Introduction to Coordination Chemistry

Inorganic Chemistry

A Wiley Series of Advanced Textbooks
ISSN: 1939-5175

Previously Published Books in this Series

Chirality in Transition Metal Chemistry
Hani Amouri & Michel Gruselle; ISBN: 978-0-470-06054-4

Bioinorganic Vanadium Chemistry
Dieter Rehder; ISBN: 978-0-470-06516-7

Inorganic Structural Chemistry, Second Edition
Ulrich Müller; ISBN: 978-0-470-01865-1

Lanthanide and Actinide Chemistry
Simon Cotton; ISBN: 978-0-470-01006-8

Mass Spectrometry of Inorganic and Organometallic Compounds: Tools – Techniques – Tips
William Henderson & J. Scott McIndoe; ISBN: 978-0-470-85016-9

Main Group Chemistry, Second Edition
A. G. Massey; ISBN: 978-0-471-49039-5

Synthesis of Organometallic Compounds: A Practical Guide
Sanshiro Komiya; ISBN: 978-0-471-97195-5

Chemical Bonds: A Dialog
Jeremy Burdett; ISBN: 978-0-471-97130-6

Molecular Chemistry of the Transition Elements: An Introductory Course
François Mathey & Alain Sevin; ISBN: 978-0-471-95687-7

Stereochemistry of Coordination Chemistry
Alexander Von Zelewsky; ISBN: 978-0-471-95599-3

Bioinorganic Chemistry: Inorganic Elements in the Chemistry of Life – An Introduction and Guide
Wolfgang Kaim; ISBN: 978-0-471-94369-3

For more information on this series see: www.wiley.com/go/inorganic

Introduction to Coordination Chemistry

Second Edition

Paul V. Bernhardt
The University of Queensland
Brisbane, Australia

Geoffrey A. Lawrance
The University of Newcastle
Callaghan, Australia

WILEY

This edition first published 2025
© 2025 John Wiley & Sons Ltd

Edition History
© 2010, 2025 by John Wiley & Sons Ltd.

Registered Offices
John Wiley & Sons, Inc., 111 River Street, Hoboken, NJ 07030, USA
John Wiley & Sons Ltd, New Era House, 8 Oldlands Way, Bognor Regis, West Sussex, PO22 9NQ, UK

For details of our global editorial offices, customer services, and more information about Wiley products visit us at www.wiley.com.

Library of Congress Cataloging-in-Publication Data

Names: Bernhardt, Paul V., author. | Lawrance, Geoffrey A., author. | John
    Wiley & Sons, publisher.
Title: Introduction to coordination chemistry / Paul V. Bernhardt, Geoffrey
    A. Lawrance.
Description: Second edition. | Hoboken : Wiley, 2025. | Includes
    bibliographical references and index.
Identifiers: LCCN 2024041135 (print) | LCCN 2024041136 (ebook) | ISBN
    9781394190713 (paperback) | ISBN 9781394190720 (adobe pdf) | ISBN
    9781394190737 (epub)
Subjects: LCSH: Coordination compounds.
Classification: LCC QD474 .L387 2025 (print) | LCC QD474 (ebook) | DDC
    341/.2242–dc23/eng/20241017
LC record available at https://lccn.loc.gov/2024041135
LC ebook record available at https://lccn.loc.gov/2024041136

Cover Design: Wiley
Cover Image: © Prof. Paul Bernhardt

Set in 10/12pt TimesLTStd by Straive, Chennai, India

Contents

Preface

This textbook is written with the assumption that readers will have completed an introductory tertiary-level course in general chemistry or its equivalent, and thus be familiar with basic chemical concepts including the foundations of chemical bonding. Consequently, no attempt to review these in any detail is included. Further, the intent here is to avoid mathematical and theoretical detail as much as practicable, and rather to take a more descriptive approach. This is done with the anticipation that those proceeding further in the study of the field will meet more stringent and detailed theoretical approaches in higher-level courses. This allows those who are not intending to specialise in the field or who wish to simply supplement their own separate area of expertise to gain a good understanding largely free of a heavy theoretical loading. While not seeking to diminish aspects that are both important and central to higher-level understanding, this is a pragmatic approach towards what is, after all, an introductory text. Without doubt, there are more than sufficient conceptual challenges herein for a student. Further, as much as is practicable in a chemistry book, you may note a more relaxed style which we hope may make the subject more approachable; not likely to be appreciated by the purists, perhaps, but then this is a text for students.

The text is presented as a suite of sequential chapters, and an attempt has been made to move beyond the pillars of the subject and provide coverage of synthesis, physical methods, and important biological and applied aspects from the perspective of their coordination chemistry in the last four chapters. While it is most appropriate and recommended that they be read in order, most chapters have sufficient internal integrity to allow each to be tackled in a more feral approach. Each chapter has a brief summary of key points at the end. Further, a limited set of references to other publications that can be used to extend your knowledge and expand your understanding is included at the end of each chapter. Topics that are not central to the thrust of the book (nomenclature and symmetry) but nevertheless are important in this field are presented as appendices. Self-assessment of your understanding of the material in each chapter has been provided for through assembly of a set of questions (and answers). However, to limit the size of this textbook, these have been provided on the supporting website at wiley.com/go/coordinationchemistry2e.

We have employed colour in figures in this revised edition to assist understanding. Figures and drawings herein employed *ChemDraw*® and *Chem3DPro*®, as well as the *Adobe*® suite of software. The coordinates for structure drawings come from the Cambridge Structural Database, curated by the Cambridge Crystallographic Data Centre (CCDC) – www.ccdc.cam.ac.uk – and appear courtesy of the CCDC. X-ray crystal structure images were created using the *Mercury* program (CCDC).

The visual appeal of transition metal complexes and their variety of colours is generally the first thing that students recall, as we did, and the inclusion of pictures of compounds in the solid state and in solution in this book hopefully piques the reader's interest. We hope that our continuing fascination and curiosity for coordination chemistry that we have attempted to convey in this book serves as a catalyst for those about to begin their journey.

Acknowledgements

I personally thank a number of colleagues for their technical assistance and enthusiastic cooperation with aspects of this book. Dr Masnun Naher (UQ) collected original NMR spectra that were presented in Chapter 7, and Dr Tri Le (UQ) also provided expert technical advice on these experiments. Associate Professor Jeff Harmer (UQ) provided a critical and valuable review of the section on electron paramagnetic spectroscopy also in Chapter 7.

Finally, I thank my co-author Geoff Lawrance for inviting me to collaborate on the second edition of a project that he initiated and completed single-handedly in its first iteration. Some 40 years ago, Geoff introduced me to the then mysterious and fascinating world of coordination chemistry during the second-year laboratory component of my chemistry BSc degree, and this mentorship continued as my PhD supervisor. We have co-authored research papers and reviews over the years, but this project has been a different kind of collaboration, and (again) I have learnt a lot in the process. It has been an opportunity to explore and reflect on the things that initially drew me towards coordination chemistry as an undergraduate chemistry student.

Paul V. Bernhardt
Brisbane, Australia
September 2024

For all those who have trodden the same path as me from time to time over the years, I thank you for your companionship; unknowingly at the time, you have contributed to this work through your influence on my path and growth as a chemist. I was particularly pleased to gain Paul Bernhardt as co-author, and be able to tap his younger and sharper brain, and numerous skills. This book has been written against a background of informal discussions with a number of colleagues on various continents at various times. Several have contributed their time in reading and commenting on draft sections of this book in both its first and second editions that have enhanced its structure and clarity. The publication team at Wiley have also done their usual fine job in production of the textbook. While this collective input has led to a better product in this new edition, we remain of course fully responsible for both the highs and the lows in the published version.

Most of all, and on behalf of both of us, we could not possibly finish without thanking our wives and families for their support over the years and forbearance during the writing of this book.

Geoffrey A. Lawrance
Newcastle, Australia
September 2024

Preamble

Coordination chemistry in its 'modern' form has existed for over a century. To identify the foundations of a field is complicated by our distance in time from those events, and we can do little more than draw on a few key events; such is the case with coordination chemistry. Deliberate efforts to prepare and characterise what we now call coordination complexes began in the 19th century, and by 1856 Wolcott Gibbs and Frederick Genth had published their research on what they termed 'the ammonia–cobalt bases', drawing attention to '*a class of salts which for beauty of form and colour ... are almost unequalled either among organic or inorganic compounds*'. With some foresight, they suggested that '*the subject is by no means exhausted, but that on the contrary there is scarcely a single point which will not amply repay a more extended study*'. In 1875, the Danish chemist Sophus Mads Jørgensen developed rules to interpret the structure of the curious group of stable and fairly robust compounds that had been discovered, such as the one of formula $CoCl_3 \cdot 6NH_3$. In doing so, he drew on immediately prior developments in organic chemistry, including an understanding of how carbon compounds can consist of chains of linked carbon centres. Jørgensen proposed that the cobalt invariably had three linkages to it to match the valency of the cobalt, but allowed each linkage to include chains of linked ammonia molecules and or chloride ions. In other words, he proposed a carbon-free analogue of carbon chemistry, which itself has a valency of four and formed, apparently invariably, four bonds. At the time this was a good idea and placed metal-containing compounds under the same broad rules as carbon compounds, a commonality for chemical compounds that had great appeal. It was not, however, a great idea. For that the world had to wait for Alfred Werner, working in Switzerland in the early 1890s, who set this class of compounds on a new and quite distinctive course that we know now as coordination chemistry. Interestingly, Jorgensen spent around three decades championing, developing and defending his concepts, but Werner's ideas that effectively allowed more linkages to the metal centre, divorced from its valency, prevailed and proved incisive enough to hold essentially true up to the present day. His influence lives on; in fact, his last research paper actually appeared in 2001, being a determination of the three-dimensional structure of a compound he crystallised in 1909! For his seminal contributions, Werner is properly regarded as the founder of coordination chemistry.

Coordination chemistry is the study of coordination compounds or, as they are often defined, coordination complexes. These entities are distinguished by the involvement, in terms of simple bonding concepts, of one or more coordinate (or dative) covalent bonds, which differ from the traditional covalent bond mainly in the way that we envisage they are formed. Although we are most likely to meet coordination complexes as compounds featuring a metal ion or set of metal ions at their core (and indeed this is where we will overwhelmingly meet examples herein), this is not strictly a requirement, as metalloids may also form such compounds. One of the simplest examples of formation of a coordination compound comes from a now venerable observation – when BF_3 gas is passed into a liquid trialkylamine, the two react exothermally to generate a solid which contains equimolar amounts of each precursor molecule. The solid formed has been shown to consist of molecules $F_3B–NR_3$, where a routine covalent bond now links the boron and nitrogen centres. However, what is peculiar to this assembly is that electron book-keeping suggests that the boron commences with an empty valence orbital, whereas the nitrogen commences with one lone pair of electrons in an orbital not involved previously in bonding. It was reasoned that the new bond must form by the two lone pair valence electrons on the nitrogen being inserted or donated into the empty orbital on the boron. Of course, the outcome is well known – a situation arises where there is an increase in shared electron density between the joined atom centres, or formation of a covalent bond. It is helpful to reflect on how this situation differs from conventional covalent bond formation; traditionally, we envisage covalent bonds as arising from two atomic centres each providing an electron to form a bond through sharing, whereas in the coordinate covalent bond one centre provides both electrons (the donor) to insert into an empty orbital on the other centre (the acceptor); essentially, you can't tell the difference once the coordinate bond has formed from that which would arise by the usual covalent bond formation. Another very simple example is the reaction between ammonia and a proton; the former can be considered to donate a lone pair of electrons into the empty orbital of the proton. In this case, the acid–base character of the acceptor–donor assembly is perhaps more clearly defined for us through the choice of partners. Conventional Brønsted acids and bases are not central to this field, however; more important is the Lewis definition of an acid and base, as an electron pair acceptor and electron pair donor, respectively.

Today's coordination chemistry is founded on research in the late 19th and early 20th centuries. As mentioned above, the work of French-born Alfred Werner, who spent most of his career in Switzerland at Zürich, lies at the core of the field, as it was he who recognised that there was no required link between metal oxidation state and number of ligands bound. This allowed him to define the highly stable complex formed between cobalt(III) (or Co^{3+}) and six ammonia molecules in terms of a central metal ion surrounded by six bound ammonia molecules, arranged symmetrically and as far apart as possible at the six corners of an octahedron. The key to the puzzle was not the primary valency of the metal ion, but the apparently constant number of donor atoms it supported (its 'coordination number'). This 'magic number' of six for cobalt(III) was confirmed through a wealth of experiments, which led to the Nobel Prize in Chemistry for Werner in 1913. While his discoveries remain firm, modern research has allowed limited examples of cobalt(III) compounds with coordination numbers of five and even four to be prepared and characterised. As it turns out, Nature was well ahead of the game, since some metalloenzymes discovered in recent decades contain metal ions with a low coordination number, which contributes to their reactivity. Metals can show an array of preferred coordination numbers, which vary not only from metal to metal, but can change for a particular metal with formal oxidation state of a metal. Thus, Cu(II) has a greater tendency towards five-coordination than Mn(II), which prefers six-coordination. Moreover, unlike six-coordinate Mn(II), Mn(VII) prefers four-coordination. Behaviour in the solid state may differ from that in solution, as a result of the availability of different potential donors resulting from the solvent itself usually being a possible ligand. Thus, $FeCl_3$ in the solid state consists of Fe(III) centres surrounded octahedrally by six Cl^- ions, each shared between two metal centres; in aqueous acidic solution, '$FeCl_3$' is more likely to be met as separate $[Fe(OH_2)_6]^{3+}$ and Cl^- ions.

Inherently, whether a coordination compound involves metal or metalloid elements is immaterial to the basic concept. However, one factor that distinguishes the chemistry of the majority of metal complexes is an often incomplete d (for transition metals) or f (for lanthanoids and actinoids) shell of electrons. This leads to the spectroscopic and magnetic properties of members of these groups being particularly indicative of the compound under study and has driven interest in and applications of these coordination complexes. The field is one of immense variety and, dare we say it, complexity. In some metal complexes, it is even not easy to define the formal oxidation state of the central metal ion since electron density may reside on some ligands to the point where it alters the physical behaviour.

What we can conclude is that metal coordination chemistry is a demanding field that will tax your skills as a scientist. Carbon chemistry is, by contrast, comparatively simple, in the sense that essentially all stable carbon compounds have four bonds around each carbon centre. Metals, as a group, can exhibit coordination numbers from 2 to 14, and formal oxidation states that range from negative values to as high as 8. Even for a particular metal, a range of oxidation states, coordination numbers, and distinctive spectroscopic and chemical behaviour associated with each oxidation state may (and usually does) exist.

Because coordination chemistry is the chemistry of the vast majority of the Periodic Table, the metals and metalloids, it is central to the proper study of chemistry. Moreover, since many coordination compounds incorporate organic molecules as ligands, and may influence their reactivity and behaviour, an understanding of organic chemistry is also necessary in this field. Further, since spectroscopic and magnetic properties are central to a proper understanding of coordination compounds, knowledge of an array of physical and analytical methods is important. Of course coordination chemistry is demanding – but it rewards the student by revealing a diversity that can be at once intriguing, attractive and rewarding. Welcome to the wild and wonderful world of coordination chemistry – let's explore it.

About the Companion Website

This book is accompanied by a companion website:

www.wiley.com/go/coordinationchemistry2e

The website includes Supplementray materials

1 The Central Atom

1.1 Key Concepts in Coordination Chemistry

The simple yet distinctive concept of the *coordinate bond* (also sometimes called a *dative bond*) lies at the core of coordination chemistry. Molecular structure, in its simplest sense, is interpreted in terms of covalent bonds formed through shared pairs of electrons. First introduced by Gilbert N. Lewis in 1916, the concept of a covalent bond formed when two atoms share an electron pair remains as a firm basis of molecular chemistry, giving us a basic understanding of single, double and triple bonds as well as of non-bonding (lone) pairs of electrons on an atom. Evolving from these simple concepts came valence bond theory from Linus Pauling, an early quantum mechanical theory that expressed the concepts of Lewis in terms of wavefunctions. Pauling would later have the unique distinction of being awarded the Nobel Prize in both Chemistry (1954) and Peace (1962). These basic bonding concepts are applicable to both organic and coordination chemistry. The coordinate bond, however, arises not through an equal contribution of electrons from the paired atoms, as occurs in organic chemistry, but from donation of a lone pair of electrons on one atom to an empty orbital on what will become its partner atom.

In coordinate bond formation, the bonding arrangement between electron pair acceptor (designated as A) and electron pair donor (designated as :D, where the pair of dots represent the lone pair of electrons) can be represented simply as:

$$A + :D \rightarrow A : D \qquad (1.1)$$

The product alternatively may be written as A←:D or A←D, where the arrow denotes the direction of electron donation, or, where the nature of the bonding is clearly understood, simply as A–D. This latter standard representation is entirely appropriate since this bond, once formed, is a two-electron covalent bond like any other (other representations involving an arrow are now discouraged, as they could imply a special character to the covalent bond once it is formed that is inappropriate). The bond formation process should be considered reversible in the sense that, if the A–D bond is broken, the lone pair of electrons originally donated by :D remains entirely with that entity.

Coordination compounds have always been an integral part of chemistry since its recognition as a distinct scientific discipline in the middle of the 18th century, although at that time nothing was known about molecular structure or bonding. Key early developmental work employed the metal cobalt (isolated first only in 1735 by Georg Brandt) and simple reagents such as ammonia and chloride ion. French chemist B.M. Tassaert reported in 1798 the reaction between the mineral

Introduction to Coordination Chemistry, Second Edition. Paul V. Bernhardt and Geoffrey A. Lawrance.
© 2025 John Wiley & Sons Ltd. Published 2025 by John Wiley & Sons Ltd.
Companion website: www.wiley.com/go/coordinationchemistry2e

cobaltite (CoAsS) and excess ammonia, which yielded an uncharacterised yellow compound. With the benefit of hindsight, it is apparent that Tassaert had accidentally prepared the chloride salt of an ammonia complex of cobalt(III), now known to be $[Co(NH_3)_6]^{3+}$ but did not appreciate its significance at the time; Leopold Gmelin reported the oxalate salt of the same complex in 1822. Although many chemists pursued variations of this reaction in the first half of the 19th century, most credit is given to Oliver Wolcott Gibbs and Frederick Augustus Genth for their now classic 1856 review '*Researches on the Ammonia-cobalt Bases*'. This paper describes the synthesis, composition and properties of numerous compounds including the colourfully named, but structurally misunderstood, praseocobaltic chloride (formulated as $CoCl_3\cdot4NH_3$), purpureocobaltic chloride ($CoCl_3\cdot5NH_3$), roseocobaltic chloride ($CoCl_3\cdot5NH_3\cdot H_2O$) and luteocobaltic chloride ($CoCl_3\cdot6NH_3$) (Figure 1.1). Suffice to say that this nomenclature soon ran out of steam (or at least colours). The French team of Louis Bernard Guyton de Morveau, Antoine-Laurent de Lavoisier, Claude Louis Berthollet and Antoine François Fourcroy had published in 1787 a forward-looking treatise on nomenclature that set under way adoption of a binary notation for salts (such as potassium carbonate, rather than an array of names in existence, including salt of tartar) still in use today. The names used by Gibbs and Genth and other early coordination chemists do follow the binary notation concept in part, but the names of the complex cations, not surprisingly, lack any real precision. An unfortunate tendency to, alternatively, name new compounds after the discoverer only compounded the problem. Modern nomenclature is based on sounder structural bases, demonstrated in Appendix 1.

These coordination compounds are historically important as they provided the impetus for various theories proposed throughout the 1800s to explain their appearance, physical properties and composition, and, in particular, the nature of the attraction between cobalt and ammonia. Thomas Graham had introduced the ammonium theory where ammonia gained a metal substituent instead of a hydrogen ion as in an ammonium salt. Jöns Jacob Berzelius' conjugate theory noted a special pairing (coupling) between metal and ammonia, while Karl Klaus recognised that ammonia once bonded to a metal became 'passive' and lost its basic properties. All these observations contributed to a clearer picture of what we now know as the coordinate bond.

In most coordination compounds, it is possible to identify the central or core atom or ion that is bonded not simply to one other atom, ion or group through a coordinate bond but to several of these entities at once. The central atom is an acceptor, with

Figure 1.1

Four classical coordination compounds reported by Gibbs and Genth in 1856, originally named: (a) praseo-cobaltic chloride ($CoCl_3\cdot4NH_3$); (b) purpureocobaltic chloride ($CoCl_3\cdot5NH_3$); (c) roseocobaltic chloride ($CoCl_3\cdot5NH_3\cdot H_2O$) and (d) luteocobaltic chloride ($CoCl_3\cdot6NH_3$).

the surrounding species each bringing (at least) one lone pair of electrons to donate to an empty orbital on the central atom, and each of these electron pair donors is called a *ligand* when attached. The central atom is a metal or a metalloid (from the *p*-block) and the compound that results from bond formation is called a coordination compound, coordination complex, or often simply a *complex*. We shall explore these concepts further below.

The species providing the electron pair (the electron pair donor) is thought of as being coordinated to the species receiving that lone pair of electrons (the electron pair acceptor). The coordinating entity, the ligand, can be as small as a monoatomic ion (e.g. F^-) or as large as a protein—the key characteristic is the presence of one or more lone pairs of electrons on an electronegative donor atom (i.e. one with a tendency to draw shared electron density towards itself). Donor atoms often met are heteroatoms like N, O, P and S as well as halide ions, but this is by no means the full range. Moreover, most existing organic molecules can act as ligands or else can be converted into molecules capable of acting as ligands. As mentioned earlier, a classical and well-studied ligand is ammonia, NH_3, which has one lone pair (Figure 1.2). Isoelectronic (same number of electrons) with ammonia is the methyl carbanion CH_3^- which can also be considered a ligand under the simple definition applied; even hydrogen as its hydride, H^-, has a pair of electrons and can act as a ligand. It is not the type of donor atom that is the key but rather its capacity to supply an electron pair.

The acceptor with which a coordinate covalent bond is formed is conventionally either a metal or metalloid. With a metalloid such as boron, which is also electron deficient, covalent bond formation is invariably associated with an increase in coordination number, and simple electron counting based on the donor–acceptor concept can account for the number of coordinate covalent bonds formed. In this case, boron attains an octet of electrons by coordinate bond formation. With a genuine metal ion (from the *s*-, *p*-, *d*- or *f*-block), this model is less applicable since the metal ion usually carries its preferred number of ligands into a reaction. Consequently, the coordination number tends to remain constant, with an incoming ligand not simply adding to the coordination sphere but replacing an existing ligand in a *substitution* reaction; a more sophisticated model needs to be applied and will be developed herein. What is apparent with metal ions in particular is the strong drive towards

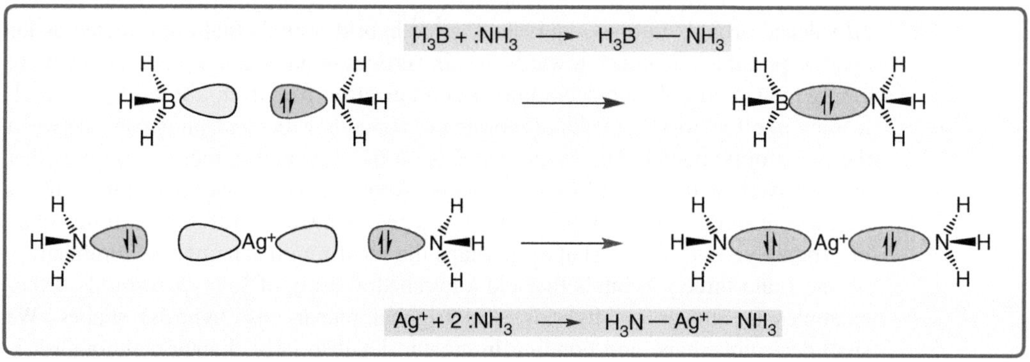

Figure 1.2

A schematic view of ammonia acting as a donor ligand to a metalloid acceptor (boron, shown here with the BH$_3$ *sp*^{3} hybridised like the ammonia) and to a metal ion acceptor (silver(I), shown here as the 'naked' cation for simplicity) to form coordinate bonds.

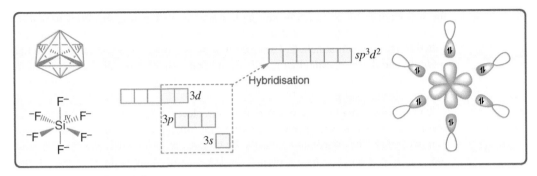

Figure 1.3

The octahedral $[SiF_6]^{2-}$ molecular ion and a simple valence bond approach to explaining its formation. Overlap of a p-orbital containing two electrons on each of the six fluoride anions with one of six empty hybrid orbitals on the Si(IV) cation, arranged in an octahedral array, generates the octahedral shape with six equivalent covalent σ-bonds.

complexation—'naked' ions are extremely reactive, and even in the gaseous state, complexation will occur. It is a case of the whole being better than the sum of its parts, or, put more appropriately, coordinate bond formation is energetically favourable.

A more elaborate example than those shown above is the anionic compound SiF_6^{2-} (Figure 1.3), which adopts the shape of a classical octahedron (a polyhedron with eight equilateral triangular faces, sets of four of which meet at each of six vertices) that we will meet also in many metal complexes. Silicon lies below carbon in the periodic table, and there are some limited similarities in their chemistry. However, simple valence bond theory and the octet rule that works so well for carbon (and other elements of the second period) cannot deal with a silicon compound exhibiting six equivalent covalent bonds. One way of viewing this molecular species is as being composed of a Si^{4+} or Si(IV) centre with six F^- anions bound to it through each fluoride anion donating an electron pair ($:F^-$) to an empty orbital on the central Si(IV) ion, which has lost all of its original four valence electrons in forming the Si^{4+} ion. Using traditional valence bond theory concepts, a process of hybridisation is necessary to accommodate the outcome where all six bonds formed are equivalent (Figure 1.3). The generation of the shape arises through asserting that the silicon arranges a combination of one $3s$, three $3p$ and two of five available $3d$ valence orbitals into six equivalent sp^3d^2 hybrid orbitals that are directed as far apart as possible and hence towards the six vertices of an octahedron. Each identical empty hybrid orbital then accommodates an electron pair from a fluoride ion, each leading in effect to six covalent (coordinate) σ-bonds with electron density between the two atomic nuclei. The shape depends on the type and number of orbitals that are involved in the hybridisation process. Above, a combination resulting in an octahedral shape (sp^3d^2 hybrids) is developed. However, different combinations of orbitals yield different shapes, perhaps the most familiar being the combination of one s and three p orbitals to yield a tetrahedral array of four sp^3 orbitals. Other examples are linear (sp hybrids) and trigonal planar (sp^2 hybrids) shapes. We shall deal with shape and bonding in greater detail in later chapters, although it is worth noting the experimental complexities involved in defining the actual shape of molecules, which draw on an array of modern instrumental methods discussed in Chapter 7. Theories and models of bonding exist to interpret what we 'see' at the molecular level but do not lose sight of the key word—they are models.

A central atom or ion with vacant valence atomic orbitals accepting lone pairs of electrons from ionic or neutral atoms (ligands) is the classic requirement for formation of what we have termed coordinate bonds, leading to a coordination compound. The very basic valence bonding model described above can be extended to metal ions, as we will see, but with some adjustments due to the presence of electrons in the *d*-orbitals; more sophisticated models are really required. Of all the developed approaches, molecular orbital theory is the most sophisticated and is focussed on the overlap of atomic orbitals of comparable energy on different atoms to form molecular orbitals to which electrons are allocated. While providing accurate descriptions of molecules and their properties, it is relatively complicated and time-consuming, and somewhat difficult to comprehend for large complexes; consequently, simpler models still tend to be used.

In the simple theory based on Lewis' concepts exemplified above, the key aspects are an empty orbital on one atom and a filled orbital (with a pair of electrons present, the lone pair) on the other. Many ligands such as ammonia are also bases in the classical Brønsted–Lowry concept of acids and bases (as proton donors or acceptors) since these species are able to accept a proton. However, in the description we have developed here, no proton is involved, but the concept of donation to an electron-deficient species does apply. The broader and more general concept of an electron pair donor as a base and an electron pair acceptor as the acid evolved, and these are called a *Lewis base* (electron pair donor) and a *Lewis acid* (electron pair acceptor). Consequently, $H_3B–NH_3$ is traditionally considered a coordination compound, arising through coordination of the electron deficient (or Lewis acid) H_3B and the electron lone pair containing (or Lewis base) compound $:NH_3$ (Figure 1.2). It is harder, in part as a result of entrenched views of covalent bonding in carbon-based compounds, to accept $[H_3C–NH_3]^+$ in similar terms as an assembly of H_3C^+ and $:NH_3$. This need to consider and debate the nature of the assembly limits the value of the model for non-metals and metalloids. With metal ions, however, you tend to know where you stand–almost invariably, you may start by considering the metal ions as the electron pair acceptors in forming coordination compounds; perhaps it is not surprising that coordination chemistry is focussed mainly on compounds of metals and their ions.

Coordination has a range of consequences for the new assembly. It leads to structural change, seen in terms of change in the number of bonds, bond angles and distances. This is inevitably tied to a change in the physical properties of the assembly, which differ from those of its separate components. With metal atoms or ions at the centre of a complex, even changing one of a set of ligands will be reflected in readily observable changes in physical properties, such as colour. With growing sophistication in both synthesis and our understanding of physical methods, properties can often be 'tuned' through varying ligands to produce a particular result, such as a desired reduction potential.

It should also be noted that a coordination compound adopts one of a limited number of basic shapes, with the shape determined by the nature of the central atom and its attached ligands. Moreover, the physical properties of the coordination compound depend on and reflect the nature of the central atom, ligand set and molecular shape. Whereas only one central atom occurs in many coordination compounds (a compound we may thus define as a monomer), it should also be noted that there exist a large and growing range of compounds where there are two or more metal atoms present either of the same or different types. These metals may be linked together through direct atom-to-atom bonding or else are linked by particular ligands that as a result are joined to at least two metals at the same time. This latter arrangement, where one or even several ligands are said to 'bridge' between central atoms, is the more common of these two options. The resulting species can usually be thought of as a set of monomer units linked together, leading to what is formally a polymer or, more correctly when only a small number of units are linked, an oligomer. We shall concentrate largely on simple monomeric species herein but will introduce examples of larger linked compounds where appropriate.

Although, as we have seen, the *p*-block metalloid elements can form molecular species that we call coordination compounds, the decision on what constitutes a coordination compound is perhaps more subtle with these than is the case with metals, as discussed earlier. Consequently, in this tale of complexes and ligands, it is with *d*-block metals and particularly their cations as the central atom that we will almost exclusively meet examples.

1.2   A Who's Who of Metal Ions

The periodic table of elements is dominated by metals. Moreover, it is a growing majority, as new elements made through the efforts of nuclear scientists are invariably metallic. If the periodic table was a parliament, the non-metals would be doomed to be forever the minority opposition, with the metalloids a minor third party who cannot decide which side to join. The periodic table is basically a system of classification, and as such reflects in part the interests and prejudices of the assemblers and the level of acceptance by the scientific community.

John Dalton's proposal in 1806 to classify elements via their relative atomic weights formed a basis for later development of the periodic table. Historically, it has changed over time, and in one of the earliest forms, proposed by John Newlands in 1864, consisted of eight columns. Several other proposals also appeared at around that time (including a remarkable rotatable cylinder carrying a spiral periodic table built by Alexandre-Émile Béguyer de Chancourtois in 1862, which showed the presence of a periodic repetition of chemical properties). Dmitri Mendeleev, in 1869, arranged all the 63 chemical elements known by the late 1860s in order of increasing atomic weight and by chemical groupings, and noticed a recurring pattern (a periodicity) of properties within groups of elements in the resulting table. His periodic table was sophisticated enough that he was able to leave gaps in places where he expected unknown elements should occur and even predicting the properties of some potential elements (two of which, Ga and Sc, were discovered in just the next decade). The later detection of 'missing' elements that he had predicted established Mendeleev as the founder of the modern periodic table, although there have been many approaches and proposals over the years. At essentially the same time as Mendeleev was developing his table, Lothar Meyer also produced a periodic table (eventually published in 1870) reflecting related trends in physical properties and with elements arranged in order of increasing atomic weights. Moreover, post-Mendeleev, development of new proposals did not stop. For example, William Crookes produced a three-dimensional spiral system in 1888, elegant but bulky. In fact, as many as 700 versions of the periodic table appeared in the century or so after 1869 (although many placed the emphasis on artistry rather than chemistry). There is a continuing general interest in the table, with 2019 (150 years after Mendeleev published his version) declared by UNESCO as the *International Year of the Periodic Table of the Chemical Elements*.

After Mendeleev's first attempt, the periodic table evolved into an arrangement of elements of effectively 18 vertical groups, which became the normal representation by the 1920s as early quantum mechanics was applied to explain the form that had developed from experimentally observed behaviours and similarities. The 18 columns relate to 2 associated with the *s*-block, 6 with the *p*-block and 10 with the *d*-block, reflecting the maximum numbers of electrons that can be accommodated by the orbitals in these sub-shells. Furthermore, the so-called inner transition elements, or *f*-block, are usually arranged as a separate block that is 14 elements wide, associated with the maximum number of electrons that can be accommodated by the set of seven *f*-orbitals. However, there is an expanded 32 column format of the Table that incorporates the *f*-block within the main body of the table along with the rest of the columns (sometimes called the 'long form', for an obvious reason). However, there is still some debate today about the current representation, focussed on Group 3 membership, with one proposal leading to a 15-element wide *f*-block. Various options proposed for *f*-block membership revolve around these elements lying in a part of the Table where assignment of electronic configurations becomes more problematical. The debate reflects a concern of specialist quantum mechanics practitioners rather than general scientists, and are usually put aside in the periodic table that we generally see and use, on the basis that quantum mechanics provides a good but not exact explanation of the Table. Notably, the International Union of Pure and Applied Chemistry (IUPAC) does not officially support any particular form of the periodic table, although their ambivalent approach does not prevent them publishing their version!

Consequently, the position of elements in the traditional form of the periodic table that we use today depends on their basic electronic configuration (Figure 1.4) and their chemistry is related to their position. Nevertheless, there are common features that allow overarching concepts to be developed and applied. For example, a metal from any of the *s*-, *p*-, *d*- or *f*-blocks behaves in a common way—it usually forms cations, and it overwhelmingly exists as molecular coordination compounds through combination with other ions or molecules. Yet the diversity of behaviour underlying this commonality is both startling and fascinating and at the core of this journey.

The inherent difficulty in isolating and identifying metallic elements meant that, for most of human history, their chemistry was unknown. Up until around the mid-18th century, only gold, silver, copper, iron, tin, lead and mercury had been isolated in their pure metallic forms. However, in an extraordinary period from 1735 to 1844, all but two of the naturally existing *d*-block elements were isolated in their pure metallic form and characterised. Along with this burst of activity in the identification of new elements came the foundations of modern coordination chemistry, building on this new-found capacity to isolate and identify metallic elements and subsequently explore their reactivity. The especially rare and elusive elements hafnium and rhenium were only isolated in 1922 and 1928, respectively. Nuclear synthesis and isolation of radioactive technetium in 1939 completed the first three rows of the *d*-block. In almost 200 years, what was to become the largest (and arguably most interesting) block of the periodic table was cemented in place. Sustained efforts of scientists in the second half of the 20th century using nuclear fusion created the so-called superheavy (but also short-lived) elements, and a fourth row of transition (*d*-block) elements, and a seventh row of the periodic table, was completed. There are now 118 known elements. Where will it end?

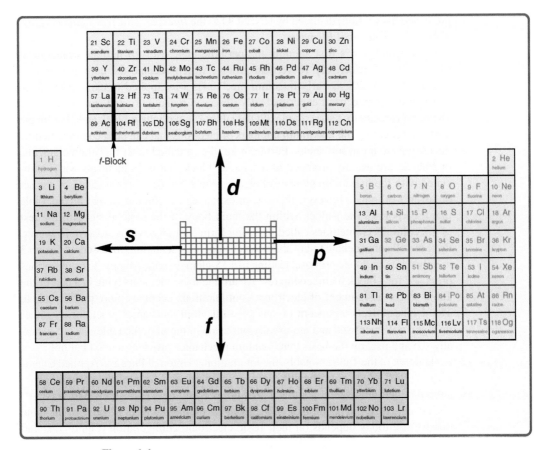

Figure 1.4
The location of metals in the periodic table of the elements. (Non-metallic elements are in faint type to distinguish them from the metals.).

Almost all metals have a commercial value because they have found industrial or medical applications. It is only the more exotic and very short half-life 'superheavy' synthetic elements that have as yet no real commercial valuation. The isolation of a metal in its elemental form can be the starting point for applications, and this is the domain of metallurgy and materials science. The chemistry of metals is overwhelmingly that of their oxidised forms. This is evident even in Nature, where metals are rarely found in their elemental state due to spontaneous oxidation by atmospheric oxygen. There are a few exceptions, of which gold is the standout example, and it was this accessibility in the metallic state that largely governed its adoption and use in antiquity. Dominantly, but not exclusively, a metal is found in a positive oxidation state, that is as a cation. These metal cations form, literally, the core of coordination chemistry; they lie at the centre of an array of neutral or anionic ligands. Nature employs metal ions in various ways, including making use of their capacity to bind to organic molecules and their ability to exist, at least for many metals, in a range of oxidation states.

The origins of a metal in terms of its periodic table position has a clear impact on its chemistry, such as the reactions it will undergo and the type of coordination compounds that are readily formed. These aspects are reviewed in Chapter 6.2, after important background concepts have been introduced. At this stage, it is sufficient to recognise that, although each metallic element is unique, there is some general chemical behaviour that relates to the block of the periodic table to which it belongs that place both limitations on, and some structure to, reactions in coordination chemistry.

1.2.1   Common and Uncommon Metals

Because we meet them daily in various forms, we tend to think of metals as common. However, 'common' is a relative term—iron may be more common than gold in terms of availability in the Earth's crust, but gold is itself more common than rhenium. Even for the fairly well-known elements of the first row of the *d*-block, abundance in the Earth's crust varies significantly, from iron (41,000 ppm) to cobalt (20 ppm); moreover, what we think of as 'common' metals, like copper (50 ppm abundance) and zinc (75 ppm abundance), are really hardly that. Availability of an element is not driven by how much is present on average in the Earth's crust, of course, but by other factors such as its existence in sufficiently high concentrations in accessible ore bodies and its commercial value and applicability (see Chapter 9). Iron, more abundant than the sum of all other *d* block elements, is mined from exceedingly rich ore deposits and is of major commercial significance. Rhenium, the rarest transition metal naturally available, is a minor by-product of some ore bodies where other valuable metals are the primary target; in any case, it has limited commercial applications at present. Nevertheless, our technology has advanced sufficiently that all metallic elements occurring naturally (on Earth) are commercially available in some amount or form for those wishing to explore their chemistry or for which applications already exist. Even some synthetic elements are available and applicable. Complexes of technetium, the only *d*-block element with no stable isotopes, are now of great significance in medical imaging yet the element itself can only be found in Nature in trace quantities near uranium ores as a daughter product of uranium fission. Even its name is a signal to its general absence, being based on the Greek $\tau\epsilon\chi\nu\eta\tau o\varsigma$ (technitos, meaning artificial). The once extinct technetium is now making a comeback and is made via nuclear decay of a radioactive isotope of neighbouring element molybdenum (^{99}Mo) produced by neutron bombardment (^{98}Mo $+ n \rightarrow$ ^{99}Mo $\rightarrow$ ^{99m}Tc $+ \beta^-$). Technetium is a vital component of medical γ-ray imaging (see Chapter 9.2.1). In fact, sufficient technetium is produced so that it may be considered as abundant as its rare naturally available partner element rhenium, although still a radioactive and inherently hazardous element. As another example, an isotope of the synthetic *f*-block actinoid element americium formed, until recently, the core of the ionisation mechanism operating in the sensor of household smoke detectors.

These observations have one obvious impact on coordination chemistry–every metallic element in the periodic table is accessible and in principle able to be studied, and each offers a suite of unique properties and behaviour. As a consequence, they are in one sense all now 'common'; what distinguishes them are their relative cost and the amounts available. In the end, it has been such commercially driven considerations that have led to a concentration on the coordination chemistry of the more available and applicable lighter elements of the transition metals, from titanium to zinc. Of course, Nature made similar choices much earlier, as most metalloenzymes employ light transition elements at their active sites, although molybdenum and tungsten are interesting outliers that will be mentioned later in Chapter 8.

1.2.2 An Alternative Concept of Common Metals

Apart from availability, discussed in the previous section, there is another more chemical approach to commonality that we should dwell on, an aspect that we have touched upon already. This is a definition in terms of oxidation states. With the most common of all metals in the Earth's crust, the main group element aluminium, only one oxidation state is found in Nature–Al(III). However, for the most common transition metal (iron), both Fe(II) and Fe(III) are common, whereas other higher oxidation states such as Fe(IV) are known but are highly reactive. With the rare element rhenium, the reverse trend holds true, as the high oxidation state Re(VII) is common but Re(III) and Re(II) are rare. What is apparent from these observations is that each transition metal can display one or more 'usual' oxidation states, a unique feature of the *d*-block elements. The main group (*s*- and *p*-block) and lanthanoid/actinoid (*f*-block) metals are much less capable of existing in more than one stable ionised form.

What allows us to observe different oxidation states in the *d*-block elements is their particular ligand environment, and in general, there is a close relationship between the donor atoms coordinated to a metal and the oxidation states it can sustain, which we will explore later. The definition of 'common' in terms of metal complexes in a particular oxidation state is an ever-changing aspect of coordination chemistry, since it depends in part on the amount of chemistry that has been performed and reported; over time, a metal in a particular oxidation state may change from 'unknown' to 'very rare' to 'uncommon' as more chemists beaver away at extending the chemistry of an element. At this time, a valid representation of the status of elements of the first row of the *d*-block with regard to their oxidation states is shown in Figure 1.5. As is clear from this figure, oxidation states two and three are the most common. Notably, hydrated transition metal ions of charge greater than 3+ (that is oxidation state over three) are not stable in water, so higher oxidation state species invariably involve other ligands apart from water. Differences in the definition of what amounts to a common oxidation state leads to some variation, but the general trends remain constant.

Immediately apparent from Figure 1.5 is that most transition metals offer a wealth of oxidation states, with the limit set by simply running out of *d*-electrons (i.e. reaching the $4s^0 3d^0$ arrangement that corresponds to the argon noble gas electron configuration) or else reaching such a high redox potential that stability of the ion is severely compromised (that is, it cannot really exist, because it involves itself immediately in oxidation–reduction reactions that return the metal to a lower and more common stable oxidation state). Notably, it gets harder to 'use up' all *d*-electrons on moving from left to right across the periodic table, associated with both the rising number of *d*-electrons and lesser screening from the charge on the nucleus. Still, you are spoilt for choice as a coordination chemist!

The standard redox potential (E^o) (formally the reduction–oxidation potential) provides a measure of the stability of a metal in a particular oxidation state. Although this will be covered in greater detail in Chapter 7, the E^o value (in Volts) is an intrinsic property of the coordination compound in its oxidised and reduced forms comprising the so-called redox couple. Put simply, for systems with large positive E^o values, the reduced form of the couple is more stable while a very large negative redox potential indicates the oxidised form is more stable. Most metals in their elemental state undergo spontaneous oxidation by atmospheric oxygen and are found in the Earth's crust as oxide and sulfide minerals. Coordination chemistry underpins

Sc	Ti	V	Cr	Mn	Fe	Co	Ni	Cu	Zn
Scandium	Titanium	Vanadium	Chromium	Manganese	Iron	Cobalt	Nickel	Copper	Zinc
		0 d^5	0 d^6	0 d^7	0 d^8	0 d^9	0 d^{10}		
		1 d^4	1 d^5	1 d^6	1 d^7	1 d^8	1 d^9	1 d^{10}	
	2 d^2	2 d^3	2 d^4	2 d^5	2 d^6	2 d^7	2 d^8	2 d^9	2 d^{10}
3 d^0	3 d^1	3 d^2	3 d^3	3 d^4	3 d^5	3 d^6	3 d^7	3 d^8	
	4 d^0	4 d^1	4 d^2	4 d^3	4 d^4	4 d^5	4 d^6		
		5 d^0	5 d^1	5 d^2	5 d^3	5 d^4			
			6 d^0	6 d^1	6 d^2				
				7 d^0					

Figure 1.5

Oxidation states met amongst complexes of transition metal elements; *d*-electron counts for the particular oxidation states of a metal appear below each oxidation state. [Oxidation states that are relatively common with a range of known complexes are in red, others in green.]

the mining of many metals, especially some of the more precious metals, as will be shown in Chapter 9.

Yet another way of defining commonality with metal ions relates to how many ligand donor groups may be attached to the central metal. This was touched on in the Preamble, and we will use and expand on the same example again. Cobalt(III) was shown decades ago to have what was then thought invariably six donor groups or atoms bound to the central metal ion or a *coordination number* of six. While this is still the overwhelmingly common coordination number for cobalt in this oxidation state, there are now stable examples for Co(III) of coordination number five and even four. In its other common oxidation state, as Co(II), there are two 'common' coordination numbers, four and six; it is hardly a surprise, then, that more and more examples of the intermediate coordination number five have appeared over time. Five-coordination has grown to be almost as common as four- or six-coordination for another metal ion, Cu(II), illustrating that our definitions of common and uncommon do vary historically. That is a problem with chemistry generally–it never stands still. The number of research papers published with a chemical theme each year continues to grow at such a rate that it is impossible to read a single year's complete offerings in a decade, let alone that year.

1.3   Metals in Molecules

Metals in the elemental form typically exhibit bright, shiny surfaces–what we tend to expect of a 'metallic' surface. In the atmosphere, rich with oxygen and usually containing water vapour, these surfaces may be prone to attack, depending on the E^o value; this leads to the bright surface changing character as it becomes oxidised. Although a highly polished steel surface is attractive and valued, the same surface covered in an oxide layer (better known in the case of iron as rust) is hardly a popular fashion statement. Yet it is the formation of rust that is perfectly natural, with the shiny metal surface the unnatural form that needs to be carefully and regularly maintained to retain its initial condition. What we are witnessing with rust formation is a chemical process governed by thermodynamics (attainment of an equilibrium defined by the stability of the reaction products compared to the reactants) and kinetics (the rate at which change, or the chemical reaction, proceeds to equilibrium under the conditions prevailing). While the outcome may not be aesthetically appealing (unless one wants to make a virtue of rusted steel as a 'distressed' surface with character), chemistry is not given to making allowances for the sake of style or commerce–it is a demanding task to 'turn off' the natural chemistry of a system. Some metals, such as titanium, are less wilful than iron; they undergo surface oxidation but form a tight monolayer of oxide that is difficult to penetrate and thus is resistant to further attack. This fact is of great comfort to recipients of titanium metal implants following orthopaedic surgery.

Of course, were the metal ions that exist in the oxidised surface to undergo attack in a different way, through complexation by natural ligands and subsequent dissolution, this would result in fresh metal surface being exposed and hence available for attack. Such a process, occurring over long periods, would suggest that free active metals would increasingly end up dissolved as their ions in the ocean, and this is clearly not so–most metal ion concentrations in the ocean are very low (apart from alkali metal ions, being <0.001 ppm). In reality, because reactive elemental-state metals are made mainly through human action, the contribution to the biosphere even by reversion to ionic forms will be small. Most metals are locked up as ions in rocks—particularly as highly insoluble oxides, sulfides, sulfates or carbonates that will dissolve only with human intervention, through reaction with strong acids or ligands. Even if they enter the biosphere as soluble complex ions, they are prone to chemistry that leads to re-precipitation. The classic example is dissolved iron(II), which readily undergoes aerial oxidation to Fe(III) and precipitation as an hydroxide, followed by dehydration to an oxide, all occurring below neutral pH.

Thus, in the laboratory, we tend to meet almost all metals in a pure form as synthetic cationic salts of common anions. These tend to be halides, nitrates or sulfates, and it is these metal salts, hydrated or anhydrous, that form the entry point to almost all of metal coordination chemistry. In Nature it is no accident that the more abundant transition metals ions tend to have found vital roles, mediated of course by their chemical and electrochemical properties. Iron is ubiquitous in and essential to all life forms not only because it is common but also due to its versatility. It commonly forms stable complexes with biomolecules in both its oxidised and reduced forms (Fe(III)/(II)) and is arguably the most biologically important transition metal of all.

1.3.1 Metals in the Natural World

Most metals in the Earth's crust are located in highly inorganic environments—as components of rocks or soils on land or under water. Where metals are aggregated in local high concentrations through geological processes, these may be sufficient in amount and concentration to represent an ore deposit, which is really an economic rather than a scientific definition. In addition, metals are present in water bodies as dissolved cations; however, as mentioned earlier, their concentrations are in only a very few cases substantial, as is the case with sodium ion in seawater. Complex anions, usually of metals in high oxidation states, also are a component of our oceans. Molybdenum is, perhaps unexpectedly, the most abundant transition metal in surface seawater (although at only about $10\,\mu g/L$); vanadium is also relatively abundant, at about $2\,\mu g/L$. Molybdenum's unique advantage stems from the fact that it is stable in its highly oxidised and extremely soluble Mo(VI) form as $[MoO_4]^{2-}$; no other transition metal can match this combination of stability in a high oxidation state (low redox potential) and solubility in water. Other more common terrestrial elements such as iron, manganese and titanium in their highly oxidised forms cannot remain in solution and precipitate immediately to soon form charge neutral insoluble oxides, e.g. Fe_2O_3, MnO_2 or TiO_2. However, even if present in very low concentration, as for gold in seawater, the size of the oceans means that there is a substantial amount of gold (and other metals) dispersed in the aquatic environment; one estimate suggests that there are around 5×10^{16} tonnes of minerals dissolved in the oceans.

The other location of metals is within living organisms, where, of the transition metals, iron, zinc and copper predominate in humans and other vertebrates. On rare occasions, the concentration of another metal may be relatively high; this is the case in some plants that tolerate and concentrate particular metal ions, such as nickel in *Hybanthus floribundus*, native to Western Australia, which can be hyper-accumulated at up to ~50 mg per gram dry weight. Levels of metal ions in animals and in particular plants vary with species and environment. Generally, the *d*- and *p*-block metals are present in Nature in only trace amounts (Table 1.1), including potentially harmful elements absorbed from our environment (e.g., Cr, Cd and Pb). High levels of most metal ions are toxic to living species; for example, ryegrass displays a toxicity order Cu > Ni > Mn > Pb > Cd > Zn > Al > Hg > Cr > Fe, with each species displaying a unique trend.

Metals eventually were recognised as having a presence in a range of biomolecules. Where metal cations appear in living things, their presence is rarely if ever simply fortuitous. Rather, they play a particular role from simply providing an ionic environment through to being at the key active site for reactions in a large enzyme. Notably, it is the lighter alkali, alkaline earth and transition elements that dominate the metals present in living organisms. Of transition metals, although iron, copper and zinc are most dominant, other first-row transition elements play some part in the functioning of organisms. Keys to metal ion roles are their high charge (and charge density), capacity to bind organic entities through strong coordinate bonds and ability in many cases to vary their oxidation states. We shall return to look at metals in biological environments in more detail in Chapter 8.

Table 1.1 Typical concentrations (ppm) of selected metals ions in nature.

Metal	Earth's crust	Oceans	Plants (ryegrass)	Animals (human blood)
Na	23,000	10,500	1000	2000
K	21,000	1620	28,000	1600
Mg	23,000	1200	2500	40
Ca	41,000	390	12,500	60
Al	82,000	0.0005	50	0.3
Sc	16	0.000006	>0.01	0.008
Ti	5600	0.00048	2.0	0.055
V	160	0.001	0.07	<0.0002
Cr	100	0.00018	0.8	0.008
Mn	950	0.00011	130	0.005
Fe	41,000	0.0001	240	450
Co	20	0.000001	0.6	0.01
Ni	80	0.0001	6.5	0.03
Cu	50	0.00008	9.0	1.0
Zn	75	0.00005	31	7.0
Mo	1.5	0.01	1.1	0.001
Cd	0.11	0.0000011	0.07	0.0052
Pb	14	0.00002	2.0	0.21
Sn	2.2	0.0000023	<0.01	0.38
Ce	68	0.000002	<0.01	<0.001

1.3.2 Metals in Contrived Environments

What defines chemistry over the past century has been our growing capacity to design and construct molecules. The number of new compounds that have been synthesised now number in the tens of millions, and that number continues to grow at an astounding pace, along with continuing growth in synthetic sophistication; we have reached the era of the 'designer' molecule. Many of the new organic molecules prepared can bind to metal ions or else can be readily converted to other molecules that can do so. This, along with the diversity caused by the capacity of a central metal ion to bind to a mixture of ligands at one time, means that the number of potential metal complexes that are not natural species is essentially infinite. Coordination chemistry has altered irreversibly the composition of the world.

It is possible to assign a role for coordination chemistry in some ancient technologies, such as mordants (metal salts) used to fix dyes to fabrics (a technique perhaps 9000 years old), but there was no understanding of the molecular level in the ancient world. Even discovering when the first synthetic metal complex was deliberately made and identified is not as easy as one might expect, because so much time has passed since that event. One popular candidate is *Prussian blue*, a cyanide complex of iron discovered accidently in the first decade of the 18th century, and developed as a commercial artist's colour within a few years of the discovery (and known to have been used in a 1709 painting); it is still a commercially important pigment today. Preparation and isolation of what we now know are gold(III) and platinum(IV) halide complexes and reaction of the latter with ammonia were reported by W. Lewis in 1763. As we have seen earlier, hexaamminecobalt(III) chloride, discovered serendipitously by Tassaert in 1798, set under way a quest to understand how different combinations of cobalt, ammonia and chloride could lead to such an array of compounds. As new compounds evolved, it was at first sufficient to identify them simply through their maker's name. Thus came into being species such as *Magnus' green salt* ($PtCl_2 \cdot 2NH_3$) and *Erdmann's salt* ($Co(NO_2)_3 \cdot KNO_2 \cdot 2NH_3$). This early attempt at nomenclature was doomed by profligacy, but as many compounds isolated were coloured, another way of identification arose based on colour (as mentioned above for the series of salts obtained from combination of cobalt, ammonia and chloride and shown in Figure 1.1), prior to the 19th and 20th century development of the systematic nomenclature that is used today.

While some may quail at the outcomes of all the extravagant molecule building, what remains a constant are the basic rules of chemistry. A synthetic metal complex obeys the same basic chemical 'rules' as a natural one. 'New' properties result from the character of new assemblies, not from a shift in the rules. As a classic example of how this works, consider the case of Vitamin B_{12}, isolated by Karl Folkers and a team of researchers at Merck & Co in 1947, and structurally characterised by Dorothy Crowfoot Hodgkin at Oxford University using small bright red crystals supplied by Lester Smith of Glaxo Laboratories, who isolated the compound at about the same time as the Merck group. The structural solution by X-ray crystallography, particularly challenging at this time, took 13 years. Crowfoot Hodgkin, who also was responsible for the structural elucidation of the groundbreaking antibiotic penicillin, was awarded the Nobel Prize in Chemistry in 1964. Vitamin B_{12} is distinguished by being one of a limited number of biomolecules centred on cobalt and was discovered to exist with good stability in three oxidation states, Co(III), Co(II) and Co(I). Moreover, it was found to be one of a rare few natural organometallic (metal–carbon bonded) compounds, involving a C–Co(III) bond. At the time of these discoveries, examples of low-molecular weight synthetic cobalt(III) complexes also stable in both Co(II) and Co(I) oxidation states were few if any in number, nor had the Co(III)–carbon bond been well defined. Such observations lent some support to a view that metals in biological entities were 'special'. Of course, time has removed the distinction, with synthetic Co complexes stable in all of the (III), (II) and (I) oxidation states well established, and examples of the Co(III)–carbon bond reported even with quite simple companion ligands in other sites around the metal ion. The 'special' nature of metals in biology is essentially a consequence of their usually very large and specifically arranged macromolecular environments embedded within a protein. While it is demanding to reproduce such natural environments in detail in the laboratory, it is possible to mimic them at a sufficient level to reproduce aspects of their chemistry.

Of course, the synthetic coordination chemist can go well beyond nature, by making use of facilities that don't exist in Earth's natural world. This can include even re-making 'extinct' elements such as radioactive technetium and promethium. The element boron has given rise to a rich chemistry based on boron hydrides, most of which are too reactive to have any geological existence. Some boron hydrides as well as mixed carbon-boron compounds (carboranes) can bind to metal ions. Nitrogen forms a vast array of carbon-based compounds (amines) that are excellent at binding to metal ions; Nature also makes wide use of these for binding metal ions, but the construction of novel amines has reached levels that far exceed the limitations of Nature. After all, most natural chemistry has evolved at room temperature and pressure in near pH-neutral aqueous environments–limitations that do not apply in a chemical laboratory. What the vast array of synthetic molecules for binding metal ions provides is a capacity to control molecular shape and physical properties in metal-containing compounds not envisaged as possible a century ago. These have given rise to applications and technologies that are limited only by our imagination.

1.3.3   Natural or Made-to-Measure Complexes

Metal complexes are natural–expose a metal ion to ligands capable of binding to that ion and complexation almost invariably occurs. Dissolve a metal salt in water, and both cation and anion are surrounded by water molecules, as separate aquated ions. In particular, the metal ion acts as a Lewis acid and water as a Lewis base, and a structure of defined coordination number with several $M^{n+}-OH_2$ bonds results. The coordinate bond exemplified in these aquated cations is at the core of all natural and synthetic complexes.

While metals are usually present in minute amounts in living organisms, techniques for isolation and concentration have been developed that allow biological complexes to be recovered. The array of metalloproteins now offered commercially by chemical companies in gram quantities is evidence of this capacity to isolate biological complexes at scale. Relying on natural sources for some compounds was once both limiting and expensive, but advances in biotechnology have led to large-scale bioreactors being used to produce common drugs like the antibiotic penicillin and the hormone insulin on industrial scales, with Nature doing the complicated chemistry for us. Other designer-made synthetic drugs must be prepared by humans and at a cost that is economically sustainable. Although less sophisticated and potent, simple across-the-counter compounds of metals find regular medical use; zinc supplements, for example, are actually usually supplied as a simple synthetic zinc(II) amino acid complex. Aspects of biological coordination chemistry are covered in Chapter 8.

Isolation of pure metals from ores by hydrometallurgical (water-based) processing often relies on complexation as part of the process. For example, gold recovery from ore currently employs oxygen as oxidant and cyanide ion as ligand, leading selectively to a soluble gold(I) cyanide complex. Copper(II) ion dissolved from ore is recovered from an aqueous mixture by solvent extraction as a metal complex into kerosene, followed by decomposition and back extraction into aqueous acid. Recovery of the parent metal is then readily achieved electrochemically by reduction to its elemental state. Electrochemical processes are also in regular industrial use; aluminium and sodium are recovered via electrochemical processes from molten salts. Historically, the emergence of electrochemistry as a technique in the early 1800s was the catalyst for the isolation of dozens of new and highly reactive metallic elements from their oxidised forms. Pyrometallurgical (high temperature) processes for isolation of metals, on the other hand, usually rely on reduction reactions of oxide ores at high temperature. An overview of applied coordination chemistry is covered in Chapter 9.

1.4   The Road Ahead

Having identified the important role of metals (usually but not exclusively as cations) as the central atom in coordination chemistry, it is appropriate at this time to recognise that the metal has partners and to reflect on the nature of the partnership. The partners are, of course, the ligands. A coordination compound (or complex) can perhaps be thought of as the product of a molecular marriage—each partner, metal and ligand, brings something to the relationship, and the result involves compromises that, when made, means the product of the union is distinctly different (in terms of chemical and physical properties) from the prior independent parts. While this analogy may be taking anthropomorphism to the extreme, it is nevertheless not a bad analogy and not so unreasonable an outlook to think of a complex as a 'living' combination. After all, as we shall touch on later, it is not a totally inert combination. Metal complexes undergo ligand exchange, and their preferred shape may change depending in part on the oxidation state and in part on the ligands' preferences. Furthermore, the stability of the metal–ligand assembly is something that we can actually measure experimentally. It is no doubt stretching the analogy to talk of a perfect match, but the concept of fit and misfit between metals and ligands (particularly where the ligand offers more than one donor group) has been developed. These metaphors should alert you to the core aspects of coordination chemistry–partnership, and compromise.

The compounds formed from coordination of ligand molecules and ions to a central atom adopt particular shapes, of which a common shape for complexes with six attached ligands is octahedral, already introduced in Figure 1.3. The early research of Albert Werner and his contemporaries involved a range of metal complexes that adopt this common geometry. An aspect of coordination chemistry that should be addressed at the outset is the 'language' we use, that is how to represent a coordination complex so that others can understand to what you are referring. As mentioned earlier, naming a compound after yourself, or anyone else, is of limited value to the rest of the scientific community. A more productive way forward is to adopt a systematic approach by using a unique chemical name, a formula or a drawing. Naming (nomenclature) is pursued in Appendix 1, and it may soon become apparent that, while formal nomenclature leads to precise identification, interpretating lengthy formal names that often include coordinated organic molecules as components of the full name, as well as devices to identify the mode of coordination of molecules, can be as daunting as compound German nouns are to those with English as a native language. A formula is often simpler but may not be sufficiently clear; for example, it may not define shape or even bonding mode. Consequently, the use of drawings of molecules developed and, if well presented, these pictures provide an unambiguous solution to identity. Chemical structure drawings have grown to play a major role in the field, just as they have in organic chemistry. There are a number of representations commonly met in drawings of coordination compounds, illustrated in Figure 1.6. The space-filling drawing, based on the covalent radii of the components, reflects the space occupied by the electron clouds of each atom within the assembly but lacks clarity due to the bulk of the components. A ball-and-stick representation, where the component atoms are 'shrunk' to yield a more open view, is clearer. The ball-and-stick representation

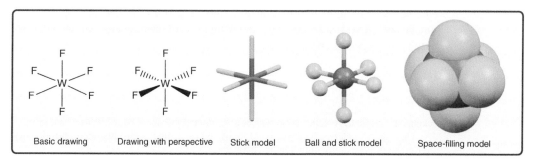

| Basic drawing | Drawing with perspective | Stick model | Ball and stick model | Space-filling model |

Figure 1.6

Drawings of the tungsten(VI) hexafluoride complex in (from right to left) space-filling, ball-and-stick, stick, perspective line drawing and basic line drawing modes.

shown in Figure 1.6 has two-coloured boundaries that reflect the covalent radii in the space-filling model. However, this is not strictly necessary, and a single-coloured bond is both sufficient and as appropriate in terms of the nature of the coordinate bond, where the bonding electron pair originates from one bond partner. As it happens, this type of drawing has similarities in terms of appearance to the views of complexes shown in structures determined experimentally by X-ray crystallography (see Chapter 7.3.8), where the atomic positions are defined in terms of probabilities (Heisenberg's Uncertainty Principle strikes again); this technique will be addressed in Chapter 7. More commonly, line drawings are used to represent a complex, either drawn with perspective to thus address both molecular bonding and shape, or else drawn as simple representations where bonds are defined but shape must be inferred. You shall meet several approaches to drawing complexes in this book but remember that they are simply conventional and related ways of defining a complex species.

In the rest of this book, we will be examining in more detail ligands, metal–ligand assembly and the consequences. These include molecular shape, bonding models, stability, properties and how we can measure and interpret these. Furthermore, we will look at metal complexes that occur in Nature and find application in commerce and speculate on future developments. Overall, the intent is to give as broad and deep an overview as is both reasonable and proper in an introductory text. Pray continue.

Concept Keys

A coordination complex consists of a central atom, usually a metal ion, bound to a set of ligands by coordinate covalent bonds.

A coordinate covalent bond is distinguished by the ligand donor atom donating both electrons (of a lone pair) to an empty orbital on the central atom to form the bond.

A ligand may be considered a *Lewis base* (electron pair donor), the central atom a *Lewis acid* (electron pair acceptor).

A 'common' metal may be defined simply by its geo-availability, but from a coordination chemistry perspective, it is more appropriate to define 'common' in terms of aspects such as preferred oxidation state, number of coordinated donors, or even preferred donor types.

Metal ions may exist and form complexes in a number of oxidation states; this behaviour is particularly prevalent in the *d*-block.

First row *d*-block metal ions are found dominantly in the M(II) or M(III) oxidation states. Heavier members of the *d*-block tend to prefer higher oxidation states.

Regardless of origin, natural or synthetic, the basic rules of chemistry applying to metal complexes are a constant, supporting measurement, interpretation and modelling of properties and behaviour.

Further Reading

Beckett, M. and Platt, A. (2006). *The Periodic Table at a Glance*. Oxford: Wiley-Blackwell.
    This short undergraduate-focussed book gives a fine well-illustrated introductory coverage of periodicity in inorganic chemistry.
Blackman, A., Bottle, S., Schmid, S., Mocerino, M., and Wille, U. (2022). *Chemistry*, 5th ed. Melbourne: Wiley.
    An introductory general chemistry textbook appropriate for reviewing basic concepts, that covers all fields of chemistry.

Gillespie, R.J. and Popelier, P.L.A. (2002). *Chemical Bonding and Molecular Geometry*. Oxford: Oxford University Press.

A venerable but still appropriate coverage from the fundamental level upward of various models of molecular bonding.

Housecroft, C.E. and Sharpe, A.G. (2012). *Inorganic Chemistry*, 4th ed. Harlow, UK: Pearson Education.

Of the large and sometimes daunting general advanced textbooks on inorganic chemistry, this is an enduring, finely written and well-illustrated example, useful as a resource book.

Jackson, T. (2020). *The Periodic Table. A Visual Guide to the Elements*. London: Aurum Press.

This is an entertaining and easy-to-read book by a science writer, rather than a scientist, and is an accessible entry level route to basic information.

2   Complexes

2.1   The Central Metal Ion

The first chapter introduced aspects of coordination chemistry from the perspective of a central atom (usually a metal ion) and a ligand. This chapter will expand on their interrelationship. A key observation of coordination chemistry is that the properties and reactivity of a bound ligand differ substantially from those of the free ligand. Concomitant with this outcome is a change in the properties of the central metal ion. This gross modification of the character of the molecular components upon binding of ligand to metal ion is at the heart of the importance of the field. Thus, understanding metal–ligand assemblies is vital.

We have considered variable oxidation states in Chapter 1 and should now be aware that many metal ions, especially those from the d-block, can exist in more than one oxidation state. Let us recall how this can arise by looking at manganese, which has the atomic electronic configuration $[Ar]4s^2 3d^5$. With the $4s$ and $3d$ electronic energy levels close together, electrons in these levels can be considered together as valence electrons. Thus, there are seven electrons 'available', and successive removal of electrons will take us from oxidation state Mn(I), on removal of the first electron, through to Mn(VII), on removal of the seventh electron. Each successive removal costs more energy; successive ionisation enthalpies for manganese rise from $0.724\,MJ\,mol^{-1}$ for the first through 1.515, 3.255, 4.95, 6.99, 9.20, and finally $11.514\,MJ\,mol^{-1}$ for the seventh. By this stage, the electron configuration has reached that of the inert gas argon, and breaking into this inert core of electrons (requiring $18.77\,MJ\,mol^{-1}$ of energy) is not practical. Therefore, the highest accessible oxidation state for manganese is Mn(VII), or Mn^{7+} as the free ion. Note that as more valence electrons are removed from Mn (and the positive charge rises), the energy cost to remove the next electron becomes higher; the seventh ionisation energy is 16 times greater than the first. Therefore, it is no surprise to find that, only the early d-block elements from the fourth row of the periodic table may attain the $3d^0 4s^0$ electronic configuration of [Ar]; i.e. Sc(III), Ti(IV), V(V), Cr(VI) and Mn(VII) (but Fe(VIII) does not form). However, despite common electronic configurations, their chemistry is quite different. While Sc(III) and Ti(IV) are quite unreactive, Cr(VI) and Mn(VII) are powerful oxidants and seek to temper this reactivity by the re-acquisition of electrons into their empty $3d$-orbitals. The difference between Sc(III) and Mn(VII), of course, is the additional nuclear charge of Mn, and this is ultimately the brake that prevents the removal of any more electrons for the following elements; as a result, Fe(VIII) is not a viable oxidation state even though it would have the stable electron configuration of argon. Apart from the ionisation energy limitations illustrated above, theoretically, if any more than seven valence electrons were removed from the later d-block elements

Introduction to Coordination Chemistry, Second Edition. Paul V. Bernhardt and Geoffrey A. Lawrance.
© 2025 John Wiley & Sons Ltd. Published 2025 by John Wiley & Sons Ltd.
Companion website: www.wiley.com/go/coordinationchemistry2e

(Fe, Co, Ni, Cu or Zn), then the metal would become such a powerful oxidant that it would oxidise practically any molecule it met, including solvents. This means they can have no stable existence; i.e. you will not find such compounds in a bottle. As a consequence, very high oxidation states are not often met, at least with the first transition metal ion series. It is only with heavier elements of the periodic table, such as the second and third rows of the *d*-block metals, that an oxidation state of +VIII is encountered (e.g. RuO_4, OsO_4), an outcome that may be related to the greater number of core electrons shielding the more exposed outer valence shell electrons from the nuclear charge, but we will not explore this in any greater depth. Higher oxidation states than +VIII are unknown.

From experimental evidence, the first-row transition metals prefer oxidation states +II and +III. Higher and lower formal oxidation states, even negative ones (such as –I, –II), are known, although oxidation states less than +I are almost exclusively found with organometallic compounds (i.e. those with at least one metal–carbon bond) discussed later in Section 2.4. Oxidation states in organometallic compounds are sometimes hard to define. This arises where it may be difficult to decide whether an electron resides exclusively in an orbital that belongs to the metal or the ligand, or even is shared (delocalised). Most reported oxidation states are what can be termed spectroscopic oxidation states; i.e. defined by an array of spectroscopic properties as most consistent with a particular metal *d*-electron set. The key point is that the chemistry of every metal is oxidation state dependent to the point that it takes on a completely new character upon oxidation or reduction. For example, Co(III) reacts slowly in its ligand substitution reactions and shows little variation in its coordination geometry, whereas Co(II) reacts rapidly and exhibits a wide variety of structures and ligand preferences; you would not think they were the same element!

Because we are usually dealing with ionic salts that are water-soluble when we begin working with transition metal ions, the complexes formed in pure water, where water molecules are also the ligand set, are a common starting point for initiating other coordination chemistry. Most first-row transition metals in the two commonest oxidation states form a stable complex ion with water as a ligand, usually of the formulation $[M(OH_2)_6]^{n+}$ ($n = 2$ or 3), commonly abbreviated as M_{aq}^{n+}. These ions are usually coloured (see Figure 7.5 for the known M_{aq}^{2+} ions), and display a wide variation in the redox potentials for both the M(III)/(II) and M(II)/(0) couples, as shown in Table 2.1. This suite of observations will form the background to later

Table 2.1 Colours of the M_{aq}^{2+} and M_{aq}^{3+} ions of the first row of the *d*-block, along with their standard reduction potentials.[a]

Parameter	Sc	Ti	V	Cr	Mn
$[M(OH_2)_6]^{3+}$ colour	Colourless	Violet	Green	Violet	Red-brown
$[M(OH_2)_6]^{2+}$ colour	—	—	Violet	Blue	Very pale pink
$E°(M^{III/II})$V in aq. acid	—	—	−0.26	−0.42	+1.56
$E°(M^{II/0})$V in aq. acid	—	−1.60	−1.13	−0.89	−1.18

Parameter	Fe	Co	Ni	Cu	Zn
$[M(OH_2)_6]^{3+}$ colour	Very pale purple	Blue	—	—	—
$[M(OH_2)_6]^{2+}$ colour	Pale green	Pink	Green	Blue-green	Colourless
$E°(M^{III/II})$V in aq. acid	+0.77	+1.84	+2.05	>+2	≫+2
$E°(M^{II/0})$V in aq. acid	−0.44	−0.28	−0.24	+0.34	−0.76

[a]Entries marked by a dash are unknown and or inaccessible.

discussion where water ligands are replaced by other ligands, with an associated change in physical properties such as colour and redox potential. Theories that we develop must be able to accommodate these changes.

2.2 Metal–Ligand Assembly

2.2.1 The Coordinate Bond

'Bond – Coordinate Bond'. Maybe it doesn't exactly roll off the tongue, but it's hard to avoid this adaption of the personal introduction used by perhaps our best-known and most enduring screen spy to introduce this section – it serves its purpose to remind us of the endurance and strength of bonds between metals and ligands, which at a rudimentary level we can consider as a covalent bond. Moreover, it is not just any bond, but a specially-constructed coordinate bond – hence the name of this field, coordination chemistry. Although the coordinate bond arises in the simple 'classical' model through donation of an electron pair from a ligand donor atom to an empty orbital on the central atom (usually a metal ion), the result is essentially a covalent bond. Unfortunately, the simple covalent bonding concepts applied to organic chemistry do not provide valid interpretations for all the physical properties of coordination complexes, and more sophisticated theories are required. We shall examine a number of bonding models for coordination complexes in this chapter.

2.2.2 The Foundation of Coordination Chemistry

Before we travel too far, however, it may be valuable to look back to the beginnings of coordination chemistry. While examples of coordination compounds were slowly developed during the 19th century, as discussed in Chapter 1, it was not until the 20th century that the nature of these compounds was understood and the bonding theories that explained their existence were developed. They were at a very early stage named 'complex compounds', a reflection of their unexplained structures, and we still call them 'complexes' today. Around the beginning of the 20th century, the wealth of instrumental methods we tend to take for granted today simply did not exist. Chemists employed chemical tests, including elemental analyses, to probe formulation and structure, augmented by limited physical measurements such as solubility and conductivity in solution. Analyses defined the components, but not their structure. As a consequence, it became usual to represent them in a simple way, as illustrated in Chapter 1 (Figure 1.1).

Simple tests of halide-containing compounds, involving gravimetric analysis of the amount of silver halide precipitated upon addition of silver ion to a solution and comparison with the known total amount of halide ion present from a more robust microanalysis method, identified the presence in some cases of both reactive and unreactive halide. For example, $CoCl_3 \cdot 6NH_3$ and $IrCl_3 \cdot 3NH_3$ precipitated three and zero chlorides, respectively, with the addition of silver ion, meaning only one type of halide was present in each; one where all 3 chlorides were 'free' (from Co) and the other where they were all tightly bound (by Ir). The compound $CoCl_3 \cdot 4NH_3$ precipitated only one of the three halides readily, so that there were clearly two types of chloride present in the one compound. It was surmised that one type is held firmly in the complex and so is unavailable, whereas the other is readily accessed and behaves more like a dissociated chloride ion in a solution of the ionic compound

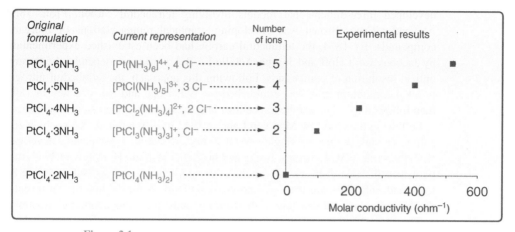

Figure 2.1

Identifying ionic composition from molar conductivity experiments for a series of platinum(IV) complexes of ammonia and chloride ion. The 'current representation' above refers to the cationic or neutral complex only, as counter-ions are shown separately to clearly define the number of ions, central to the interpretation of the conductivity results.

sodium chloride. We now know these two classes as a coordinated chlorido ligand (held via a Co–Cl coordinate bond in this example) and ionic chloride (present as simply the Cl$^-$ counter-ion to a positively-charged complex cationic species).

The availability of equipment to measure the molar conductivity of solutions was turned to good use. It is interesting to note that coordination chemists still make use of physical methods heavily in their quest to assign structures – it is just that the extent and sophistication of instrumentation has grown enormously in over a century. What conductivity could provide for the early coordination chemist was some further information about the apparently ionic species inferred to exist through the silver ion precipitation reactions. This is best illustrated for a series of platinum(IV) complexes with various amounts of chloride ion and ammonia present (Figure 2.1). From comparison of measured molar conductivity with conductivities of known compounds, the number of ions present in each of the complexes could be determined. We now understand these results in terms of the modern formulation of the complexes as octahedral platinum(IV) compounds with coordinated ammonia, where coordinated chloride ions make up any shortfall in the fixed coordination number of 6. This leaves in most cases some free ionic chloride ions to balance the charge on the complex cation.

This type of what we now consider simple experiments provided a foundation for coordination chemistry as we now know it. What became compelling following the discovery of these intriguing compounds was to develop a theory that would account for these observations. One of the key aspects relates to shape in these molecules – and it is a wondrously variable area to behold.

2.2.3 Complex Shape

It is now over 200 years since the realisation that chemical compounds have a distinctive three-dimensional shape was expressed by William Wollaston in a paper published in 1808. As early as 1860, Louis Pasteur suggested a tetrahedral grouping around carbon, and both August Kekulé (1867) and Emanuele Paterno (1869) developed three-dimensional models involving tetrahedral carbon. From 1864, Alexander Crum Brown was developing graphical representations of inorganic compounds. By 1874, the tetrahedral carbon had been established experimentally by Jacobus van't Hoff and Joseph Achille Le Bel, working independently, through optical resolution of compounds. Following this approach, the tetrahedral nitrogen in alkylammonium salts was established by W.J. Hope and S.J. Peachey in 1899, then followed by other tetrahedral centres and subsequently other stereochemistries.

Following the report by Max von Laue in 1912 of diffraction of X-rays by a zinc sulfide crystal, the use of single-crystal X-ray diffraction subsequently developed from the work of W. Lawrence Bragg and his father William H. Bragg. In 1913, they published the first-ever crystal structures of simple salts such as NaCl, KCl, KBr, CaF_2 and ZnS. This was truly ground-breaking work as for the first time it revealed the three-dimensional structure of these compounds (including coordination number and geometry) and provided the first measurement of interatomic distances, i.e. bond lengths. The Braggs also published the crystal structure of diamond (an allotrope of elemental carbon) in 1913, confirming van't Hoff and Le Bel's findings that the four substituents on a saturated carbon atom were oriented towards the vertices of a tetrahedron. X-ray crystallography was then established as the supreme method for determination of three-dimensional shape of all chemical compounds (molecular and ionic) in the solid state and remains so today.

Shapes of transition metal compounds received limited attention until late in the 19th century. Historical aspects introduced in Chapter 1 are expanded somewhat below. From 1875 for several decades, S.M. Jørgensen championed graphical chain formulae for metal complexes, building on an original proposal of C.W. Blomstrand of 1871. For example, with cobalt(III), three direct bonds to the cobalt were assumed by asserting a relationship between the metal oxidation state and the number of bonds to the metal, with other groups in the molecule linked in chains (such as $Co–NH_3–NH_3–NH_3–Cl$) reminiscent of the $–CH_2–$ chains in organic compounds that had been developed a little earlier; another key assumption was that Co–Cl bonds were considered unable to ionise in solution, whereas NH_3–Cl bonds could do so. The theory made too many assumptions, with insufficient experimental evidence in support. Although chain theory was simply wrong, it was the belief that there should be a close link between structure in organic compounds and coordination compounds that drove the proposal. It failed since it could not explain experimental observations of the era; for example, the non-electrolyte $CoCl_3 \cdot (NH_3)_3$ could not be fitted at all to this model. Although he drew on Jørgensen's careful and extensive experimental results, Alfred Werner rejected this model, and in a series of studies between 1891 and 1893 introduced two important concepts to coordination chemistry: stereochemistry and 'affinities'. His formulation for $CoCl_3 \cdot (NH_3)_3$ as a compound with three Co–NH_3 and three Co–Cl bonds satisfies the non-electrolytic nature of the compound and takes the 'modern' six-coordinate octahedral shape, which has now been proven by structural studies.

Werner's coordination theory is acknowledged as the birth of modern coordination chemistry. His proposal of *primary* valency (oxidation state) as distinct from *secondary* valency (coordination number) laid the foundations for our understanding of coordination compound structure and reactivity. Further, Werner recognised and proposed that these 'secondary valencies' would be distributed around the metal in fixed positions in space in such a way as to lead to the least repulsion between the groups or atoms attached; a three-dimensional view, remarkable at the time. Subsequently, he devoted himself to finding experimental proof for his theory, using the limited methods available at the time, which included conductivity and resolution of complexes into optical isomers. Werner's model interpreted all experimental observations of the time correctly, including being able to explain how $CoCl_3 \cdot 4NH_3$ can exist in two forms of what Werner showed were geometric isomers, species with the same formula but different spatial dispositions of groups around the cobalt (see Chapter 4 for a detailed discussion of shape). Only for the octahedral shape are two isomers predicted for $CoCl_3 \cdot 4NH_3$, matching the experimental observations. Using either a flat hexagonal shape or else a trigonal prismatic shape, three isomers are predicted, supporting (but not proving) the octahedral shape for six-coordination. The proof came from Werner's success in resolving some six-coordinate complexes into optical isomers – something that could occur if they exist in the octahedral shape but not some other proposed shapes. His suite of studies of six-coordinate compounds, which he argued should exist in many cases as geometric and or optical isomers, convincingly supported his predictions. By isolating optical isomers of a complex without any carbon atoms present, he also put to rest the view that optical activity was a property of carbon compounds alone.

Werner also proposed in 1912 the concept of a second (outer) coordination sphere through directed but weaker interactions of groups in the first (inner) coordination sphere with species arranged in an outer shell around the main coordination sphere. These effects, involving what we would now call specific hydrogen bonding interactions, were later shown to exist by Paul Pfeiffer (1931) through his work on optically active substances interacting with an optically inactive central metal complex, where an effect was induced in the metal complex despite it not being covalently bonded to the optically active species.

We now know conclusively that Werner was correct with his theories, or at least nearly so. Interestingly, although his octahedral shape is certainly dominant in six-coordinate complexes, we now know that there are also trigonal prismatic six-coordinate shapes – albeit with ligand systems to which Werner had no access. What sophisticated modern physical methods tell us conclusively is that coordination complexes do not just happen upon a shape. Each different coordination number known, and these vary from 2 to 14, supports a limited number of basic shapes. The shape, or stereochemistry, depends both on the metal and its oxidation state and on the form of the ligand – each contributes in different ways.

In 1940, Nevil Sidgwick and Herbert M. Powell proposed a simple approach to inorganic stereochemistry based on the concept that the broad features of molecular shape could be related to the distribution achieved by all bonding and non-bonding electron pairs (bond pairs and lone pairs) around a central atom as a result of minimised repulsion. A basic but very useful model for shape evolving out of this is the *Valence Shell Electron Pair Repulsion* (VSEPR) model developed by Ronald S. Nyholm and Ronald J. Gillespie. This, an extension of Lewis' ideas, is based on a simple concept – it considers the electron pairs as point charges placed on a spherical surface with the central atom in the centre of the sphere. These charges of identical type will distribute themselves on the surface so as to minimise repulsive interactions. In other words, this is a simple *electrostatic* model. For any set of like charges, there is usually either only one lowest energy arrangement or a very limited set of arrangements of the same or at least very similar energy. This very simple model has proven to be of great value as a predictor of polyatomic molecular shape for p-block compounds, since in these compounds lone pairs often are present and play an important structural and directional role in defining molecular shape. The concepts can be adapted to some extent for d-block metal complexes and were first applied by Nyholm and Gillespie in the late 1950s to coordination complexes, although the strong structural influence of lone pairs found in *p*-block chemistry is absent in the *d*-block.

For *d*-block metal ions in particular, it was known that a series of structurally common $[M(OH_2)_6]^{2+}$ complexes form for metal ions with different numbers of *d*-electrons. Although from a VSEPR viewpoint, one would expect their structures to vary since the electronic configurations differ, the *d*-electrons are not playing a major structural role. Recognising this, David L. Kepert amended the VSEPR model for use with transition metals by ignoring *d*-electrons altogether, and considering only the set of ligand donor groups, treating them as point charges on a spherical surface as in the VSEPR model and considering repulsions between them. For two to six point charges, the lowest energy outcomes are represented in Figure 2.2; only for five-coordination are two geometries of essentially equal energy predicted. These dispositions of point charge on the surface then can be taken to represent the location of the donor atoms in the attached ligands, joined by coordinate bonds to the metal that is placed at the centre of the sphere. Once bonds are inserted, this then represents the basic shape of the complex (Figure 2.2). The solid lines shown in the right half of Figure 2.2 represent the actual coordinate covalent bonds; the dashed lines in the left half of the figure only define the shape of the polyhedral framework and do not relate to bonding at all.

The Kepert model has been applied to coordination numbers from 2 to at least 9, although the higher numbers are met less frequently. As with any simple model, it is not universally correct, and for each coordination number, there may be alternate shapes. For example, the predicted shape for four-coordination is *tetrahedral*, but we

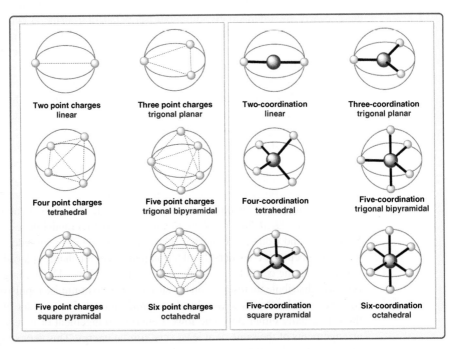

Figure 2.2

Distributions predicted for from two to six point charges on a spherical surface (left), and shapes of complexes evolving from this point-charge model, when a central atom is placed at the core of the sphere and considered to bond to each donor atom located where a point charge occurs.

know that one well-known shape found for four-coordination in metal complexes is *square planar*, where all four donor atoms lie in the same plane as the metal, disposed in a square arrangement with the metal ion at the centre. Obviously, there are other electronic influences on the shape or stereochemistry of complexes. We shall look at real outcomes for each coordination number in turn in Chapter 4, where shape is examined in detail, but at this stage, it is sufficient to recognise that there are several basic shapes, most of those being defined perfectly well by the simple model depicted in Figure 2.2.

For coordination numbers above 6, VSEPR predicted and observed shapes also match reasonably well. Shapes predicted for seven-, eight- and nine-coordination are shown in Figure 2.3. What is notable about higher coordination number structures is that several can be thought of as derived from shapes for lower coordination numbers. For example, from the common octahedral shape, adding an extra metal–ligand bond in a plane in an octahedron that includes the metal where four bonds lie produces five bonds in that plane, now rearranged in a pentagonal rather than the original square shape around the plane (yielding the seven-coordinate *pentagonal bipyramidal* shape). The nine-coordinate *tricapped trigonal prismatic* shape is based on the other rarer shape of six-coordination, the trigonal prism; here, an additional ligand is attached through the centre of each of the three rectangular faces of the trigonal prism. The eight-coordinate *square antiprismatic* shape can be viewed adequately as two square planes of four ligands set on either side of a central metal, with the top square rotated 45° compared to the other square. Whereas the three shapes depicted in Figure 2.3 are the lowest repulsion geometries under the Kepert model, several other low repulsion geometries are predicted (capped octahedral and capped trigonal prism for seven coordination, dodecahedral and bicapped trigonal prism for eight coordination, and capped square antiprism for nine-coordination). Although few examples of high coordination numbers will be met in this introductory book, it is important to be aware that many complexes, particularly with heavier central metals, adopt these geometries.

It is also appropriate at this stage to recognise that the way we represent or draw molecules for display and discussion is important for communication of concepts. Technology has provided a number of ways in which molecules can be illustrated (as shown earlier in Figure 1.6). For everyday use and discussion between chemists, it is

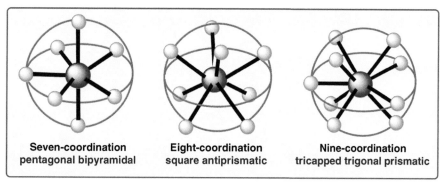

| Seven-coordination | Eight-coordination | Nine-coordination |
| pentagonal bipyramidal | square antiprismatic | tricapped trigonal prismatic |

Figure 2.3

The lowest repulsion shapes predicted for seven-, eight- and nine-coordination: pentagonal bipyramidal (seven-coordinate, e.g. ZrF_7^{3-}), square antiprismatic (eight-coordinate, e.g. ReF_8^-) and tricapped trigonal prismatic (nine-coordinate, e.g. ReH_9^{2-}).

most likely that simple basic drawings will be used, as they can be hand or computer sketched rapidly. Views with perspective, or else ball and stick models, tend to be met in formal presentations, as will be the case in this book. Space-filling models, while giving a view closer to the actual situation for the molecular assembly, are difficult to visualise, even for very simple molecules, because atoms at the front tend to obscure those behind. As a consequence, the ball and stick models are met more often in formal presentations. With the extensive use of single-crystal X-ray crystallography in defining coordination compounds, it has become common to meet structural views developed from structure solutions using this technique. Molecules characterised by X-ray structure analysis appear throughout the chapters, with the instrumental methodology introduced in Chapter 7.

2.3   The Nature of Bonding in Metal Complexes

2.3.1   Orbital Symmetry

Having established the basic structural concepts of coordination complexes, it is now time to understand how these complexes are held together, or bond. To pursue this aspect, we need to develop models for bonding that not only provide a satisfactory basis for dealing with the array of shapes that exist but also can provide interpretation of the spectroscopic and other physical properties of this class of compounds. It is useful to introduce the core concepts and models that we use to interpret observation immediately, as they pervade discussion throughout the field.

Before we travel too far, it is perhaps valuable to remind ourselves of the nature of the valence s, p and d orbitals that are intimately involved in transition metal coordination chemistry (Figure 2.4). The spherical s orbital and the dumbbell-shaped p orbitals aligned with the x, y and z axes should already be familiar to you. Within each transition metal series, there are also five nd orbitals (d_{xy}, d_{yz}, d_{xz}, $d_{x^2-y^2}$ and d_{z^2}, $n = 3$–6) which have different spatial orientations and involve two different basic shapes (Figure 2.4). To begin to understand how they differ, notice that $d_{x^2-y^2}$ and d_{z^2} have lobes lying *along* the x, y and z axes of the defined coordinate system, whereas the d_{xy}, d_{yz}, d_{xz} orbitals lie *between* the axes. It is these two classes of spatial

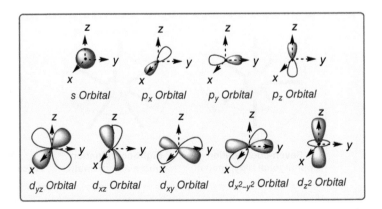

Figure 2.4

The shapes, phase and symmetry relationships of the s, p and d orbitals. The position of the atomic nucleus is indicated by a black dot.

arrangement that are the key to much of the discussion that will follow in coming sections – suffice to say that this spatial difference will be significant.

It is notable that, for elements of the f-block, there are seven f orbitals ($f_{xyz}, f_{z(x^2-y^2)}, f_{y(z^2-x^2)}, f_{x(z^2-y^2)}, f_{x^3}, f_{y^3}$ and f_{z^3}) which offer three different basic shapes and also, like the d orbitals, occupy quite different spatial positions. While we will not be discussing f orbitals in any detail, it is useful to be reminded that they do many of the things that d orbitals do, except that the $4f$ orbitals of the lanthanoids, for example, are less 'exposed' due to being screened by larger $5s$, $5p$ and $6s$ shells, which leads to weaker interactions with their surrounding ligands and very similar chemistry across the series. The nd orbitals are more exposed, and hence more involved in their element's chemistry.

We also must address the concept of *phase* and *symmetry* as it applies to atomic orbitals. These features are derived from quantum mechanics (where the three-dimensional shapes of orbitals come from), but we will explain this in a visual and non-mathematical way. We are familiar with the idea of waves having peaks (positive amplitude) and troughs (negative amplitude), like a sine wave. To represent the phase relationships of an orbital in Figure 2.4, we use different colours and shading to discriminate positive phase from negative phase. On all p orbitals (each oriented with an x, y or z axis), we note that one lobe is positive while the opposite lobe has a negative phase, a so-called *antisymmetric* arrangement. For the d orbitals, each one is symmetric so that *opposite* lobes have the same phase while adjacent lobes have opposite phases. All s orbitals are also symmetric as shown. We will return to this feature in Section 2.3.3 when we consider what happens when these orbitals overlap, but for the time being, we will only be concerned with their basic shapes, which define the space where an electron may be found.

The metals of the d-block characteristically exist in a limited range of stable oxidation states for which the nd subshell has only partial occupancy by electrons. This differs from the situation with main group metal ions, for which it is common or at least more usual to have fixed oxidation states associated with s- and p-subshells that are completely empty. Importantly, the chemical and physical properties characteristic of the transition elements are determined by the partly filled nd subshells (likewise in the f-block, but to a lesser extent, by partly filled nf subshells). In the simple atomic model of the first-row d-block elements, however, the set of $4s$, $4p$ and $3d$ orbitals lie close in energy and can be considered the metal's valence orbitals. This provides the luxury of nine atomic orbitals (one s, three p and five d) that may, in principle, be used to account for coordination numbers of up to 9 using a valence bond model approach.

Valence bond theory, which describes bonding in terms of bringing together valence (hybrid) atomic orbitals on different atoms to form bonding electron pairs, evolved from the 1927 Heitler–London model for the covalent bond. This in turn evolved from the original 1902 concept of covalency proposed by Gilbert Lewis and published in 1916. This was later augmented to include the concepts of hybridisation and resonance by Linus Pauling. While it remains the most accessible and popular bonding model in organic chemistry, it has never really found favour in coordination chemistry, and we can demonstrate why this is the case with a counter-example.

Consider the well-known complex cation $[Ni(NH_3)_6]^{2+}$ as an example, deconstructed as a bare Ni^{2+} cation (electronic configuration $3d^8 4s^0 4p^0$) and six neutral NH_3 ligands. In accord with the electron pair donor–acceptor concept of coordinate bonding, each ammonia ligand will donate a lone pair of electrons to an empty acceptor orbital on the metal, so we need six acceptor orbitals on the metal. To arrive at the octahedral geometry in valence bond theory we mix (hybridise) one 4*s*, three 4*p* and two 3*d* orbitals to create six new equivalent (and energetically identical, or degenerate) sp^3d^2 hybrid orbitals, each pointing towards a vertex of an octahedron leaving three 3*d* atomic orbitals unchanged. The problem is immediately apparent from the d^8 electronic configuration of Ni(II). The six sp^3d^2 hybrid orbitals on Ni^{2+} must be empty to each accept a pair of electrons from each ammonia ligand but we have 8 valence electrons to accommodate on the Ni ion in only three remaining 3*d* orbitals ($3d_{xy}$, $3d_{xz}$ and $3d_{yz}$), which is impossible. Indeed, valence bond theory fails to explain the bonding in any octahedral complex with more than six *d*-electrons. We can just about deal with the d^6 cobalt(III) cation, through a somewhat complicated process of ion formation, electron rearrangement, orbital hybridisation and filling of empty hybrid orbitals, as depicted in Figure 2.5(a), but not beyond there. Workarounds were proposed in the 1950s by 'borrowing' unoccupied 4*d* orbitals for hybridisation to accommodate the metal's valence electrons in the 3*d* subshell, the so-called *outer shell* and *inner shell* model (Figure 2.5(b)). However, the much higher energy 4*d* orbitals are not valence orbitals, which violates one of the fundamental principles of valence bond theory. The limits of valence bond theory illustrated starkly here led to its demise as the premier model to describe bonding in transition metal complexes, although it still sees some use due in part to its simplicity.

When a model reaches its limit in explaining experimental observations (as illustrated in Figure 2.5), it becomes time to set it aside and look for alternative, perhaps more sophisticated models. After all, the only value of a model is that it explains observations – otherwise, it is about as useful as a chocolate teapot. Despite deficiencies, valence bond theory nevertheless yielded some concepts that still see use today. The focus on the nine 3*d*–4*s*–4*p* orbital set suggested to early researchers that there may be some special stability associated with systems that employ only these nine inner valence shell orbitals and hence can accommodate no more than 18 electrons. Thus arose the *18-electron rule*, which suggests that coordination complexes whose

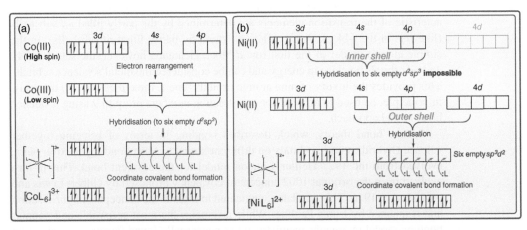

Figure 2.5

(a) A simple valence bond description of bonding for the d^6 complex ion $[Co(NH_3)_6]^{3+}$. Beyond six *d*-electrons in a low-spin configuration, this approach fails. (b) The attempted valence bond theory description for an octahedral Ni(II) complex. The six sp^3d^2 hybrid orbitals are shown for coordinate bonding but the d^8 electron count of Ni(II) means that some of them are partially occupied and unable to accept electrons from the donor atoms of the ligands, driving the use of non-valence higher energy 4*d* orbitals ($4d_{x^2-y^2}$ and $4d_{z^2}$) in this inner shell–outer shell model.

total number of valence electrons approaches or equals 18 (but does not exceed it) are more stable assemblies, with those actually achieving 18-electron sets (like an inert gas) being the most stable. This simple rule works best for organometallic compounds, which are more covalent in character, with more ionic character Werner-type compounds tending to break the rule more often and consequently it is very much less used for these. We shall develop this rule later in the chapter when we discuss organometallic complexes.

The covalent bonding model for complexes has a rival, an ionic bonding model, as we shall soon see. It is at first sight an eccentricity of metal coordination compounds that we can invoke either a covalent bonding model or an ionic bonding model. However, if we think of ionic and covalent bonds as the two extremes of a bonding continuum, the apparent contradiction becomes more acceptable. It is almost a state of mind; we would rarely consider CH_4 as anything but covalently bonded, but at the same time recognise that representation as CH_3^- and H^+ (or CH_3^+ and H^-) at least provides a way of understanding some nucleophilic and electrophilic organic reactions. At first sight, we are likely to be reasonably comfortable in considering a neutral compound such as $[Mo(CO)_6]$, composed formally of uncharged Mo(0) bonded to the carbon atoms of neutral CO molecules, as covalently bonded. When presented with a highly ionic assembly, such as $[CrF_6]^{3-}$, we can at least accept there is some wisdom in thinking of this as a Cr^{3+} ion with six fluoride anions attached by strong ionic bonds. What is common to both approaches to bonding in coordination chemistry is that the bonding models tend to be holistic in nature rather than focused on individual atom-to-atom bonding; this is probably a better description of systems where a set of entities are assembled to form a much larger whole. Two sophisticated theories have been developed to explain the properties of metal complexes. For the former example above, the *molecular orbital theory* represents an appropriate treatment of holistic covalent bonding, whereas for the latter example, the *crystal field theory* may be invoked to explain *d*-electron configurations. Each model has its strengths and weaknesses. Our models are developed to not only allow some level of understanding of physical and chemical properties of complexes but also to provide a means of prediction of properties where changes or new species are involved. A model is only useful if it performs the tasks that we ask of it.

2.3.2 Introducing Crystal Field Theory – *d*-Electrons Versus Ligand Lone Pairs

All bonding, covalent or ionic, has some component of electrostatic attraction between electrons and nuclei on different atoms. The simplest way of viewing coordinate bonding is to invoke a purely electrostatic (ionic) model. In this case, we recognise that the metal at the centre of the complex is usually a cation or at least more electropositive than its ligand donor atoms and that donor atoms are either anionic or highly electronegative elements carrying a full or partial negative charge. Thus, we can conceive of a situation where a set of donors is arranged around a metal ion centre in an array defined mostly by their electrostatic attraction to the metal ion and electrostatic repulsion towards other ligands. Kepert's model will dictate that, for example, six ligands will occupy the six vertices of an octahedron.

Crystal field theory (CFT) emerged from Hans Bethe's work published in 1929; focused on a group theory approach to the influence on the energy levels of an ion or atom upon lowering symmetry in a ligand environment, it is based on geometric and symmetry relations. First developed to interpret paramagnetism (the presence of unpaired electrons), it was subsequently applied more widely in the development of the field of coordination chemistry in the 1950s. Although it is sometimes incorrectly referred to as such, CFT is *not a bonding model*, as it relies on the brave assumption that we have just invoked that the attraction between metal cation and ligand negative charges is purely electrostatic (ionic).

We now need to consider what influence the ligand negative charges have on the metal d-electron orbital energies. As we have seen in Figure 2.4, the five d orbitals have distinctly different orientations: the d_{xy}, d_{xz} and d_{yz} orbitals point between x, y and z axes while the $d_{x^2-y^2}$ and d_{z^2} orbitals are oriented along the axes and directly at the six negative charges. To help with our explanation, we will choose as an example the Cr(III) complex $[CrF_6]^{3-}$ (Figure 2.6) which has three $3d$-electrons. In the free Cr^{3+} ion, devoid of any ligands, the three electrons can move freely from one d orbital to the other, all being energetically equivalent, so it does not matter which one is which at this stage. However, Hund's rule tells us that if you have a set of five degenerate $3d$ orbitals and three electrons, then they will remain unpaired and as far apart from each other as possible due to their mutual repulsion. Now let us introduce a spherical 'cloud' of negative charge equivalent to the charges of the six ligands but distributed randomly around the metal. This negatively charged cloud does not discriminate, and the electrons in the three $3d$ orbitals feel this electrostatic repulsion equally regardless of where they are, so all $3d$ orbital energies increase by the same amount. However, when this negatively charged cloud is focused into six negative point charges (e.g. six F^- ions) arranged along the x, y and z axes at the vertices of an octahedron (the most common shape met in coordination chemistry), an entirely different outcome to the spherical field situation results. The $3d_{x^2-y^2}$ and $3d_{z^2}$ orbitals, with lobes pointing along the axes, interact head-on with the ligand point charges, and so their electronic energies (when occupied) are raised by an equal amount. The repulsions between the point charges and the $3d_{xy}$, $3d_{xz}$ and $3d_{yz}$ electrons are minimal – less, in fact, than felt in a spherical field so that they drop in energy versus the spherical field. The outcome of the introduction of a defined arrangement of point charges in an octahedral field is that the degeneracy of the five $3d$ orbitals is removed, and the orbitals arrange themselves into two sets of differing energy. The formation of

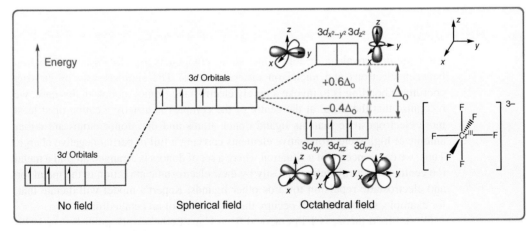

Figure 2.6

Crystal field influences for a set of d orbitals in an octahedral field exemplified by the Cr^{3+} ion (as in $[CrF_6]^{3-}$).

a higher energy doubly degenerate and lower energy triply degenerate set of d orbitals is characteristic of the octahedral crystal field. Because it is the spatial location of the set of point charges that is significant in generating this outcome, it will hardly come as a surprise to find that *every different coordination geometry (arrangement of point charges) will lead to a different d-orbital energy sequence* – but more on that later.

We pause to introduce some shorthand symbols that encapsulate both the symmetry and the energy degeneracy of an orbital. These symbols are from mathematical group theory, but this formalism is credited to Nobel Laureate Robert Mulliken who developed molecular orbital (MO) theory, which we will come to shortly. In terms of degeneracy, the Mulliken symbols are a = non-degenerate, e = doubly degenerate and t = triply degenerate. For orbital symmetry, a subscripted 'g' = symmetric (*gerade* is German for even) and 'u' = antisymmetric (*ungerade*).

In an octahedral complex, all of the d orbitals are in a symmetrical environment, so they all carry the 'g' subscript. The lower energy set of three orbitals (d_{xy}, d_{yz} and d_{xz}) is called the t_{2g} set, whereas the higher energy set of two orbitals ($d_{x^2-y^2}$ and d_{z^2}) is called the e_g set. The energy difference between these levels Δ_o is called the *crystal field splitting energy*, where the subscript 'o' is an abbreviation for 'octahedral'. An energy balance between the two sets of orbitals is struck, so that the three lower levels are $-0.4\Delta_o$ lower and the upper two levels $+0.6\Delta_o$ higher than the spherical field position set as the zero reference point. The language of chemistry contains several dialects, and you will find some older texts refer to the energy gap Δ_o as $10D_q$ instead (with the diagonal and axial sets lowered by $4D_q$ and raised $6D_q$, respectively, in this case; the D_q symbol arises from quantum mechanics as the product of two terms, and we shall not pursue its mathematically confronting origins). Do not be confused, as the same general conceptual model is being applied. Using the Δ_o (or Δ_t for tetrahedral) symbolism is more appropriate and has better applicability to other geometries beyond octahedral and tetrahedral, and we will employ it exclusively.

Electrons occupy the lowest energy orbitals available so the final decision is clear; the three valence electrons of Cr^{3+} will occupy the t_{2g} orbitals while the e_g orbitals remain empty. As already mentioned, the $3d_{xy}$, $3d_{yz}$ and $3d_{xz}$ orbitals have an identical shape but different orientations, so not surprisingly they remain separate and degenerate in an octahedral complex. However, it is notable (Figure 2.6) that the $3d_{x^2-y^2}$ and $3d_{z^2}$ orbital pairing also is degenerate despite their apparent differences in shape. How can this be? While it could be said, simply, that it just works out that way, this is hardly a satisfying outcome. The answer lies deep within the mathematical derivation of the atomic model and orbital functions, and mathematical detail is something we are doing our best to avoid here. However, there is a certain simplicity and logic here that we can inspect without recourse to mathematics. It turns out that the $3d_{z^2}$ orbital is a combination of two orbital functions, $d_{z^2-y^2}$ and $d_{z^2-x^2}$, that are shaped just like the separate $d_{x^2-y^2}$ orbital but lying in the xy and xz planes. It is then less surprising that the composite we see as the d_{z^2} orbital is energetically identical to the $d_{x^2-y^2}$ orbital. But if you are now wondering why these orbital functions need to be combined in the first place – well, that is a story you simply will not find here.

For the best known of the four-coordination geometries, tetrahedral, the ligand arrangement is distinguished by none of the point charges lying along the x, y and z axes that we employed for the octahedral case. The tetrahedron is based on a cube, with pairs of opposite corners removed. Thus, in the tetrahedral ligand field, the x, y and z axes bisect the angles formed at the metal centre by its four ligands. It becomes clear that the orbitals lying along these axes ($d_{x^2-y^2}$ and d_{z^2}) will interact least with the tetrahedral point charges, the *opposite to the behaviour in an octahedral field*

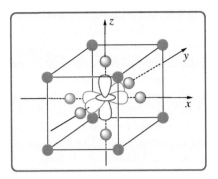

Figure 2.7
A view representing the interaction of the d_{z^2} and $d_{x^2-y^2}$ orbitals with a cubic field (dark olive circles) and an octahedral field (light green circles). The orientation of the d-orbital lobes directly towards the octahedral set contrasts with the orientation for the cubic set.

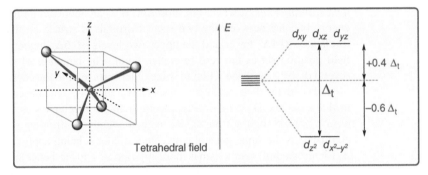

Figure 2.8
The d-orbital splitting diagram for a tetrahedral field. Note how the x, y and z axes perfectly bisect the coordinate angles.

(Figure 2.7). Conversely, the d_{xy}, d_{yz} and d_{xz} orbitals (not shown here for clarity) that lie between the x, y and z axes are disposed towards, *but not directly at*, the corners of the cube and will thus interact most.

The tetrahedral crystal field and the resulting d-orbital energy splitting are shown in Figure 2.8. As we anticipated, the interactions between the point charges and the d orbitals are reversed from the situation in the octahedral field, and it is no surprise that the orbital energy sequence is also reversed for the tetrahedral field. In this case, the symmetry labels change slightly, now simply e and t_2, the removal of the 'g' symbols being associated with the loss of symmetry – a tetrahedral crystal field is neither symmetric (g) or antisymmetric (u) according to our definitions.

Geometric arguments relating to relative distances between point charges and orbital lobes allow calculation of the relative strength of the orbital energy splitting in the tetrahedral case (called, now, Δ_t) compared with the octahedral case (Δ_o), which shows that $\Delta_t = \frac{4}{9}\Delta_o$ when identical ligands (as point charges) are placed equidistant from the metal centre in each case (Figure 2.9). Without going into details, the smaller value of Δ_t is due to there being only four point charges, as opposed to six in an octahedral crystal field, as well as the d-orbital interactions with the point charges in a tetrahedral field being slightly misdirected relative to the octahedral case. For the related cubic field, where there are twice as many point

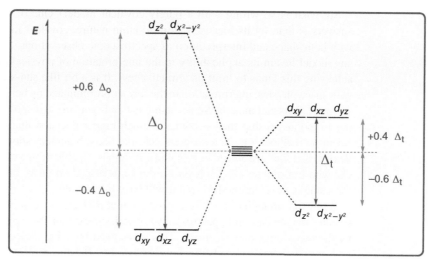

Figure 2.9

Comparison of the *d*-orbital splitting for octahedral and tetrahedral fields ($\Delta_t = {}^4/_9\Delta_o$).

charges present compared with the tetrahedral field, the splitting energy is twice the tetrahedral value, or $\Delta_{cubic} = {}^8/_9\Delta_o$; this is a rarely-met field (but present, for example, in [$Zr(CO)_8$]), so we will not develop it here.

There are many other geometries that we could cover with the CFT model, but octahedral and tetrahedral are the most encountered and will suffice for now. We must return to the nature of the coordinate bond.

2.3.3   Ligand Field Theory – Molecular Orbitals to the Rescue

It is perhaps a bit confusing to assert that such a thing as covalent coordinate bonds exist in Chapter 1, then introduce a theory (CFT) based on purely ionic interactions. While there is an array of experimental outcomes from a range of methods that can be, at least qualitatively, fitted to the *d*-orbital splitting model of CFT, quantitative calculations based on this electrostatic model tend to fail to match experimental values. However, the key failing is that a purely ionic description of bonding in coordination complexes is not consistent with the now abundant experimental evidence. Put simply, electrons in *d* orbitals, which are fully localised therein in the CFT model, in fact spend part of their time in ligand orbitals (they are 'delocalised', or there is covalency in the bonding). The spectroscopic method of electron paramagnetic resonance (see Chapter 7.3.6), which 'maps' unpaired electron density, indicates that *d* orbitals in fact can have their electrons appreciably delocalised over the ligands, whereas evidence for donation of ligand electron density to the metal comes from the nephelauxetic effect of electronic spectroscopy. The term 'nephelauxetic' means 'cloud expanding', and the effect relates to the observation that electron pairing energies are lower in metal complexes than in 'naked' gaseous metal ions. This suggests that interelectronic repulsion has fallen on complexation, which equates with the effective size of metal orbitals increasing. The effect varies with the type of ligand bound to the metal. Even clawing back some of the concept by asserting mixed bonding character seems a grudging concession.

So, what is so wrong with a purely covalent model? One of the most obvious answers is that, by its localised atom-to-atom nature, covalent models deal poorly with both shape and interpretation of spectroscopic observations. Since the value of any model lies in its applicability in the interpretation of physical observations, not achieving this tends to limit its attractiveness. It is a bit like sun-worship compared with space physics; interpretation in the former is dominated by belief, in the latter by scientific facts and models. Sun-worship is fine if you are so disposed, except changing to moon-worship then becomes a much bigger decision than simply applying scientific methodology to a different celestial body. Scientists should be sufficiently flexible that they are able to accept that a system is capable of sustaining several models, whose usefulness depends on the demands placed on them. Thus, there is room for a simple covalent model – but it will have limitations.

While the strength of CFT is its prediction of *d*-electronic configurations (encompassing magnetism and spectroscopy), it fails to explain the essential role played by the ligand lone pair electrons. This is resolved by using molecular orbital (MO) theory, which may already be familiar to you. Briefly, molecular orbital theory involves mixing different combinations of atomic orbitals from two (or more) atoms making the covalent (or coordinate) bond. As a holistic model, it is based on the recognition that it is not essential to limit bonding to linkages between only two atoms at a time. Rather, it allows a molecular orbital to spread out over any number of atoms in a molecule, from two to even all atoms. This is a complex and mathematically intensive theory, which is not explored here in depth. It applies mathematical *group theory* to determine the allowed combinations of metal orbitals and ligand orbitals (or orbital overlap) that lead to bonding situations. One example of a combination involving many centres is the overlap of a metal *s* orbital with all six of the lone pair orbitals from a set of six ligands at once, and this will be described shortly.

The phase and symmetry relationships introduced in Section 2.3.1 are largely of academic interest only *until the atomic orbitals from different atoms come together to form new molecular orbitals*. Then, like waves, their positive and negative phases will either reinforce or cancel each other. If the overlapping regions have the same phase (like with like), there will be reinforcement resulting in electron density *between* the two atoms which is known as a *bonding molecular orbital* (BMO). If the phases of the overlapping atomic orbitals are opposed, then cancellation occurs, and no electron density remains between the two atoms which typifies an *antibonding molecular orbital* (AMO). Note that you must take the good with the bad, as two atomic orbitals will inevitably create one BMO and one AMO.

To illustrate MO theory, we shall start with the simplest example, H_2, comprising the combination of two H atoms each bearing a singly occupied $1s$ orbital (Figure 2.10). It is helpful to think about this as a chemical reaction between two hydrogen atoms forming the H_2 molecule. The colours of the orbitals reflect their phase, for which we can assign blue to positive and colourless to negative (this could be reversed with the same outcome). Also note that reversing the phase of an orbital and adding it to another orbital is the same as subtracting one orbital from another, i.e. $a - b = a + (-b)$. That is as mathematical as this will get!

When the H atoms come together, both combinations (+/+ [blue/blue] and +/– [blue/white]) of their $1s$ atomic orbitals result in one BMO and one AMO. The BMO has electron density *between* the two hydrogen nuclei which leads to attraction between the electrons and both positively-charged H nuclei at the same time. The MO symbol σ_{1s} is self-explanatory as it defines a σ-bond where electrons are mostly

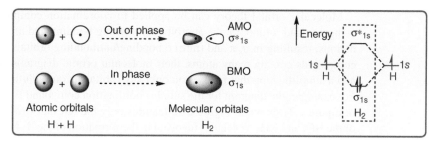

Figure 2.10
The construction of H_2 molecular orbitals from the two combinations of $1s$ atomic orbitals on each H atom. Both electrons are found in the bonding molecular orbital (BMO), and none in the antibonding orbital (AMO).

between the nuclei, and the subscript tells us which atomic orbitals were used. When the phases are opposed, there is cancellation, and the resulting AMO has no electron density between the nuclei. The '*' symbol in $\sigma_{1s}{}^{*}$ is reserved for antibonding orbitals in general. There is balance in the orbital energies on the right of Figure 2.10 where the BMO energy is lower than those of the $1s$ atomic orbitals while the AMO is higher by the same amount. Both electrons in H_2 occupy the BMO as it is of lower energy. Note that electrons in this σ_{1s} BMO now belong to the H_2 molecule; they have lost their original identity as $1s$-electrons. Once the atomic orbitals are used, there is no going back! We can also see (Figure 2.10) that we have gained electronic energy (stability) in forming H_2 from two isolated H atoms as the two electrons now move to a lower energy (molecular) orbital. This can be equated to the bond formation/dissociation energy.

Before pushing on to MOs in coordination complexes, specifically known as *Ligand Field Theory*, which use combinations of $(n+1)s$, $(n+1)p$ and nd orbitals, we should define some symmetry rules that must *never* be broken. It is all or nothing when sets of atomic orbitals on different atoms are brought together, as overlap is either all in phase (BMO) or all out of phase (AMO). This places some symmetry restrictions on the way atomic orbitals can interact. For example, an s and a p orbital from different atoms may overlap in but one way. A 'head-on' interaction is permitted creating a σ_{sp} BMO and a σ_{sp*} AMO (Figure 2.11). However, a side-on interaction is not allowed as the symmetries are mismatched; half of the antisymmetric p orbital presents a positive phase while the other half presents a negative phase, and the totally symmetric s orbital is confused! We call this a *non-bonding* situation, where the orbitals do not mix and retain their atomic orbital character.

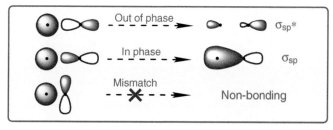

Figure 2.11
Symmetry-allowed and forbidden combinations of s and p orbitals to form molecular orbitals.

Molecular orbital theory can be applied to coordination compounds in general, using *s*, *p* and *d* (and, rarely, *f*) orbitals in making bonding and antibonding combinations, resulting in σ, π and (rare) δ bonding/antibonding orbitals. As coordination compounds contain many atoms, their molecular orbital diagrams are complicated, and a simplification that focuses on only key orbitals is one solution you may meet. For example, the diagram for the simple coordination compound $F_3B–NH_3$ (first met in Figure 1.2) shown in Figure 2.12 features only the relevant two 'frontier' orbitals of the BF_3 and NH_3 molecules involved in the formation of the B–N coordinate bond, despite them having 16 and 7 molecular orbitals, respectively. Without reference to their origin or character, we can define these from the complete MO diagrams as *the lowest unoccupied molecular orbital* (LUMO) on the precursor molecule BF_3 and the *highest occupied molecular orbital* (HOMO) on the precursor molecule NH_3. These are the two orbitals you can consider specifically involved in forming the B–N coordinate bond; think of it, if you must (albeit inappropriately), as the equivalent of one occupied orbital donating electrons to another empty orbital in valence bond theory. As presented, Figure 2.12 is similar in form to the diagram for H_2 in Figure 2.10 (except that now the two precursor orbitals are not identical and hence of different energies). It is clearly possible, not only simply for ease of visualisation but also because frontier bonding orbitals are important in the theory, to sometimes restrict a molecular orbital diagram to solely depicting 'orbitals of interest'. In the present case (Figure 2.12), the pair of electrons occupy the resultant BMO in the $F_3B–NH_3$ assembly, leading to a more stable situation (as the electrons now occupy a lower energy orbital compared to the energy of the orbitals on the original separate NH_3 and BF_3 centres), and hence a stable bonding outcome occurs. This frontier orbital view, with its focus on two core molecular orbitals in combination, is somewhat contrived, although it does show how the theory presented for the simplest diatomic molecule has wider application. For transition metal complexes, however,

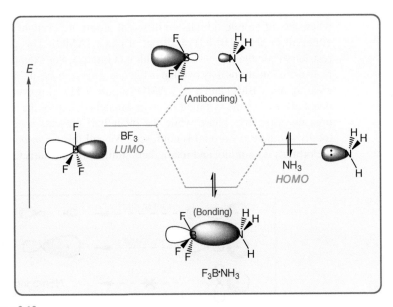

Figure 2.12
A simplified molecular orbital diagram for $F_3B–NH_3$, reporting only the orbital interaction of the lowest BF_3 unoccupied molecular orbital and the highest occupied NH_3 molecular orbital.

we need to switch our focus to molecular orbitals formed from combinations of several ligand atomic orbitals that may interact with orbitals on the metal.

Now, having reviewed and illustrated the basics of MO theory, we can move to bonding in transition metal complexes, which is within the scope of *Ligand Field Theory* (LFT). Our first step is to recognise that we have nine valence atomic orbitals on our metal. If we are in the first transition metal series, these comprise one $4s$, three $4p$ and five $3d$ atomic orbitals. The most common coordination geometry is six-coordinate octahedral, so let us start there. Each of the six ligands will contribute two electrons from their lone pair orbital. We cannot continue the bilateral arrangements of pairs of orbitals brought together as in H_2 or $F_3B–NH_3$, and we need to switch to combinations of ligand orbitals, called *ligand group orbitals* (LGOs), and they will have phase and symmetry, just as the metal AOs do. Each ligand group orbital 'collective' interacts in unison with a metal-based atomic orbital, just in the same way we saw with H_2, but the symmetry rules dictated by the metal AOs must be followed. This means that all we need to know are the symmetries of the $4s$, $4p$ and $3d$ orbitals (see Figure 2.4), and we can construct LGOs that are compatible with each of them.

The $4s$ atomic orbital is totally symmetric so all six ligands may contribute to the LGO if they all have the same phase. Two combinations are possible (Figure 2.13, top left) leading to a totally symmetric BMO (using Mulliken's symbols, a_{1g}, all in phase) and an AMO (all out of phase, $a_{1g}*$). Moving to the antisymmetric metal $4p_z$ orbital, its corresponding LGO must also be antisymmetric. As we have seen in Figure 2.11 for an s orbital, a ligand lone pair orbital cannot approach a $4p_z$ orbital side-on, only head-on. So, in this case, none of the four ligands in the xy plane can contribute to this LGO. Guided by the antisymmetric $4p_z$ orbital, the LGO must comprise one sole pair of ligands oriented along the *z-axis* with opposite phases leading to the BMO and AMO in Figure 2.13 (centre left). The situation is analogous for the $4p_y$ and $4p_x$ orbitals. As they have the same shape and size, but different orientations we can group them into sets of threefold degenerate, antisymmetric MOs with Mulliken symbols t_{1u} (BMOs) and $t_{1u}*$ (AMOs).

You may be wondering about the symmetrical $3d$ orbitals, as until now we have not mentioned them. The $3d_{x^2-y^2}$ orbital lies exclusively in the xy plane pointing towards the ligands (remember what we saw in CFT) so we immediately know that no LGO contributions along the z-axis are allowed from symmetry. Instead, the LGOs follow the planar symmetry of the $3d_{x^2-y^2}$ orbital with contributions along the x- and y-axes of opposite phase to give the BMO and AMO as shown in Figure 2.13 (centre right). Finally, the symmetric d_{z^2} orbital is directed mainly along the z-axis with a small contribution in the xy plane. The LGO must follow this lead and the resulting BMO and AMO are shown in Figure 2.13 (lower right). These two d-based BMOs have the same energy and symmetry, so they have the symbol e_g while the antibonding pair of orbitals are grouped together as e_g*.

Three valence atomic orbitals were not used, namely symmetric $3d_{xy}$, $3d_{xz}$ and $3d_{yz}$. It is clear that their orientation with respect to the six σ-bonding ligands does not lead to any LGOs that are permitted by symmetry. This is illustrated in Figure 2.14 for the $3d_{xz}$ orbital. It does not matter from which direction the LGO frontier orbitals approach (along the x, y or z axes), as in each case they meet symmetry mismatched lobes of the d orbital, and this is not permitted. The same situation occurs for the $3d_{xy}$ and $3d_{yz}$ orbitals, so these three remain *non-bonding* orbitals in octahedral complexes and retain their atomic orbital character. Their threefold degeneracy and symmetry

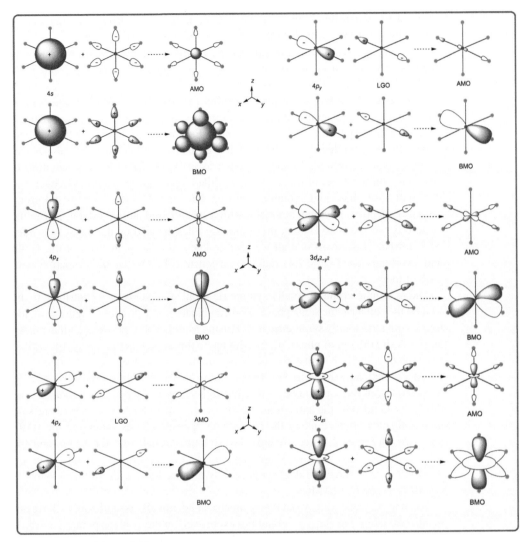

Figure 2.13

The linear combinations of valence atomic orbitals that lead to the six bonding and six antibonding orbitals in an octahedral complex, shown as two parts. (LGO refers to a ligand group orbital, BMO represents a bonding molecular orbital and AMO an antibonding molecular orbital.)

are in accord with their t_{2g} symbol for the set, which is no surprise, as they are the same three orbitals grouped together in our CFT model.

Let us take stock. We have created six BMOs and six AMOs (Figure 2.13) while three of the metal's atomic orbitals remain non-bonding orbitals ($3d_{xy}$, $3d_{yz}$ and $3d_{xz}$, Figure 2.14). So now we have 15 orbitals in total. The relative energies of these orbitals are shown in Figure 2.15. As we saw with H_2, the BMOs are stabilised relative to the non-bonding atomic orbitals by the same amount that their paired AMOs are destabilised. Predicting the energy sequence of the BMOs is non-trivial but some qualitative observations are worth making. The a_{1g} BMO (derived from the 4s atomic orbital) is the most stable as it is privileged by having six metal–ligand pairings within the one orbital: i.e. an electron in this orbital feels the attraction

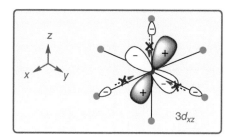

Figure 2.14
The mismatched symmetry of the $3d_{xz}$ orbital for head-on (σ-type) bonding in an octahedral complex.

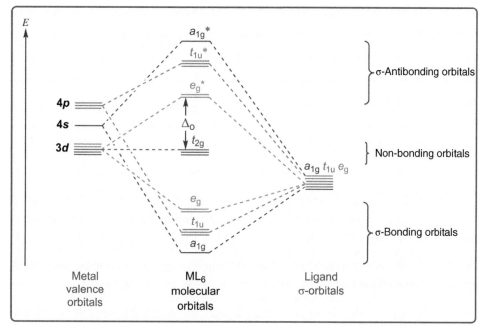

Figure 2.15
The MO energy diagram for an octahedral complex.

of seven different positively-charged nuclei at once. The strongly bonding t_{1u} set (each oriented along the x, y and z axes) is next in energy, with the two e_g BMOs derived from the $3d_{x^2-y^2}$ and d_{z^2} orbitals completing the set of six. The AMOs follow the same pattern, in reverse for symmetry, in our diagram. This energy diagram is perfectly general for *any* octahedral complex with the same six ligands. If we return to the CFT example $[CrF_6]^{3-}$ and instead apply LFT, the electrons from the fluorido ligands ($6 \times 2 = 12$) go into the six lowest energy σ-BMOs while the remaining three valence electrons from Cr^{3+} occupy the degenerate t_{2g} orbitals, one in each orbital. Reference to the shapes of these six σ-BMOs (Figure 2.13) shows that the electron pairs are not distributed in the same way and their energies are not all the same. They are *not* simply localised between six discrete pairs of Cr^{3+} and F^- ions.

We also see a familiar pattern. The set of triply degenerate $3d_{xy}$, $3d_{xz}$ and $3d_{yz}$ (t_{2g}) non-bonding orbitals lie slightly lower in energy than the degenerate pair of AMOs derived from the $3d_{x^2-y^2}$ and $3d_{z^2}$ orbitals ($e_g{}^*$). The energy separation Δ_o is the same

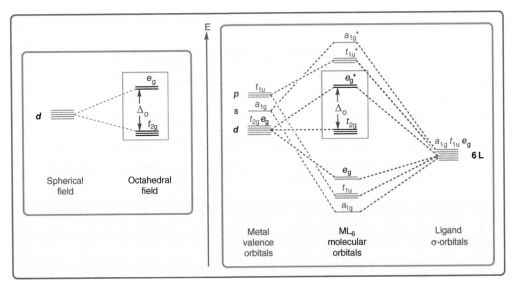

Figure 2.16

Comparison of CFT (ionic; at left) and LFT (molecular orbital; at right) development of the *d*-orbital splitting diagram for octahedral systems. Both reduce to the equivalent consideration of insertion and location of metal-dominated *d*-electrons in two degenerate sets of orbitals separated by a relatively small energy gap (Δ_o).

value defined by CFT (Figure 2.16). It is comforting to know that the more elaborate LFT is in accord with the simpler CFT model that we know accurately predicts *d*-electronic configurations, so we have lost nothing and gained a complete understanding of how the six ligand lone pairs combine to stabilise six-coordinate bonds.

One of the strengths of LFT (and MO theory in general) is that the same MO energy diagram can be used for systems with different numbers of electrons, i.e. metals with various *d*-electronic configurations. As we can now see, in *any* octahedral transition metal complex bearing σ-bonding ligands, the six lowest energy molecular orbitals will be fully occupied and *any six donor atoms with a lone pair will fulfil this requirement*, e.g. combinations of F^-, H_2O, NH_3, etc. It is the ligands that provide the 'glue' to hold the complex together – no *d*-electrons are required. The five t_{2g} and e_g* MOs may accommodate no more than 10 electrons, after which strongly antibonding (high energy) molecular orbitals start to be populated. As soon as the antibonding e_g* MOs are populated with electrons the coordinate bonds are weakened, as the electrons in these orbitals are unable to draw the metal and donor atom nuclei together. Again, making the connection to CFT (Figure 2.16), the e_g* AMOs were formerly described there as pure $d_{x^2-y^2}$ and d_{z^2} atomic orbitals. LFT provides a more accurate picture, as with the e_g* AMOs shown in Figure 2.13, there is a small amount of electron density on the donor atoms, but the remaining electron density around the metal does look very much like the original $d_{x^2-y^2}$ and d_{z^2} atomic orbitals. You will often see these t_{2g} and e_g* MOs and their electrons loosely referred to as *d* orbitals and *d* electrons, as a throwback to CFT. No harm is done if you understand the distinction.

After the e_g* MOs are full (a d^{10} configuration in CFT language), no more electrons can be accepted unless they are placed in much higher energy, strongly antibonding MOs (see Figure 2.15), which then would destabilise the coordinate bonding interactions we have tried so hard to create, so we will not go there! The transition metal series stops abruptly when the electronic configuration reaches nd^{10}, as if any more electrons are added, then they must be accommodated by different molecular orbital interactions that are the domain of p-block chemistry and well outside the scope of this book.

The same LFT model may be set up involving combinations of LGOs and d orbitals in a tetrahedral complex. We will not present this in this book, but the reader may be assured that the end result is what we would expect; the four lowest energy BMOs house the electrons that form the four coordinate bonds while the 'inverted' d-orbital splitting pattern seen in the CFT model for tetrahedral complexes (Figure 2.8) is also reproduced.

Regardless of how we got there (CFT or LFT), the signature pattern of d-based orbital splitting energies characteristic of an octahedral or tetrahedral field remains. One key question that begs an answer at this stage is what factors govern the size of Δ_o or Δ_t? Answering this should give us the satisfaction of being able to predict and explain changes in spectroscopic properties like colour or magnetic properties. Obviously, since we are involving both metal and ligand in our complex, one can anticipate that both have a role to play. If we fix the metal's identity, then we can focus on the ligands and their properties that influence $\Delta_{o/t}$.

For octahedral complexes of most first-row transition metal ions at least, the presence of colour tells us that some part of the visible (white) light spectrum has been filtered out leaving an apparent colour: e.g. if we filter out the violet part of the spectrum, but nothing else, we will see the complementary colour yellow. Filtering out a particular colour means the energy associated with that light has been used in some way, as energy must always be conserved. On a molecular scale, this energy is harnessed to promote electrons from lower energy orbitals to higher energy orbitals, so-called *electronic transitions*.

It is nothing more than a happy accident, but the many and varied colours of most first-row transition metal ions are due to their Δ_o values coinciding with the energies of light that is visible to our eyes ($\sim$400 to $\sim$700 nm). In the examples shown in Figure 2.17, four six-coordinate vanadium ions in different oxidation states but with essentially the same ligands (H_2O or O^{2-}) are shown along with their associated d-electronic configurations. For the d^1, d^2 and d^3 ions, absorption (filtering, if you like) of a component of the visible spectrum occurs, and the absorbed energy is used to force an electron from the lower energy t_{2g} set to the higher energy $e_g{}^*$ set, a d–d electronic transition. The colours that were not absorbed, and *complementary* to the colour absorbed (red/green, violet/yellow, blue/orange) define the colour we see. These electronic transitions evidently occur at different wavelengths resulting in their blue, green or purple colours due to the different charges and electronic configurations of the ions. They take on a completely different appearance simply due to the changes in d-electrons. The d^0 V(V) ion is pale yellow but this cannot be due to a d–d transition (as no d-electrons are present!). In this case, the absorption of light in the purple (near UV) region is due to a ligand-to-metal charge transfer electronic transition promoting an electron from an O-based orbital to a V(V) $3d$-orbital. These features will be covered in more depth in Chapter 7.

Simply by monitoring the change in colour as ligands are swapped for others, we can determine the energy gap $\Delta_{o/t}$ for any ligand set. The stronger the ligand field, the larger $\Delta_{o/t}$ and hence the more energy required to promote an electron, leading to a decrease in the wavelength of absorbed light (higher energy). This allows us to rank particular ligands in terms of their capacity to separate the t_{2g} and $e_g{}^*$ energy levels.

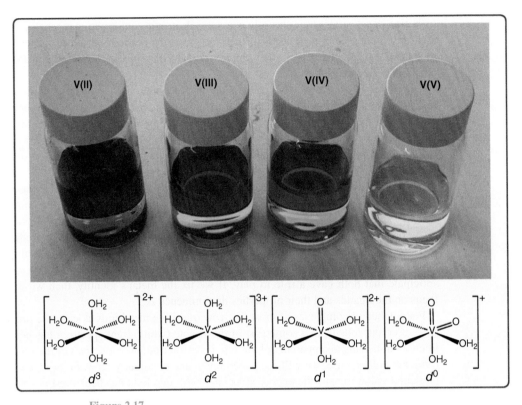

Figure 2.17
The variety of colours exhibited by stable vanadium metal ion complexes of different oxidation states (and hence number of *d*-electrons) in aqueous solution.

This has been done to produce what is called the *ligand spectrochemical series*. For some common ligands that bind as monodentates to a single site, this order in terms of increasing ability to split the t_{2g} and $e_g{}^*$ orbital energy levels is of the form:

$$\text{I}^- < \text{Br}^- < \underline{\text{S}}\text{CN}^- < \text{Cl}^- < \text{S}^{2-} < \text{F}^- < \text{HO}^- < \text{O}\underline{\text{N}}\text{O}^- < \text{OH}_2 < \underline{\text{N}}\text{CS}^-$$

$$< \text{NH}_3 < \text{O}\underline{\text{N}}\text{O}^- < \text{PR}_3 < \underline{\text{C}}\text{N}^- < \underline{\text{C}}\text{O}.$$

Notably, this order is just about independent of metal ion, oxidation state and coordination geometry, and it applies equally well to Δ_o or Δ_t. Note that this series does not sit comfortably with the crystal field (ionic bonding) model. For example, the series indicates that an anion like Br^- is far *less* effective than a neutral molecule like CO at splitting the *d*-orbital set, which is at odds with a model based on electrostatic repulsions, where one would expect a charged or highly electronegative ligand atom to be more effective than a neutral one when the actual order is reversed.

Not surprisingly the metal (its position in the periodic table and its charge) also plays a role in the magnitude of Δ_o. For the same ligand donor set, the variation of Δ_o follows the *metal spectrochemical series* which is dictated by two properties. The first is the formal charge (oxidation state). As the formal charge increases, the coordinate bonds decrease due to a greater ionic attraction between the cation and its electronegative donor atoms. According to our LFT model (Figure 2.16), this coordinate bond shortening will raise the energy of the antibonding $e_g{}^*$ orbitals. It follows that in the metal spectroelectrochemical series we have the ordering (+4 > +3 > +2).

The second effect is due to the transition series in which the metal lies ($3d$, $4d$ or $5d$). The magnitude of Δ_o increases down any group (triad) with the same set of ligands resulting in the ordering ($5d > 4d > 3d$). Metal ions of the same charge within the same row of the periodic table are essentially equivalent within this series (shown in brackets here):

$$Pt^{4+} > Ir^{3+} > Rh^{3+} > (Co^{3+}, Cr^{3+}, Fe^{3+}) > (Fe^{2+}, Co^{2+}, Ni^{2+}, Cu^{2+})$$

Until now, our focus has been on coordinate bonds involving a ligand lone pair (or LGO) directed between the metal and donor atom, namely, a σ-bond. We already know from organic chemistry that π-bonding is encountered in compounds with double and triple bonds. The coordinate bond can also accommodate π-bonding within LFT if the all-important symmetry rules are obeyed. The first point to make is that σ-bonds and π-bonds can be discriminated well by reference to the x, y and z axes of a coordinate system. We have seen that σ-bonds in an octahedral complex cannot involve the d_{xy}, d_{xz} or d_{yz} (t_{2g}) orbitals (Figure 2.14). If we now include π-bonding, then things change quite dramatically. The example we choose is the ligand Br^- which is replete with lone pairs of electrons. Its electronic configuration is $4s^2 4p_x^2 4p_y^2 4p_z^2$. If we bring the $4p_z$ orbital within bonding range of a transition metal, viewed here only in the xz plane, the d_{xz} orbital is symmetry-matched with the occupied $4p_z$ orbital so a new π-BMO is created as well as a new π-AMO; this will be replicated with the other two p orbitals and appropriate d orbitals in other metal-ligand pairings (Figure 2.18). In organic chemistry, π-bonds are always accompanied by σ-bonds and a similar feature occurs with coordinate bonds. The same σ-BMOs (and LGOs) illustrated in Figure 2.13 may augment the π-bonds formed, using some of the remaining atomic orbitals on the bromide anion and creating a stronger coordinate bond overall. The σ-bond that may be created by MOs aligned with the x-axis are not shown in Figure 2.18, but their position is indicated by the dashed line.

The situation in Figure 2.18 applies when the ligand provides the lone pair of electrons via a $4p$ orbital. This is analogous to the donor–acceptor concept developed before for σ-donor coordination, so we describe Br^- in this context as a π-*donor ligand*. In any MO interaction, the orbital energies change. The electrons in the three

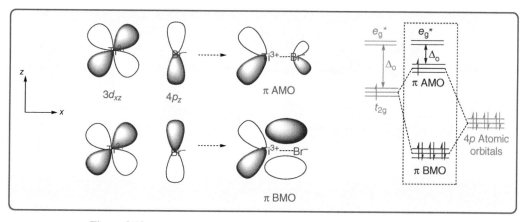

Figure 2.18

The π-donor MO interactions presented by Br^- ions in a complex with the d^1 metal ion Ti^{3+}. The formation of one of three π-MOs is illustrated at left, with the molecular orbital diagram including all three presented at right.

4p orbitals of Br⁻, formerly non-bonding lone pairs, all are stabilised by overlap with the d_{xy}, d_{xz} and d_{yz} orbitals along the three different axes (only one is shown here). The out-of-phase AMO that shifts to higher energy is where the single Ti^{3+} valence electron resides. It is clear that Δ_o *decreases* as a result of this π-donor interaction. The decrease in Δ_o means that any ligand acting as a π-donor will appear lower in the spectrochemical series relative to a ligand with the same donor atom but one without orbitals of appropriate symmetry for π-MO formation. The most notable example is the pairing of ⁻OH (an effective σ- and π-donor) and H₂O (a pure σ-donor), which are well separated in the spectrochemical series.

Now that we have established that the d_{xy}, d_{xz} and d_{yz} orbitals are capable of being involved in π-bonding, we need to address a new and interesting situation where the roles are reversed, i.e. the metal provides the electrons for the coordinate bond and the ligand accepts them. This places some demands on both metal and ligand. The ligand must have empty π-orbitals in which electrons can reside and the metal must have electrons in d orbitals of appropriate symmetry. In terms of the ligand, this type of bonding is impossible for a halide anion (like bromide), as this example already has its full complement of 18 valence electrons ($4s^2 3d^{10} 4p^6$). Indeed, the ligands that are most capable of accepting electrons are those equipped with empty π-antibonding (π^*) MOs. The most celebrated example is CO, a remarkable molecule which can be drawn as a combination of Lewis (resonance) structures in valence bond theory, none that are truly representative of its character. The MO description is much more realistic and accounts for the multiple bond character of the C–O bond, as shown in Figure 2.19. A lone pair formally resides on the C-atom so it may function as a σ-donor just like NH₃ or H₂O, but CO has much more to offer. The bonding in CO is effectively a triple bond comprising a σ-bonding MO and two orthogonal π-bonding

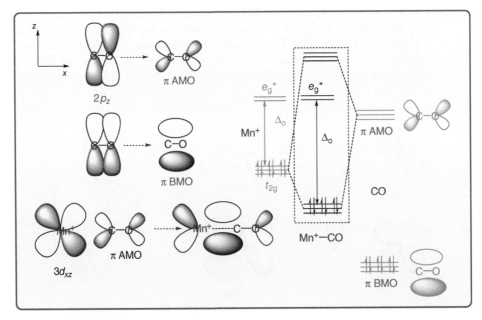

Figure 2.19

MO diagram for CO acting as a π-acceptor from Mn⁺ via its π-antibonding MO. One of the orthogonal π-AMOs of CO is shown at the top left. The molecular orbital energy diagram for the interaction of a π-AMO of CO with the Mn⁺ $3d_{xz}$ orbital in the complex [Mn(CO)₆]⁺ is shown at right.

MOs; one of these is shown in Figure 2.19. The BMOs are occupied and at very low energy, so they are incapable of accepting electrons from the metal, in this case Mn^+ in $[Mn(CO)_6]^+$ The higher energy empty π^* AMO of CO has the correct symmetry to overlap with the filled $3d_{xz}$ orbital of Mn^+, and it is these $3d$-electrons that are donated back to CO, hence the term for this interaction as *back-bonding*. The CO ligand is acting as a π-acceptor (as well as a σ-donor). As with the π-donor case in Figure 2.18, the driving force for this process is that electrons (originally non-bonding in the t_{2g} set of Mn^+) now find themselves in strongly bonding molecular orbitals (attracted to both the Mn and C nuclei), and in this case, the value of Δ_0 increases.

Back-bonding is a common feature of metal coordination to a molecular ligand that includes double or triple bonds. The outcome, as seen in Figure 2.19, is a stabilisation of the t_{2g} set and an increase in the size of the splitting (or of Δ_0). The experimental observation that CO complexes tend to be spin paired (low-spin) is consistent with the larger Δ_0 predicted in the model, which also equates with π-acceptor ligands lying at the top of the spectrochemical series as they split the d orbitals by the greatest amount.

LFT, with its more comprehensive treatment of bonding, superseded CFT as it was able to recognise that both σ and π bonds can occur in complexes as in pure organic compounds. The two models are sufficiently closely related in terms of their core operations involving the d orbitals for them to be often used (albeit inappropriately) interchangeably. While deficient in some aspects, CFT remains a useful theory for qualitative and limited quantitative interpretation. What is perhaps most remarkable about this theory is that it works at all, given the simplicity and level of assumptions – but it does. Further, it has stood the test of time to remain adequate despite the vast changes in the field over the decades since its evolution.

However, it is LFT that is more sophisticated and applicable. The capacity to model and account for the interaction of ligands capable of additional π-bonding with metals provides an enabling mechanism for dealing with the experimental spectrochemical series positioning of ligands. Of course, not all ligands will have the capacity to undergo further π-type interactions with metal orbitals. For example, ammonia has but one lone pair directed to σ-bonding, and otherwise three internal N–H σ-bonds; it cannot undertake any further bonding. Alternatively, carbon monoxide ($C\equiv O$) has an array of π-bonds resulting from its multiple bond character and is a clear candidate for further π-type interactions, as described earlier. However, such ligands must have low lying empty orbitals of the right symmetry to participate. The scene has now changed, so that we have a greater capacity to understand and predict the effects of ligands on the chemistry of metal complexes – this is the strength of the LFT and molecular orbitals.

What we have in the end is a reasonably consistent set of models (CFT and LFT) that differ in their focus and assignment of importance to electrostatic and covalent character. What is the 'real' situation, and how can we effectively assess the contribution of each component? A key and indeed simple approach to this came forward from Linus Pauling, who asserted that metals and ligands would adopt, as far as practicable, nett charges close to zero, through metal–ligand bond polarisation or π-type metal-to-ligand or ligand-to-metal electron density donation. As a consequence of this concept of *electroneutrality*, a metal ion in a high oxidation state with a high formal positive charge would seek to attract ligands that neutralised its charge or promoted ligand-to-metal charge donation, whereas one in a low oxidation state may do the reverse. Support for this concept, experimentally, is the observation that a high-charged metal ion such as Mn^{7+} prefers dominantly O^{2-}

ligands (as in the well-known permanganate ion, $[MnO_4]^-$), whereas lower-charged Mn^{2+} is satisfied with neutral H_2O ligands in $[Mn(OH_2)_6]^{2+}$.

What we have seen so far is that bonding between metals placed centrally in an array of ligands that participate in coordinate bonding can be represented by models of eventually high sophistication. Most of our understanding and interpretation of physical observables draws on these models, and it is their success in allowing us to interpret what we see that keeps the models in use. But a model is imperfect because it is a model – and so higher levels of sophistication will reinvent or replace these models as time goes by. At present, at least, and for an introduction to the subject, they are more than sufficient – indeed they have proven remarkably resilient and effective, despite the extraordinary growth in coordination chemistry over the decades since their development.

2.3.4   *d*-Electronic Configurations and Magnetism

The derivation of a model that manipulates the *d*-orbital set is incomplete without consideration of the placement of *d*-electrons into the orbitals. Because the simple CFT model we have developed is ionic in character, one of the advantages is that we do not have to worry about any additional electrons coming from attached groups; LFT can deal with that. The only electrons of concern are the *d*-electrons themselves, which will vary from 0 (a trivial case) to a maximum of 10 (all that the five *d* orbitals can accommodate) depending on the metal and its oxidation state. At first, we can apply simple rules for filling atomic energy levels – fill in the order of increasing energy – and apply Hund's rule (which states that for degenerate levels, electrons add to each orbital separately maintaining parallel spins before pairing up commences). The outcome is the set of electron arrangements shown for the octahedral case in Figure 2.20. What this model provides is support for complexation, since in most cases the energy of the assembly in the presence of the octahedral field is lower than

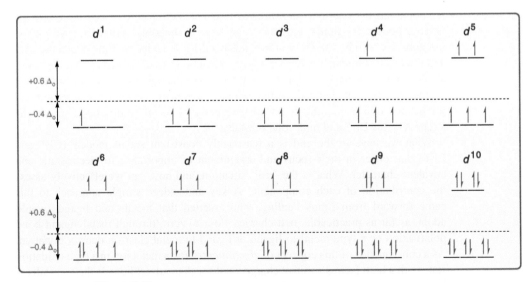

Figure 2.20

Electron arrangements in the t_{2g} and $e_g{}^*$ orbitals in an octahedral field with a maximum number of unpaired electrons for each configuration.

the spherical field situation. To clarify this, consider the simple d^1 case. Here, an electron resides only in a t_{2g} orbital, which is more stable by $-0.4\Delta_o$ than the spherical field zero-point shown by the level of the dotted line. The energy of this stabilisation is called the crystal field stabilisation energy (CFSE). For d^2, with electrons in different t_{2g} orbitals, stabilisation is $2 \times (-0.4\Delta_o) = -0.8\Delta_o$.

This set is only part of the story, however, since we must now consider the consequence of the relatively small energy gap (Δ_o) between the t_{2g} and e_g^* set, as a decision has to be made regarding more than one possible electronic configurations in some cases. For the d-electronic configurations in an octahedral field, with its two sets of degenerate levels, there is no case for consideration with d^0, d^1, d^2, d^3, d^8, d^9 and d^{10}, since there is only one option (assuming Hund's rule applies – i.e. every orbital in a subshell is occupied by a single electron of the same spin before any pairing up occurs).

However, let us consider the example of the d^4 electronic configuration in an octahedral complex. The four d-electrons can be accommodated in two different ways, all electrons in the t_{2g} set (t_{2g}^4), the so-called *low-spin* (or partially spin paired) configuration, or all unpaired and spread over both sets ($t_{2g}^3 e_g^{*1}$), the *high-spin* (or spin free) configuration (Figure 2.21). By the same method as employed earlier, we can calculate the CFSE as $-1.6\Delta_o$ [from $4 \times (-0.4\Delta_o)$] for the low-spin case and only $-0.6\Delta_o$ [from $3 \times (-0.4\Delta_o) + 1 \times (+0.6\Delta_o)$] for the high-spin configuration. The choice of

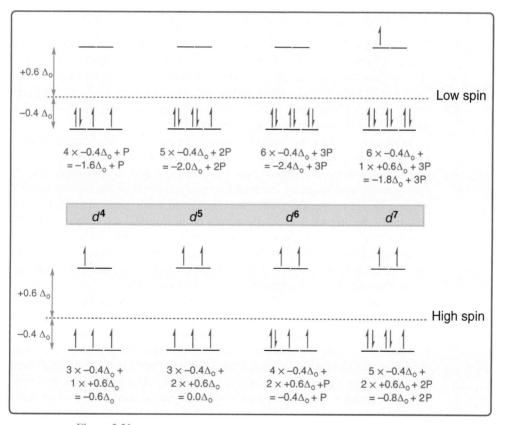

Figure 2.21

Comparing the high-spin and low-spin arrangements for d^4–d^7 in an octahedral field.

the former seems clear, but putting all four electrons into the three lowest energy available orbitals comes at an energy cost arising from forcing two electrons into the same orbital, a penalty defined as the spin pairing energy P (a positive number).

If the two electronic configurations have *the same energy*, then we have the following equality: low-spin d^4 $(-1.6\Delta_0 + P)$ = high-spin d^4 $(-0.6\Delta_0)$, which simplifies to $\Delta_0 = P$. Now P is essentially constant (independent of the type of orbital), but the value of Δ_0 is highly dependent on the ligands present (their position in the spectrochemical series) as well as the metal oxidation state. For a large Δ_0 enforced by strong-field ligands high in the spectrochemical series and high metal oxidation states (where consequently $\Delta_0 > P$), low-spin is clearly the more stable arrangement, whereas for a small energy gap with weak-field ligands (where now $\Delta_0 < P$), high-spin is preferred. In circumstances where P and Δ_0 are similar, the situation is non-trivial and in fact an equilibrium between the two states can exist.

What is important about this model (based on CFT) is that it can be used to understand experimental observations (as any viable model should, of course). It allows us to understand how the colour (dependent on Δ_0 and electronic configuration) and magnetism (governed by the number of unpaired electrons) of a complex can change even where the central metal ion (including its oxidation state) is unaltered, simply as a result of a changing ligand environment perturbing the size of Δ_0 – but more on that later.

2.3.5   Coordination Geometries Beyond Octahedral and Tetrahedral

Coordinate bonding has been accounted so far for by the molecular orbitals created within LFT for an octahedral complex. However, the same principles can be applied to all other coordination geometries. To show all of them would be labouring the point, so we return to the focus of all transition metal chemistry, the d-based orbitals in the middle of each MO diagram and their electronic configurations, which are central to understanding the properties of any complex.

As we have seen for the octahedral and tetrahedral cases, the electrostatic field of the ligand lone pairs perturb the central ion environment, and every d orbital will be raised in energy relative to the prior free-ion situation. A spherically symmetrical electrical field raises all d-orbital energies equally then the distribution of this negative charge to specific points where the ligand donor atoms are found leads to a splitting of the original fivefold degeneracy of the d-orbital set. Conventionally (and conveniently), it is common to represent only the effect of the crystal field specific component, defined relative to the orbital energy of the degenerate set in the spherical field (see broken horizontal lines in Figure 2.16). Since we are commonly dealing with differences between levels, this is entirely appropriate.

A key requirement is that when a set of degenerate d orbitals splits due to a change in ligand field symmetry, the net sum of the energy changes (positive and negative) is zero; i.e. the average energy level (sometimes loosely referred to as the barycentre due to it being analogous to a centre of gravity) equals that of the orbitals in the spherical field alone. Of course, an empty orbital (a virtual orbital) is a rather odd concept to deal with, since we tend to consider orbitals as only meaningful when they contain electrons. Perhaps, grasping for a macroscopic example, think of a set of orbitals like a set of shelves – the individual shelves only become important to you when they contain an item, yet you are certainly aware of the full set available, whether full or empty.

We have already seen that the ligand spectrochemical series is ordered according to the value of Δ_o (or Δ_t) from spectroscopy measured as the wavelength (and energy) of light required to promote a d-electron from lower to higher energy orbital (explored in later chapters). The metal also plays a role in the values of Δ_o and Δ_t. The observed splitting between the d-orbital sets increases with increasing oxidation state of the metal. The ionic radius decreases as ionic charge (which equates to the oxidation state of the metal) increases, because the positively-charged nucleus draws the remaining electrons, in the minority, closer. As this electron cloud shrinks, the charge density (i.e. charge divided by volume) increases substantially, which is associated with the contraction of metal–ligand bond distances (r) and an increase in Δ_o (which varies empirically with r^{-5}). In every known coordination geometry, the unique arrangement of donor atom lone pairs presents a different electrostatic field to the electrons primarily housed in metal-based d orbitals and purely geometric arguments can be used to explain d-orbital energy splitting in all coordination geometries within the CFT model. However, all our discussions to date have focused on only two geometries, albeit both common, octahedral and tetrahedral.

That different shaped complexes exist is a fact, so our model developed to date must be flexible enough to accommodate other shapes. One of the best ways to understand how this works is to start with our octahedral field and look at what happens as we distort it. One well-known type of structural distortion is where there is one axis where the bond distances are longer than along the other two axes; it still has the basic octahedral shape but is 'stretched' along one axis direction – like a stretched limo still being, at heart, a standard car. By convention, the 'stretched' axis is defined as the z-axis. This will be observed when the ligands on the z-axis are different to those in the xy plane. In the extreme, these stretched bonds get so long that they effectively do not exist, meaning that there are only four coordinate bonds left, in a plane around the central metal. As such, it is the orbitals which point in the z-direction that feel the effect of change most – easy to identify, from their names, as d_{z^2}, d_{xz} and d_{yz}. As we pull the two ligands away from the metal, their influence on these three orbitals diminishes, and the orbitals fall in relative energy (or are stabilised); this removes the degeneracy of the t_{2g} and $e_g{}^*$ sets of an octahedral complex as at least one of the d_{z^2}, d_{xz} and d_{yz} orbitals resides in each set. The outcome is shown in Figure 2.22. This trend continues to the extreme case where they are removed completely (and

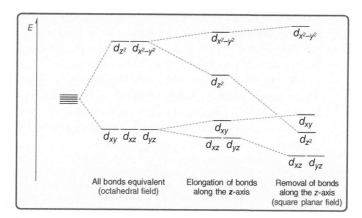

Figure 2.22

Variation in the d-orbital splitting diagram of an octahedral complex as a result of elongation of bonds along the z-axis and shortening of the bonds along the x- and y-axes.

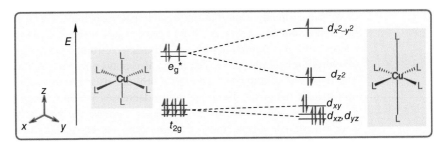

Figure 2.23

Electronic configurations of a Cu(II) complex with six identical ligands (L) showing the crystal field stabilisation energy gain through spontaneous elongation of its axial ligands.

square planar four-coordination is achieved), with increasing separation of the two types of orbitals in each originally degenerate set.

The $d_{x^2-y^2}$ orbital (strongly antibonding in LFT terms) is raised very high in energy compared with the other four (Figure 2.22), which favours the d^8 configuration since all electrons fill the four low energy levels (all essentially non-bonding in LFT terms), leaving the highest energy one empty, providing an energetically favourable arrangement. The experimental observation that nearly all square planar complexes have a d^8 electronic configuration is consistent with this model. The energy gain also increases with increasing ligand field strength; this occurs for the heavier ($4d$ and $5d$) transition elements like Pd(II) and Pt(II), which are usually found in square planar geometries.

One classic, and unusual, example of bond elongation is the behaviour of d^9 Cu(II) six-coordinate complexes with *identical ligands*, which typically display a spontaneous elongation of M–L bond lengths along one axis direction. This is known as the *Jahn-Teller effect*. Put simply for this particular situation, if there is an *odd number of electrons* in a degenerate higher energy d-orbital set (here $3d_{z^2}$ and $3d_{x^2-y^2}$), then there will be a *spontaneous* distortion (elongation or compression) that removes this degeneracy. The effect is more significant in the e_g^* orbitals which are directed towards the ligands on the x, y and z axes, whereas the essentially non-bonding t_{2g} sets are hardly affected, as shown in Figure 2.23. Why would a complex spontaneously adopt a lower symmetry geometry? The answer lies in noting that two electrons in the $3d_{z^2}$ orbital are stabilised by stretching the bonds along the z-axis and only one electron in the $3d_{x^2-y^2}$ orbital is destabilised, an energetically advantageous outcome: a case of two steps forward and one step back. This is routinely observed for d^9, low-spin d^7 and high-spin d^4 six-coordinate complexes, with particularly strong evidence coming from structural studies of d^9 Cu(II) complexes.

What the above shows is that *the d-orbital energies are responsive to molecular shape*, which is one of its key attributes. In principle, it can be manipulated to yield an energy level arrangement for any shape, and we will encounter other coordination geometries later on where relative d-orbital orientation can again be used to provide explanations of structure, bonding and shape.

2.4 Metal–Ligand Bonding in Organometallic Chemistry – Beyond Classical Werner Complexes

The field of *organometallic chemistry* is vast and covers complexes specifically with metal–carbon bonds. This distinction may appear artificial. What is so special about carbon anyway? In some organic ligand families comprising simple σ-donor ligands such as alkyl anions, e.g. CH_3^- which is structurally identical to ammonia and isoelectronic, organometallic complexes can be described with the same classical bonding models already described for complexes of ligands such as NH_3 or H_2O. Organometallic coordination chemistry, however, becomes more intriguing when the ligands possess π-molecular orbitals.

2.4.1   π-Complexes – No Lone Pairs Needed

Now that we have established molecular orbital theory describes the way ligand lone pairs become incorporated with the metal's outer shell 3*d*, 4*s* and 4*p* orbitals, we can venture further and explore ligands without any lone pairs! An example is the simple hydrocarbon ethene (C_2H_4). In 1830, the Danish chemist William Christopher Zeise reacted the Pt(IV) compound $PtCl_4$ with ethanol and isolated a highly flammable compound with the empirical formula $Pt(C_2H_4)Cl_2$. Zeise first published his findings (in Latin) in the University of Copenhagen's 1830 Annual report. The compound was given the descriptive name *Chloridum platinæ inflammabile*, which we now know was the dimer $[Pt(C_2H_4)Cl_2)]_2$. Recrystallisation by addition of KCl yielded a yellow crystalline compound of formula $K[Pt(C_2H_4)Cl_3]·H_2O$, which became commonly known as Zeise's salt. Their molecular structures, which were determined by X-ray crystallography more than a century later, are shown in Figure 2.24.

After Zeise published his work, and several other papers during the 1830s, many years of vigorous debate and controversy followed involving luminaries of chemistry at that time (Berzelius, Liebig, Dumas and Wöhler, to name a few) to account for what Zeise had observed. The presence of ethene in the compound was even challenged, as this weighed into hotly contested theories of organic chemistry at the time, including the etherin (ethene) theory which stated that compounds like ethanol and diethyl ether were formally derived from ethene, and Zeise's observations seemed to support this.

In 1861, 14 years after Zeise's death, ethene was confirmed independently as being a constituent of Zeise's salt. Nevertheless, the field was no closer to understanding the structure of this compound. Ideas of three-dimensional structure and stereochemistry were still being developed, so theory had to catch up with experiment. Even Alfred Werner's coordination theory, published at the end of the 19th century, failed to explain Zeise's discovery as ethene has no lone pairs, unlike ammonia, water and halide anions, so there is no obvious point of attachment.

The story continued into the 20th century when theory finally caught up with experiment. Double bonds in organic chemistry were poorly understood until the 1930s when theoretician Erich Hückel showed that all double bonds in organic compounds can be described as a combination of a σ-bond and a π-bond with electrons placed in

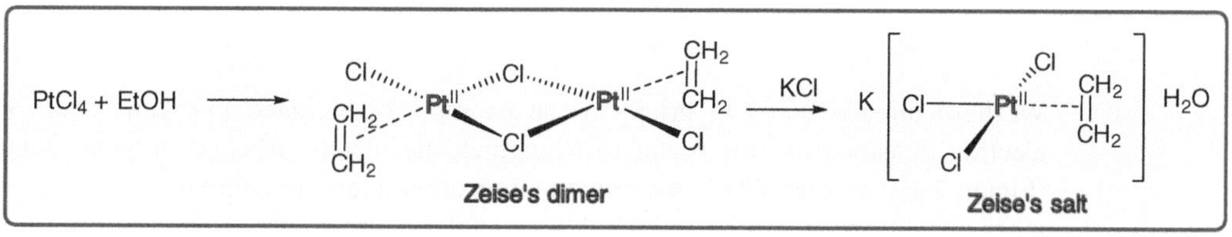

Figure 2.24
The preparation of Zeise's dimer and Zeise's salt, the first non-classical coordination compounds.

orbitals of different symmetry. At that time, molecular orbital theory was an emerging field. Both MO theory and Pauling's alternative resonance model (within valence bond theory) were viewed with suspicion by many experimental chemists of the day.

The conundrum of alkene coordination was finally resolved in 1951 by Michael J.S. Dewar, another theoretical chemist, who published a MO theory model predicting that ethene could and would coordinate to a metal in a side-on fashion. The example he chose was Ag^+, but we now know that this can be applied generally. To understand this model, we first need to look at the bonding in ethene with molecular orbital theory.

There are 12 molecular orbitals in even a simple molecule like ethene (two for each atom in its formula). Ten are involved in σ-bonding and antibonding interactions within the plane of the molecule but play no part in bonding to a metal, so they are omitted here. The remaining two are shown in Figure 2.25 and are both π MOs, one bonding which is the highest occupied MO (HOMO), and the other antibonding which is the lowest unoccupied MO (LUMO) of ethene.

The occupied π BMO of ethene (the π bond component of its double bond) has the correct symmetry to interact with a metal $d_{x^2-y^2}$ orbital (illustrated in this case with Pt^{2+} for historical relevance) along the x-axis, as drawn, to create a new type of donor–acceptor coordinate bond. This is equally as valid as the types of donor–acceptor interactions seen earlier for a classical coordinate bond. However, on its own, this would not be enough to make a stable coordinate bond as ethene is a poor electron pair donor (Lewis base), so this interaction is reinforced by back-bonding from an occupied metal d_{xy} orbital into the empty π AMO.

This proposal by Dewar was quickly supported by Joseph Chatt and L.A. Duncanson (then working for the British chemical company ICI) who published their own findings including vital spectroscopic data in 1953, and rightly credited Dewar with the original bonding model. The story was finally over more than 120 years after Zeise's original synthesis was published, which is surely history's longest-running saga concerning chemical bonding.

This synergistic bonding arrangement that pervades much of organometallic chemistry is referred to as the Dewar–Chatt–Duncanson model. Nothing is ever simple, it seems, though. Extensive data are now extant on hundreds of alkene complexes that include accurate crystallographic measurement of C=C bond lengths and infrared (IR) vibrational spectroscopy reporting the vibrational frequencies of the C=C bond. These show a wide spectrum of behaviour ranging from normal C=C double bonds to those so weakened that they approach a C–C single bond. This reflects the degree of back-bonding from the metal, and once again shows us that the diversity of properties found across the transition metals is always going to present challenges for a 'one size fits all' model. Fortunately, MO theory allows us to provide a holistic picture of this unusual coordinate bonding interaction by considering the relative contributions of the two bonding orbitals shown in Figure 2.25.

2.4.2   Carbon Monoxide and the Carbonyl Ligand

We have already met CO briefly as an example of a ligand that can accept π electron density from the metal to strengthen the metal–carbon coordinate bond (Figure 2.19). In fact, CO is a common companion ligand in complexes containing other metal–carbon bonds, and $M(CO)_n$ complexes are common precursors in the synthesis of other organometallic compounds. The back-bonding transfer of electron density from the metal to a π antibonding molecular orbital on CO helps stabilise

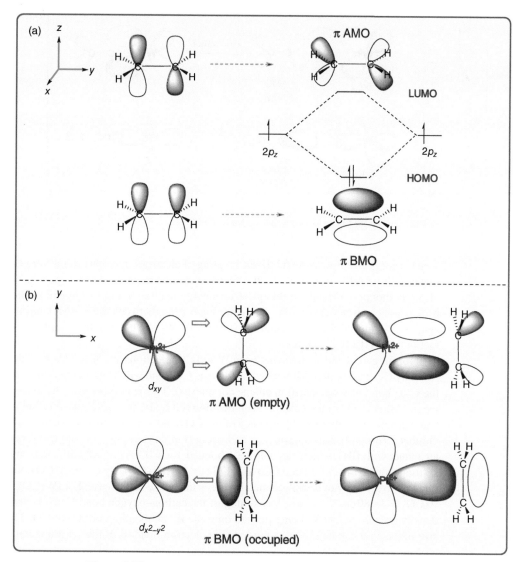

Figure 2.25

(a) The MO description of ethene focusing on its π MOs and (b) symmetry-allowed overlap with d orbitals in the xy plane (illustrated here with Pt^{2+}).

complexes of formally low-valent (even zero-valent) and d-electron-rich metals. Zero-valent metals have no affinity for simple σ-donor ligands such as ammonia or water. In concert, these observations explain why zero-valent $[W(NH_3)_6]$ has not been prepared, whereas $[W(CO)_6]$ is quite stable, even against oxidation by air, as a result of highly efficient back-bonding.

The CO ligand, bound through its C-atom, has the capacity to act as both a σ-donor and a π-acceptor ligand in what may be described in an analogous way by the Dewar–Chatt–Duncanson model although the orientation of the ligand is rotated by 90°. This is shown in detail in Figure 2.26 with a W–CO coordinate bond as our focus. Two (of several) MOs of CO are shown, one a BMO with σ-character but where a significant amount of electron density sits on the C-atom. The other we have

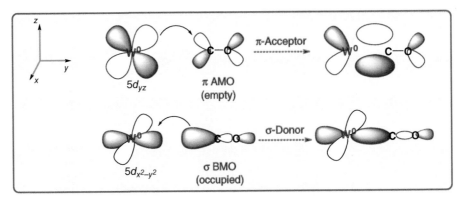

Figure 2.26
MO diagrams for the CO ligand and its interaction with W(0).

seen earlier is an empty π-AMO. Back-bonding of electrons into this orbital from the metal with its fully occupied t_{2g} set effectively results in some electron population of an antibonding orbital. In the quest for charge neutrality, the CO ligand pushes back some electron density onto the metal via overlap between the σ-BMO and the symmetry-matched $5d_{x^2-y^2}$ orbital. A very stable W–CO bond is the result, six in fact in the [W(CO)$_6$] complex.

In forming a coordinate bond between W and CO, electrons are shared by W with a formally CO antibonding MO, which means the CO bond will be weakened (just like the C=C bond was weakened in ethene complexes). This is in fact shown by experiment, as this C–O bond weakening can be observed by IR spectroscopy, a technique that measures the vibrational frequency of the CO bond, reported as $\bar{v}$ (CO), the wave number $\bar{v}$ (proportional to frequency and energy) in units of cm^{-1}. When CO forms an adduct with BH$_3$, which can only be σ-bonded, the $\bar{v}$ (CO) is hardly affected, only shifting from 2,149 cm^{-1} (free CO) to 2,178 cm^{-1} (H$_3$B–CO). By contrast, CO coordination to a variety of transition metals causes this value to decrease. For [W(CO)$_6$], $\bar{v}$ (CO) = 1,980 cm^{-1} due to a weakening of the carbon–oxygen bond from electron density on zero-valent W being pushed onto the π^* antibonding orbital of CO. The size of the shift can be used as a guide to the electron richness of the central metal.

2.4.3 The Metallocenes

Once the possibilities of different kinds of metal–ligand bonding modes had been understood by Dewar's MO bonding model reported in 1951, the organometallic chemistry field was ready for more unusual compounds to emerge. This happened quickly, as in the same year Thomas J. Kealy and Peter L. Pauson reported an unexpected result from the reaction between FeCl$_3$ and the Grignard reagent cyclopentadienyl magnesium bromide. The target of their synthesis was dihydrofulvalene (C$_{10}$H$_{10}$, effectively a dimer of cyclopentadiene); instead, they produced a crystalline yellow compound of formula FeC$_{10}$H$_{10}$ to which they assigned the simple two-coordinate structure shown in the upper section of Figure 2.27 with each ligand acting as a conventional σ-donor. This compound showed extraordinary stability against chemical reagents such as 10% NaOH, concentrated HCl and even heating. Organometallic iron complexes analogous to the well-known but reactive magnesium Grignard regents were not known at that time, so this lack of reactivity

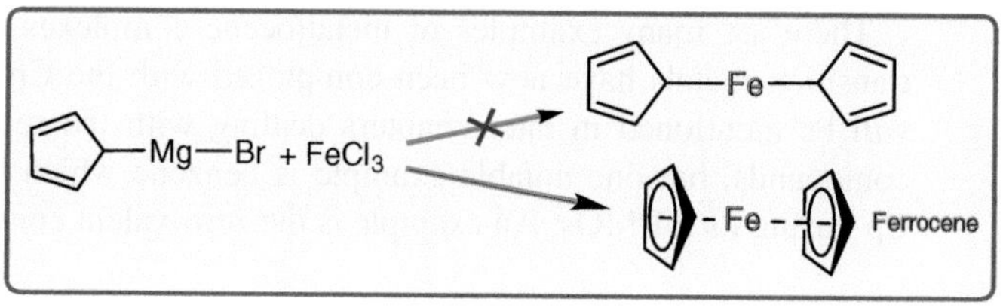

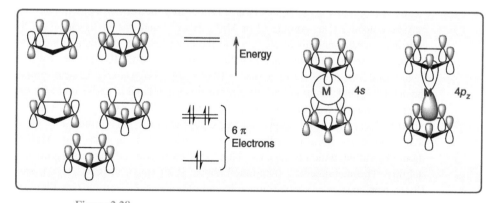

Figure 2.27

The accidental synthesis of the first so-called sandwich complex ferrocene in 1951 and its originally incorrect structure assignment. Its true symmetrical structure is shown in the lower part of the figure.

seemed extraordinary, and this report created considerable interest. The true structure was something more remarkable. In a flurry of papers from different groups, all published in 1952, the compound was assigned the highly symmetric structure at the bottom of Figure 2.27 and given the trivial name ferrocene. A combination of physical methods was employed: X-ray crystallography revealed its three-dimensional structure, magnetic measurements showed it had a low-spin diamagnetic electronic configuration, the dipole moment measured was zero, and a single C–H vibrational peak in its infrared spectrum indicated that all C–H bonds were identical. The groups of Geoffrey Wilkinson and Ernst Otto Fischer were major contributors to this discovery, and these two chemists would share the 1973 Nobel Prize in Chemistry for their contributions. More of their work will appear in later chapters.

With bonding models now on a firmer footing, we can understand the unusual side-on bonding of the cyclopentadienyl anion by recourse to MO theory. With 5 carbon and 5 hydrogen atoms, there are 25 possible MOs derived from the valence atomic orbitals for the $C_5H_5^-$ cyclopentadienyl anion (Cp$^-$ for short). Of this total, 20 are involved in σ-bonding/antibonding MOs leaving 5 that are involved in π-MO formation, and these are shown in Figure 2.28. There are several possible symmetry-matched combinations with the metal atomic orbitals, but for now, we show just two, each involving the same Cp$^-$ LGOs overlapping with either the $4s$ or $4p$ orbitals of the metal. Both are strongly bonding within the Cp$^-$ ligand and between the metal and ligand (Figure 2.28).

Figure 2.28

(Left) The three π bonding and two antibonding MOs of cyclopentadiene and (right) two selected symmetry-matched bonding MOs involved in the structure of a bis(cyclopentadienyl)metal complex (or metallocene).

There are many examples of metallocene complexes like this and most of the transition metals have now been complexed with the Cp^- ligand. Other examples will be mentioned in later chapters dealing with the reactivity of organometallic compounds, but one notable example is benzene which binds in a similar way to Cp^- using its π BMOs. An example is the zero-valent complex $[Cr(C_6H_6)_2]$.

2.4.4 The 18 and 16 Electron Rules

Unlike classical Werner coordination compounds, where all *d*-electronic configurations (d^0–d^{10}) and coordination numbers (2–9) are encountered, organometallic compounds tend to fall into a much narrower range of stable electron configurations and coordination numbers. There is a reason for this that again can be explained by MO theory. The similar energies of the metal *nd*, $(n+1)s$ and $(n+1)p$ orbital sets means that these are the only orbitals available for coordinate bond formation. In classical Werner coordination compounds, this leads to an upper limit of nine coordinate bonds arising from nine BMOs, all fully occupied. This is hardly ever met and, as we know, six-coordination is much more likely, requiring only 12 bonding electrons.

In organometallic complexes, especially those that bear π-bonded ligands like alkenes and aromatic hydrocarbons, assigning formal oxidation states becomes less obvious in cases where electrons are donated by the metal towards the ligand (back-bonding). That is, the relatively clear distinction between metal-centred *d*-electrons and those provided by the ligand lone pairs apparent in the MO diagram for an octahedral purely σ-bonded complex (Figure 2.15) becomes difficult. In organometallic complexes bearing ligands that participate in some of the more exotic bonding modes seen in the previous sections, the only consideration is the *total electron count* from both the metal and its ligands as a single number. Remarkably, these total electron counts are in most cases 18 or 16; the latter associated with formally square planar complexes. Unsurprisingly, this is known as the *18-electron rule* (or 16-electron rule, depending on your preferred geometry).

As an example, consider the simple organometallic compound, octahedral $[Mn(CO)_5I]$. There are two methods of electron counting: the closed shell (or oxidation state) method and the neutral ligand method. Both exist because accounting for electrons is not always straightforward. Employing the former method, $[Mn(CO)_5I]$ may be considered as composed of Mn^+, five CO and one I^- ligand. Each ligand contributes one lone pair of electrons to the complex, and the metal contributes its valence electrons. This amounts to 6 $3d$-electrons for Mn^+, 10 electrons for five CO ligands and 2 electrons for I^-, a total of 18 electrons. For a charge-neutral compound, the neutral ligand method treats all components (metal and ligands) as charge neutral so in this example we arrive at Mn (7 electrons: $4s^2 3d^5$), $5 \times CO$ (10 electrons: 5 lone pairs) and I (1 electron: as opposed to 2 electrons for I^-). The total of 18 electrons is of course the same regardless of the method. This suggests that $[Mn(CO)_5I]$ should be stable, which is the case. This 'rule' more correctly acts as a guideline for stability. The metallocenes mentioned above, $[Fe(Cp)_2]$ and $[Cr(C_6H_6)_2]$, are also 18-electron species as each aromatic ligand provides 6 π-electrons (see Figure 2.28, left), while the metal has formally 6 $3d$-electrons.

Returning to Zeise's salt $K[Pt(C_2H_4)Cl_3]$, we have formally Pt^{2+} ($5d^8$), three Cl^- lone pairs (6 electrons) and the ethene ligand which shares its 2 π-bonding electrons, totalling 16 electrons and again consistent with its square planar geometry. The 18/16 electron rule will return in later chapters when we consider oxidation and reduction reactions where the total electron count is linked to changes in coordination number and geometry.

2.5   Coupling – Polymetallic Complexes

Most examples met so far are based on simple monomeric complexes, containing just one central metal. This might suggest that these are the sole or dominant type – not so. It is quite common to meet complexes that feature two or even more metal centres linked together. Complexes may be monomeric, dimeric (two metals involved), oligomeric (several metals involved) or polymeric (many metals involved). The metal centres may be linked directly, yielding a metal–metal bond, or else may be linked or bridged by a ligand or ligands. This latter class can arise in a number of different ways:

- linkage via a small neutral or anionic ligand that binds simultaneously to both metals through a common donor atom (e.g. Cl^-);
- linkage via a small ligand that binds to both metals through two different donor atoms (e.g. NCS^-, using both the N and S atoms);
- linkage via a polydentate ligand that spans across two (or more) metals and provides some donor groups for each (e.g. $^-OOC–CH_2–COO^-$, using the two carboxylate groups);
- incorporation of two or more metal ions in a large cavity in a polydentate ligand, each metal binding to several of a large set of potential donors (a process called encapsulation);
- formation of a polyhedron by a set of metal ions (a *cluster*) linked by bridging ligands with additional ligands bound to satisfy the coordination spheres of the metals.

Moreover, the complex may involve all metals of one type and in the same oxidation states; all metals of the same type in several oxidation states; two or more different types of metals in a common oxidation state; or two or more different types of metals in a range of oxidation states – that is, all combinations are feasible!

Complexes that are oligomeric (multiple metal centres but discrete molecules) and roughly spherical are often called *clusters*. Clusters may involve several layers of metal ions from a core outward and, perhaps not surprisingly, metals in different environments (core or surface, for example) behave differently, even if inherently the same type and in the same oxidation state – in other words, 'environmental' effects contribute to the metal's behaviour. The shapes of clusters known are extensive in number and may show some bond angle and bond length deformation due to the complexity of the structure and the associated demands of the ligands. In clusters with more than one metal type involved, an array of isomers may exist resulting from differing placements of the metals in the cluster framework. Perhaps the most demanding class of clusters is the double-shell (cage within cage) architecture, where sets of metals form cages of different sizes and geometries, one inside the other, and strongly interconnected by covalent bridging units. For example, a double-shell structure with a small internal cluster involving 5 vanadium atoms and a large outer cluster having 12 vanadium atoms has been reported. Suffice it to say that clusters are an enticing, but difficult to control, area of coordination chemistry.

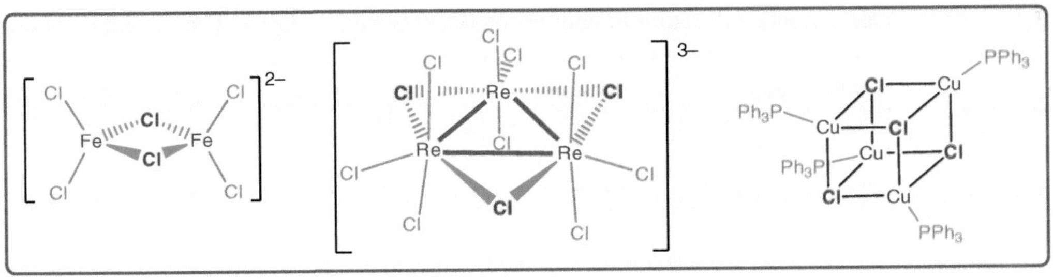

Figure 2.29

Examples of oligomeric complexes of metal ions with halides. Both bridging and terminal halides can be identified in the figure; bridging halides are shown in bold type.

Nevertheless, we shall very briefly examine one of the classes identified above a little more carefully, with some examples. One of the simplest families of related compounds is the halido metal clusters. Many metal ions can exist with a set of halogen anions as common donors in monomers of the general formula MX_n^{q-}; an example is the $[CuCl_4]^{2-}$ ion. In some cases, oligomers can form, of which the first is the dimer $M_2X_n^{q-}$; an example is $[Fe^{II}_2Cl_6]^{2-}$, distinguished by having two bridging halides in addition to four terminal halides (Figure 2.29, left). Higher oligomeric clusters are also known; trimers, such as $[Re^{III}_3Cl_{12}]^{3-}$, have the three metal centres disposed in an equilateral triangular arrangement, with both bridging and terminal halides (Figure 2.29, centre), whereas tetramers, such as $[Cu^I_4Cl_4L_4]$, adopt a slightly distorted 'cubane' shape (Figure 2.29, right). These may also involve metal–metal bonding; for example, $[Re^{III}_3Cl_{12}]^{3-}$ has Re–Re bonds of similar length to the Re–Cl bonds, which further strengthen the cluster.

The above examples are relatively simple since they contain just two types of atoms and adopt symmetrical shapes. It should be obvious that many examples prove much more complicated in structure, addressed in Chapter 3.2.2. Although polymeric coordination compounds are growing in number and importance, we shall limit our exploration of these, leaving this to more advanced textbooks.

2.6　Complex Lifetimes and Complexation Consequences

Metal complexes are not usually unchanging, everlasting entities. It is true that many, once formed, do not react readily. However, the general observation is that coordinate bonds are able to be broken with more ease than cleaving a covalent carbon–carbon bond. The strength of metal–ligand bonds is typically less than most bonds in organic molecules, but much greater than non-covalent hydrogen bonds. We recognise carbon–carbon bonds as strong and usually unchanging and hydrogen bonds as weak and easily broken; the coordinate covalent bond lies in the middle ground, albeit nearer the carbon–carbon end of the park. Breaking a metal–ligand bond is almost invariably tied to a follow-up process of making another metal–ligand bond in its place, so as to preserve the desired coordination number and shape of the complex. When the same type of ligand is involved in each sequential process, *ligand exchange* is said to have occurred. Where the ligands and solvent are one and the same, there is no opportunity for any alternative reaction. Even at equilibrium, metal complexes in solution display a continuous process of ligand exchange, where the rate of the exchange process is driven by the type of metal ion, ligands and/or solvent involved. From the perspective of the central metal, ligand exchange can vary with metal ion from extremely fast (what we refer to as *labile* complexes) to extremely slow (termed *inert* complexes).

Lability and inertness are *kinetic* terms (focused on transition *to* equilibrium). You need to be aware, further, that *thermodynamic* considerations (focused on the situation *at* equilibrium) also play a role in reactions. Suffice to say that kinetics and thermodynamics, in combination, govern all chemical reactions, but obviously tempered by what exactly is available chemically to react. We shall examine the stability and reactions of coordination complexes in detail in Chapter 5, and the application of these in chemical synthesis is met in Chapter 6.

In complex formation, we are bringing together metals, that are usually positively-charged, and donor atoms and groups that are partially or fully negatively-charged to assemble a tightly bound molecule. Inevitably, the ligands affect the core element (metal) which in turn affects the ligands. Whereas the bonding models developed earlier in the chapter interpret the behaviour successfully, we can also see this in experimentally measurable effects. Changes with complete or even partial ligand exchange in the shape, colour, redox potential, magnetic properties and wider physical properties are characteristic behaviours for coordination complexes; we shall develop this in later chapters.

That the position of an element in the periodic table influences its chemistry is inevitable, and a consequence of the Periodic Law which follows from the electronic configurations of the elements. However, it is less clearly recognised that elements within the same column (and hence with the same valence shell electron set) differ in their chemistry. It is instructive to overview the chemical impact of the central atom to illustrate both similarities and differences. These, along with specific examples of synthetic coordination chemistry, appear in Chapter 6, whereas shape and stability aspects are detailed in Chapters 4 and 5, respectively. Complexes *are* coordination chemistry, and the introduction to structure and bonding met in this chapter sets a foundation for what is to follow.

Concept Keys

Complexes form a limited number of basic shapes associated with each coordination number (or number of ligand donor groups attached). Six-coordinate octahedral is a particularly common shape.

A simple covalent σ-bonding model employing five *d* orbitals and the next available *s* and three *p* orbitals, with appropriate hybridisation to match a particular shape, allows a limited interpretation of bonding in coordination complexes.

CFT, an ionic bonding model, is focused on the *d*-orbital set, and the way this degenerate set of five orbitals on the bare metal ion is split in the presence of a set of ligands into different energy levels. It provides a fair understanding of spectroscopic and magnetic properties.

A holistic MO theory description of bonding in complexes provides a more sophisticated model of bonding in complexes, leading to LFT, which deals better with ligand influences. Both CFT and LFT reduce to equivalent consideration of *d*-electron location in a set of five core *d* orbitals.

Variation in the splitting of the *d*-orbital set into different energy levels is dependent on complex shape.

Arrangement of a set of electrons in the *d*-orbital energy levels, as a result of relatively small energy gaps, can in some cases occur with different arrangements where unpaired electrons are either minimised (low-spin) or maximised (high-spin). Properties differ as a result.

Certain ligands can employ their π-orbitals for additional interaction with the central metal, either for increasing (π-donor) or decreasing (π-acceptor) electron density on the central metal. This alters the energies of levels associated with the *d* orbitals, influencing physical properties.

Some ligands are capable of binding to more than one metal ion at once, acting as bridging groups in polynuclear assemblies.

Further Reading

Cotton, F.A., Wilkinson, G., Bochmann, M., and Murillo, C. (1999) *Advanced Inorganic Chemistry*, 6th ed. New York: Wiley Interscience.

    Viewed by many as the classical advanced textbook in the broad inorganic chemistry field, this lengthy book includes a detailed coverage of *d*-block chemistry, with a leaning towards organometallic chemistry.

Kauffman, G.B. (Ed.) (1994). *Coordination Chemistry: A Century of Progress*. Washington: American Chemical Society.

    This is a readable account of the rise of coordination chemistry, celebrating the centenary of Werner's key work, for those with a historical bent; puts coordination chemistry in its historical context.

Kent, B. (2019). *Advanced Inorganic Chemistry*. Forest Hills, NY: NY Research Press.

    With an extensive coverage of descriptive and theoretical aspects across the breadth of the field, this may suit those seeking to extend their understanding beyond an introductory level, as it is approachable and fairly compact despite being at an advanced level.

Ribas Gispert, J. (2008). *Coordination Chemistry*. Wiley-VCH.

    This is a large advanced text for graduate students, but undergraduates seeking an extended view of aspects of the field should find it useful.

Wilkinson, G., Gillard, R.D., and McCleverty, J.A. (Eds) (1987). *Comprehensive Coordination Chemistry: The Synthesis, Reactions, Properties, and Applications of Coordination Compounds*. UK: Pergamon; McCleverty, J.A. and Meyer, T.J. (Eds) (2004). *Comprehensive Coordination Chemistry II: From Biology to Nanotechnology*. UK: Elsevier, Pergamon; Constable, E.C., Parkin, G., and Que, L. Jr. (Eds) (2021). *Comprehensive Coordination Chemistry III*, UK: Elsevier.

    These three multi-volume sets provide coverage at an advanced level from ligand preparation through their detailed coordination chemistry and beyond; they provide, through independent chapters, a fine though sometimes daunting in-depth resource.

3 Ligands

3.1 Membership: Being a Ligand

A ligand is an entity that binds strongly to a central species. This broad general definition allows extension of the concept beyond chemistry – you will meet it in molecular biology and biochemistry, for example, where the 'complex' formed by a 'ligand' with a biomolecule involves relatively weak noncovalent interactions like ionic and hydrogen bonding. The origin of the name ligand may be traced to the Latin word *ligare*, meaning to bind. In coordination chemistry, a ligand is a molecule or atom (charged or neutral) carrying suitable groups capable of *binding* (i.e. coordinating covalently) to a central atom. Electrons are a prerequisite for any type of covalent bonding, including coordinate bonding, so electronic structure and molecular orbitals are central to any proper understanding of coordination chemistry, as described in Chapter 2. The central atom that is the focus of ligand coordination is most commonly a metal that may be from any of the *s-*, *p-*, *d-* or *f*-blocks.

The term ligand first appeared early in the 20th century, with German chemist Alfred Stock credited with its introduction in the 1917 published version of a research lecture given in late 1916, but it achieved popular use only by mid-century. Strictly speaking, a molecule or ion does not become a ligand until it is bound, and this is reflected in nomenclature (Appendix 1) where, for example, water becomes an aqua ligand once coordinated to a metal. Prior to that, it is formally a proligand, although this distinction is rarely used. For an atom or molecule, being a ligand is a lot like a plant being green – surprisingly common. That a metal atom or ion is almost invariably found with a tied set of companion atoms or molecules has been known for a long time, but it was only from around the beginning of the 20th century that a clear concept of what a ligand is and how it binds to a central atom began to develop.

3.1.1 What Makes a Ligand?

The range of chemical species that can bind to metal ions as ligands is diverse, but elements from the *p*-block (plus hydrogen) assembled into charge-neutral or negatively charged molecular forms dominate. With metal–metal bonds also well known, one could argue in a simplistic sense that metals themselves can act as ligands, but this is not a direction we shall take. We have already met with central metal ions, metal–ligand assemblies (complexes) and the coordinate bonds involved in the first two chapters; in this chapter, we will focus on the character of the ligand itself.

Introduction to Coordination Chemistry, Second Edition. Paul V. Bernhardt and Geoffrey A. Lawrance.
© 2025 John Wiley & Sons Ltd. Published 2025 by John Wiley & Sons Ltd.
Companion website: www.wiley.com/go/coordinationchemistry2e

Metal ions are hardly ever found naked. They are always clothed with ligands.

Figure 3.1
An anthropomorphic view of being a metal coordination complex.

We will start with the simplest, and most encountered, form of metal–ligand bonding, within so-called 'classical' coordination compounds. In this context, the metal–ligand coordinate bond has but one basic requirement – in the simple valence bond model, the metal-free ligand bears at least one lone (nonbonding) pair of electrons, as indicated in the cartoon in Figure 3.1. The atom that carries the lone pair needed for coordinate bonding is termed the *donor atom*; where it is part of a well-recognised functional group (like an amine, $R–NH_2$, or carboxylate, $R–COO^-$), we speak of a *donor group* but must recognise that it is typically one particular donor atom of the group that is bound to the metal. In its initial development, the general view of a ligand was that its role, beyond electron pair donation, is somewhat passive – a spectator rather than a player, if you wish. We now know that ligands influence the central metal ion significantly and can display reactivity and undergo chemistry of their own while still bound to the metal ion, which we shall return to in Chapter 6; this does not alter the fundamentals of their behaviour as ligands, however.

Ligands (often represented by the general symbol L) that present a single donor atom with one lone pair for coordination to a metal ion may only occupy one coordination site, $M^{n+}–L$, and are then referred to as *monodentate* (i.e. 'one-toothed'). Classical examples of monodentate ligands include NH_3, CH_3^- and $P(CH_3)_3$. As a general rule, elements from groups 16 and 17 (e.g. O, S, Cl and Br) carry more than one lone pair of electrons, which also enables them to bind to more than one metal simultaneously as a *bridging ligand* (see Section 3.2.2). Recognising the presence of these donor atoms is the first step when deciding if and how a molecule may become a ligand. Of course, location, local environment and relative orientation of potential donors in a larger molecule play a role in how many nominally 'available' donor groups may bind collectively to a single metal ion. Some simple molecules can contain two different donor atoms that are each inherently capable of coordination, but where bonding of both to the same metal is forbidden due to their orientation. A classic example is the linear anion thiocyanate (SCN^-), which can coordinate as a monodentate ligand either through its S or N atom. It is termed an *ambidentate* ligand; there are only a limited number of these extant. Such ligands do offer another mode of coordination, in which they employ both donor atoms to simultaneously bind to two different metal ions; in this mode, they function as what are termed bridging ligands.

3.1.2   Making Attachments – Coordination

Only in the gas phase is a ligand likely to meet a 'naked' metal or its ion, and this is a highly contrived situation. In solution, where coordination chemistry is most commonly studied, a potential ligand will normally be confronted by a metal already carrying a set of ligands within its *inner coordination sphere*. In the absence of any competition, these ligands are usually solvent molecules – but no less legitimate as ligands simply because they can serve two roles, as ligand or solvent. Most simple counter anions (halides and oxoanions such as sulfate, nitrate and phosphate) of common metal ion salts can also coordinate. Binding of a new ligand to a metal ion in solution usually requires that it replace an existing one, a process termed *ligand substitution*. What drives this process we shall deal with mainly in Chapter 5.

The strong drive towards complexation of otherwise 'naked' metal ions is readily observed. For example, if a simple hydrated salt like copper(II) sulfate ($CuSO_4 \cdot 5H_2O$), with its characteristic blue colour (Figure 3.2), is dried in a vacuum oven to the point where no water molecules remain, the very pale anhydrous salt $CuSO_4$ is obtained. If this solid is dissolved in water, a blue solution is immediately formed as the metal ion is rehydrated, a process which simply involves a set of water molecules rapidly binding to the metal ion as ligands. An enthalpy change is associated with this process – the *enthalpy of hydration*. If the solvent is then evaporated at atmospheric pressure without excessive heating, the original blue solid $CuSO_4 \cdot 5H_2O$ is recovered. That most of the molecules are coordinated to the copper ion can be shown by simply measuring the weight change as temperature is slowly increased. What is observed is that the water in $CuSO_4 \cdot 5H_2O$ does not evaporate simply by heating to 100°C (the boiling point of water) but is only removed fully following heating to over 200°C for an extended period, with recovery after that stage of the anhydrous species. In its solid-state crystal structure, four of the five water molecules in $CuSO_4 \cdot 5H_2O$ are coordinated to the Cu ion in the solid state, whereas the fifth is not and forms only hydrogen bonds to the SO_4^{2-} anions, which also coordinate to the copper ion (Figure 3.2, left). In the anhydrous solid, all Cu–O bonds arise from sulfate oxygens (Figure 3.2, right).

Application of an array of advanced experimental methods allows us to observe the species in solution also; not only can we observe the presence of separate cations and anions, but also the size, shape and environment of the ions can be elucidated. This confirms that the metal ion exists with a well-defined sheath of water molecules, the inner coordination sphere, where each water molecule is attached to the central metal through a coordinate covalent bond via an oxygen lone pair. When this

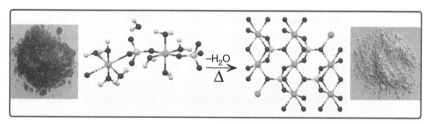

Figure 3.2

The different polymeric solid state forms of the simple salt copper sulfate in its pentahydrate (left) and anhydrous (right) forms. The water (aqua) ligands lead to its characteristic blue colour in the solid state and in solution. In each case, the copper(II) ions (green) coordinate to six oxygen donors (red).

entity is charged, as is the case for copper(II), this complex in aqueous solution is surrounded by a partially ordered *outer* (or *secondary*) *coordination sphere* where polar water molecules attracted to the complex cation are hydrogen-bonded to the inner-sphere ligated water molecules; a third and subsequent sheath surrounds the second layer, the process continuing until the layers become indistinguishable from the bulk water. The various layers moving outwards from the centre experience a successively diminishing electrostatic influence of the metal ion.

It is also important to think about the lifetime of a particular complex ion. For a metal ion in aqueous solution surrounded by water (aqua) ligands, it is not correct to assume that, once formed, an aquated metal ion remains with the same set of water molecules. In solution, it is possible (indeed usual) for water molecules in the outer coordination sphere to change places with water molecules in the inner coordination sphere. Obviously, this is a difficult process to observe since there has been no real change in the metal environment when one water molecule replaces another – a little like taking a cold can out of a refrigerator and replacing it with a warm one of the same kind so that no one can tell (unless they pick up the warm can). At the molecular level, one can adopt the cold/warm can concept to probe what is called *ligand exchange* by adding water with a different oxygen isotope (e.g. $H_2^{18}O$ instead of more common $H_2^{16}O$) and following its uptake into the coordination sphere. The facility of this water exchange process varies significantly with the type and oxidation state of the metal ion. Moreover, the rate of exchange varies not only with the metal but also with the ligand – to the point where longevity of a particular complex can indeed be extreme, or the coordination sphere is fixed for all intents and purposes. We shall return to the concept of ligand exchange again in Chapter 5.

What other observable physical consequences of complex formation arise? As one might expect, it is the properties of specifically the donor group of a ligand, directly bound to the metal, that are changed most substantially upon coordination to a metal ion. This can be illustrated by looking at molecules where the donor group carries protons. Acidity of coordinated water groups increases substantially on coordination – from a pK_a of ~14 for free bulk water to a pK_a of 5–9, typically for coordinated water molecules. Consequently, aqua–metal complexes are often only fully protonated in acidic solution. As charge on the metal ion increases, aqua ligand acidity increases to the point where deprotonation occurs readily, to form the hydroxido (M–OH), and to even in some cases form the oxido (M=O) ligand:

$$M^{n+}-OH_2 \rightleftharpoons M^{n+}-OH^- + H^+$$

$$M^{n+}-OH^- \rightleftharpoons M^{n+}=O^{2-} + H^+$$

High-valent (high-oxidation state) metal ions tend to promote deprotonation to form oxido-metal complexes (see Figure 2.17) and further desolvation to eventually yield simple metal oxides. Thus, pure aqua complexes are found mainly with metals in oxidation states III and below. The influence of the central metal ion on the ligand falls off rapidly with distance of components from the central metal. For example, in a molecule featuring coordination of a diol through one of the two alcohol groups, such as $M^{n+}-O(H)-CH_2-CH_2-OH$, the alcohol O-atom bound to the metal has a pK_a of ~7, whereas the unbound −OH group on the end of the chain has a pK_a of >14, essentially the same as for the free ligand. Even where two heteroatoms are joined directly, as in hydrazine ($M^{n+}-NH_2-NH_2$), it is the metal-bound −NH_2 group that is more acidic. These concepts are developed further in Section 3.2.

Ligand substitution (as distinct from ligand exchange) is another reaction that leads to clear changes in a complex, as seen when commencing with aqua metal ions. When the water ligand is substituted by other ligands (common donors are halide ions and O-donor, N-donor, S-donor and P-donor groups in monodentate or polydentate ligands), observable physical changes occur. *Colour* and *redox potential* change with ligand substitution – for example, $[Co(OH_2)_6]^{3+}$ is light blue, with a redox potential E^o of $+1.8$ V, making it unstable in solution as it oxidises the solvent water slowly to become the more stable $[Co(OH_2)_6]^{2+}$. However, yellow $[Co(NH_3)_6]^{3+}$ has an E^o value of about 0 V, making it stable in its oxidised form in solution. Sometimes, the change of ligand can lead to a gross change in molecular shape also; light pink, octahedral $[Co(OH_2)_6]^{2+}$ changes to blue, tetrahedral $[CoCl_4]^{2-}$ when water ligands are replaced completely by chloride ions. Complete or partial ligand substitution generally leads to an array of physical changes, which we will explore in later chapters.

3.1.3 Chelation by Didentate Ligands

The classic simple monodentate ligand is ammonia since it offers but one lone pair of electrons and thus cannot form more than one coordinate covalent bond (Figure 3.3). A water molecule has two lone pairs of electrons on the oxygen, yet it also usually forms one coordinate covalent bond. If one looks at the arrangement of lone pairs, this is hardly surprising; once one coordinate bond is formed, the remaining lone pair points in the wrong direction to allow it to become attached to the same metal ion – only through attachment to a different metal ion could this second lone pair achieve coordination (a *bridging* situation for the ligand).

Let us try to make it a bit easier for two lone pairs to interact with a single metal ion by putting them on different atoms and examining the result. A simple example is the carboxylate group (R–COO⁻), which can coordinate in at least three ways – to one metal through one oxygen, bridging two metals with each bound to a different oxygen, or as a *chelating* (claw-like) ligand to one metal via *both* oxygen atoms (Figure 3.4). In this mode, the ligand binds as a *didentate* (i.e. 'two toothed'). The number of times a ligand binds to a central metal may be greater than two if sufficient donor atoms are available, and *denticity* is the general term used, addressed further in Section 3.1.6.

In situations where the one ligand employs two different donors to attach to the same metal, we have achieved *chelation* – a *chelate ring* has been formed. The name

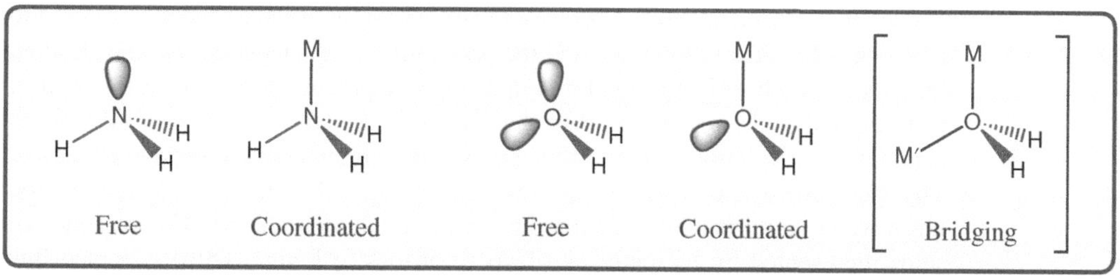

Free   Coordinated   Free   Coordinated   Bridging

Figure 3.3

Free and coordinated ammonia and water molecules. The second lone pair of water is oriented in a direction that prohibits its interaction with the same metal centre as occurs for the first lone pair. However, water does have the potential, in principle, to use this lone pair to bind to a second metal centre in a *bridging* coordination mode. (Other groups bound to the metals are left off to simplify the views.)

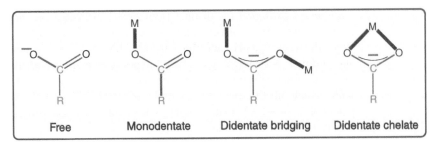

Figure 3.4

Metal ion binding options for a carboxylate group, featuring various monodentate and didentate coordination modes.

derives from the concept of a lobster using both claws (Greek χηλη *chele*) to get a better grip on its prey, put forward by Gilbert T. Morgan and Harry D.K. Drew in a research paper in 1920; not a bad analogy, given that chelates form much stronger complexes than an equivalent pair of simple monodentate ligands. A chelate ring is defined formally as the cyclic system that includes the two donor atoms, the metal ion, and the part of the ligand framework joining the two coordinated donors. The size of the chelate ring is then obtained by simply counting the number of atoms linked covalently in the continuous ring, starting and ending at the metal, and including it. For example, the chelated carboxylate in Figure 3.4 (at right) involves the sequence M–O–C–O–(M), with the last O returning us to the M, so four atoms participate in the continuous ring, meaning it is a four-membered chelate ring.

Just in the way that ring structures of a certain range of sizes in organic compounds are inherently stable, chelation leads to enhanced stability in metal complexes for chelate rings of certain sizes. For the organic compound formally named ethane-1,2-diamine (but also called ethylenediamine and abbreviated as 'en'), chelation leads to a five-membered chelate ring (Figure 3.5(a)). It is immediately apparent that this ring is more flexible than the rigid chelating carboxylate in Figure 3.4. The free ligand itself is flexible, and rotation about C–C and C–N bonds is facile. This allows it to move around to adopt shapes that suit it best for coordination and chelation.

When a free ethane-1,2-diamine molecule and an aquated metal ion react, the most stable product contains a chelate ring; this *overall* reaction is represented by the solid brown arrow in Figure 3.5(b). This representation is not an indication of how the process occurs, however, and it is very unlikely that the two amines bind to the metal ion at once in a concerted process because the organisation of the molecules required to permit this to occur is too great and the probability of it happening too low. Rather, the process of chelation occurs through a stepwise process. First, one nitrogen must form a bond to the metal, and then the remaining lone pair must be brought to an appropriate orientation as the nitrogen approaches the metal to complete chelation. The anchoring of the first nitrogen to the metal means the second one cannot be too far away in any orientation, facilitating its eventual coordination. The sequence of steps represented by the pathway of green arrows in Figure 3.5(b) is a plausible *mechanism* for the reaction. Along the way, an *intermediate* compound presumed to form is the monodentate ethane-1,2-diamine complex (albeit short-lived, because it reacts rapidly). This is a viable entity, as this mode of coordination has been seen in some isolable complexes where only one accessible binding site exists around a metal ion due to other ligands already being firmly and essentially irreversibly bound in other sites or when the metal prefers a lower coordination number than is offered by all of the N-donor atoms, as in the case of the unusual but experimentally observed five-coordinate $[Cu(en)_2(N\text{-}en)]^{2+}$ complex (Figure 3.5(c)).

As dissolution of simple metal salts in water typically produces metal ions with a coordination sphere initially consisting of water molecules, as discussed earlier,

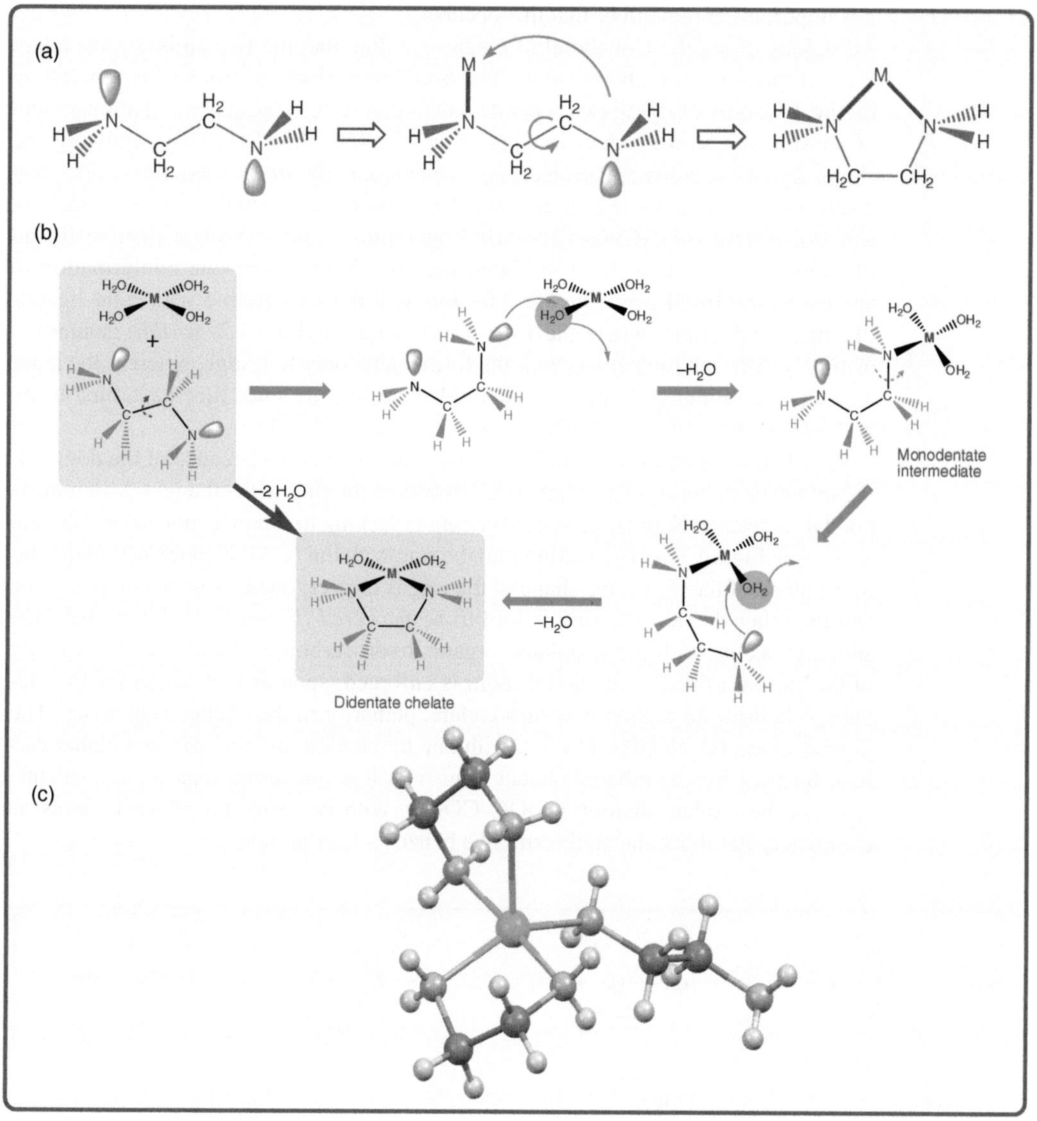

Figure 3.5

(a) Outline of the stepwise process for chelation of ethane-1,2-diamine (en). (b) A detailed plausible mechanism of the chelation process, drawn for clarity around a square-planar metal but appropriate to a range of geometries. This features initial monodentate coordination of en through replacement of an aqua ligand, rearrangement and orientation of the second lone pair and its subsequent binding to form the chelate ring. A single concerted step (brown arrow) from reactants to products is *not* favoured, but rather a stepwise process (green arrows) is involved. (c) The crystal structure of the unusual complex $[Cu(en)_2(N\text{-}en)]^{2+}$ is shown, where one of the en ligands binds as a monodentate, while the other two are didentate. This supports the mechanism presented above.

the process of chelation inevitably involves sequential replacement (substitution) of these metal-bound water molecules (now aqua ligands) by the donor atoms or groups of the chelating ligand. A process of substitution occurs because the metal centre has a strong preference for a particular shape and number of donors and cannot simply increase the number of donors in its coordination sphere. This essential and inevitable process is often left out of simple representations of the mechanism of chelation, but it is important to remember that this occurs.

Looking along the C–C bond of ethane-1,2-diamine, the two amines must adopt a *cis* disposition for chelation; in the *trans* conformation (shown at top left in Figure 3.6), only bridging two separate metals can result. The staggered arrangement of substituents shown when looking along the C–C bond (known, from organic chemistry, as a Newman projection) is energetically more favourable than one where the substituents are in alignment (eclipsed), as shown both by modelling and experiment (see $[Cu(en)_2(N-en)]^{2+}$ in Figure 3.5(c)). With a flexible ligand like this, rotation about the C–C bond permits change from one conformation to another in the free ligand (Figure 3.6); this will not be possible with rigid ligands like benzenediamine, where the 1,4- (*para*) isomer and the 1,2- (*ortho*) isomer are distinctly different molecules, with the former able only to bridge, whereas the latter may chelate (although both are called didentate ligands since they each bind using both of their two nitrogen donors).

The chelate ring formed with benzene-1,2-diamine is flat because of the dominating influence of the rigid aromatic ring. However, the ring with ethane-1,2-diamine is not flat since each N and C centre in the ring is seeking to retain its normal tetrahedral shape (see Figure 3.5(c)). Looking into the ring with the N–M–N plane perpendicular to the plane of the paper, the shape of the ring is clearer; one C is up above this plane and the other down – the ring is said to be *puckered* (Figure 3.7). Much the same situation occurs with the nonplanar organic hydrocarbon cyclopentane. If planarity of the carbon joined to the donor atom is enforced, such as is the case for the planar sp^2-hybridised carbon in a carboxylate, planarity in the chelate ring arises. The glycine anion ($H_2N–CH_2–COO^-$), with one tetrahedral and one trigonal planar carbon, forms a five-membered chelate ring with less puckering than diaminoethane, whereas the oxalate dianion ($^-OOC–COO^-$), with two trigonal planar carbons, is completely flat in its chelated form, like benzene-1,2-diamine.

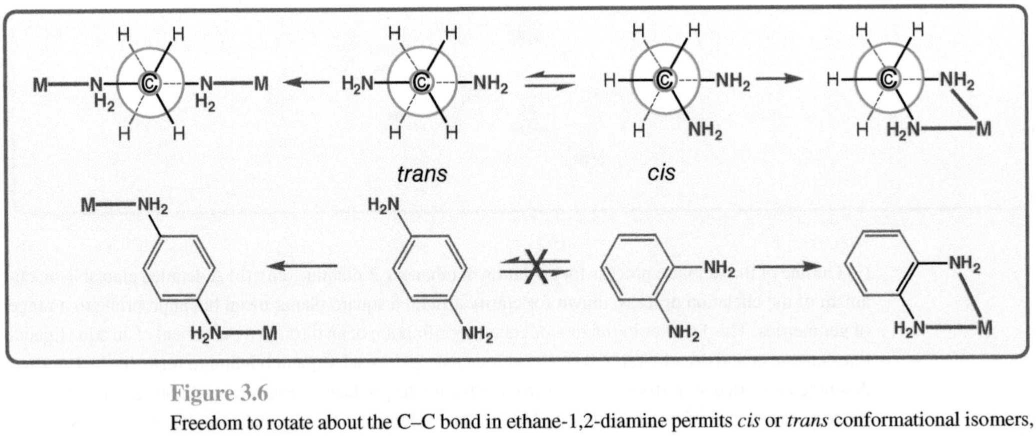

Figure 3.6

Freedom to rotate about the C–C bond in ethane-1,2-diamine permits *cis* or *trans* conformational isomers, capable of chelation and bridging, respectively (top). For rigid benzenediamine (bottom), rearrangement is not possible, and the two isomers shown have exclusive, different coordinating modes as didentate ligands.

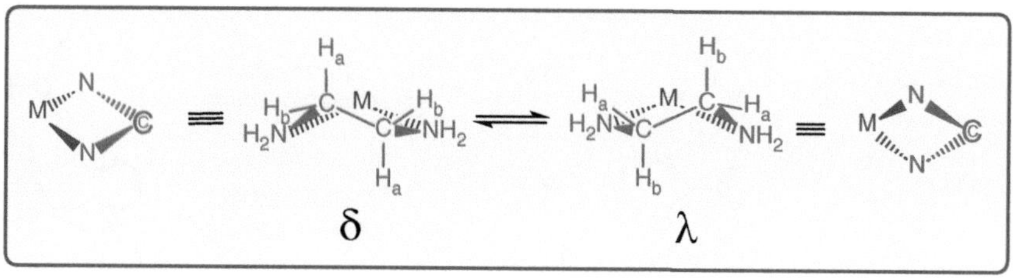

Figure 3.7

Chelate ring conformations in chelated ethane-1,2-diamine (en), designated as δ and λ. In views (figure centre) looking into the N–M–N plane and considering H atoms bound to C atoms, for the δ conformer, H_a is axial (perpendicular to the plane) and H_b is equatorial, whereas in the λ conformer, these positions are interchanged. On the left and right, these chelate rings are viewed along the C–C bond (H atoms removed for clarity). Ready interconversion between the two conformations (which are mirror images) is possible, as only a small energy barrier exists between them.

For the puckered ethane-1,2-diamine, there are some further observations to make. The chelate ring is more rigid than the freely rotating unbound ligand so that the protons on each carbon are in nonequivalent environments, as one points essentially vertically (axial) and the other sideways approximately parallel (equatorial) to the N–M–N plane (Figure 3.7). Nevertheless, it is sufficiently flexible that it can invert – one carbon moving upwards while the other moves downward to yield the other form and the H-atoms exchange axial/equatorial positions. These two forms are examples of two different *conformational isomers* or *conformers*; one is called λ and the other δ, by convention; they are mirror images of each other. Any chelate ring that is not flat may have such conformers.

A vast array of didentate chelates exist so that this one type alone can be daunting because of the variety. However, there are a number of essentially classical and popular examples, many of which tend to form flat chelate rings rather than puckered ones as a result of the shape of the donor group or enforced planarity of the whole assembly due to conjugation. A selection of common ligands appears in Figure 3.8, along with 'trivial' or abbreviated names often used to identify these molecules as ligands. One aspect of the set of examples is that the chain of atoms linking the donor atoms can vary – they do not all lead to the same chelate ring size; however, it is notable that a four-atom chain leading to five-membered chelate rings is most common. This aspect is addressed in the next section. Although most of the chelate ligands you will encounter have a carbon backbone and heteroatoms like nitrogen, oxygen or sulfur as donors, this is not essential. This can best be illustrated by comparing the two pairs of chelated systems in Figure 3.9, which have identical ring sizes and shapes, yet one of each pair contains no carbon atoms at all. There are a range of ligand systems without carbon chains known, but they need no special consideration. They can be treated at an elementary level just like the more common carbon-backboned ligands. Herein, it is this latter and most common type upon which we shall concentrate.

3.1.4 Chelate Ring Size

As the size of the chain of atoms linking a pair of donor atoms increases, the size of the chelate ring formed by the molecule increases. This affects the '*bite*' of the chelate, which is the preferred separation of the donor atoms in the chelate,

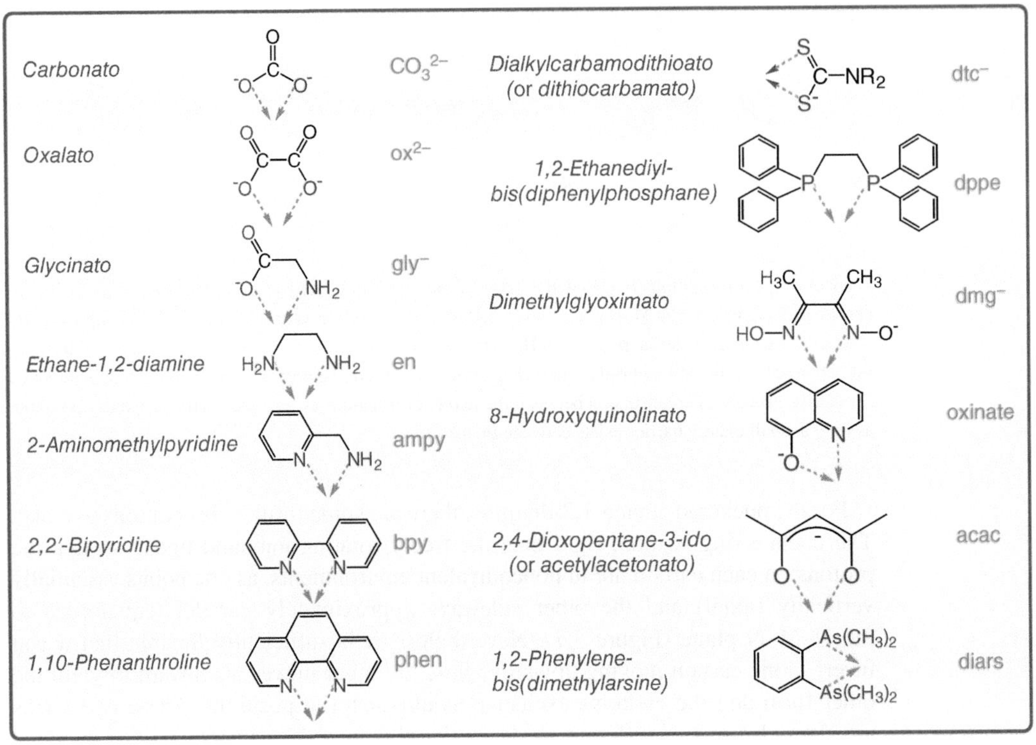

Figure 3.8

Some common didentate chelating ligands. Common abbreviations used for the ligands are given to the right of each line drawing. The arrows show the orientation of the donor atom lone pairs.

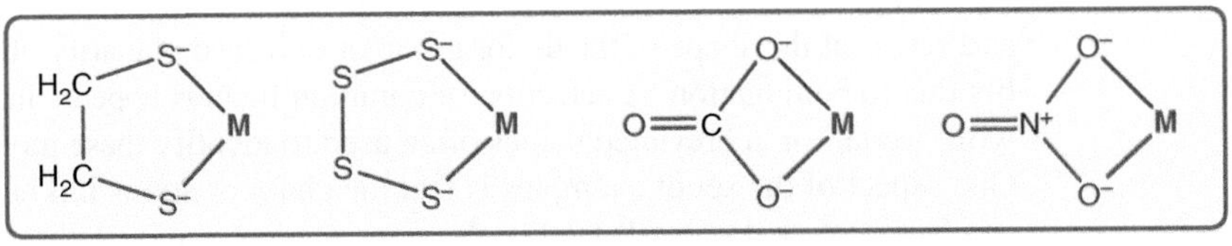

Figure 3.9

Simple pairs of chelate rings with common donor atoms but different chain atoms.

with concomitant effects on the stability and strength of the assembly. Altering bond distances and angles in the organic framework of a ligand involves a greater expenditure of energy than is the case with distances and angles around the metal, and so adjustments observed are often greater around the metal centre. Where a mismatch occurs in chelation, varying the M–L bond length or L–M–L angles (or usually both in concert) is the dominant way adjustment is made to deal with the nonideal 'fit' of metal and potential chelate (Figure 3.10).

Nevertheless, there is inherently no real upper limit on chelate ring size, except that as the chains between donors elongate, the size of the ring is such that it imparts no special stability on the complex, and thus no benefit is obtained – nor is it as easy for chelation to occur since the second donor may be located well away from the anchored first donor and thus not in a preferred position for binding. Examples of three- to seven-membered chelate rings appear in Figure 3.11; note how the experimentally

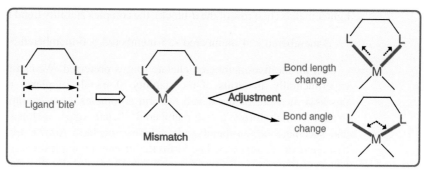

Figure 3.10
Chelate ring formation may not be ideal in terms of the 'fit' of the ligand to the metal. Potential mismatch involves, in large part, (but not exclusively) adjustment in the metal–donor distances and the angles around the metal.

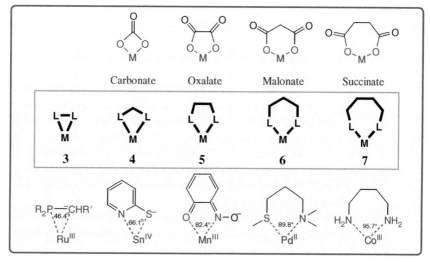

Figure 3.11
Examples of from three- to seven-membered chelate rings. Note how ring size impacts on the L–M–L intraligand angle, shown in the lower set.

measured L–M–L angle (where L represents the donor atom or group) changes with ring size. You may note that the smallest possible ring, three-membered, is missing from the top row of simple O-donor chelates. This is because there may be concern arising from bonding complexities about admitting the side-on binding of the simple diatomic molecule O_2 (termed η^2 binding) as a genuine simple chelate. In general, three-membered chelate rings are extremely rare. One example, with a compressed C–Ru–P angle of only 46.4°, is included in Figure 3.11. Since you are very unlikely to come across a three-membered chelate ring, it is appropriate to think of four-membered rings as the minimum size commonly met.

What we find is that there is a preferred chelate ring size; as the ring size increases, there is an increase in stability of the assembled complex and then a decrease as the ring continues to grow. This trend depends on a number of factors, such as the metal ions, donor groups and ligand framework involved. Nevertheless, for the common lighter metals (first row of the *d*-block), the complex stability trend is fairly consistent:

3-membered < 4-membered < 5-membered > 6-membered > 7-membered.

Overall, the five-membered chelate ring is preferred. We can measure this trend experimentally, in terms of the stability of metal complexes (you can consider this as a measure of the desire to form a complex). For example, comparing the oxalate (which forms a five-membered chelate ring), malonate (six-membered) and succinate (seven-membered) didentate chelates (drawn in Figure 3.11), the five-membered oxalate chelate forms the strongest complexes across a range of divalent transition metal ions (see Figure 5.2 in Chapter 5). Further, in all cases, the trend 5-membered > 6-membered > 7-membered ring is preserved. The size of the metal ion and its preferred bond lengths also play a role – for example, while the trend is consistent for the lighter metals, larger metal ions may prefer a different ring size. There is another stability trend observed for these complexes that we will return to later in Chapter 5 – dependent only on the metal, irrespective of the ligand.

3.1.5   Donor Group Variation

What should already be obvious is that there can be different types of donors since we have by now introduced examples of molecules where N, O, S, P and even C atoms bind to the metal (Figure 3.8, for some examples). In the same way that different ethnic groups are all members of the same human species, molecules with different donors are all members of the same basic category – in this case ligands. You expect different groups of people to have differentiating features, and the same applies to ligands with different donors. Because the donor atom is directly attached to the central metal, the metal feels the influence of the donor atom much more than any other atoms and groups bonded to that donor atom. This is reflected in the chemical and physical properties of a complex.

Most donor atoms in ligands are members of the *p*-block of the periodic table. Of these, common simple ligands you are likely to meet are the following where R is an organic group:

R_3N	R_2O	F^-
R_3P	R_2S	Cl^-
R_3As	R_2Se	Br^-
R_3Sb	R_2Te	I^-

Those ligands with N, O, P and S donors, as well as the halide anions, are particularly common. The presence of these donor atoms covers the majority of ligands that you are ever likely to meet, including in natural biomolecules. There are others, of course; even carbon, as H_3C^-, for example, is an effective donor, and there is an area of coordination chemistry (organometallic chemistry) devoted to compounds that include M–C bonds, introduced in Chapter 2 and addressed later in Section 3.6.

Very often, as illustrated already, ligands contain more than one potential donor group. Where these are not identical, we have a *mixed-donor* ligand. These are extremely common, indeed the dominant class; classic examples are the naturally occurring α-amino acid anions {$H_2N–CH(R)–COO^-$}, which present both N (amine, –NH_2) and O (carboxylate, –COO^-) donors. Where there are choices of donor groups available to a metal ion, it is hardly surprising that some preference may exist, which we will address in due course.

3.1.6 Denticity

So far, we have restricted our discussion to ligands that attach to a metal at one or two coordination sites. While these are very common, they by no means represent the limit. It is possible to have ligands that can provide sufficient donor groups in one molecule to satisfy the coordination sphere of a metal ion fully. Indeed, they may have sufficient donor atoms to bind several metal ions. As already introduced, *denticity* is the term defining the number of coordinated donors of a ligand. Ligand denticity is expressed by the appropriate Greek prefix: i.e. *monodentate* (one), *didentate* (two), *tridentate* (three), and *tetradentate* (four coordinated donor atoms). This nomenclature extends to even higher denticity ligands, reported in Appendix 1. As a class, these ligands are termed *polydentate* (see Appendix 1). Denticity represents the way we regularly discuss ligand binding in spoken form; it provides brevity at least, as we can then say 'a didentate ligand' rather than 'a ligand bound through two donor groups'.

It is important to recognise that denticity and the number of *potential* donor groups of a ligand are not always the same thing. We can see this with a very simple example, the amino acid anion glycinate, which can bind to a metal via only one O of the carboxylate group, only the amine group, or as a chelate through both O and N at once; in fact, chelation using both O atoms of the carboxylate is also possible. The first two modes are examples of the molecule acting as a monodentate ligand, with the last two of the molecules acting as a didentate chelating ligand (Figure 3.12). For a *polydentate* ligand, the number of sites at which it binds to a central metal can be expressed as the symbol κ^n, where n indicates the number of attachment sites. A related term to denticity that is met in organometallic compounds is *hapticity* (η^n), relating to the number of adjoining atoms in a ligand that are considered to contribute as a donor unit in binding to a metal. This concept was introduced in Chapter 2 for ligands binding in a side-on manner, like η^2–$H_2C=CH_2$. Here, we have strayed into the field of nomenclature (see Appendix 1), which we will promptly leave.

The groups that bind to a metal are also influenced by the shape of the ligand and their location on the ligand framework. Some further examples of ligands of different denticity are given in Figure 3.13. To retain our focus on the concept at this stage, the examples involve coordination to a simple two-dimensional square planar geometry that we shall meet again later; of course, other shapes are able to accommodate polydentate ligands as well. Note that some ligands undergo deprotonation to form an anion in achieving complexation, consistent with the anion being a better donor for a cationic metal ion than a neutral ligand on purely electrostatic grounds. Not all donor groups can readily deprotonate, as this ability depends on the acidity of the group.

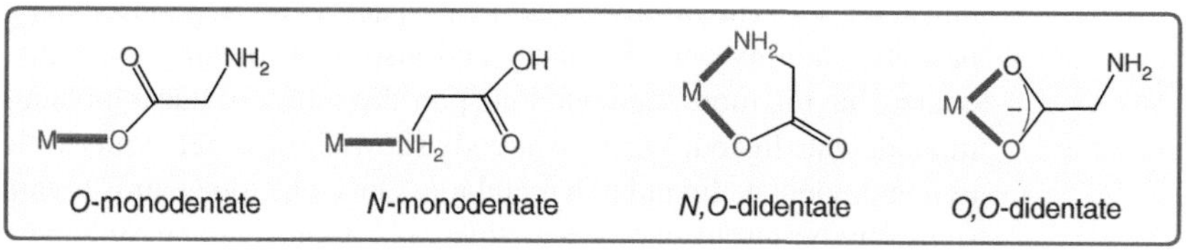

Figure 3.12

Possible coordination modes for the amino acid ligand glycine to a single metal centre.

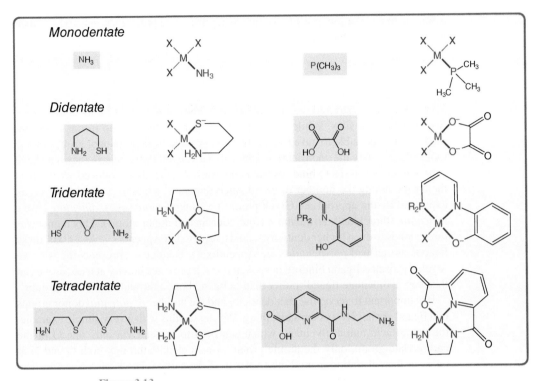

Figure 3.13

Examples of ligands of different denticity. Several undergo deprotonation prior to coordination, as indicated by the binding of an anionic form. X-groups inserted in other available metal coordination sites on the square-planar metal used for illustration here identify those sites not used by the target ligand.

Further, some ligands contain groups with more than one heteroatom capable, in principle, of acting as a donor atom; simple examples in Figure 3.13 are carboxylate ($-COO^-$, with potentially two O donors) and amido ($-N^--CO-$, with both N and O potential donors). Coordination of both donors to the one metal ion at once may be forbidden by the shape of the ligand molecule or else occurs under only special synthetic conditions. We shall explore polydentate ligands in more detail in Section 3.4.

3.2 Ligand Basicity

3.2.1 Basic Binders

The simplest class of ligands offers a single lone pair for binding to a central atom or ion. The ammonia molecule (NH_3) is the classical example, which we have introduced earlier. The ability to act as a ligand is related to a molecule's own inherent basicity, stability and chemistry and also to it meeting the right partner metal. Of course, having more than one lone pair does not exclude a molecule from acting as a monodentate ligand. We have already shown how water, with two lone pairs, can bind as a monodentate ligand to a metal ion. Once one lone pair is bound, as the other subsequently points away in space, this second lone pair can only be employed, if at all, by a different metal ion. It is most usual to meet water as a simple monodentate ligand. In the extreme, a halogen ion has four lone pairs, but as all point to different corners of a tetrahedron, only one pair at a time can effectively interact with one particular metal ion. Again, the other lone pairs are available for binding (through bridging) to other metal ions, and this is a known but not a necessary outcome. All of the ligands R_3N, R_2O and X^- (X = halogen), with one, two and four lone pairs respectively, are met as monodentate ligands. The latter two types can also, in principle, bridge other metal ions using their extra lone pairs, an option that ammonia does not have. However, even ammonia can gain another lone pair by removing a proton to form the amido ($^-NH_2$) ion, which is then capable of bridging two metal ions. Acting as an acid is not our usual view of ammonia because it is a base, and losing a proton is not easily achieved; ammonia's more usual behaviour is addressed below. Since bridging ligands bring two or more metal cations reasonably close together, it is common for bridging ligands to be anionic to offset some of the metal–metal electrostatic repulsion. Clearly, ligand basicity and or the capacity to undergo deprotonation is important in complex formation generally; the latter aspect is addressed further in Section 3.2.2.

Ammonia is weakly basic and able to be protonated to form the ammonium ion, with its base strength defined by its pK_a of 9.3 for the conjugate acid (NH_4^+), an experimentally measurable parameter ($pK_a = -\log_{10}K_a$, where K_a is the acid dissociation constant). In this case, at pH 9.3, the concentrations of NH_3 and NH_4^+ will be equal, and the higher the pK_a of the conjugate acid, the stronger the base. If base strength is considered a measure of affinity for a proton, it seems reasonable to assume that this is also a reasonable measure of affinity for small, positively charged metal ions. This works reasonably well as a gross measure when the metal ion is small and highly charged. For example, NH_3 is much more basic than PH_3, which can thus account for ammonia being a better ligand for first row transition metal ions than phosphine. Further, the oxygen analogue H_3O^+ (oxonium) has one lone pair remaining, but it is an extremely poor base and is *not* able to be protonated to H_4O^{2+}. The oxonium ion is also totally ineffective as a ligand to metal ions. We are now beginning to define some ligands as 'better' than others on the basis of certain chemical and physical properties – this is a move towards understanding selectivity and preference or why molecules dissolved in a solvent can compete with that solvent as ligands for a metal ion.

Another aspect of coordination of note is that all ligands with H atoms become stronger acids once coordinated to a metal. The acidity of free water (pK_w 14.0) increases dramatically when coordinated to a metal ion, and the charge of the metal has the greatest influence: e.g. Ag^+ (pK_a 12.0), Zn^{2+} (9.0), Fe^{3+} (2.2) and Zr^{4+} (−0.3). In general, both the ligand acidity and reactivity are affected by coordination to the metal.

The ligand basicity is also affected by organic substituents. For example, ammonia (NH_3, pK_a 9.3) and methanamine (CH_3NH_2, pK_a 10.6) carry a common N donor atom and act as monodentate ligands, but the introduction of an electron-donating methyl group in place of an H-atom is responsible for its increase in basicity. Interestingly, NH_3 and CH_3NH_2 both form complexes of similar stability, as the methyl group on CH_3NH_2 creates a destabilising steric (size) effect, which opposes the greater basicity of its N-atom.

This can be carried further by considering a secondary amine such as $(CH_3)_2NH$ and a tertiary amine such as $(CH_3)_3N$ (Figure 3.14). Both can, in principle, bind to metals, but again there are opposing effects of additional steric crowding at the N-atom counterbalanced by the stronger basicity of the N-atom from more electron-donating methyl groups. Further, there are other types of nitrogen donors; for example, pyridine (C_6H_5N), has an sp^2-hybridised nitrogen in a benzene-like

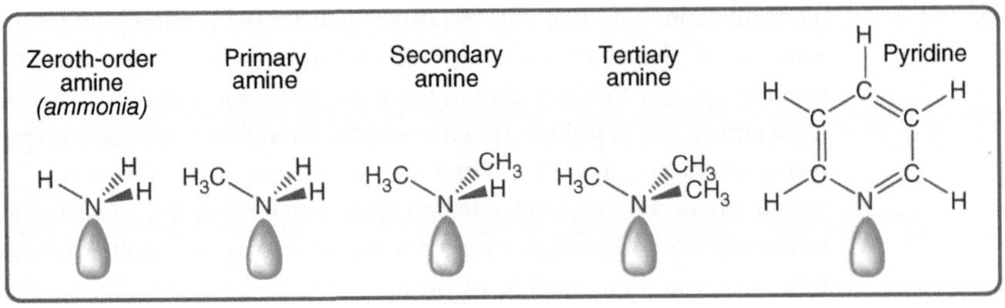

Figure 3.14

Examples of the N-donor family of monodentate ligands.

aromatic ring (Figure 3.14). In an amide group (R–HN–CO–R′), the N-atom has no coordinating properties due to the N-atom lone pair being conjugated with the carbonyl group, but in its deprotonated amido form (R–$^-$N–CO–R′), it becomes an effective N-donor ligand. Each type of N donor has characteristics as a ligand that do not equate with other N donors – variety is indeed the spice of life in coordination chemistry.

3.2.2 Bridging Ligands

We have identified ammonia (NH_3) as the classical example of a monodentate ligand, with only a single lone pair. However, as already mentioned, if ammonia is stripped of a proton to form the anion $^-NH_2$, it now offers two lone pairs. Although only accessible in nonaqueous solution (as it is more basic than HO^-), it is known for its capacity to bind efficiently to two different metal ions at once. This is an example of how the creation of more lone pairs associated with forming a new anion can expand access to more metal ions, through these ligands making additional attachments to other metal ions – the process of *bridging*. This is illustrated for ammonia and its anion in Figure 3.15. Bridging groups are represented by the Greek letter μ (mu) as a prefix in formulae (see Appendix 1).

Water can lose protons sequentially to form hydroxide (HO^-) and oxide (O^{2-}). As the protons are stripped off, the number of lone pairs increases, with hydroxide offering three and oxide offering four. However, just having more than one lone pair is no guarantee of forming more bonds. Binding as a monodentate ligand to a single metal occurs for H_2O, HO^- and O^{2-}, in all cases involving the oxygen as the donor atom. However, these ligands can, in principle, expand their linkages by making

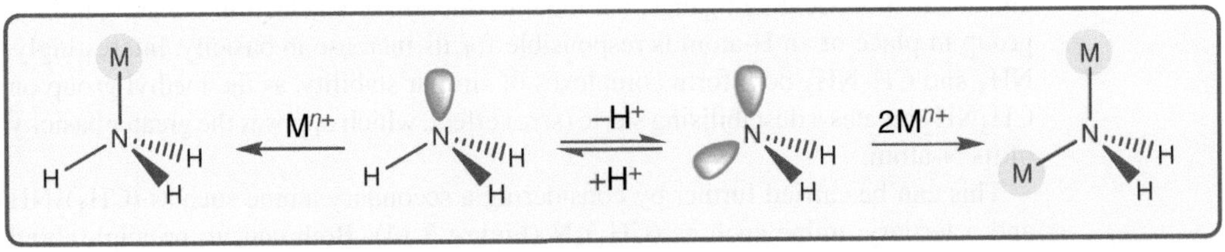

Figure 3.15

Ammonia coordinates as a monodentate ligand to one metal. When a proton is removed, it exhibits the capacity to attach the resultant additional lone pair (orbital shown in green) to a second metal ion in a *bridging* mode. Charges are not shown in this diagram.

additional *bridging* attachments to other metal ions. We see this with the anions, but less commonly with water itself. It is useful to examine why neutral H_2O does not normally bridge two metal ions. When a coordinate bond is formed with a positively charged metal ion, there is a strong tendency for electron density to 'shift' from the donor atom towards the metal ion; we can visualise this as a diminution in the size of the remaining lone pair, making it less accessible and attractive to a second metal ion.

If a proton is removed from the coordinated water, creating a second free lone pair, there is a significant increase in electron density on the oxygen, and so its affinity for a second metal cation is increased. The negative charge on the HO^- ligand acts to partially balance the positive charges on the two metal ions forced to be located closely together (Figure 3.16). So, when H_2O acts as a bridging ligand, it is more often met between metal ions of low charge where its pK_a is still high (e.g. alkali metal ions such as $Na^+-H_2O-Na^+$). When more positively charged metals are involved, the decrease in the pK_a of the bridging water ligand is more significant through the influence of the adjacent positive charges, and the $M^{n+}-HO^--M^{n+}$ or even $M^{n+}-O^{2-}-M^{n+}$ modes are more stable.

An extension of this deprotonation process is illustrated in the lower part of Figure 3.16 (ignoring nonbonding or so-called 'spectator' lone pairs) starting from water bound in its usual form as a monodentate ligand to one metal. As each proton is successively removed, the capacity to attach the resultant lone pair to a second and then even, with the next deprotonation, to a third metal ion arises. The O donor is acting as a bridging monodentate ligand to more than one metal at once. This is well-known behaviour. What is not shown (for simplicity) are the other bonds to the metals that can lead to large clusters in some cases, such as the box-like framework shown at bottom right in Figure 3.16, nor those additional ligands not involved in bridging. The oxido-ligand O^{2-} can even bind to four metal ions at once.

Thus, ligands can accommodate two metal ions in two distinct ways: by using two different donor atoms on the one ligand to bind to two separate metal ions or by using two different lone pairs on the same donor group to bind to two separate metal ions (Figure 3.17). The former leads to greater distance between the like-charged metal centres and is thus likely to be a less demanding process. The latter brings the metal ions much closer together but can only occur, of course, with donor groups that have the potential to provide at least two lone pairs and that are usually anionic to help two

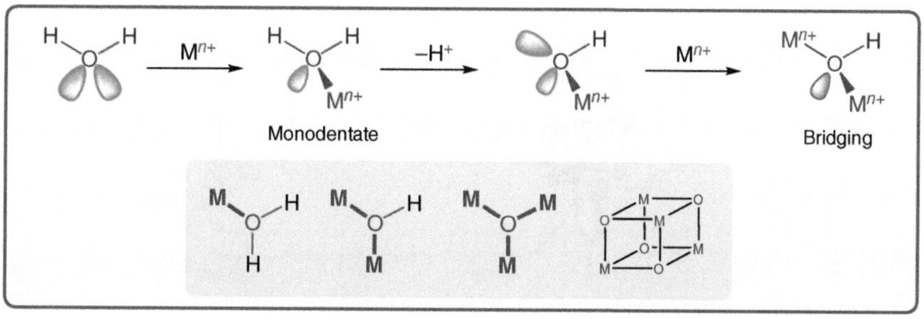

Figure 3.16

Deprotonation of water enhances the prospect of bridging between two metal ions by increasing electron density on the O atom. Successive deprotonation can permit multiple bridging to metal ions, resulting in small metal–oxide clusters (exemplified at bottom right).

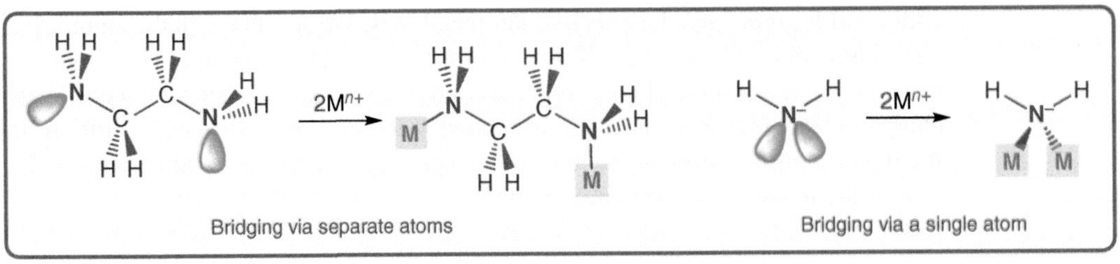

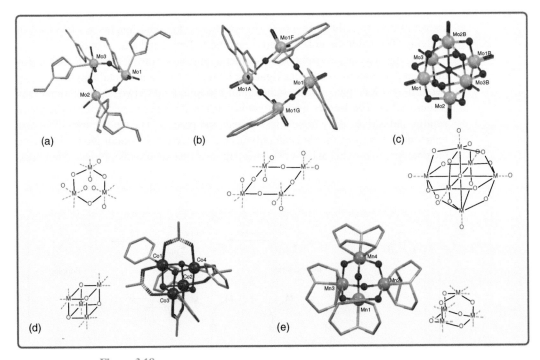

Figure 3.17
Modes through which bridging between two different metals can arise, illustrated for simple N-donor ligands ethane-1,2-diamine and the amide ion.

like-charged metal centres approach each other closely. Systems where several metal ions are bound in proximity employ the same basic concepts discussed here.

As mentioned, one consequence of coordination by a bridging ligand is obviously bringing two like charged metal centres into proximity. With several bridging ligands participating, more than two metal centres may be assembled near each other. This was illustrated in Chapter 2, where a few simple examples with exclusively bridging and terminal halide ligands were presented (Figure 2.29). As mentioned in Chapter 2, assemblies of three or more metal ions in proximity (in essentially closed polyhedra) are called *clusters*. Examples where the metal ions are linked by O^{2-} bridging ions are termed *metal-oxido clusters*, and some example shapes are shown in Figure 3.18.

Figure 3.18
Examples of mainly metal-oxido clusters with various number of metal ions involved. The metal-containing core in each case is of the type: (a) M_3O_9; (b) M_4O_{12}; (c) M_6O_{18}; (d) M_4O_4 cubane and (e) M_4O_6 adamantoid.

While O^{2-} is a common bridging entity, others such as hydroxide, sulfide and halides are known. Apart from the bridging ligands, an array of other ligands is bound to the several metals involved in the cluster and arranged around the cluster core; each cluster metal acquires a coordination sphere quite like that it would acquire as an independent complex. Clusters with many metal ions have been reported, and these assemblies can display physical properties of commercial interest; synthetic clusters are discussed in more detail in Chapter 6. Clusters are also met in biological systems, and examples will be examined in Chapter 8. There is another category of clusters that involve direct bonding between metal centres or at least significant metal–metal interaction, often met in organometallic compounds.

3.3   Making Choices

3.3.1   Selectivity

Of all the donors in all the ligands in all the world, why this one? This question takes us back to a movie analogy again (*Casablanca*), but we are not going to take it too far. Suffice to say that, in the same way that there is a vast number of 'gin joints' in the world to choose from, there is an even vaster range of potential ligands for a metal ion to select. Why, then, does selection happen (or choice arise) at all?

What we do know from experimental evidence is that metal ions do not interact with potential ligands purely on a statistical basis. A simple example will suffice. Pure water, the most common solvent, is present at a concentration of $\sim$55 M. If ammonia is dissolved in water to a concentration of $\sim$0.5 M, there is a 100-fold excess of water molecules over ammonia molecules. If a small concentration of copper ion is introduced into the solution, the vast majority of the copper ions bind to the ammonia – even though each cation 'meets' many more water molecules. Why is there this clear preference for ammonia over water as a ligand in this case? This is but one example of a general observation.

3.3.2   Preferences

Ligand preference for and affinities of metal ions present themselves as experimentally observable behaviour that is not easily reconciled. As a general rule, when a metal (M) is mixed with equimolar amounts of ligands A and B, the result is *not* usually equimolar amounts of MA and MB. Both metal ions (*Lewis acids*) and ligands (*Lewis bases*) show preferences.

In seeking an understanding of this phenomenon, we can sort ligands according to their preference for forming coordinate bonds to metal ions that exhibit more ionic rather than purely covalent character. Every coordinate covalent bond between a metal ion and a donor atom will display some polarity since the electronegativities of the two atoms joined are not equivalent. The extreme case of a polar bond is the ionic bond, where formal charge separation (cation–anion) rather than electron sharing occurs; coordinate bonds show differing amounts of ionic character. Small, highly charged entities will have a high charge density, and their bonds will have significant ionic character. In a sense, we can think of these ions or atoms as 'hard' spheres since their electron clouds tend to be drawn inward more towards the core and thus are more compressed and less efficient at orbital overlap. A large, low-charged entity tends to have more diffuse and expanded electron clouds, making it less dense or 'soft' and better suited to orbital overlap. This concept, applied to donor atoms and groups, leads to us defining 'hard' donors as those with significant ionic character in their bonds and 'soft' donors as those forming coordinate bonds with a high degree of covalent character. Note that there is a continuum from 100% ionic to 100% covalent character, and the extrema are rarely met. An important observation is that the 'hard' donor atoms are also the most electronegative elements (i.e. they avidly draw shared electrons or electron density towards their own nucleus). Therefore, arranging donor ions or groups in terms of increasing electronegativity provides us with a trend from 'soft' to 'hard' ligands:

$$\text{'SOFT'} \quad - \textit{increasing electronegativity} \rightarrow \quad \text{'HARD'}$$

$$\underline{C}N^- \quad S^{2-} \quad RS^- \quad I^- \quad Br^- \quad Cl^- \quad NH_3 \quad {}^-OH \quad OH_2 \quad F^-$$

The ligands at each 'end' show distinct preferences for different types of metal ions. The F^- ligand binds strongly to Ti^{4+}, whereas CN^- binds strongly to Au^+, and it is clear that these metal ions differ in terms of both ionic radius and charge. Any individual metal ion will display preferential binding when presented with a range of different ligands. This means that metal ions can be graded and assigned 'hard–soft' character like ligands; however, by convention, we define the least electronegative as 'hardest' and the most electronegative as 'softest'. For the metals, they grow increasingly 'soft' from left to right across the periodic table and also increase in softness down any column of the table. This definition for metals allows us to apply a simple 'like prefers like' concept – *'hard' ligands (bases) prefer 'hard' metals (acids)* and *'soft' ligands (bases) prefer 'soft' metals (acids)*. This principle of hard and soft acids and bases (HSAB), developed by Ralph Pearson in 1963, is a simple but surprisingly effective way of looking at experimentally observable metal–ligand preference. To see the concept in action, we should look at a few examples of how it allows 'interpretation' of experimental observations; yet without a strong theoretical basis, the concept is somehow unsatisfying.

One example involves the long-established reaction of Co^{2+} and Hg^{2+} together in solution with SCN^-. The ligand has both a 'soft' donor (S) and a 'hard' donor (N) available; although like-charged, Hg^{2+} is larger than Co^{2+} and further across and down the periodic table and thus 'softer'. This reaction results in a polymeric crystalline solid $[Co(NCS)_4Hg]$, with each metal in a tetrahedral coordination geometry of thiocyanate donors, with four S atoms bound to each Hg^{2+} and four N atoms bound to each Co^{2+} (Figure 3.19). Only this thermodynamically stable product, where the softer S bonds to softer Hg^{2+}, and harder N bonds to harder Co^{2+}, is known. If the 'wrong' end of a ligand like thiocyanate binds initially to a mismatched metal ion, it will usually undergo a rearrangement reaction to reach the stable, preferred form. It is possible to apply the concepts exemplified above to reaction outcomes generally. In all cases, it is a comparative issue; for example, a ligand defined as 'harder' versus one other donor type (such as OH_2 versus Cl^-) may be considered 'softer' if compared with another type of ligand (such as NH_3 versus OH_2); this aspect will also be true for metal ions.

Definition of hard/soft character is the result of empirical observations and trends in measured stability of complexes. For example, hard acids (such as Fe^{3+}) tend to bind the halides with the order of complex strength being $F^- > Cl^- > Br^- > I^-$ and

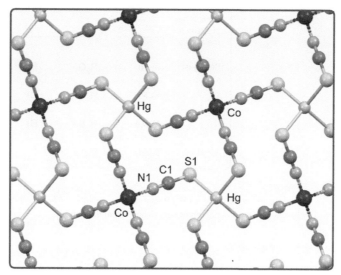

Figure 3.19

Crystal structure of the compound [Co(NCS)$_4$Hg] showing preferred coordination of nitrogen to Co^{II} and sulfur to Hg^{II}. The structure is polymeric and extends in all dimensions off the page.

Table 3.1   Selected examples of hard/soft Lewis acids and bases.

Character	Lewis acids	Lewis bases
Hard	H$^+$, Li$^+$, Mg^{2+}, Al^{3+}, Cr^{3+}, Ti^{4+}	F$^-$, HO$^-$, H$_2$O, H$_3$N, CO$_3$$^{2-}$, PO$_4$$^{3-}$
Intermediate	Fe^{2+}, Co^{2+}, Ni^{2+}, Cu^{2+}, Zn^{2+}	Br$^-$, $\underline{N}$O$_2$$^-$, SC$\underline{N}$$^-$
Soft	Cu$^+$, Ag$^+$, Au$^+$, Hg$^+$, Cd^{2+}, Pt^{2+}	I$^-$, CN$^-$, CO, H$^-$, $\underline{S}$CN$^-$, R$_3$P, R$_2$S

soft acids (such as Hg^{2+}) in the reverse order of stability. However, as with any model with just two categories, there will be a 'grey' area in the middle, where borderline character is exhibited. This is the case for both Lewis acids and Lewis bases. Selected examples are collected in Table 3.1; a more complete table appears later in Chapter 5.

The 'hard'–'soft' concept is, from a perspective of the metal ion, often recast in terms of two classes of metal ions as follows:

'Class A' Metal Ions ('*hard*') – These are small, compact and not very polarisable; this group includes alkali metal ions, alkaline earth metal ions and lighter and more highly charged metal ions such as Ti^{4+}, Fe^{3+}, Co^{3+} and Al^{3+}. They show a preference for ligands (bases) that are small and less polarisable.

'Class B' Metal Ions ('*soft*') – These are larger and more polarisable; this group includes heavier transition metal ion such as Hg^{2+}, Pt^{2+} and Au$^+$, as well as low-valent metal ions including formally M(0) centres in organometallic compounds. They exhibit a preference for larger, polarisable ligands. This leads to a preference pattern outlined in Figure 3.20.

There is also a changing preference order seen across the rows of the periodic table, with *Class A* showing a trend from weaker to stronger from left to right and *Class B* showing the opposite trend of stronger to weaker, defined in terms of the measured stability constant of ML complexes formed.

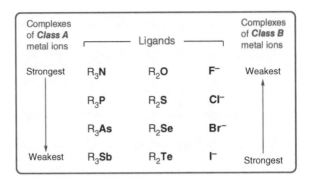

Figure 3.20

Metal–ligand preferences for key ligands. In any column, a *Class A* metal ion prefers ligands from the top, whereas a *Class B* metal ion prefers ligands from the bottom.

3.4   Polydentate Ligands

The concept of a chelate ring has been introduced earlier in this chapter. The simplest chelating ligands consist of just two potential donors attached to a (usually, but not necessarily) carbon chain in positions that permit both donors to attach to the same metal ion through bonding at two adjacent sites around that central metal. There is no compulsion for such a ligand to chelate, but the fact that they usually do is an indication that such arrangements are especially favourable, or the complex assembly is more stable as a result of chelation. Ligands may offer more than two potential donor groups, but the step-by-step process of chelation described earlier for the simple didentate ethane-1,2-diamine is believed to be the usual way all polydentate ligands go about coordinating to metal ions in general – 'knitting' themselves onto the metal ion 'stitch by stitch' (or, more correctly, undergoing *stepwise coordination*). It also provides a way whereby compounds can achieve only partial coordination of a potential donor set, when there are insufficient coordination sites for the full suite of potential donor groups available on a ligand.

As a general observation, the more donor atoms by which a molecule is bound to a metal ion, the stronger will be the assembly – think of it as more teeth giving a stronger bite. There is basic common sense in this, of course, since more donors bound means more coordinate bonds linking the ligand to the metal ion. To separate the metal and ligand, all these bonds would need to be broken, which one would obviously anticipate costing more energy than breaking just a single bond to separate a metal ion from a monodentate ligand. A case of more is better or greed is good. There is a conventional nomenclature to identify the number of bonds formed between a ligand and a central metal ion, shown in Table 3.2. Here, as mentioned earlier (Section 3.1.6), *denticity* defines the number of bound donor groups; this can equal the number of potential donor groups available (as in examples in the table) or be a lesser number if not all groups available are coordinated. See Appendix 1 for some other notes on polydenticity.

Table 3.2 identifies molecules that can bind at up to six coordination sites. This is by no means the limit, as inherently a molecule (such as a polypeptide) may offer a vast number of potential donors. It is not uncommon for polydentate ligands to use only some and not all of their donors in forming complexes; this may be driven by the relative location of the potential donors on the ligand (and the overall shape and folding of the ligand), which affects their capacity to all bind at once, and the number of donors, as more may be offered than can be accommodated around the metal ion.

Table 3.2 Naming of different numbers of donor groups bound to a single metal ion (denticity), with examples of typical simple linear ligands of each class.

Number of bound donors	Ligand denticity	Examples of ligands (donor atoms highlighted in bold)
One	Monodentate	$\mathbf{N}H_3$ $O\mathbf{H}_2$ $\mathbf{F}^-$
Two	Didentate	$H_2\mathbf{N}-CH_2-CH_2-\mathbf{N}H_2$ $H_2\mathbf{N}-CH_2-\mathbf{COO}^-$
Three	Tridentate	$H_2\mathbf{N}-CH_2-CH_2-\mathbf{N}H-CH_2-CH_2-\mathbf{N}H_2$ $^-\mathbf{OOC}-CH_2-H\mathbf{N}-CH_2-\mathbf{COO}^-$ $H_2\mathbf{N}-CH_2-CH_2-\mathbf{S}-CH_2-CH_2-\mathbf{P}(CH_3)_2$
Four	Tetradentate	$H_2\mathbf{N}-CH_2-CH_2-\mathbf{N}H-CH_2-CH_2-\mathbf{N}H-CH_2-CH_2-\mathbf{N}H_2$ $^-\mathbf{OOC}-CH_2-^-\mathbf{N}-CO-CH_2-CO-\mathbf{N}^--CH_2-CH_2-\mathbf{N}H_2$
Five	Pentadentate	$H_2\mathbf{N}-CH_2-CH_2-\mathbf{N}H-CH_2-CH_2-\mathbf{N}H-CH_2-CH_2-\mathbf{N}H-CH_2-CH_2-\mathbf{N}H_2$ $^-\mathbf{OOC}-CH_2-\mathbf{N}H-CH_2-CH_2-\mathbf{S}-CH_2-CH_2-\mathbf{N}H-CH_2-CH_2-\mathbf{O}^-$
Six	Hexadentate	$^-\mathbf{OOC}-CH_2-\mathbf{N}H-CH_2-CH_2-\mathbf{N}H-CH_2-CH_2-\mathbf{N}H-CH_2-CH_2-\mathbf{N}H-CH_2-\mathbf{COO}^-$

As an example, a ligand like $H_2N-CH_2-CH_2-NH-CH_2-CH_2-NH_2$ has three amine groups, all of which are capable of coordination. However, it may bind through one, two or all three of these donors, acting, in turn, as a mono-, di- and tri-dentate ligand, respectively. Where more are available than required, the set that eventually binds is usually the one that produces the thermodynamically most stable assembly (i.e. the lowest energy state). Some molecules may easily accommodate more than one metal ion; a polymer with an array of donors would be an obvious example, with different parts of the chain binding to different metal ions – we shall look at this prospect soon.

3.4.1 Introducing Ligand Shape

High denticity ligands need not be simply composed of linear chains like the examples in Table 3.2. The shape of the ligand has an important role to play in the manner of coordination and the strength of the complex, but there is no particular shape that is forbidden to a ligand as they are simply organic molecules that follow conventional bonding models (valence bond theory – hybrid orbitals). These ligands may be linear, branched, podal, cyclic, polycyclic, helical, as well as having aliphatic or aromatic components, or mixtures of these – the over-arching requirement is simply a set of donor groups with lone pairs of electrons. In some cases, they may be dinucleating or polynucleating, that is, capable of binding to more than one metal at the same time. Some fairly simple examples and their family name appear in Figure 3.21. We shall deal more with shape later since ligand shape or topology can have an important influence on coordination, but it may be useful to introduce some concepts now.

Let us consider two different types of hexadentate ligands: a linear polyamine and a branched polyaminoacid. To coordinate a metal ion and use all six donors, the linear molecule must wrap around the metal ion like a ribbon, whereas the branched molecule wraps up the metal ion in a different way (Figure 3.22). On a macroscopic scale, the latter is like holding a ball in the palm of your hand with the fingers wrapped tightly around it – a nice, secure way of holding it. On the molecular scale, branching can likewise lead to a more 'secure' complex, which makes the branched polyaminoacid $EDTA^{4-}$ (ethylenediamine tetraacetate), represented in its anionic form in Figure 3.22 (left), such an effective, strong ligand for binding most metal ions.

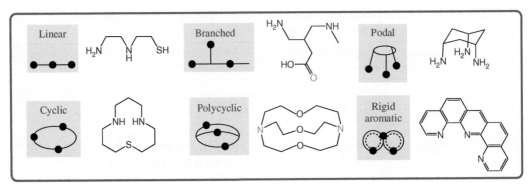

Figure 3.21
Examples of polydentate ligands of selected different basic shapes. Potentially tridentate examples appear.

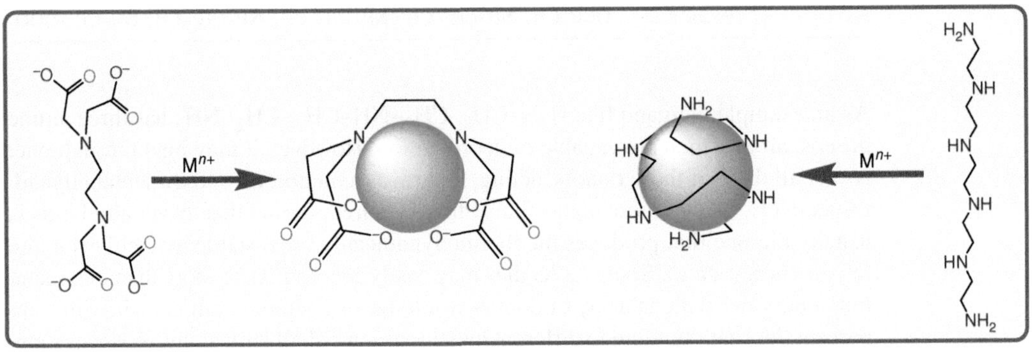

Figure 3.22
The process of complexation of polydentate ligands involves an aspect of 'wrapping' around the metal ions, usually with gross rearrangement of the ligand shape. At left is the EDTA^{4-} polyaminoacid anion and at right the linear polyamine pentaethylenehexamine (called 'penten').

The shape of the ligand places demands on how it can coordinate to a metal ion. For a many-branched ligand, where the branches or 'fingers' are terminated with donor groups capable of binding to a metal ion, the ligand can achieve a firm grip by employing these 'fingers'. Looking at the way such ligands wrap around a metal ion, you may also get an idea of how they can be removed – by detaching each 'finger' in turn and then taking the metal ion out of the 'palm' of the ligand. Initial coordination is, in effect, the reverse of this process – step-by-step binding to eventually achieve the final complex. Of course, because the metal ion is not 'naked' in the first place but carries a suite of usually simple ligands when it encounters the polydentate, each step of attachment of a donor of the polydentate involves departure of a simple ligand; it is a *substitution* process.

3.5 Polynucleating Ligands

3.5.1 Dinucleating Ligands

Small ligands like didentate ethane-1,2-diamine will not satisfy the coordination sphere of any metal ion so that, provided sufficient ligand is available in the reaction mixture, the metal may choose to attach more than one ligand, leading not only to ML species but also to ML_2 and ML_3 or even more. One consequence of working with big ligands that carry many potential donor groups is that they may decide to do the exact opposite – use their suite of donor groups to attach to more than one metal ion at once. Thus, instead of a simple ML species, they may form M_2L or even M_3L or higher.

The whole situation is about getting satisfaction (by achieving a thermodynamically stable complex) through meeting the desires of both the ligand and metal. In this partnership of the metal and ligand, there is usually some form of give and take to reach an agreeable (lowest energy) outcome. The metal is seeking to maintain its desired number of metal–donor bonds, and the ligand is seeking to employ as many of its donor groups as practicable, while preserving its preferred conformation. They are pursuing the same general outcome but come to assembly (complexation) with different demands and expectations. Normally, they reach a satisfactory match, which may be just a simple 1:1 assembly; however, it often really is a case of one is not enough. Compromise (through reaching the most stable assembly) may require either the use of more ligands than metals or the exact opposite.

Let us look more closely at a situation where more than one metal ion could be accommodated, with a relatively simple set of ligands. The polyaminoacid anion $EDTA^{4-}$ on the left in Figure 3.23 we have met already above, and it has a flexible $-CH_2-CH_2-$ linker between the two $-N(CH_2COO^-)_2$ groups. Thus (as described above), it can bind a single metal ion in a hexadentate manner effectively, forming a 1:1 ML complex species with two N-donors and four O-donors. However, the molecule on the right has the flexible linker replaced by a rigid, flat aromatic ring. Now, the two nitrogen atoms cannot bend around to coordinate to two adjacent sites

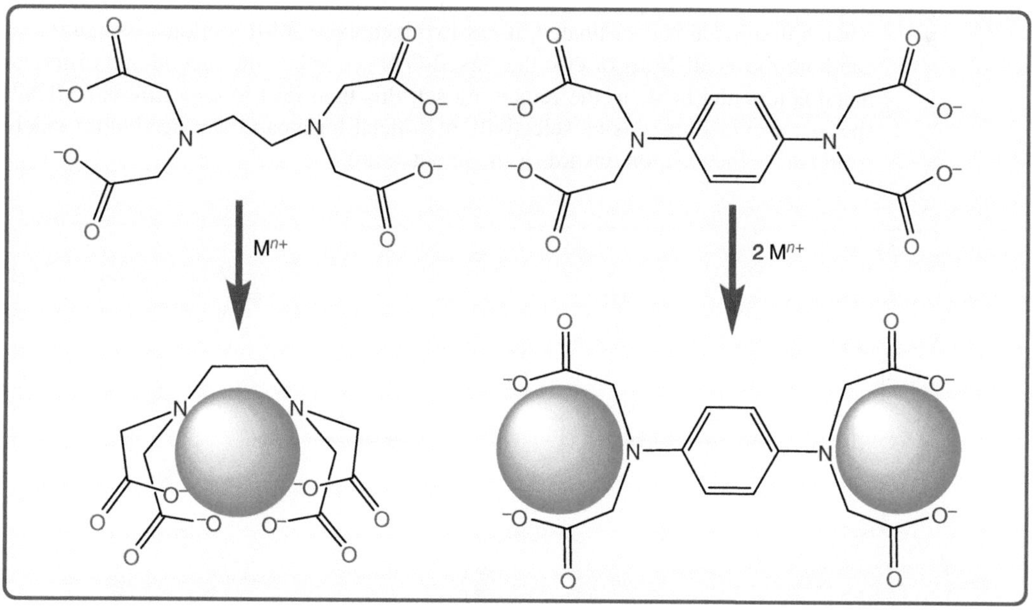

Figure 3.23

How ligand flexibility influences the process of complexation of polydentate ligands to metal ions. The rigid aromatic link in the right-hand example prohibits 'wrapping' of the two arms around a single metal, and coordination to two separate metal ions may be preferred. (Additional ligands to complete the coordination sphere in the example at right are required, but not shown.)

of the metal and thus cannot form the initial central chelate ring. Consequently, this ligand has more success binding a metal ion in each of its two $-N(CH_2COO^-)_2$ head units; this leads to less donor groups coordinated to each metal (which will therefore seek additional other ligands to satisfy its coordination sphere), but this is the sole way that all six donor groups can be employed with a minimum number of metal ions, leading to an M_2L complex.

Complexation of the ligand on the right in Figure 3.23 provides the ligand with a choice; it may form a 1:1 M:L complex and leave some donor groups uncoordinated or form a 2:1 M:L complex and employ all donor groups. In reality, both forms may exist in solution at once, depending on the relative concentrations of the ligand and metal, and we will explore those types of situations separately in Chapter 5.

3.5.2 Mixed-metal Complexation

As mentioned already, the way a polydentate ligand goes about binding several metal ions is inevitably a stepwise process, as the probability of all metal ions attaching to the ligand at once is so low that it can be ignored. This suggests that one may be able to not only see some intermediate species, as they could be inherently stable, but also that one should be able to intercept these species and insert a different metal ion into an empty cavity to produce, not an M_2L species, but an MLM' species (Figure 3.24). The formation of the mixed-metal compound is indirect evidence for a stepwise coordination process. Because MLM' carries different metal ions, it will differ in chemical and physical properties from an M_2L or M'_2L species with the same metal ions in each site and thus can be detected readily.

We can carry that concept a little further and examine a slightly different ligand set (Figure 3.25). Here, each of the two ligand head units is different. This means that when a first metal is coordinated, it has two inequivalent ML options (isomers) – it can bind to the all-N set or else the mixed-O,N set. When the second and different metal enters and binds to the remaining set, this then lead to two different MLM' species, provided there is no selectivity of a metal for a particular head unit (which would drive the reaction towards a single outcome).

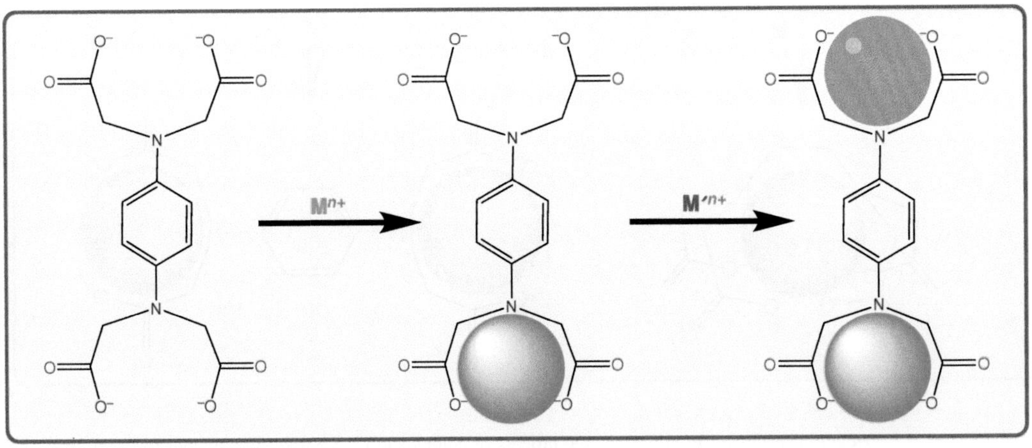

Figure 3.24

How the process of sequential complexation of metal ions may be applied to permit coordination to two different metal ions. (Additional ligands to complete the coordination spheres are required, but not shown.)

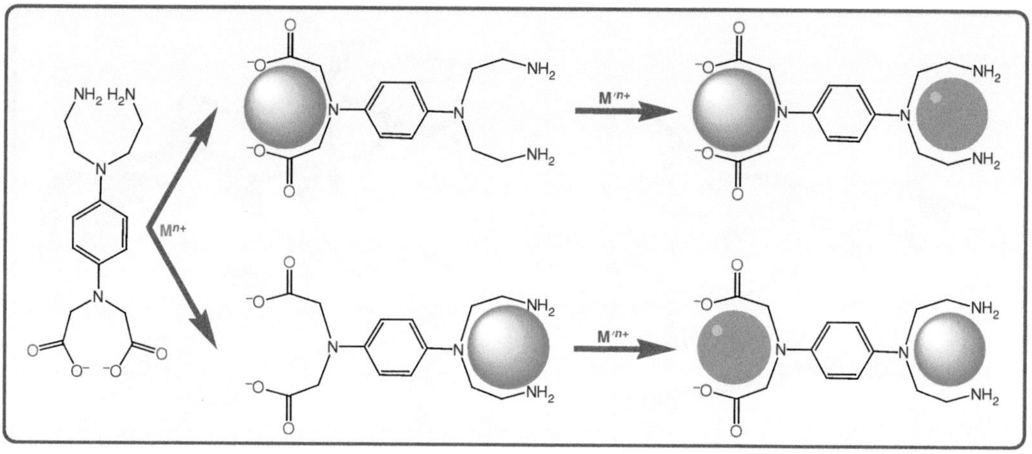

Figure 3.25

How the process of choice and selection during complexation of metal ions to an unsymmetrical ligand may lead to two differing MLM′ forms. (Additional ligands to complete the coordination spheres are required, but not shown.)

Thus, with this unsymmetrical ligand, there are two isomeric complexes possible with different metal ions inserted in the two different compartments. This simple example is sufficient to illustrate the often vast range of options that become available when complicated ligands bind to metal ions. However, not all these options will be equally probable, as the thermodynamic stability of each of the various options will differ, concepts that we will also deal with further in Chapter 5.

3.5.3 Binding to Macromolecules

There is, in principle, no limit to the size of a molecule that may bind metal ions. What we often see is that as molecules get larger, they become less soluble, which places a limit on their capacity to coordinate to metal ions in solution. However, what we also now know is that even solids carrying groups capable of coordinating to metal ions can adsorb metal ions from a solution with which they are in contact onto the solid surface, effectively removing them from solution through complexation (in this context by a process called *chemisorption*). Hence, complexation is not restricted to the liquid state and will occur in the solid state; as mentioned earlier, it can also occur in the gas phase.

Large, 'supersized' molecules (usually termed *polymers*) are common. Many biomolecules are composed of complex polymeric units (peptides or nucleotides), and most have the potential to coordinate metal ions; in fact, the metal ions may contribute to the shape and chemical activity of the polymer. So size is no object. Proteins, for example, typically bind transition metal ions via N-, O- and S-donors on their side chains. As an example, the amino acids histidine (an N-donor), methionine and cysteine (S-donors) appear as ligands in Figure 3.26 and can in some cases chelate more than one metal ion in proximity; biomolecules are addressed in detail in Chapter 8. What remains true even in these huge molecules is that the rules of coordination do not change, and any particular type of metal ion tends to demand the same type of coordination environment or number of donors irrespective of the

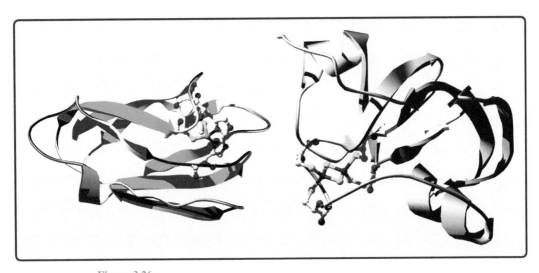

Figure 3.26
The structures of two metalloproteins: (left) spinach plastocyanin with its Cu ion coordinated by two histidine N-donors and cysteine and methionine S-donors; (right) the Fe/S protein spinach ferredoxin with its 2Fe–2S cluster core and four cysteine amino acid side chains acting as terminal ligands.

size of the ligand. This means that, if a metal ion prefers to form six bonds to six donor groups, it will usually achieve this whether only one donor group is offered by a molecule (by binding six separate molecules) or whether 100 donor groups are offered by a molecule (by binding selectively only 6 of the 100 donors available).

3.6 Nonclassical Ligands in Organometallic Chemistry

We have mentioned earlier the prospect of the carbanion H_3C^- being an effective ligand since it offers an electron pair donor in the same way that ammonia does. To emphasise this aspect, simple compounds like $[Co(CH_3)(NH_3)_5]^{2+}$ have been prepared in recent decades. There has grown a vast area of chemistry featuring transition metal–carbon bonds called, not surprisingly, organometallic chemistry. It is also conventional practice to include hydride (H^-) and phosphine (PR_3) ligands in this category as they are common companion ligands. However, these still fall under the 'classical' description proposed by Werner of a donor atom bearing a lone pair of electrons forming a σ coordinate bond and all the features already discussed in this chapter apply equally well to these ligands.

The bonding that makes organometallic chemistry a special field worthy of its own name is exemplified by the coordination of 'nonclassical' ligands such as alkenes and arenes met in Chapter 2. Lacking any lone pairs, they coordinate to mostly transition metals in a side-on fashion through overlap between their π molecular orbitals and the metal's *d*-orbitals. Our understanding of bonding in alkene and arene complexes emerged in the 1950s (more than half a century after Werner's coordination theory for 'classical' coordination compounds) in the form of the Dewar–Chatt–Duncanson model, which was covered in Chapter 2. In contrast to classical σ-bonded complexes, when a ligand like ethene ($H_2C=CH_2$) bonds in a side-on η^2 mode, this generally involves a synergistic overlap between filled π-bonding molecular orbitals on the

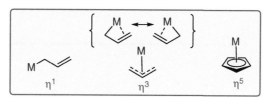

Figure 3.27

Binding of allyl as an η^1-ligand and an η^3-ligand, and relationship of the latter to cyclopentadienyl.

ligand and unoccupied metal d-orbitals, supplemented by π^*-antibonding molecular orbitals on ethene accepting electrons from symmetrically related occupied d-orbitals on the metal.

Unlike the usually charged classical Werner-style compounds, many nonclassical arene and alkene complexes are neutral, low melting point compounds that dissolve in organic solvents and display greater reactivity. Further, they usually feature metals in low (sometimes zerovalent) oxidation states and in polymetallic systems are more likely to involve direct metal–metal bonds. The low oxidation state is important so that the metal can provide d-electrons for back-bonding to augment the M−C bonds.

Another type of ligand somewhat related to ethene is the allyl anion, $^-CH_2-CH=CH_2$, the conjugate base of propene. When drawn in its simplest representation, this molecule presents different types of carbon centres; the allylic carbon ($^-CH_2-$) is sp^3-hybridised, whereas the two carbons of the $-CH=CH_2$ (vinyl) entity are sp^2-hybridised. This ligand is ambidentate and can bind to a metal in two ways: as an η^1-ligand or an η^3-ligand (Figure 3.27). For η^1-bonding, which occurs without any participation of the double bond, it is behaving as a simple monodentate σ-bonded alkyl. Alternatively, when η^3-bound through all three conjugated carbons, it can be considered to exist as two resonance forms, often represented in its true delocalised form (with dashed bonds). Related to this versatility in bonding, it has been observed that η^3-allyl complexes are often highly reactive and react with both electrophiles and nucleophiles under certain circumstances. In the simplest metal–ligand bonding view, the plane of the allyl ligand should be aligned parallel to the plane of the metal d_{xy} orbital and normal to the z-axis, but is in reality tilted slightly, considered to be the outcome of achieving maximum orbital overlap. Allyl binding as an η^3-ligand can be considered a fragment of the aromatic η^5-bound cyclopentadienyl ligand.

Indeed, the 'side-on' bonding situation seen with ethene and the allyl anion is exaggerated further when coordinating the cyclopentadienyl ion, $C_5H_5^-$ (Cp$^-$), with 6π electrons in its bonding molecular orbitals (Chapter 2, Figure 2.28). This organic anion is formed by deprotonation of neutral cyclopentadiene; in an electronically localised format it could be considered to have one carbon with a lone pair, which could bind to a metal ion through this single carbon in a simple η^1 σ-bonded mode. However, this is not the only, or even most usual, form of coordination. Rather, it commonly binds to a metal ion 'side-on' (η^5 mode) with all five carbon atoms equidistant from the metal and with C−C bond lengths in the C_5 ring identical, in what can be considered a π-bonding mode of the aromatic anion. In this mode of coordination to what would normally be a six-coordinate octahedral centre, it effectively occupies a face of the octahedron (and thus three standard coordination sites), in a somewhat similar manner to the way a cyclic, saturated triamine ligand can do so.

In Figure 3.28 (where H atoms are usually not shown for simplicity), a comparison is made between a metal ion bound to two cyclic triamine ligands, one each of

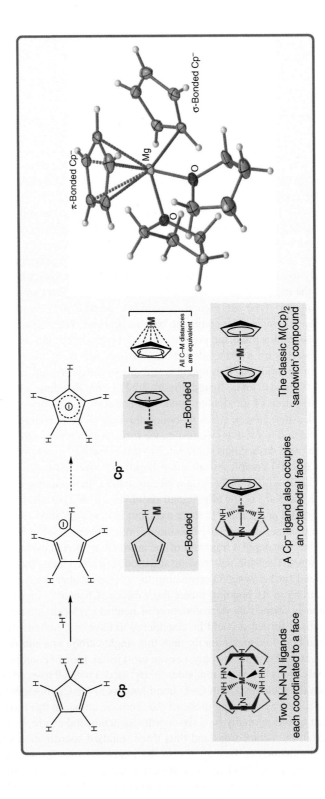

Figure 3.28

Formation and coordination of $H_5C_5^-$ (Cp^-) as a σ-bonded ligand and as a π-bonded ligand. In the latter mode, the ligand occupies the face of an octahedron in the same way that a conventional cyclic triamine does. The crystal structure of the complex $[Mg(\eta^5\text{-}Cp)(\eta^1\text{-}Cp)(THF_2)]$ shown (at right) is an example showing both classical (η^5) and nonclassical (η^1) coordinate bonding of Cp^-.

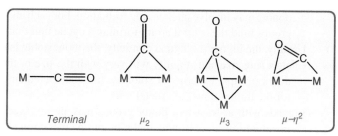

Figure 3.29
Coordination of CO as a monodentate and bridging ligand.

a triamine and a cyclopentadienyl anion, and two cyclopentadienyl anions. You may see from the figure how the Cp⁻ anion effectively replaces three traditional σ-donor groups on a 'face', leading eventually with two anions bound to a [$M^{n+}(Cp^-)_2$] compound called, for obvious reasons, a 'sandwich' complex (also known as a metallocene). The Cp⁻ and other unsaturated carbon-bonding ligands can exhibit variable coordination through the use of different numbers of available carbons (in the same way that a traditional σ-bonding polydentate ligand may not employ all of its potential donor groups in coordination). The mode with a single M−C σ bond is a η^1 bonding form, whereas the usual facially bound π-bonded form is an η^5 mode. Both modes can even occur in the one compound, as shown in Figure 3.28 for the Mg(II) complex [$Mg(\eta^5\text{-Cp})(\eta^1\text{-Cp})(THF)_2$], where THF is the cyclic ether tetrahydrofuran.

Apart from unsaturated alkenes, cycloalkenes and arenes acting as ligands, the most common ligand of organometallic chemistry is carbon monoxide, where bonding can again be described by the Dewar–Chatt–Duncanson model as a combination of both σ and π orbital overlap comprising the M−C coordinate bond. The M−C≡O atoms are typically colinear in this array. Interpretation of more elaborate bridging modes (Figure 3.29) can also be accommodated by molecular orbital theory, but we will not bother with this pursuit here. We also should mention the example of a rare cationic ligand, the nitrosonium (nitrilooxonium) ion NO⁺, isoelectronic with CO (and also with anionic cyanide, CN⁻). The related compound carbon dioxide may also act as a ligand, with CO_2 able to coordinate through only the carbon atom, or both a carbon and one oxygen atom, to one metal centre or by bridging between two metal centres, albeit involving bonding complexities beyond the scope of this book.

With organometallic chemistry being such an extensive and demanding field in its own right and given the potentially thousands of organic ligands that may coordinate via carbon atoms in several ways, we have deliberately limited our discussion of these somewhat unusual ligands and their compounds in this introductory textbook. Nevertheless, there are aspects of their mode of coordination and spectroscopic properties that have driven extension of bonding models, the key reason why this has been addressed in Chapter 2.

Concept Keys

Metal ions are almost invariably met with a set of coordinated ligands, each of which provides one or more lone pairs of electrons to make one or more covalent bonds. Denticity defines the number of donor groups of a ligand that are coordinated. Apart from one (monodentate), those providing two (didentate) or more (polydentate) donors may involve chelate ring formation. For each chelate ring, two linked donor groups bind the central atom, forming a cyclic unit that includes the central atom.

Usually, the higher the ligand denticity, the more stable is the complex formed.

Chelate ring size and stability will vary with the size of the chain linking the pair of donor groups. In terms of stability, five-membered rings are usually most stable for the lighter transition metal ions.

Ligands with at least two donor groups may be involved in bridging between two metal ions as an option to chelation of one metal ion.

The type of donor atom influences the stability and properties of complexes.

Preference exists between ligands and metals undertaking complexation. Ligands and metals can be categorized as 'hard' or 'soft' bases/acids; a like-prefers-like situation operates.

The properties of both the ligand and the metal play a role in complex formation, with the physical properties of both being altered as a result of complex formation.

The shape of a ligand influences the mode of coordination and stability of formed complexes.

Organometallic complexes are those in which a metal–carbon bond is involved. The organic ligand may be σ-bonded (as occurs for $^-CH_3$ and CO) or π-bonded (as occurs for $H_2C=CH_2$ and $C_5H_5^-$).

Even simple ligands like CO can adopt a number of binding modes in organometallics.

Further Reading

Constable, E.C. (1996). *Metals and Ligand Reactivity: An Introduction to the Organic Chemistry of Metal Complexes*. Germany: VCH, Weinheim.

> Although at a more advanced level, the early chapters give a useful introduction to metal–ligand systems and some key aspects of their chemistry.

Constable, E.C., Parkin, G. and Que, Jr. L. (2021). *Comprehensive Coordination Chemistry III*. Elsevier.

> A nine-volume set that does exactly what the label says – this is a comprehensive resource focused on metal–ligand interactions with a systematic coverage of the chemistry. Although at a more advanced level, the reader with access to this set may be able to use this to expand their knowledge of particular topics.

Crabtree, R.H. (2019). *The Organometallic Chemistry of the Transition Metals*, 7th ed. Wiley.

> A popular, advanced comprehensive coverage of this adjunct field to coordination chemistry, with a fine clear introduction to the principles of the field that students may find revealing.

Powell, P. (2011). *Principles of Organometallic Chemistry*, 2nd ed. Springer.

> A valuable undergraduate-level introduction to the broad field of organometallic compounds, with sets of problems also provided; beyond the focus of the present text, but useful for those who wish to extend their knowledge.

Scott, R.A. (Editor-in-Chief) (2011). *Encyclopaedia of Inorganic and Bioinorganic Chemistry*. UK: Wiley.

> An on-line resource, regularly updated, that provides over 2,000 short, readable articles on specific topics, mostly at an approachable level, by leading specialists; topics stretch from ligands to coordination, organometallic, and bioinorganic chemistry, and beyond. A good resource set for this and other chapters.

4 Shape

4.1 Getting in Shape

Coordination complexes adopt a limited number of basic shapes. In Chapter 2, we developed a predictive set of molecular shapes evolving from an electrostatic model of the distributions predicted for a range of point charges (two to nine) dispersed on a spherical surface. The model can also be applied as a predictive aid for a larger number of point charges on a surface. These shapes, evolving from a modification of the Valence Shell Electron Pair Repulsion (VSEPR) model developed initially for compounds where a p-block element is the central atom, are satisfactory as models for many of the basic shapes met experimentally for complexes across the periodic table. The original VSEPR model and its electron counting rules have limited predictive value for shape in complexes of d-block elements compared with its application for p-block elements. This relates to the defined directional properties of lone pairs in p-block elements, whereas in transition elements non-bonding electrons play a much-reduced role in defining shape. Notably, the complexes $[M(H_2O)_6]^{n+}$ are all octahedral for M = V^{3+} (d^2), Mn^{3+} (d^4), Co^{3+} (d^6), Ni^{2+} (d^8) and Zn^{2+} (d^{10}) despite the electronic configurations of the central metal ion varying. Rather, it is simply the number of donor groups bound about the metal that is the key to shape in transition metal complexes. This is recognised in the Kepert model introduced in Chapter 2 – a variation of the VSEPR concept developed for transition elements. This model ignores d-electrons and considers only the set of donor groups represented as point charges on a surface. This essentially electrostatic model has limitations, as shape is influenced by other factors such as inherent ligand shape and steric interactions between ligands, as well as the size and valence electron set of the central metal ion.

The actual shape of complexes can be determined readily in many cases. The advent of X-ray crystallography (described in Chapter 7.3.8), now at a level where highly automated instruments allow rapid determination of accurate and absolute three-dimensional structures of coordination complexes in the crystalline form, has been a boon for the chemist. Provided a complex can be crystallised, its structure in the solid state can be accurately defined. Although it is important to recognise that, for a coordination complex, solid-state structure and structure in solution can differ, it is nevertheless true that they often are essentially the same, so we have at our fingertips an exceptionally fine method for structural characterisation. This is a technique that can define angles with an error approaching 0.01° and bond distances (which are typically between 100 and 300 pm) with an error as low as 0.1 pm. A simple example structure is shown in Figure 4.1, in which all atom locations were accurately defined.

What crystallography has now shown clearly is that the predicted shapes we developed earlier are often observed but usually not achieved ideally – we notice that bond angles are often not exactly those anticipated, and, at times, geometries occur that are clearly better described in terms of a different basic shape. There are obviously many factors influencing the outcomes we see experimentally.

Introduction to Coordination Chemistry, Second Edition. Paul V. Bernhardt and Geoffrey A. Lawrance.
© 2025 John Wiley & Sons Ltd. Published 2025 by John Wiley & Sons Ltd.
Companion website: www.wiley.com/go/coordinationchemistry2e

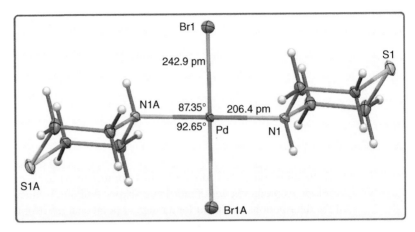

Figure 4.1

A view of the structure determined by X-ray crystallography of the neutral complex *trans*-[PdBr$_2$(thiomorpholine)$_2$]. The non-hydrogen atom locations are represented by probability surfaces, with the smaller the size the better defined is the atom.

There are some basic ideas we can immediately introduce to explain the experimental observations. Consider the simple cation [Co(NH$_3$)$_6$]$^{3+}$. As an ML$_6$ compound, we would predict initially (see Figure 2.2) that this complex cation will be octahedral in shape. From X-ray crystallography, this is exactly what is found; all N–Co–N angles between neighbouring ammonia groups are (or at least are very close to) 90°, and in addition, all six M–N bond distances are identical within very small margins of error. Experiment has justified the use of our simple point-charge model. Now consider what happens if just one ammonia ligand is replaced by a bromide ion, to form [CoBr(NH$_3$)$_5$]$^{2+}$. The basic shape is still octahedral, but there are some changes found. First, the Co–N distances are different to the Co–Br distance. This is reasonable, since we are, after all, linking different species of different sizes, charges and shapes (also see Figure 4.1 for another example). However, the intraligand angles also change, with N–Co–Br angles opening out a little to be greater than 90° and some N–Co–N angles closing in a little to be less than 90°. This implies that the interactions 'sideways' between types of ligands differ – an ammonia and a bromide interact in a non-bonding manner (or push against each other) differently to two neighbouring ammonia molecules. We can define these effects between neighbouring ligands as *ligand–ligand repulsion forces*, called generally a *steric effect*. Simply, two ligands cannot occupy the same space, and compromises must be reached which involve bond angle distortion and bond length variation. In addition, there are other effects that are electronic in character. It is apparent that the Co–NH$_3$ bond length directly opposite the bromide ion (*trans*) differs from the Co–NH$_3$ distances adjacent to the bromide ion (*cis*). This could arguably be related to steric effects for *cis* groups differing from *trans* groups, but may also be related to electronic effects; in the simplest view, consider the latter as reflecting the way different ligands compete differently for *d* orbitals, or 'push' or 'pull' electron density to or from the metal centre. Regardless of the identification of the effects, one obvious outcome is that the 'octahedral' [CoBr(NH$_3$)$_5$]$^{2+}$ ion is not an ideal octahedron. In a similar vein, in Figure 4.1, the *cis* Br–Pd–N angles of the square-planar complex are close to, but not exactly, the ideal 90°. As will be seen later in Chapter 7, the structures we observe by X-ray crystallography are also influenced by the way the molecules pack together within the crystal. Sometimes these distortions vanish when the complexes are studied in solution.

If we replace the cobalt(III) centre in $[Co(NH_3)_6]^{3+}$ by a different metal ion such as Pt(IV) or Ni(II), what we discover is that the octahedral shape is retained, but the average M–N distance changes. Obviously, each metal has a preferred distance between its centre and the donor atom. We can understand this in terms of both the size of the metal ion – the larger, the longer the bonds – and in terms of the charge on the metal ion. With variation in charge (or oxidation state) two factors operate; first, the number of *d*-electrons differs, which can influence both shape preference and repulsive terms related to the number of valence shell electrons; second, the charge changes, influencing simple electrostatic interactions. The effects are exemplified by examining one metal in two oxidation states; for example, Co(III)–N bonds are invariably shorter than Co(II)–N bonds. Because a coordinate bond links two different atomic centres, it is reasonable to expect that the preferred metal–donor distance changes with donor also. We have seen already how a Pd(II)–N bond distance differs from a Pd(II)–Br bond distance. This is a universal observation – the preferred metal–donor distance varies with the type of donor even when the metal is fixed. In fact, the effect is quite subtle, as the Co(III)–NH_3 distance differs from that found for Co(III)–NH_2CH_3, even though they both form a Co–N bond. This can be accounted for by two factors: first, ammonia is a smaller, less bulky molecule than methylamine; second, the basicity of ammonia and methylamine differs, affecting their relative capacities to act as a lone pair donor (Lewis base) in forming a coordinate covalent bond to the metal ion (Lewis acid). The size and also the shape of ligands can force much more dramatic effects upon their complexes, as we shall see, so it is apparent that ligand geometry and rigidity are important. We can illustrate this in part for three simple diamines, all of formula $C_3H_{10}N_2$. Chelation of each of these in turn would produce a six-, five- and four-membered ring (Figure 4.2). With five-membered chelate rings usually being more stable (less strained) than six- or four-membered rings, it is not surprising to find that shapes of complexes with the ligand forming a five-membered chelate ring are less distorted than those with the other ligands. In fact, the four-membered chelate ring shown in Figure 4.2 is much less stable than the other two. Further, if we join the terminal carbon and nitrogen atoms in this ligand to form a small five-membered heterocyclic ring with a loss of

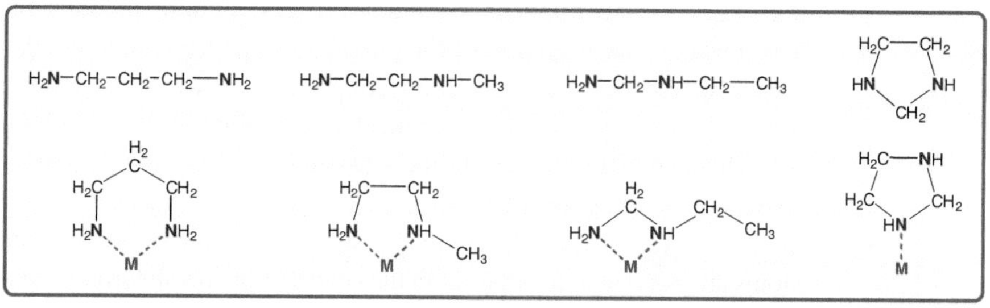

Figure 4.2

Chelation modes for simple linear diamines of common empirical formula $H_{10}C_3N_2$. Also included for contrast is a related molecule where the linear diamine to its immediate left has the five atoms in the chain confined in an organic ring; it can adopt only monodentate coordination to the metal.

two H atoms, the ligand shape is now such that only one of the two amine groups can bind to a single metal centre at any one time (Figure 4.2), the other having its lone pair directed in an inappropriate direction for chelation.

There are in fact several effects that contribute to the outcome of metal–ligand assembly. Overall, stereochemistry and coordination number in complexes appear to depend on four key factors:

1. central metal–ligand electronic interactions, particularly influenced by the number of d (or f) electrons of the metal ion;
2. metal ion size and preferred metal–ligand donor bond lengths;
3. ligand–ligand repulsion forces;
4. inherent ligand geometry and rigidity.

The simple amended VSEPR point-charge model discussed in Chapter 2 is based in effect on the third factor given above, but despite this provides a basis for predicting shape. However, it must, because of its limitations, be deficient in predicting shape in all metal complexes. We shall see as we explore actual shapes in the following that it is, nevertheless, a good starting point.

4.2   Coordination Number and Shape

Molecules are certainly greatly varied. Even carbon-based compounds can be considered as unlimited in number, despite the fact that they almost exclusively involve four bonds around each carbon centre in a very limited number of shapes. When we move to coordination compounds, the range of coordination numbers and shapes expands considerably, so that coordination complexes live up to their name – they are inherently complex molecular forms. Fortunately, as we have shown, we can identify a number of basic shapes and even some system that governs outcomes – that is, there *is* a predictive aspect to the shape in coordination complexes. We shall examine complexes from the perspective of coordination numbers in following sections. As the coordination number rises, it is notable that complexes may be described in terms of a three-dimensional polyhedral shape; *polyhedra* have a surface composed of a finite number of plane faces (or polygonal surfaces), with vertices (or corner points) and edges (or lines connecting pairs of vertices) defining the joins between the faces. The metal sits at the centre of the polyhedron, with covalent bonds to donor groups located at vertices; the edges connecting vertices may be present in some molecular drawings but serve merely to define an overarching shape and play no role in bonding. This can be confusing for the new student. For example, the simple octahedral shape has 8 triangular plane faces, 12 edges, and 6 vertices. It is the *vertices* that interest us most since these are the sites of six ligand donor atoms. These donor atoms are bonded to a metal set in the centre of the octahedron by six covalent bonds, none of which define the polyhedral shape but have spatial dispositions as a result of the overarching shape. In other words, a polyhedral shape only enables our view of a three-dimensional complex, particularly where high coordination numbers apply. In many cases, it is not necessary to invoke their use in drawings of coordination complexes – shapes are confusing enough without having both bonds and polyhedral edges present!

4.2.1   One-Coordination (ML)

This unlikely coordination number suffers from the fact that a single donor bound to the metal would still leave the metal highly exposed, a situation that would most likely invite additional ligands to bind and thus increasing the coordination number. It is nevertheless prudent to describe it as extremely rare, because there is a small possibility that a suitably bulky and appropriately shaped ligand may achieve one-coordination. It may be more practicable in the gas phase under high dilution conditions, where metal–ligand encounters are limited.

Consequently, the number of genuinely one-coordinate complexes is very limited. A series of monovalent Group 13 complexes (Al(I) to Tl(I)) have been prepared with the isopropyl substituted terphenyl ligand (Figure 4.3, left). They all possess a single M–C bond from a σ-bonded phenyl carbanion that carries two bulky substituted phenyl substituents in *ortho* positions; these partially block the approach of other potential donors to the metal cation. Transition metals have also shown this rare type of coordination, in this case, with less crowded aromatic ligands (Figure 4.3, right). In any case, this coordination number is trivial in the sense that it can have only one shape – a linear M–L arrangement.

4.2.2   Two-Coordination (ML$_2$)

Two-coordination is the lowest stable coordination number that is well represented. It also presents the first opportunity for variation from the shape predicted by VSEPR in Figure 2.2. We expect an ML$_2$ molecule to be linear, with the two donor groups disposed as far away from each other as possible on opposite ends of a line joining them and passing through the metal centre. Deviation from prediction may arise in this case simply by *bond angle deformation*, with the usual 180° L–M–L bond angle reduced to <180° through bending (Figure 4.4).

Experimentally, **ML$_2$** complexes are overwhelmingly *linear*. Both electron pair repulsion and simple steric arguments favour this shape. If bending occurs, it brings the two ligands closer towards each other, providing greater opportunity for repulsive interaction between the ligands; this would seem both unreasonable and unlikely, yet bent molecules do occur. 'Bent' geometries are well known in *p*-block

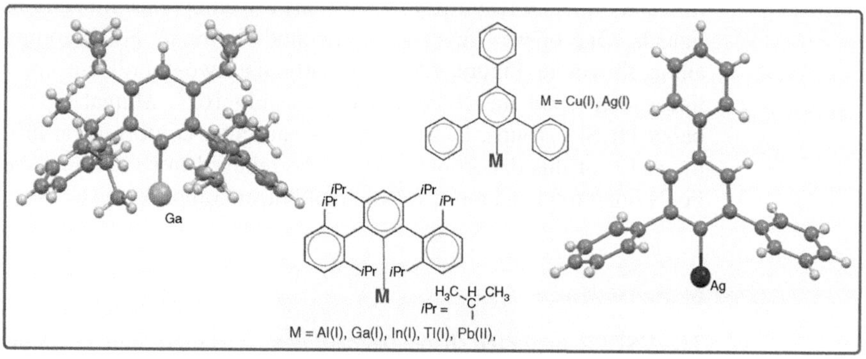

Figure 4.3

Extremely rare one-coordinate complexes with representative molecular structures. In each case, rotation about the single bonds linking the aromatic rings relieves the steric clashing and hinders the approach of an additional ligand.

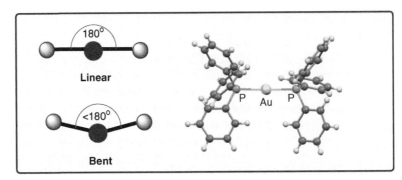

Figure 4.4
Possible shapes for two-coordination, and (at right) an example of a linear complex cation, $[Au(PPh_3)_2]^+$.

chemistry, of course, where lone pairs play an important directional role. Water is the classic example, with its two lone pairs and two bonding electron pairs around the oxygen centre leading to a bent H–O–H geometry as a result of repulsion between its inherently tetrahedral array of bonding and non-bonding electron pairs (from VSEPR concepts). Where such bending is seen in metal complexes, it can often be assigned to a higher pseudo-coordination number shape associated with a distant but still attractive metal–ligand interaction. Lone pairs are difficult to observe directly; the presence of non-bonding electron density usually can only be inferred, as a result of an apparent distortion to the shape of the molecule due to its repulsive influence on adjacent bonding electrons.

The ML_2 geometry is rare for all but metal ions rich in d-electrons, particularly d^{10} metal ions. This is a recurring theme in coordination numbers for transition metal complexes – as a very rough rule, the more electrons in the valence shell, the lower the coordination number. Complexes that are typically two-coordinate include those of the d^{10} cations Ag(I) and Au(I), for example, the $[Ag(NH_3)_2]^+$ and $[Au(CN)_2]^-$ complexes, which have linear N–Ag–N and C–Au–C cores, respectively; a related example with a bulkier PPh_3 ligand is drawn in Figure 4.4. It is also believed that the simple dihalides (MX_2) of most d-block metal ions in the gas phase tend to be linear two-coordinate species, although this is not generally the case in the solid or solution state. Examples of genuinely non-linear two-coordinate complexes are extremely rare and typically result from distortions driven by exceptionally bulky asymmetric ligands. One of very few such compounds is the d^6 Fe(II) complex of the amido anion shown in Figure 4.5. This particular two-coordinate complex approaches linear shape, but the N–Fe–N angle is only 162°. Mutual repulsion between the bulky Ph_3Si– groups, locked into this particular conformation in the solid state, is the cause of this distortion. This can be relieved in solution by the rotation of one Fe–N bond to give a more symmetrical linear complex.

4.2.3   Three-Coordination (ML_3)

The VSEPR-predicted shape, *trigonal planar*, is predominant amongst this relatively rare coordination number. **ML_3** is (like ML_2) favoured by transition metal ions with lots of d-electrons (particularly d^{10}). The only other shape of significance in metal complexes is called *trigonal pyramidal* although this is reserved for the p-block elements where stereochemically active lone pairs on the metal distort the donor atoms

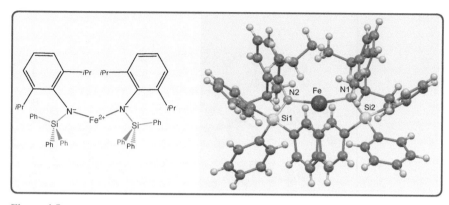

Figure 4.5

An example of a two-coordinate Fe(II) complex with a slightly bent geometry due to ligand–ligand repulsion.

away from coplanarity. The *T-shaped* geometry is also found in *p*-block compounds (such as the interhalogen compound BrF_3) where more than one lone pair is present on the central Br atom, but this is straying into main group chemistry. In terms of transition metal chemistry, T-shaped or trigonal pyramidal geometries are negligible.

These latter two uncommon shapes can be seen to arise from distortions of the ideal trigonal planar shape, as depicted in Figure 4.6. It is usual for different shapes of a particular coordination number to be able to interconvert without any bond breaking, simply through rearrangements such as those exemplified in Figure 4.6. There is an obvious outcome of this. These various shapes reported in this chapter should be considered as limiting shapes – they are the extremes or termini of change, and as a consequence, molecules may adopt a shape that is intermediate or partway along the process of changing from one basic shape to another. This is something that three-dimensional structural studies have clearly identified. However, it is both convenient and essential, if we are to have any pattern to our family of shapes, to identify

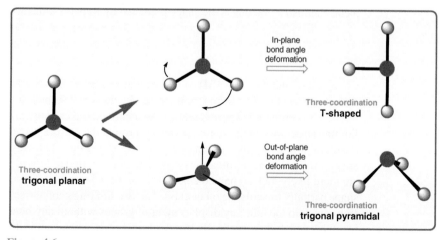

Figure 4.6

Trigonal planar, the parent shape of three-coordination, and the way it can transform to other potential geometries.

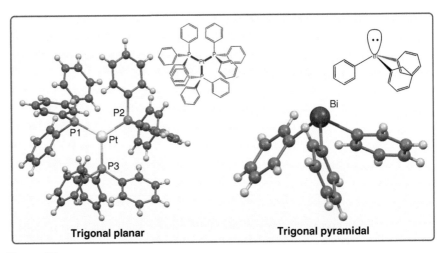

Figure 4.7

Example of a trigonal planar geometry, the neutral d^{10} complex [Pt(PPh$_3$)$_3$], and a trigonal pyramidal geometry, the *p*-block complex [BiPh$_3$]. The 3D structures from X-ray crystallography are shown as well as molecular drawings.

a coordination complex in terms of its nearest ideal (symmetrical) shape – and in most cases for three-coordination the deviation from the ideal trigonal planar shape is small, as it is inherently more stable than any alternative.

A simple trigonal planar complex ion is [HgI$_3$]$^-$, whereas a more elaborate example, also a d^{10} complex, is [Pt(PPh$_3$)$_3$], where all Pt–P bonds are equivalent (226 pm). For L–M–L angles less than the ideal 120°, the metal has moved out of the plane of the three donor atoms. While this is rarely seen in transition metal chemistry, unless very bulky and asymmetric ligands are present, this geometry is ubiquitous in metal complexes from the *p*-block. For example, trigonal pyramidal structures are common in the Group 15 metal Bi(III) where the lone pair on the central atom is stereochemically active (Figure 4.7).

Overall, trigonal planar is the dominant shape of three-coordination in transition metal chemistry, and this is the shape predicted by Kepert's amended VSEPR model.

4.2.4   Four-Coordination (ML$_4$)

Coordination number four (**ML$_4$**) is common and has two major forms, *tetrahedral* and *square planar*. The former is predicted by the VSEPR model; the latter is a different shape observed experimentally, with many examples known. These are ideal or limiting structures, in the sense that they represent the perfect shapes which lie at the structural limits for this coordination number; as mentioned already, ideal structures are relatively rare in coordination chemistry, and distorted or *intermediate* geometries are more likely met, so named since they can be achieved by distorting one or other shape partially towards the other class. The two limiting geometries can be converted from one into the other by displacement of groups without any bond breaking being involved, which is energetically less demanding, since bond angle deformations, with their lower energy demand than bond breaking, are then the dominant energy 'cost'.

The shapes of the two limiting and an intermediate geometry are shown in Figure 4.8, all based on a cubic box frame. The tetrahedral shape is defined by

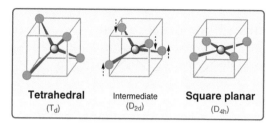

Tetrahedral
(T_d)
Intermediate
(D_{2d})
Square planar
(D_{4h})

Figure 4.8

The two limiting shapes for four-coordination, tetrahedral and square planar, along with an intermediate geometry formed in transitioning from one limiting shape to the other.

placing pairs of donors on diagonally opposite corners of a face of the cube such that no two donors are adjacent. This geometry can be converted to the square-planar shape by simply 'sliding' the top two donors down the edges of the cubic box and 'sliding' the other two bottom donors up the other two edges of the box until they are all half-way along the edges. When this occurs, the donors and the metal are all coplanar, resulting in the ideal square-planar shape. If the 'sliding' is stopped partway, then an intermediate distorted shape is achieved: a so-called flattened tetrahedral geometry. In reality, all types are well known; for example, for simple $[MCl_4]^{2-}$ ions, d^7 Co(II) is tetrahedral, d^8 Ni(II) is an intermediate geometry and d^9 Cu(II) is known in both square-planar and intermediate geometries. Many complexes described as square planar display small out-of-plane (tetrahedral) distortions that place the two pairs of donors slightly above or below the average plane including the metal ion; likewise, many tetrahedral complexes display small distortions towards square planarity. Where these distortions are minor, it is convenient to ignore them in defining the basic shape.

It is also often convenient to represent shapes in terms of their actual symmetry, expressed as the appropriate mathematical 'label' – T_d for tetrahedral, D_{4h} for square planar, and D_{2d} for intermediate geometries in this case – as this defines the shape succinctly and is appropriate for application in spectroscopy. There are simple rules for deciding the symmetry of a complex, and these are described and exemplified in Appendix 2.

Tetrahedral or distorted tetrahedral geometries are, from experimental observations, dominantly found in complexes that are overall *neutral* or *anionic*, e.g. $[TiCl_4]$ (d^0), $[CrCl_4]^-$ (d^3), $[FeCl_4]^-$ (d^5), $[CoCl_4]^{2-}$ (d^7) and $[CuCl_4]^{2-}$ (d^9). The preference for four-coordinate tetrahedral over six-coordinate octahedral geometry is again a consequence of the electroneutrality principle (Chapter 2) where the complex resists taking on any more negative charge. Although most d-electronic configurations are known to form tetrahedral complexes, the tetrahedral geometry is most common amongst d^0 complexes (such as $[MnO_4]^-$), whereas d^{10} compounds (such as $[Ni(PF_3)_4]$ and $[Ni(CO)_4]$) also lean towards tetrahedral geometry. Ligand arrangement in the tetrahedral geometry minimises inter-ligand repulsions, so negatively charged ligands prefer this shape. With a balance between favourable repulsions and unfavourable ligand field effects, it is not surprising to find that steric effects, or ligand size, are an important consideration in this geometry.

Square-planar transition metal complexes are found most commonly with a d^8 electronic configuration and occur in all three chemically explored transition metal series, e.g. Ni(II), Rh(I), Ir(I), Pd(II), Pt(II) and Au(III). Several examples are shown in Figure 4.9. One consequence of square planarity is clear in these examples – there

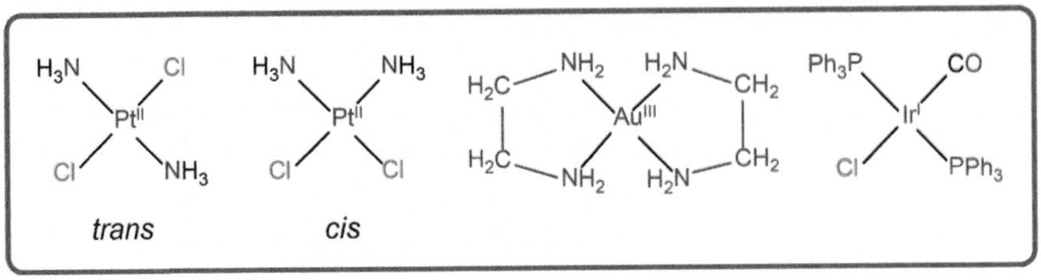

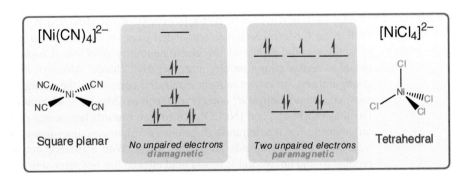

Figure 4.9

Examples of simple four-coordinate square-planar complexes.

are geometric isomers possible. Neutral $[PtCl_2(NH_3)_2]$ exists in two geometric isomeric forms, *trans* (where each pair of groups is as far apart as possible on opposite sides of the molecule) and *cis* (where each pair occupies adjacent sites). Like all geometric isomers, they display distinct chemical and physical properties, including biological properties; the *cis* isomer is otherwise known as the anti-cancer drug cisplatin (see Chapter 9.1.2), but its *trans* isomer is ineffective as a drug.

Although square planarity is common for d^8 complexes, this does not mean that the shape is not seen for other metal ions, and indeed it is known amongst metals with d^6 to d^9 electronic configurations. Some metal ions are ambivalent, with the ligand field playing a clear role in the outcome. Nickel(II) (a d^8 metal ion) is a good example, as it is known to form tetrahedral or square-planar complexes, with strong-field ligands favouring square-planar geometry, while weak-field and negatively charged ligands favour a tetrahedral shape. They are distinguished readily by their spectroscopic and magnetic properties, as these different geometries result in distinct electronic configurations and numbers of unpaired electrons (Figure 4.10).

Since square-planar complexes are essentially flat two-dimensional compounds, it is not a surprise to find other molecules coming into reasonably close but still non-bonding positions (distances >300 pm) above and below the plane in the solid state and in solution, including metal–solvent and even some cases of metal–metal interaction. The square-planar cation–anion pair $[Pt(NH_3)_4][PtCl_4]$ (historically known as 'Magnus' green salt' and first isolated in the 1830s), for example, stacks cationic and anionic complexes in alternating positions in the crystalline lattice, with a M…M separation of ~325 pm (Figure 4.11). The separation represented is small

Figure 4.10

Examples of d^8 nickel(II) complexes adopting square-planar or tetrahedral geometry, depending on the type of ligand. The square-planar complex has no unpaired d-electrons (diamagnetic), whereas the tetrahedral complex has two unpaired d-electrons (paramagnetic), allowing easy identification (*energy splitting diagrams are detailed in Figures 2.22 and 2.8, respectively*).

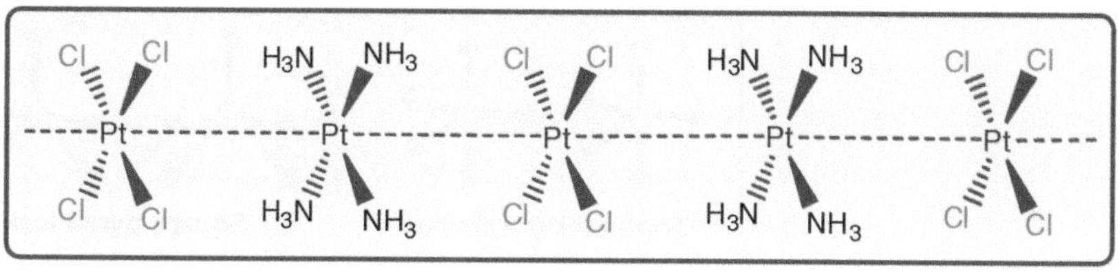

Figure 4.11

Stacking of d^8 platinum(II) complex cations and anions in the solid state leads to weak Pt···Pt interaction, influencing physical properties.

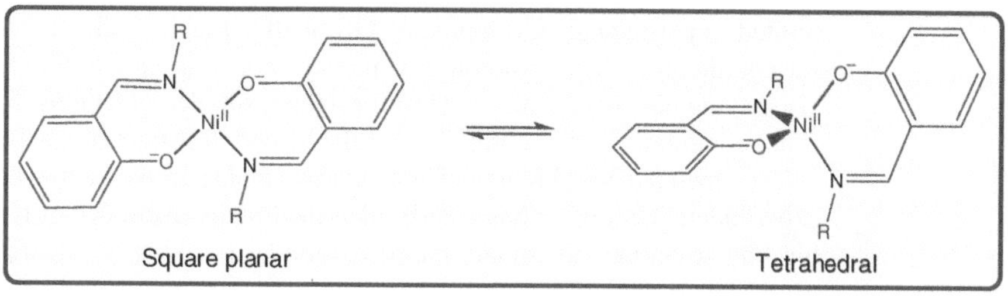

Square planar   Tetrahedral

Figure 4.12

An N,O-chelate ligand whose Ni(II) complex displays interconversion between different geometries depending on conditions.

enough to have an effect on physical properties, with the green colour indicating significant metal–metal interaction, since another form of the salt with a very long anion–cation separation of over 500 pm (no Pt … Pt interaction) is pink, the colour attributable to the $[PtCl_4]^{2-}$ anion alone, as isolated $[Pt(NH_3)_4]^{2+}$ is colourless. When metal ions come very close together, it will also affect the magnetic properties of their assembly significantly.

In describing the relationship between tetrahedral and square-planar complexes above, we used a model where interconversion could occur without bond breaking. If this is a fair representation of reality, then it should be possible to find some systems that exist as a mixture of the two isomers in a rapidly interconverting equilibrium. Fortunately, there are indeed some compounds that undergo conversion between tetrahedral and square-planar forms in solution, a situation which implies that the stabilities of the two forms are very similar. One now classic example involves the Ni(II) complex of a chelated N,O-donor ligand shown in Figure 4.12, where the position of the equilibrium between dominantly tetrahedral and dominantly square-planar forms depends on the temperature, solvent and the type of R-group attached to the coordinated imine nitrogen. A shift in the equilibrium between the two forms can be readily monitored, since their colours and absorption band positions and intensities are different.

4.2.5 Five-Coordination (ML$_5$)

Examples of **ML$_5$** are found for all of the first-row transition metal ions, as well as some other metal ions. Although once considered rare, growth in coordination chemistry has led to five-coordination becoming met almost as frequently as four-coordination. This exhibits one of the limitations of making comparisons of

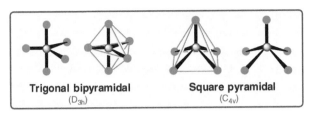

Figure 4.13
The two limiting shapes for five-coordination.

this type; rarity may not be a result of any inherent restriction but may simply reflect limited experimental development. Given that four-coordination is common and six-coordination very common, it is perhaps not surprising to find five-coordination also having matching status, at least for lighter, smaller metal ions. Five-coordination is not commonly met in complexes of the heavier transition metals, however.

The amended VSEPR model predicts two stable forms of five-coordination, and experimental chemistry has clearly identified many examples of both forms. These limiting structures are square-based pyramidal (or, simply, square pyramidal) and trigonal bipyramidal (Figure 4.13). The classic square-based pyramidal shape is formed simply by cleaving off one bond from an octahedral shape, which leaves the metal in the same plane as the four square-based ligands. In reality, almost no complexes exhibit this shape but rather adopt a distorted square pyramidal shape where the metal lies above the basal square plane, with typically the angle from the apical donor through the metal to each donor in the base around 105° instead of 90°. Considering electron pair repulsion alone, this distorted shape (distorted only in terms of the metal location; the Egyptian pyramid shape created by the donor groups is otherwise unaltered) is actually more stable than the form created by simply truncating an octahedron and is only slightly less stable than the trigonal bipyramidal geometry. As a consequence, it has become usual to regard the square pyramidal shape as that with the metal above the pyramidal base plane, and you will see it represented in this way almost exclusively.

Once again, as described for three- and four-coordination, it is possible to convert from one form to the other through bond angle changes without any bond breaking. Both geometries are common, but in practice, there are many structures that are intermediate between these two. The two limiting structures are of similar energy; as predicted, some complexes display an equilibrium between the two: for example, $[Ni(CN)_5]^{3-}$ crystallises as a salt with both structural forms of the anion present in the crystals.

If we examine the two five-coordinate shapes from a crystal field perspective, the d orbitals split in a different way to that found for octahedral, tetrahedral and square-planar shapes since the d orbitals find the ligands in clearly different locations in space. The crystal field splitting pattern for the two is shown in Figure 4.14. From this pattern, crystal field stabilisation energies (CFSEs) can be calculated, and they favour the square pyramidal geometry in all cases (apart from the trivial situations d^0 and d^{10}, and high spin d^5). This prediction differs from the outcome predicted by the VSEPR model.

Although electron pair repulsion and CFSE influences operate, it appears nevertheless that the steric or shape demands of at least polydentate ligands play a dominant influence on complex shape adopted. This is exemplified in Figure 4.15 (left), where an example of a 'three-legged' ligand shape fits best to the trigonal

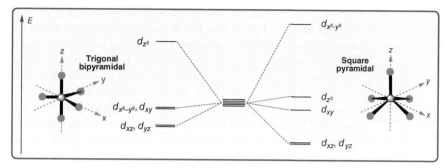

Figure 4.14
Crystal field splitting pattern for trigonal bipyramidal (left) and square pyramidal (right) ML$_5$ complexes.

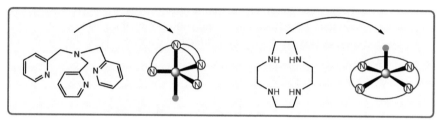

Figure 4.15
Ligand shape directing complex shape in five-coordination.

bipyramidal geometry, occupying the top four positions of the complex shape, with a fifth simple ligand then occupying the underside – the whole assembly looks a little like an open umbrella. The ligand is predisposed to this shape, with limited steric interaction when bound. Likewise, the cyclic tetraamine represented in Figure 4.15 (right) is predisposed to promote the square pyramidal shape.

The square pyramidal shape is more common overall and is favoured when steric requirements of polydentate ligands are important (Figure 4.15, right) or where π-coordinate bonding occurs, as in so-called 'vanadyl' complexes where the V=O bond occupies the axial site with four other donors in the basal plane (Figure 4.16). However, exceptions abound, reflecting the similar energies of the two forms. As an example, the simple isoelectronic complex anions $[MnCl_5]^{3-}$ and $[FeCl_5]^{2-}$ both can adopt a trigonal bipyramidal or square pyramidal geometry depending on the counter-ion present.

Whereas a limited number of complexes display equivalent lengths for all bonds, it is more common for some distortions from the regular stereochemistry to be found,

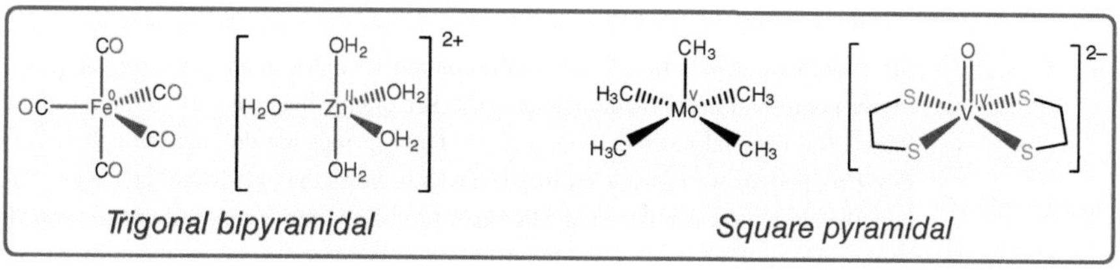

Figure 4.16
Examples of complexes adopting one of the two shapes of five-coordination.

both in terms of bond distances and angles. For example, in trigonal bipyramidal $[Co(NCCH_3)_5]^+$, the two axial Co–N bonds are 5 pm longer than the three equatorial bonds, and in $[CuCl_5]^{3-}$, axial bonds are 95 pm shorter than equatorial bonds. For square pyramidal complexes, the metal lies typically between 30 and 50 pm above the square plane of donors, close to the value of 48 pm calculated from geometry assuming an apical donor–metal–basal donor angle of 104° for 200 pm metal–donor bonds. In this stereochemistry, the single axial bond tends to be longer than the four equatorial bonds; for $[Ni(CN)_5]^{3-}$, the former is 217 pm, while the latter ones are 187 pm.

4.2.6 Six-Coordination (ML_6)

ML_6 is by far the most common coordination type exhibited by transition metal elements (seen for all configurations from d^0 to d^{10}) and also is often met for complexes of metal ions from the *s*- and *p*-blocks of the periodic table. Of the two limiting shapes (Figure 4.17), the *octahedral* geometry is by far the most common, though

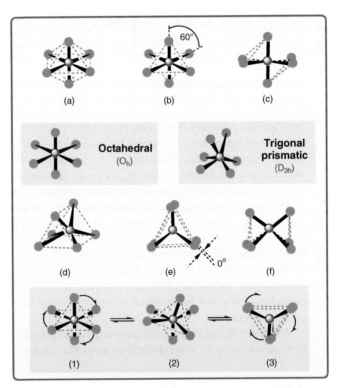

Figure 4.17

Six-coordinate geometries. Views of the common octahedral and rare trigonal prismatic shapes with (a), (d) the coordination polyhedron defined (broken orange lines) as well as the coordinate bonds (thick black lines). The trigonal faces carrying sets of donor groups are defined through views perpendicular to the face [(b), (e)], as well as side on [(c), (f)] which shows the 'sandwich' character of the L_3ML_3 sets. Note how the angle between the triangular faces projected onto a common plane varies from 60° in the octahedral case to 0° in the trigonal prismatic case. Twisting of one octahedral face (1) with the other fixed in position through an intermediate (2) to the trigonal prismatic geometry (3) provides a mechanism for interconversion without bond breaking.

examples of the other limiting shape, *trigonal prismatic*, exist. Because the six donor atoms come into closer contact in the trigonal prismatic geometry than in the octahedral geometry, trigonal prismatic is predicted by the Kepert model to be less stable. However, many structures show distortion that places them as intermediates between an ideal octahedral and the ideal trigonal prismatic forms. This distortion is best viewed in terms of the orientations of two opposite triangular faces of sets of three donors. For octahedral (which is also a special case of trigonal antiprismatic where edge and face length uniformity applies), the 'top' face is twisted around 60° versus the 'bottom' face so that they are perfectly staggered; for trigonal prismatic, the two faces are exactly eclipsed (the twist angle has been reduced to 0°). Intermediate or distorted structures show a twist angle between 60° and 0° (Figure 4.17).

The dominance of octahedral geometry may be assigned to a number of factors: it is favoured by the amended VSEPR concept, serves as an ideal shape for minimising steric clashing between donors, promotes good metal–ligand orbital overlap, and leads to favourable ligand field stabilisation energies. Distortions from pure octahedral geometry can arise from twisting of faces, as just discussed, and this effect is most likely met with chelate ligands, where the 'bite' of the chelate donor groups may be satisfied by a twisting distortion. This is typically only a modest trigonal twist, and the geometry is still best described as distorted octahedral.

Although the trigonal prismatic geometry is usually enforced by requirements of a chelating ligand, this is not always the case; and organometallic trigonal prismatic complexes of the monodentate methyl ligand $[Re(CH_3)_6]$ (d^1) (Figure 4.18) and $[Zr(CH_3)_6]^{2-}$ (d^0) are known, and their stability has been explained by molecular orbital theory considerations. More typical examples of trigonal prismatic geometries involve rigid chelates like dithiolates {$^-S–(R)C=C(R)–S^-$}. The tungsten(0) complex with six monodentate phosphinine (the P-analogue of pyridine) ligands is octahedral, whereas the analogous *tris*-biphosphinine complex is trigonal prismatic (Figure 4.18); this effectively illustrates the influence of ligand rigidity and chelation on enforcing an otherwise less preferred trigonal prismatic geometry.

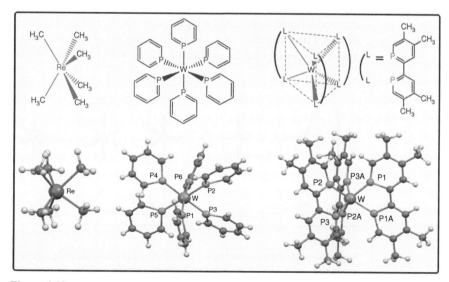

Figure 4.18

Trigonal prismatic $[Re(CH_3)_6]$ (left), octahedral $[W(phosphinine)_6]$ (centre) and its trigonal prismatic $[W(Me_4\text{-biphosphinine})_3]$ analogue (right).

Another way distortions away from ideal octahedral geometry occur is through the attachment of a mixture of ligands since each donor type has a preferred coordinate bond length. For example, although all bonds are equivalent in $[Cr(OH_2)_6]^{3+}$, when one is substituted to form $[CrCl(OH_2)_5]^{2+}$, the Cr–Cl distance is clearly longer than the Cr–O distances; moreover, even the O-donor opposite the Cl ion is slightly different in distance from the chromium(III) ion than the remainder O-donors. Differences in steric bulk and electrostatic repulsion can add to these distortions; for *cis*-$[CrCl_2(OH_2)_4]^+$, the Cl^-–Cr–Cl^- angle is opened out compared with the H_2O–Cr–OH_2 angles, leading to a distorted octahedron. In reality, it is rare to find a complex that adopts a perfect octahedral shape, since we are commonly dealing with ligands or sets of ligands with different donors. These observations, of course, apply to any basic stereochemistry.

4.2.7   Higher Coordination Numbers (ML_7 to ML_9)

Beyond ML_6 lies a range of higher coordination numbers that reach as high as ML_{12} for *f*-block metals, although examples beyond ML_9 are not known for *d*-block metals. It is appropriate to be aware of some of these, and how they may arise. Thus, we will briefly examine the relatively well-known ML_7 to ML_9. For these, a point-charge model of the distributions predicted for charges dispersed on a spherical surface can still be applied, and predicted shapes are found experimentally. Rather than elaborating on this aspect, however, we will examine another approach to understand how some of the structures arise by relating them to our lower coordination numbers already described.

If we examine selectively some shapes for two- to six-coordination already discussed, we can arrange them in such a manner that they are related by an increase in the number of groups dispersed in a symmetrical fashion around a plane including the metal; additional groups reside in each of the two axial sites in all cases (Figure 4.19). Following this trend beyond six-coordinate octahedral, where there are four donor groups around the centre in a square arrangement, it is possible to suggest that one seven-coordinate shape could arise through placing five groups around the centre (the symmetrical arrangement for which is a pentagon) or else, for eight-coordination, six groups (in a hexagonal arrangement). These would lead to a pentagonal bipyramid

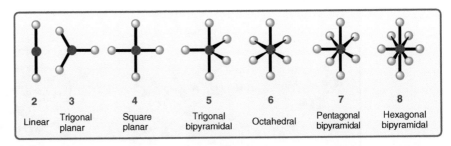

2	**3**	**4**	**5**	**6**	**7**	**8**
Linear	Trigonal planar	Square planar	Trigonal bipyramidal	Octahedral	Pentagonal bipyramidal	Hexagonal bipyramidal

Figure 4.19

Shapes for coordination numbers from 2 to 8 which reflect a trend involving a stepwise addition of groups arranged around the central plane that also includes the metal. All structures shown are actual local minima; the very rare T-shape for three-coordination, although it would preserve the axial L–M–L linearity displayed by all others, is replaced by the stable trigonal planar form (which still fits the overall trend around the central plane).

and a hexagonal bipyramid, respectively (Figure 4.19). As it happens, both these are known shapes for seven- and eight-coordination, respectively. How far this approach can be extended depends on the size of the metal ion and the size of the donor groups, as clearly steric factors will become important as an increasing number of groups are packed into the plane around a metal ion. This also introduces one of the general observations regarding complexes with high coordination numbers – they are found with larger metal ions, in other words, metal ions that exhibit long coordinate bond lengths, as this alleviates steric 'crowding' around the metal centre.

Another approach to increasing the coordination number is through expanding the layers (or planes) of donors around a metal ion. For example, if we consider either six-coordinate shape, we can visualise this as a metal ion alone in a central layer and two layers of donors above and below the central layer. We can then expand the coordination number in two ways: addition of extra donors to the upper and lower planes of donors; or addition of donors to the central plane containing the metal, analogous to that already described in Figure 4.19 except here we are starting with more than one group in 'axial' locations.

With high coordination numbers (>6) also comes a greater number of possible coordination geometries (symmetries). Seven-coordinate complexes have at least three well-represented coordination geometries and one of these, pentagonal bipyramidal, has already been introduced in Figure 4.19. Alternatively, a new seven-coordinate geometry may be made by adding a donor at the centre of a rectangular face of a six-coordinate trigonal prism, giving a *one face-centred* (or monocapped) *trigonal prismatic* structure. Similarly, adding a new donor to the trigonal face of an octahedral complex leads to the *monocapped octahedral* geometry. Pentagonal bipyramidal, monocapped trigonal prismatic and monocapped octahedral are all well represented amongst seven-coordinate complexes. To illustrate this, the simple complex anion $[ZrF_7]^{3-}$ has been observed in each of these three geometries depending on the counter-ion used to crystallise the complex (Figure 4.20). It follows that, unlike more common four- and six-coordinate systems, there is no strong preference for a particular geometry here so it is no longer realistic to be able to predict which geometry will be found.

Converting the trigonal faces of six-coordinate octahedral or trigonal prismatic structures from three to four donors effectively converts each face to a square plane of donors, leading to *cubic* eight-coordination. In practice, this perfectly eclipsed cubic geometry is of higher energy than when one of the square layers is rotated through

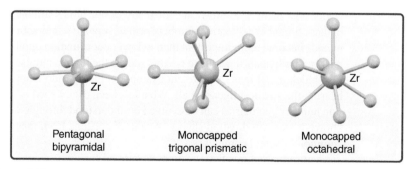

| Pentagonal bipyramidal | Monocapped trigonal prismatic | Monocapped octahedral |

Figure 4.20

The three experimentally observed seven-coordinate geometries of $[ZrF_7]^{3-}$ (each crystallised with different organic cations).

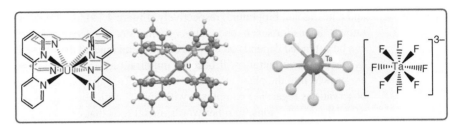

Figure 4.21

The extremely uncommon cubic coordination geometry in [U(bipy)$_4$] with the more typical square antiprismatic geometry in [TaF$_8$]$^{3-}$.

45° to produce a *square antiprismatic* (or Archimedean antiprismatic) shape. This is analogous to twisting the six-coordinate trigonal prismatic shape through 60° to produce the preferred octahedral shape, discussed earlier (Figure 4.17). In line with the expectations for relative stability, the square antiprismatic shape is experimentally much more common than the cubic. There are very few examples of molecular complexes (that maintain their structure in solution) in a cubic coordination geometry and most of these are from the actinoid (5*f*) elements where very high coordination numbers (eight and above) are the norm. An interesting case is [U(bipy)$_4$] (Figure 4.21), originally prepared in 1963 and assigned as an unusual zero-valent uranium complex. More than 50 years later, it was found to be better described as a U(IV) complex with four bipy$^-$ radical anionic ligands, a classic case of a 'non-innocent' ligand which can formally accept electrons from the metal it coordinates with. In the absence of chelation, the square antiprismatic geometry is much preferred and many eight-coordinate complexes from the *s*-, *d*- and *f*-blocks adopt this geometry. A simple example is [TaF$_8$]$^{5-}$, also shown in Figure 4.21.

Following the same 'capping' methodology as described earlier, if donor atoms are added to all three faces of a trigonal prism, the nine-coordinate *three face-centred* (or *tricapped*) *trigonal prismatic* form is obtained (Figure 4.22). Alternatively, the other commonly encountered geometry is the *monocapped square antiprismatic* structure that is well represented in lanthanoid (4*f*) complexes, particularly their nona-aqua complexes [Ln(OH$_2$)$_9$]$^{3+}$ (Ln = La–Lu). Again, there is little preference for either geometry, and the [Pr(OH$_2$)$_9$]$^{3+}$ has been identified in both its tricapped trigonal prismatic and monocapped square antiprismatic geometries (Figure 4.23) when different anions are present.

Because there is a 'crowding' of many donor atoms around the same metal ion in these higher coordination number species, repulsive interactions between adjacent ligands become more important than in lower coordination number complexes. Thus, it is smaller ligands that tend to occupy sites in the higher-number coordination sphere of *d*-block metal ions. A pertinent example is the hydrido complex anion [ReH$_9$]$^{2-}$, bearing the smallest ligand of all. What has been established is that higher coordination number geometries can be evolved or understood through extrapolation from the known lower coordination shapes. Again, distortions from these limiting shapes are common in actual complexes.

Idealised structures that we have developed for up to nine-coordination are summarised in Figure 4.24, including drawings of the polyhedra. These do not represent all of the shapes met, since apart from all these idealised structures, it is necessary to remember that bond angle and bond length distortions of these structures can occur;

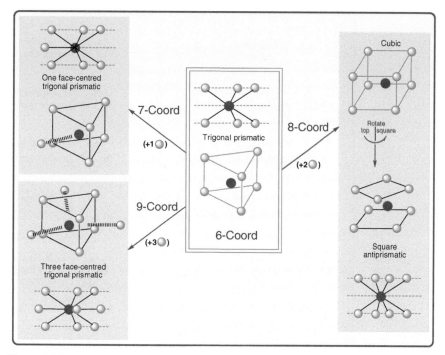

Figure 4.22
Methods of converting the six-coordinate trigonal prismatic shape, where the metal is 'sandwiched' between two layers of donors, into various seven-, eight- and nine-coordinate shapes through the addition of other groups either in the plane of the metal to form another layer of donors or else in the two existing layers of donors to expand the set of donors in those layers.

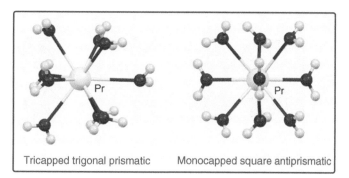

Figure 4.23
The two common nine-coordinate geometries of tricapped trigonal prismatic (left) and monocapped square antiprismatic (right) represented by the $[Pr(OH_2)_9]^{3+}$ ion.

some of the shapes resulting from these effects are themselves common enough to be represented as named shapes, and we have discussed some examples of these earlier.

Further, beyond nine-coordination, an array of additional shapes can be found, not met in *d*-block compounds but seen occasionally with *f*-block metals. We must stop somewhere, so here it will be at coordination number 9 as it has special significance in transition metal chemistry. Clearly, the options are extensive, so it may be time to find out what directs a complex to take a particular shape.

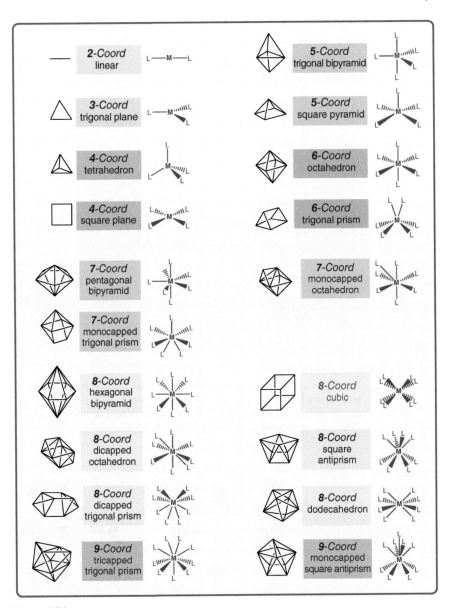

Figure 4.24

A summary of ideal polyhedral shapes for the limiting structures for coordination numbers 2 – 9. More practical metal bonding representations for complexes are also shown.

4.3 Influencing Shape

4.3.1 Metal Ion Influences

Although a coordination complex is an assembly of central atom and donor molecules or atoms, and as such is best viewed holistically, it is still instructive to identify factors influencing shape that depend on each component. If we focus first on the central atom or ion, there are two of the four key factors mentioned earlier that we can consider metal-centric. These are

- the number of *d*-electrons on the metal ion; and
- metal ion size and preferred metal ion–donor atom bond length.

Each metal in a particular oxidation state brings a unique character to play in its complexes. Examples of the way the size and bond distances vary across the *d*-block are given in Table 4.1, reporting data for N-donor, O-donor and Cl⁻ ligands. The M–L distances are averages only, as distances vary over a range of at least 20 pm, influenced by the specific type of donor group for a particular type of donor atom, the ligand shape and associated strain energy, as well as influences of other donors in the coordination sphere (through what are *trans* or *cis* effects of an electronic nature). Moreover, the spin state of the central metal plays a role (e.g. M–O distances for high spin Mn(III) are typically 20 pm longer than for low-spin compounds).

As mentioned earlier, distances vary for any particular metal ion depending on the character of the donor (carboxylate-O or alcohol-O, for example, or else amine-N versus amido-N), influences from the ligand framework itself, and influences of other ligands bound to the same complex. There are some clear trends apparent, such as the fall in M–L distance with increasing charge on central metal ions with the same number of *d*-electrons. Also, the variation in bond length M–Cl > M–N > M–O

Table 4.1 Variation of ion size and average bond distances found for first-row *d*-block metal ions in common oxidation states and with six-coordinate octahedral geometry.[a]

Metal ion	*d*-electron configuration	Free ion radius (pm)	Typical M–O[b] (pm)	Typical M–N (pm)	Typical M–Cl (pm)
Sc(III)	d^0	74.5	210	218	245
Ti(IV)	d^0	60.5	195	210	230
Ti(III)	d^1	67	185	215	235
V(IV)	d^1	58	185	205	215
V(III)	d^2	64	215	225	235
Cr(III)	d^3	61.5	195	210	235
Mn(IV)	d^3	53	185	210	230
Cr(II)	d^4	80	200	215	245
Mn(III)	d^4	64.5	195	205	230
Mn(II)	d^5	83	215	240	250
Fe(III)	d^5	64.5	190	205	230
Fe(II)	d^6	78	205	215	240
Co(III)[c]	d^6	61	185	195	225
Rh(III)[c]	d^6	*66.5*	*195*	*205*	*235*
Ir(III)[c]	d^6	*68*	*210*	*215*	*240*
Co(II)	d^7	74.5	205	220	240
Ni(II)	d^8	69	205	210	235
Cu(II)	d^9	73	200	200	225
Zn(II)	d^{10}	74	205	210	230

[a]Entries for a second- and third-row element, to exemplify differences down a column of the periodic table *d*-block, appear in italics.
[b]Distances for early transition elements have a broad range, as single and double bonds both occur.
[c]Data for common low-spin electronic configuration.

is almost universally observed. Further, there is a modest relationship between bond distances and the size of the metal ion. The increase in metal cation size from the first to the second and third row of the periodic table is accompanied by usually longer M–L distances, as exemplified for Co(III), Rh(III) and Ir(III) in Table 4.1. Overall, metal–donor distances fall within a range of ~160–260 pm, with the smaller distances found where highly charged metal ions, small anionic ligands and/or multiple bonding operating. For example, the $V^{IV}=O$ distance of ~160 pm is markedly shorter than the usual $V^{IV}-O$ distance of ~185 pm. Experimentally, every metal ion–donor atom grouping displays bond distances that vary only slightly across a usually large number of ligand systems examined, indicating that each assembly does have a preferred metal–donor distance.

4.3.2   Ligand Influences

If we focus next on the metal-bound ligand, there are also two of the four key factors mentioned earlier that we can consider ligand-centric. These are

• ligand–ligand repulsion forces; and
• ligand rigidity or geometry.

Obviously, each ligand is unique in its shape and size. The effect of repulsion between ligands, which we can term *non-bonding interactions*, can usually be recognised readily in the solid state from distortions in shape as defined by crystal structure analysis. Two PH_3 molecules may bind in a square-planar shaped complex with little preference for *trans* over *cis* geometry, whereas two very bulky $P(C_6H_5)_3$ molecules may exhibit a strong preference for coordination in a *trans* geometry, where they are much further apart. Another way that two *cis*-disposed bulky ligands can relieve ligand–ligand repulsion is for the complex to undergo distortion from square planar towards tetrahedral, which leads to the two ligands moving further apart in space; this is also observable in the solid state and, from spectroscopic behaviour, in solution. However, there are also weak structure-making effects arising from favourable hydrogen bonding interactions that can assist in stabilising a particular shape. This might occur where a carboxylate group oxygen participates in a hydrogen bond with an amine hydrogen atom, arising from the interaction:

$$R_2N-H^{\delta+} \cdots {}^{\delta-}O-CO-R$$

Such directed close contacts can actually be observed in the crystal structure also, but distances are longer than formal covalent bonding distances and the bonds much weaker.

Some ligands are structurally so rigid that they can bind to a metal ion in only one manner. The ligand at the top right of Figure 4.25 can only bind as a tridentate when all three donor atoms are coplanar with the metal, in a so-called *meridional* (*mer*) form. Replacing the pyridine ring with a simple secondary amine bridge allows the ligand to bind to the triangular face of an octahedral metal ion (the *facial* or *fac* form) while also being able to adopt the *mer* isomeric form. The aromatic porphyrin molecule (its dianion is shown at the bottom right of Figure 4.25) is completely flat, and large amounts of energy are required to distort it. Therefore, when it binds to a metal, it seeks to retain this shape and will simply use the square-shaped array of four N-donors to wrap around a metal ion in a planar manner. The porphyrin ligand is contrasted with the flexible saturated cyclic tetraamine ligand, which can bind in a flat or folded form

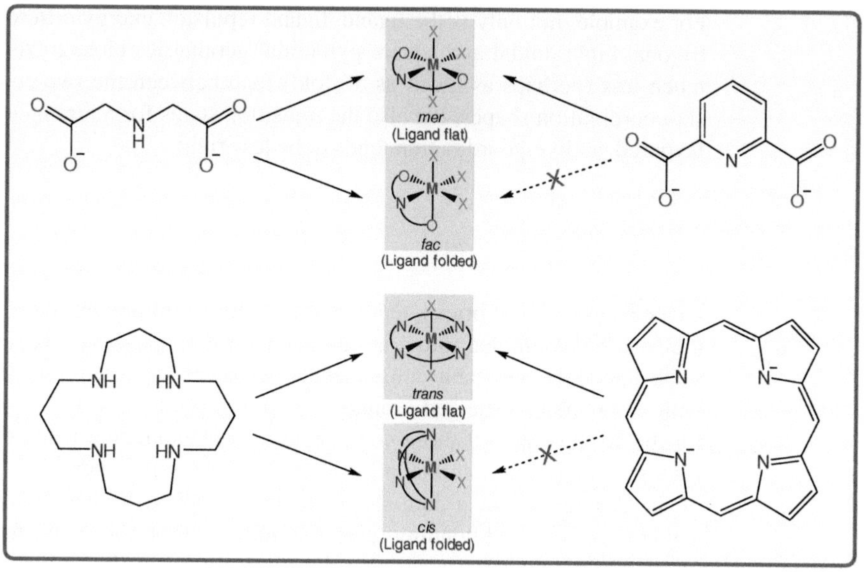

Figure 4.25

Rigidity and flexibility influence the way ligands bind to metal ions. The more rigid ligands on the right do not permit folding, whereas the more flexible ones on the left can accommodate folding and thus offer options when coordinating to an octahedral metal ion.

to octahedral metal ions. The end result is that the two remaining coordination sites (represented by X) may either be in *trans* (opposite) or *cis* (adjacent) positions.

4.3.3   Chameleon Complexes

Complexes usually arrive at a particular three-dimensional shape as the thermo-dynamically most stable form, and we expect the same outcome with a particular set of ligands each time they are made. While this is ideally the case, it should be remembered that thermodynamic stability relates to a particular set of conditions, such as temperature, solvent and counter-ions or electrolyte. Change the conditions, and the system may be perturbed enough to produce a different complex. Some changes may be readily reversible so that reverting to the initial conditions allows the complex to return to its original form – genuine chameleon-like behaviour. The most obvious example of this is where a change in conditions creates a change in geometry. We have already noted that for some coordination numbers two different shapes may be similar in stability; providing the barrier to conversion between them is not too great, then interconversion may occur as a result of, for example, changing the temperature. The tetrahedral/square-planar interconversion shown in Figure 4.12 is an example of this. Of course, conversion from one isomer to another irreversibly to achieve the more thermodynamically stable form is common, and we shall deal with this later in this chapter.

The tendency to show flexibility in shape varies with coordination number, because ligand–ligand repulsion energy differences between various shapes change, as do the heights of energy barriers in reaching transition intermediates. As a consequence, five-, seven-, eight- and nine-coordination tend to be substantially less rigid than four- and six-coordination and display more examples of chameleon-like behaviour.

For example, not only is the ligand–ligand repulsion energy difference between the trigonal bipyramidal and square pyramidal geometries close to zero and definitely much less (perhaps as much as 50-fold) than between the two common four- and six-coordination shapes, but also the transition states for rearrangement are less disfavoured, so five-coordination tends to be less rigid.

4.4 Isomerism – Real 3D Effects

Given a large pile of bricks, there exists an almost infinite number of ways in which they can be assembled into a three-dimensional shape. Likewise, on the molecular scale, one of the inevitable consequences of assembling three-dimensional molecules with components arranged around a central core atom is the existence of options for the arrangement of those components – a phenomenon that we call *isomerism*. As a starting point, note that all isomers have the same chemical formula. Because of the designation of a core shape for the complex, limitations on ligand shapes and connectivity and bonding rules, the number of options is far from infinite – yet not much less concerning for that. Defining the basic shape of the complex is the first stage, with next the task of defining options for ligand attachment needing to be examined. In very few cases is the exercise trivial because we need to think in three dimensions to identify possibilities. This can be illustrated in the case of a tetrahedral MA_3B complex, where there is only one possible outcome, yet this is not immediately obvious (Figure 4.26).

Only in the case where all ligands are identical monodentates with a single common donor atom is the situation simple and the prospect of positional isomers of any kind removed for a particular stereochemistry; this occurs for $[CoCl_4]^{2-}$ and $[Ni(OH_2)_6]^{2+}$, for example. Wherever there is more than one type of ligand bound, the possibility of isomers needs to be considered, although it is not the case that all possible isomers will necessarily exist.

4.4.1 Introducing Stereoisomers

Characteristically, metal ion complexes bearing different ligands result in several possible isomers. This is a consequence of numerous stereochemical combinations

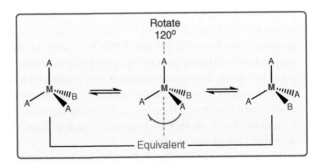

Figure 4.26

Despite first appearances, the molecules at the left and right are identical since simple rotation by 120° around the marked axis in the central view 'converts' one into the other. They are not isomers, merely different orientations of the same molecule in space. The need to consider orientation issues in three dimensions complicates decisions on isomers.

that are possible particularly in a metal with a high coordination number. The best-known examples of isomerism in complexes are geometrical isomers (such as *cis* and *trans* isomers), but these are not the sole type. We can, following a traditional approach, divide the area into two classes: constitutional (or structural) isomerism and stereoisomerism.

4.4.2   Constitutional (Structural) Isomerism

Some of the forms of isomerism have little more than historic interest now, as their significance has diminished with the rise in physical methods that make their identification and origins routine, no longer involving the demanding experiments of an earlier era to probe their form. Nevertheless, some remain important, and others at least give a flavour of the historical development of the field, and this deserves a brief discussion. Constitutional isomers are characterised by having different atom connectivity and consequently show vastly different physical properties.

4.4.2.1   *Hydrate Isomerism*

Hydrate isomerism (sometimes called solvate isomerism) was named to identify an at first puzzling observation, which was that some hydrated compounds with the same empirical formula were obviously different, in colour, charge and number of ions. A classic example is the inert compound $CrCl_3 \cdot 6H_2O$, for which three forms were detected. We know these now as the compounds $[CrCl_2(OH_2)_4]Cl \cdot 2H_2O$, $[CrCl(OH_2)_5]Cl_2 \cdot H_2O$ and $[Cr(OH_2)_6]Cl_3$, being distinguished by the groups bound as ligands to the chromium(III) ion, as shown in the views of the complex cations (Figure 4.27(a)); a fourth option, the neutral complex $[CrCl_3(OH_2)_3] \cdot 3H_2O$ shown at left in the figure, is not detected due to its reactivity in solution, converting readily to other species. The commercial form is the dark green $[CrCl_2(OH_2)_4]Cl \cdot 2H_2O$ (sufficiently venerable to have once been called Recoura's green chloride), formed by crystallisation from a concentrated hydrochloric acid solution. Upon redissolution in water, substitution reactions to release additional coordinated chloride ions commence.

4.4.2.2   *Ionisation Isomerism*

Ionisation isomerism is another case defined by recognising that an empirical formula allows some options for defining the coordination sphere of the metal. It is essentially the same situation as hydrate isomerism, but involves anionic ligands. For example, consider the inert cobalt(III) compound $CoBr(SO_4) \cdot 5(NH_3)$, which forms two different compounds, one violet, and the other red. We know these now as $[CoBr(NH_3)_5](SO_4)$ and $[Co(NH_3)_5(SO_4)]Br$, which differ in the choice of which anion occupies the coordination sphere, the other remaining as the counter-ion (Figure 4.27(b)). Early experimentalists identified their character through simple chemical reactions. For example, with silver ion, it is only $[Co(NH_3)_5(OSO_3)]Br$ that produces an immediate precipitate of AgBr, as the Co–Br bond in the other form is too strong to permit reaction readily. Likewise, reaction with barium ion causes an immediate precipitate of $BaSO_4$ only with $[CoBr(NH_3)_5](SO_4)$, again because the tightly bound SO_4^{2-} ligand in $[Co(NH_3)_5(OSO_3)]^+$ is protected from reaction with Ba^{2+}. These experiments allowed the identification of differing ionic characteristics in the two compounds.

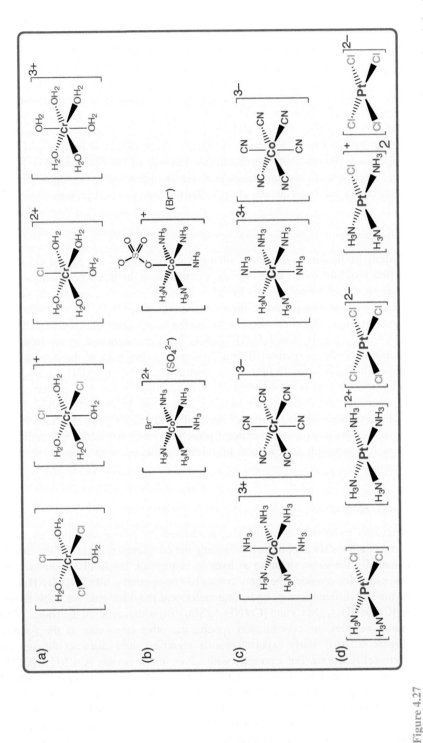

Figure 4.27

Examples of constitutional isomers. (a) *Hydrate isomers*: the octahedral complexes possible for the empirical formula $CrCl_3 \cdot 6H_2O$ (chloride counter-ions not shown). In addition, the neutral and 1+ compounds on the left may in principle exist as two geometric isomers, of which only one is shown. (b) *Ionisation isomers*: the complexes exemplified differ in which anion of the two present is coordinated to the metal in compounds of empirical formula $CoBr(SO_4) \cdot 5(NH_3)$. (c) *Coordination isomers*: the complex pairs exemplified differ in which of the two sets of ligands present is coordinated to which metal. (d) *Polymerisation isomers*: the complexes exemplified have identical empirical formula but differ in the number of replications of the empirical formula, $\{Pt(NH_3)_2Cl_2\}_n$ (n = 1, 2 or 3).

4.4.2.3   Coordination Isomerism

For complex salts where there are metal ions present in both the cation and the anion, both functioning as a complex ion, there are, with two types of ligands, a number of possible coordination forms. The simplest is where all of one type of ligand associate with one or the other metal centre; for example, $CoCr(CN)_6 \cdot 6NH_3$ can be either $[Co(NH_3)_6][Cr(CN)_6]$ or $[Cr(NH_3)_6][Co(CN)_6]$, where, effectively, metal ions 'swap' ligands, both being capable of binding to either, and the various complex cations and anions all being stable entities (Figure 4.27(c)). Mixed-ligand assemblies on each metal offer more options.

4.4.2.4   Polymerisation Isomerism

The empirical formula obtained from elemental analysis identifies the ratio of components, not their actual number. Hence an empirical formula MA_2B_2 could be considered as any of $[MA_2B_2]_n$ (for $n = 1, 2, 3, \ldots$), a series of compounds with the same empirical formula, and with the molecular formula of each some multiple of the simplest formula. In an era where we can determine the molecular mass and/or three-dimensional structure fairly readily, this is not much of an issue but was of concern in earlier times. A classic example is for the empirical formula $Pt(NH_3)_2Cl_2$, which exists not only as the $n = 1$ form $[PtCl_2(NH_3)_2]$, but also as $[Pt(NH_3)_4][PtCl_4]$ ($n = 2$, Magnus' green salt) and $[PtCl(NH_3)_3]_2[PtCl_4]$ ($n = 3$) (Figure 4.27(d)). A more unusual version of the $n = 2$ form possible is the dimeric complex $[(NH_3)_2Pt(\mu\text{-}Cl)_2Pt(NH_3)_2]Cl_2$, where two chlorido ligands bridge the two metal centres with coordinate covalent bonds. Since the syntheses of the two $n = 2$ species differ, this is more an intellectual exercise than a difficult case. However, this example does serve to remind us that oligomers (small polymers) are frequently met in coordination chemistry.

4.4.2.5   Linkage Isomerism

There are a number of anionic ligands that contain two different atoms carrying lone pairs, both capable of coordination to a metal ion. These are called *ambidentate* ligands, distinguished by their capability for binding a metal ion through either of the different donor groups (Figure 4.28). A simple example is the thiocyanate anion (SCN^-), which offers either an N-atom (SCN-*N isomer*) or an S-atom (NCS-*S isomer*) to metal ions; which one a metal ion selects depends on metal–ligand

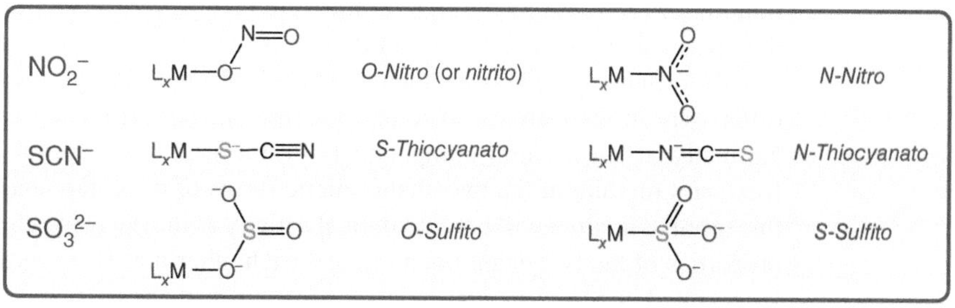

Figure 4.28

Linkage isomers: coordination modes for coordinated nitrite ion, thiocyanate ion and sulfite ion.

preferences discussed earlier. For example, the 'hard' Co(III) ion prefers to form Co–NCS complexes with the 'hard' N-donor, whereas the 'soft' Pd(II) ion prefers the 'soft' S-donor and forms Pd–SCN complexes.

Another classic example is the nitrite ion, which offers N or O atoms as donors. This example has been deeply studied, and the way it behaves is fairly well understood (see Chapter 7.3.1). The O-bound isomer converts (isomerises) to the thermodynamically more stable N-bound isomer, sometimes even in the solid state, by an intramolecular process (without the ligand departing the coordination sphere) in inert complexes. Another feature of ambidentate ligands is that they can 'bridge' two metal ions, with each of the two different donor atoms attached to one of two metal ions.

4.4.3   Stereoisomerism: In Place – Positional Isomers; In Space – Optical Isomers

Stereoisomerism is the name given to cover the more general cases of isomerism in coordination complexes. *Stereoisomers* are molecules of the same empirical formula that share identical coordination numbers, geometries and atom connectivity, but in which the ligand atoms differ in their locations in space. Put simply, everything is the same except where the ligands are positioned. Changing atomic locations in space changes physical and usually chemical properties, so the stereoisomers display at least some differences; these properties can be measured and used to identify the isomer under examination.

There are two gross classes or types with which you should become familiar. **Diastereomers** are the general class of stereoisomers which include geometric isomers (such as *cis*, *trans* isomeric forms) as a sub-class. Any particular diastereomer *may* have an **enantiomer**, which is a stereoisomer that is a *non-superimposable mirror image* of the original diastereomer (like your left hand is a non-superimposable mirror image of your right hand). Not all compounds will exist as two mirror image forms (the enantiomers), as it is a property related to the symmetry of the compound; consequently, diastereomers are not the same thing as enantiomers. For a complex molecule to have enantiomers, it must be **dissymmetric**, which in most cases means it has *neither* mirror plane symmetry *nor* a centre of inversion. However, a dissymmetric molecule *can* have an axis of rotation. A more in-depth definition of symmetry operations and the rules of symmetry are discussed in Appendix 2.

A molecule that is dissymmetric (and therefore *not* superimposable on its mirror image) is called a **chiral** compound; this means that all enantiomers are chiral. Such a compound will display **optical activity** as an individual enantiomer, which is the ability to rotate the plane of plane-polarised light (measured using a polarimeter, in which plane-polarised light is passed through a solution of a compound and any rotation of the plane of the light is measured). This is one way that we can detect the presence of an enantiomer and define its optical purity. Whereas *diastereomers* usually differ appreciably in their chemical and physical properties, *enantiomers* differ only in their ability to rotate polarised light, and related optical properties. Normally, when a chiral compound is synthesised from achiral ligands, a 50:50 (*racemic*) mixture of the two enantiomeric forms of the compound is produced, but this racemic mixture will have no optical activity as the opposing chiral spectroscopic properties of the two enantiomers cancel each other out. However, if the compound is separated into its two enantiomers (or *resolved*), each enantiomer will show optical activity; in a polarimeter, the responses of each enantiomer will differ only in the sign (or direction) of rotation of plane-polarised light.

A 50:50 (*racemic*) mixture of enantiomers is not always the outcome in synthesis. Any synthetic procedure that produces more of one enantiomer than the other is termed an *enantioselective* reaction. In the extreme when only one enantiomer forms, it is called an *enantiospecific* reaction.

Whereas chirality in tetrahedral compounds of carbon requires four different groups to be bonded to the tetrahedral carbon atom, this is not necessarily the case for other central atoms with other stereochemistries. For example, octahedral complexes have more relaxed rules. Whereas a chiral tetrahedral organic compound is, as a consequence of the rule for chirality, asymmetric (or totally lacking in symmetry), chiral octahedral complexes need not be asymmetric, but may have axes of rotation (they are then dissymmetric). The common rule for chirality is simple – a compound must have non-superimposable mirror images. For an octahedral complex, this can occur even when three different pairs of monodentate ligands are coordinated, as discussed later. We shall look a little more closely at four- and six-coordinate complexes in the following sections.

4.4.3.1 Four-Coordinate Complexes

One of the two limiting forms of four-coordination, square planar, involves at least one plane of symmetry (the plane including the metal and four donor groups), and as a result square-planar complexes cannot form enantiomers (unless a ligand itself is chiral). It can form geometric isomers (*cis* and *trans*), however, as defined in Figure 4.29. Tetrahedral complexes can have only enantiomers; they cannot, as a result of their shape, have *cis* and *trans* isomers. Chirality will occur in tetrahedral complexes with four non-equivalent ligands (although there appear to be very few known examples isolated yet) *or* with two unsymmetrical didentate ligands (Figure 4.29).

4.4.3.2 Six-Coordinate Octahedral Complexes

Six-coordination presents greater options for the location of ligands around the coordination sphere than occurs in four-coordination, as a result of the greater number of donor sites. Although there is only one form of MA_6 and of MA_5B possible, for MA_4B_2, there arise two geometric isomers, *cis* and *trans* (Figure 4.30), although there are no enantiomers as both diastereomers have planes of symmetry. When we extend

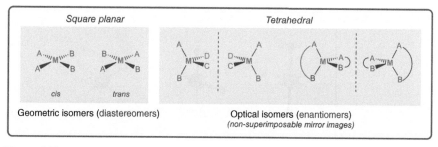

Figure 4.29

Isomers possible with four-coordination. The square-planar geometry, with a plane of symmetry, cannot exhibit optical isomerism but can display geometric isomerism, whereas tetrahedral geometry, with its symmetrical disposition of bonds, cannot exhibit geometric isomerism but may display optical isomerism.

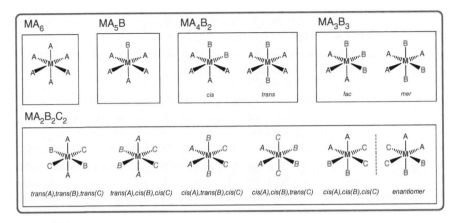

Figure 4.30

Diastereomers and enantiomers possible for octahedral complexes with various sets of monodentate ligands.

to MA_3B_3, there are two geometric isomers, *facial* (abbreviated as *fac*) and *meridional* (abbreviated as *mer*), and we briefly encountered these earlier (Figure 4.25), neither of which has an enantiomer if A and B are both monodentate ligands. If we include additional ligand types, however, the number of isomers can increase rapidly. For $MA_2B_2C_2$, there are now five diastereomers, one of which has an enantiomer (Figure 4.30). With more diverse sets of ligands, the number of possible isomers can grow even greater. In practice, complexes of the formula $MA_2B_2C_2$ are rarely met due to the inherent problems of separating so many potential isomers, so this exercise merely becomes a thought experiment.

Simple actual examples illustrating *cis/trans* and *fac/mer* isomerism are octahedral complexes of cobalt(III) formed with ammonia (or amine) and aqua ligands, namely the cations *cis-/trans*-$[Co(NH_3)_4(OH_2)_2]^{3+}$ and *fac-/mer*-$[Co(dien)_2]^{3+}$ where dien is the linear tridentate diethylenetriamine ($NH_2CH_2CH_2NHCH_2CH_2NH_2$); their X-ray crystal structures are shown in Figure 4.31. The distinctions between the *trans* and *cis* isomers should be self-evident. What should also be apparent is that in the *fac*-$[Co(dien)_2]^{3+}$ isomer, the donor atoms from each ligand are in adjacent (*cis*) coordination sites occupying one face of the octahedral coordination sphere; hence the origin of the name. In the *mer* isomer, the donor atoms are coplanar with the metal and the two ligands are perpendicular to each other.

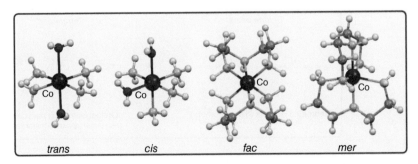

Figure 4.31

Diastereomers of the octahedral complexes *trans*- and *cis*-$[Co(NH_3)_4(OH_2)_2]^{3+}$ as well as *fac*- and *mer*-$[Co(dien)_2]^{3+}$ (dien = diethylenetriamine, a tridentate ligand).

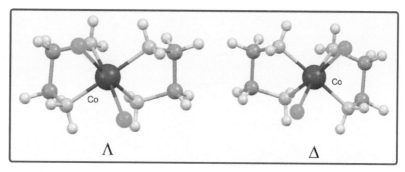

Figure 4.32

Non-superimposable mirror images (enantiomers) of diastereomeric Λ- and Δ-*cis*-[CoCl$_2$(en)$_2$]$^+$.

The introduction of chelate ligands, as just illustrated, usually acts to limit the number of geometric isomers, or at least to not extend the number. For example, M(AA)$_2$B$_2$ compounds (where (AA) refers to a symmetric didentate ligand) can adopt most of the diastereomers of its all-monodentate analogue MA$_4$B$_2$. The difference is that the introduction of chelating ligands leads to physical constraints (*cis* versus *trans* coordination sites) and lower symmetry, such that enantiomers can exist. The presence of at least two chelate rings in *cis* dispositions always leads to chiral (optically active) octahedral compounds, as the molecules become dissymmetric. A simple example is the octahedral complex *cis*-[CoCl$_2$(en)$_2$]$^+$ that has two ethane-1,2-diamine (en) ligands, and, unlike the *trans* diastereomer, has a mirror image that is non-superimposable on the original, a key requirement for optical activity (Figure 4.32). The enantiomeric complexes are distinguished by using the Δ (Greek capital delta) and Λ (Greek capital lambda) symbols. The lowercase equivalents, δ and λ, you have met in relation to chelate ring conformations (Figure 3.7). These symbols have no relationship to the *d* (dextrorotatory) and *l* (levorotatory) prefixes which only refer to chiroptical properties and carry no structural information. The *d*- and *l*-terminology once used commonly in organic chemistry was supplanted by the Cahn–Ingold–Prelog descriptors *R* and *S*, first proposed in 1966 and adopted by the International Union of Pure and Applied Chemistry in 1974.

So far, we have introduced the concept of optical activity (or chirality) in a complex arising as a result of the way ligands are arranged around the central metal. For six-coordination octahedral, there are actually several ways in which dissymmetry (chirality) can arise, namely from:

1. the distribution of monodentate ligands about the metal (which occurs with at least three pairs of different ligands, as exemplified for MA$_2$B$_2$C$_2$ in Figure 4.30);
2. the distribution of chelate rings about the metal (as occurs when at least two didentate chelates are coordinated, as exemplified in Figure 4.32);
3. coordination of asymmetric polydentate ligands (where the assembly of different donors and linkages resulting means the complex becomes dissymmetric);
4. conformations of chelate rings (where tetrahedral carbon and other atoms enforce their geometry in the chain linking donor groups, causing chelate rings to be non-planar and able to adopt conformations that have δ and λ mirror image forms – see Figure 3.7);
5. coordination of an asymmetric and hence chiral organic ligand (whereby the chirality of the part is assigned to the whole) – this chirality in the ligand may be

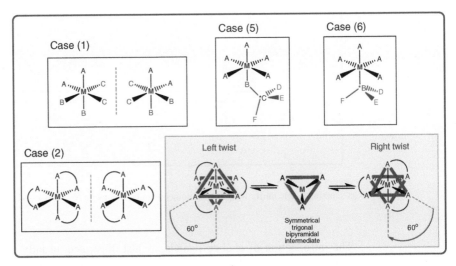

Figure 4.33

Selected examples of ways in which dissymmetry is introduced into octahedral complexes, with case numbers equating with those in the text immediately above. Enantiomers are shown for two cases only [(1) and (2)]. A process by which conversion in case (2) from one enantiomer to the other (*optical isomerisation*) may occur is illustrated, based on twisting of one octahedral face by 120° while keeping the other fixed. Each isomer must twist in opposite directions to avoid breaking the chelate chains, to generate a symmetrical intermediate, from which it may continue twisting to yield the opposite enantiomer. The 'left-twisting' isomer is called Λ and the 'right-twisting' one Δ, the symbols for the two enantiomers or optical isomers. In the examples for case (5) and (6), each chirotopic centre is identified by a star.

conventional, arising from chirotopic carbon centres of *R* or *S* chirality, or arise from binding to a helical ligand (which has *P* and *M* isomers associated with opposite helicity);

6. coordination of a donor atom that then becomes chirotopic (with four different substituents including the metal), imparting chirality to the whole complex, similar to the situation in (5) above, but with the asymmetric centre attached to the metal in this case.

Examples of several of these cases appear in Figure 4.33, but a detailed discussion will not be pursued here, this being a task for more advanced texts.

4.4.4 Isomer Preferences

Polydentate ligands present particular problems, as a range of geometric isomers, often with enantiomers, may exist in principle. This is illustrated in Figure 4.34 for a simple system where a symmetrical tetradentate is bound. This can occur in three ways (one *trans* and two different *cis* isomers) giving five isomers in total, where both *cis* isomers are chiral and produce enantiomeric pairs. What should be appreciated is that these three coordination modes do not lead to complexes of equal thermodynamic stability, meaning that the percentage of each formed is not based purely on a statistical ratio, but related to the relative stabilities of each. The isomers usually display different stability, and when there are large energy differences, one thermodynamically preferred form, or a limited number of forms, will result. This means, in effect, that in some cases only one geometric isomer could be observed experimentally if it is significantly more stable than others.

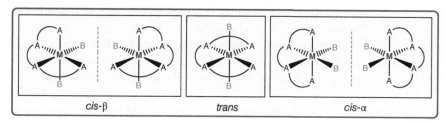

Figure 4.34

The three geometric isomers (diastereomers) for M(AAAA)B$_2$ complexes, one of which is of *trans* geometry with respect to the two simple B ligands (centre), and two of which are of *cis* geometry (*cis*-β at left, *cis*-α at right) and those cases have enantiomers, also drawn.

If we introduce mixed donors (asymmetry) into the simple tetradentate AAAA of Figure 4.34, for example to form AAAC, the number of isomers will increase. For example, now there are two different forms of the *cis*-β isomer in Figure 4.34, as illustrated in Figure 4.35. The two new isomers have the same spatial arrangement of the chelate chains, but the unique terminal groups (A and C) are located differently and so the compounds are diastereomers with different stabilities and chemical properties. An alternative reaction that increases geometric isomers is where the two common monodentate donor groups are replaced by two different monodentate donor groups (Figure 4.35, right); there are again different outcomes depending on which of two groups is replaced. There is adequate nomenclature to deal with naming these difference diastereomers, but this is not required to comprehend the outcomes when pictorial representations are available; some aspects of naming complexes appear in Appendix 1. These two simple examples suffice to illustrate the array of geometric isomers that can result when polydentate and mixed donor ligands bind to a central metal ion.

Geometric isomers are energetically inequivalent because of a suite of effects that contribute additively to the stability of each complex. When chelating ligands are present (as in Figures 4.34 and 4.35), a major factor is ligand strain energy which includes: bond length deformation (compression or extension away from an ideal distance); valence bond angle deformation (bending beyond an ideal angle);

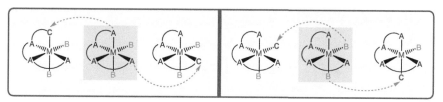

Figure 4.35

(Left) One of the geometric isomers (*cis*-β) of a M(AAAA)B$_2$ complex is transformed into two diastereomers when a mixed donor ligand replaces the common donor ligand in forming a M(AAAC)B$_2$ complex. Consider the two arising from replacement of two different terminal donors, so the product has the introduced C group either opposite a B ligand (left view) or opposite an A ligand (right view). In reality, of course, the mixed donor ligand is preformed and coordinates as a single entity. (Right) One of the geometric isomers (*cis*-β) of a M(AAAA)B$_2'$ complex is also transformed into two diastereomers when different monodentate ligands (B, C) replace the common monodentate ligand (B) in forming a M(AAAA)BC complex. Consider the two arising from replacement of two different monodentate donors, so the product has the introduced C monodentate either opposite a central A group of the tetradentate (left view) or opposite a terminal A group (right view).

torsion angle deformations (between ideally staggered and unfavourable eclipsed orientations); non-bonded repulsions; electrostatic interactions (of charged groups, both attractive and repulsive) and hydrogen bonding interactions. Models that allow the estimation of strain energies in isomers exist (see Chapter 7.3.10), providing prediction of and/or interpretation of experimental behaviour.

4.5   Sophisticated Shapes

The shapes we have described have employed, in all but the last section, simple ligands that bind at one or two sites around a metal ion. However, most ligands are more complicated in both their potential donor set and basic shape. The ultimate in complicated ligands are the natural ligands met in Chapter 8, but it is appropriate to briefly examine some examples of synthetic ligands, to illustrate the way in which they coordinate and form complexes.

4.5.1   Compounds of Polydentate Ligands

Immediately above, we have introduced some of the issues that arise when a polydentate ligand is bound rather than simple monodentate or didentate ligands. Even stepping from two to three donors increases the options in terms of ligand shape (or *topology*), and this shape will affect the way a molecule may bind to a metal ion. Some shapes for potentially tridentate ligands appear in Figure 4.36.

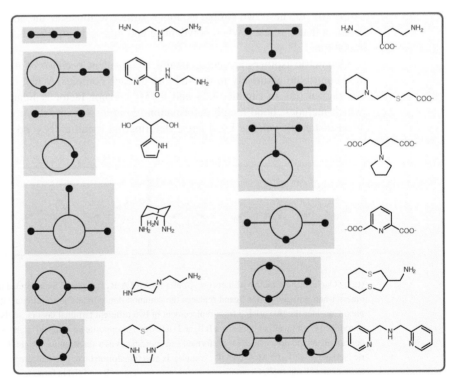

Figure 4.36
Some possible shapes for three-donor ligands, with examples.

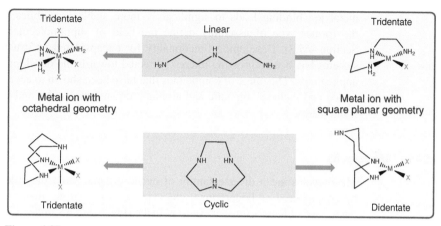

Figure 4.37

Possible coordination for three-donor linear and cyclic ligands to square-planar and octahedral metal complexes. The cyclic ligand is unable to bind all three donors in the square-planar geometry.

The shape of a ligand influences the way it can bind to a metal ion, as introduced in Chapter 3. This is illustrated for potentially tridentate ligands in Figure 4.37, comparing a linear triamine and a cyclic triamine binding to a square-planar or octahedral metal ion. Whereas the former can bind to a square-planar centre with all three donors attached, the latter cannot, as the ring is too small to permit the third donor to bind in the plane of the metal after the first two bind. This means that the cyclic ligand acts not as a tridentate but as a didentate ligand with one group not coordinated (or *pendant*). Only if there is a change to octahedral or tetrahedral geometry, where the ligand can bind to a triangular face, do all three amine groups bind. This simple example serves to highlight some of the considerations that apply when ligand topology comes into play.

As the number of donors increases, so does the diversity in shape that may be met. We tend to classify ligands in terms of their appearance or basic framework – for example, they may be described as linear, branched, cyclic and cyclic with pendant chains. Nevertheless, despite the variations possible, these types of ligands are still regarded as basic or simple in character; there exists an array of more complicated and specific shapes that we shall touch upon briefly in the following sections.

4.5.2 Preorganised Complexes

Increased sophistication in chemical synthesis has led to the development of a wide range of molecules which can 'wrap up' or 'encapsulate' metal ions or even organic and inorganic molecules or ions, although in this latter case, weaker non-bonding interactions operate rather than covalent bonding. A range of three-dimensional shapes have been devised to accommodate either metal ions with different preferred coordination numbers or small molecules. The metal ions dominantly involve coordinate bonds, while non-metallic targets use weaker non-covalent interactions (such as hydrogen bonds). A key point is that the ligand (*host*) is preorganised to accept its partner (*guest*). Encapsulation of the guest using non-covalent or coordinate bonding may be reversible or irreversible depending on the stability of the assembly. Non-covalent bonding interactions are reversible and weak while metal ion binding leads to significantly more stable assemblies. 'Guests' within the former type of assembly define the field of supramolecular chemistry (see Section 4.5.3). Developing functionality for encapsulated or structurally confined systems has been a growing interest, directed towards the development of practical applications. The range of compounds that have been shown to act as 'hosts' is quite varied and includes aliphatic and aromatic organic molecules and inorganic zeolite frameworks.

4.5.2.1 *Macrocycles*

Macrocycles are a diverse family of cyclic organic molecules (defined as having nine-membered or larger rings) characterised by a strong metal binding capacity when several heteroatoms are present in the ring. Several simple examples have been introduced earlier. This class includes the so-called 'pigments of life' – nature's aromatic macrocycles built for important biochemical use in plants (the chlorophylls) and all life forms (the hemes). Visible light is absorbed in these systems by the presence of a sequence of alternating double and single bonds to give character-istic colours. *Macromonocyclic* (single large ring) molecules including several heteroatoms represent the simplest members of the family of macrocycles. They can bind metal ions and even small molecules reasonably efficiently, depending on their size. A range of types, both aliphatic and aromatic, are known, as exemplified in Figure 4.38. In a simple sense, what is important in these cyclic systems is the matching of cavity (or hole) size to metal ion size; a good 'fit' means a stronger complex.

One well-known group of synthetic macrocyclic ligands are polyamines. Many may not be capable of 'wrapping up' the metal fully, a consequence of not having sufficient donor groups to satisfy fully its coordination sphere. However, with sufficient donor groups this can be achieved, and in a very large ring with many amine groups even more than one metal ion may be accommodated. Coordination of a saturated polyamine macrocycle, however, introduces subtle stereochemical consequences relating to the disposition of amine hydrogen atoms on complexation.

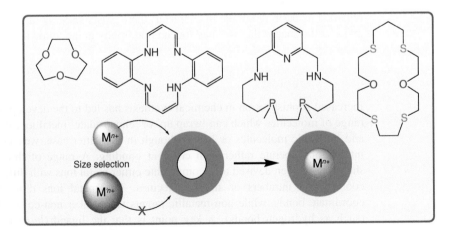

Figure 4.38
Examples of three-, four-, five- and six-donor macrocycles. Metal ion selection and binding strength are based in part on ion–hole 'fit'.

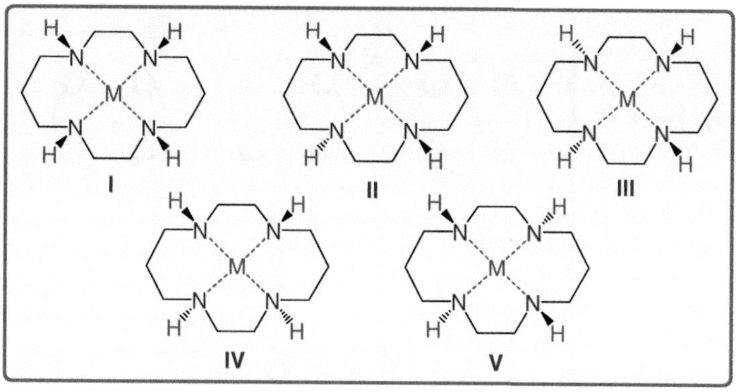

Figure 4.39

Possible dispositions of the secondary amine hydrogen atoms above and below the plane of the ring in a coordinated macrocyclic tetraamine. Each N-based diastereomer (named I–V), formed only on coordination, exhibits slightly different physical properties.

This is illustrated for a tetraamine bound to a square-planar metal ion in Figure 4.39; different amine proton dispositions possible are highlighted.

At a simple level, we can see this as 'four up' (I), 'three up and one down' (II) and three types of 'two up and two down' (III–V); other options (such as 'four down' compared with 'four up') are equivalent to one of those shown, as they are formed simply by rotating the complex by 180°. In effect, these isomers equate to a different chirality set for the four nitrogen donors, fixed upon coordination. In this case, each secondary amine RR′HN group, when coordinated, forms an RR′HN–M group and, as all four attachments are different, each N-atom is chirotopic. These N-donors are additional sources of isomerism and chirality in the molecule depending on their symmetry relationship to other equivalent centres. Although the N-based isomers may have slightly different physical properties, interconversion between these isomers can be readily achieved by raising the solution pH, which promotes N-deprotonation and exchange, leading to the formation of the thermodynamically most stable N-based isomer. Usually, this subtle N-based isomerism tends to be ignored, as it is a level of complication too far for most.

Macrocycles carrying *pendant groups* also capable of binding metal ions produce the opportunity to 'wrap up' metal ions better (Figure 4.40). These 'molecular wrappers' have pendant groups that can come on or off, so they behave as 'hinged lids'. The pendant groups may be of any type and carry any form of potential binding group – amine, carboxylic acid, thiol, alcohol, pyridine and others. These groups may themselves be further elaborated or extended using standard organic reactions, including forming dimers, linking them to biomolecules or binding them to surfaces.

4.5.2.2   *Macropolycycles*

Macropolycyclic molecules (with more than one large organic ring fused together) that include several heteroatoms can bind metal ions efficiently, and a range of types are known. One simple family is the *sarcophagines*, bicyclic (two fused large ring) amine molecules with the capacity to 'cage' metal ions. In their metal free form (Figure 4.41, centre), they have a cavity which offers six N-donors to a metal ion.

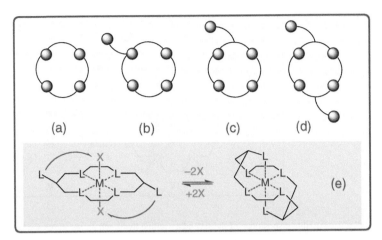

Figure 4.40

Simple macrocycles (a) may be augmented with pendant group(s) attached to either a heteroatom (b) or a carbon atom (c) of the ring. Those with two pendant groups (d) offer better opportunities for 'wrapping up' octahedral metal ions, as the two pendant arms can supply additional donors that, by being linked to the ring, enhance entrapment by 'capping' the metal, as shown in (e).

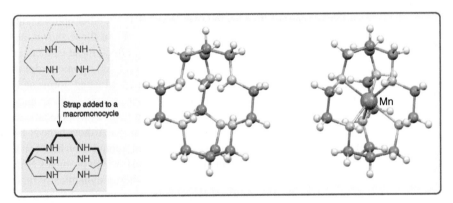

Figure 4.41

A simple macrobicycle can be considered to arise by the addition of another chain or 'strap' of atoms to a macromonocycle, to provide three chains joined at two capping carbon atoms (left), effectively two fused macromonocycles. This example (with the trivial name *sarcophagine*, with its X-ray structure shown at centre) offers six N-donor groups to a metal ion, in this case, Mn(III) (X-ray structure at right). These ligands readily form very stable octahedral complexes with many transition metal ions. Examples of ligands with mixed donor combinations of S and NH groups are also known.

This cavity is too small for anything but a single metal ion. As their name implies, they are molecular 'graves' – once a metal ion is interred in the cavity, it is trapped and can only be removed with difficulty (usually requiring very strong acid or highly competitive ligands such as cyanide). Metal ions are effectively in a prison with the frame of the macrobicycle as the bars. A typical example is shown in Figure 4.41 (right) featuring a rare Mn(III) complex where this normally reactive metal ion is stabilised within its cage. Another notable feature is that the free ligand is preorganised in a suitable conformation to bind the metal, and there is very little structural rearrangement upon metal complexation.

4.5.2.3 Cryptands

Cryptands are cyclic (mainly) polyether ligands with usually three chains linked at nitrogen 'caps' at each end of the molecule (Figure 4.42), much like the sarcophagines but with a different capping atom and different donors. While the N-donors of the sarcophagines prefer transition metals (according to the hard-soft acid/base principle), the cryptands with mostly hard O-donors were specifically designed to bind metals from the *s*-block. They can, depending on host cavity size, bind either metal ions covalently or even accommodate small charged organic molecules or ions through non-covalent attractive forces (H-bonds and electrostatic forces). All the *s*-block metal ions have been structurally characterised as complexes with cryptand ligands of various shapes and sizes. The cryptands can be effective in metal ion selection from a group of ions, which is useful in both analysis and separation of mixtures. They also can solubilise usually only water-soluble alkali metal salts in aprotic solvents and hence find wide application in organic synthesis. Their systematic names are formidable, but a simple shorthand terminology has evolved that is shown in Figure 4.42 based on the number of donor atoms in each strap. Historically the advent of the crown ethers and cryptands launched the field of *s*-block coordination chemistry which until that time was poorly understood due to a lack of suitable ligands for alkali or alkaline earth metal ions, and this work was recognised by the 1987 Nobel Prize in Chemistry shared by Donald J. Cram, Jean-Marie Lehn and Charles J. Pedersen.

4.5.2.4 Catenanes

A particularly unusual class of macrocyclic molecules are the *catenanes*, where, instead of two side-by-side covalently linked rings as in the cryptands above, two separate and interlocked rings offer donors to a single metal ion (Figure 4.43). With appropriate components in the ring, a change in oxidation state of the metal ion can lead to a ring rotating into a different position, a process that can be reversed if the

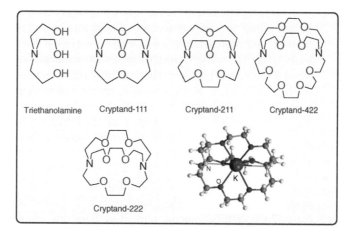

Triethanolamine   Cryptand-111   Cryptand-211   Cryptand-422

Cryptand-222

Figure 4.42

The simple non-cyclic analogue triethanolamine and macrobicyclic cryptands which have three chains joined at two capping nitrogen atoms. The cryptand trivial names reflect the number of O atoms in each linking chain. (Cryptand-222 is also shown as its eight-coordinate K$^+$ complex, the structure determined by X-ray crystallography.)

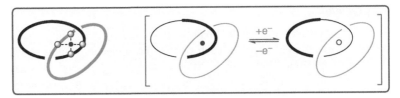

Figure 4.43

A schematic representation of a catenane complex with donors from each of two interlinked rings binding to a central metal ion, and (at right) the process by which a system with differing ring components may operate as an electrochemically driven molecular switch, involving ring rotation.

metal ion oxidation state is turned back to its original. This process of switching position may provide an electrochemically driven molecular 'switch', and this kind of orchestrated motion at the atomic scale has been dubbed a *molecular machine*. Synthesis of these interlocked systems and their precise control are not simple, so their potential application may be limited, but the pioneers of this field were also recognised by the 2016 Nobel Prize in Chemistry, awarded to Jean-Pierre Sauvage, J. Fraser Stoddart and Ben L. Feringa.

4.5.3   Host–Guest Non-covalent Molecular Assemblies

A close relative of simple coordination chemistry is *supramolecular chemistry*. Whereas coordinate covalent bonds link the ligand host and the metal guest in coordination complexes, in supramolecular compounds the guest is not a metal ion, and as such can only involve itself in weaker non-covalent interactions with the host, typically involving hydrogen bonding or electrostatic attraction for charged hosts and guests. Although hydrogen bonds are relatively strong compared with other non-bonding effects, they are still much weaker than coordinate covalent bonds. However, in certain cases, the shape and outcome are reminiscent of coordination chemistry, and it may be of value to exemplify it here in passing.

Host–guest interactions occur where a cavity in the host permits selected entry of the guest into the cavity space; it is, like the macrocycles with metal ions, a case of 'best-fit', although here it is a whole molecule or ion that is fitted into the cavity. The host–guest terminology is usually applied to molecule–molecule interactions, not to metal ion binding, however. The assembly that forms in solution (with measurable stability, as a result of spectroscopic changes that occur when the guest occupies the different environment of the host) is held together by non-covalent bonding forces – hydrogen bonding and other weaker interactions. The 'container' molecules are now many and varied, but all have the common character of acting as hosts for guest molecules or molecular ions.

Cyclodextrins, calixarenes and cucurbiturils (Figure 4.44) are well-established examples of molecular hosts. **Cyclodextrins** are naturally occurring cyclic molecules made of glucose units linked head to tail. The most common cyclodextrins have 6 (α), 7 (β) or 8 (γ) glucose units. Cyclodextrins adopt a barrel-like 3D structure with a hydrophobic central cavity while the hydroxyl groups on the outside of the ring bring water solubility, so water-insoluble non-polar guest molecules may be stabilised in an aqueous solution within the cavity. There are several related hosts that differ depending on the class of monomer units and linkages adopted in macrocyclic ring formation. **Calixarenes** (from the Latin calix = bowl) are large

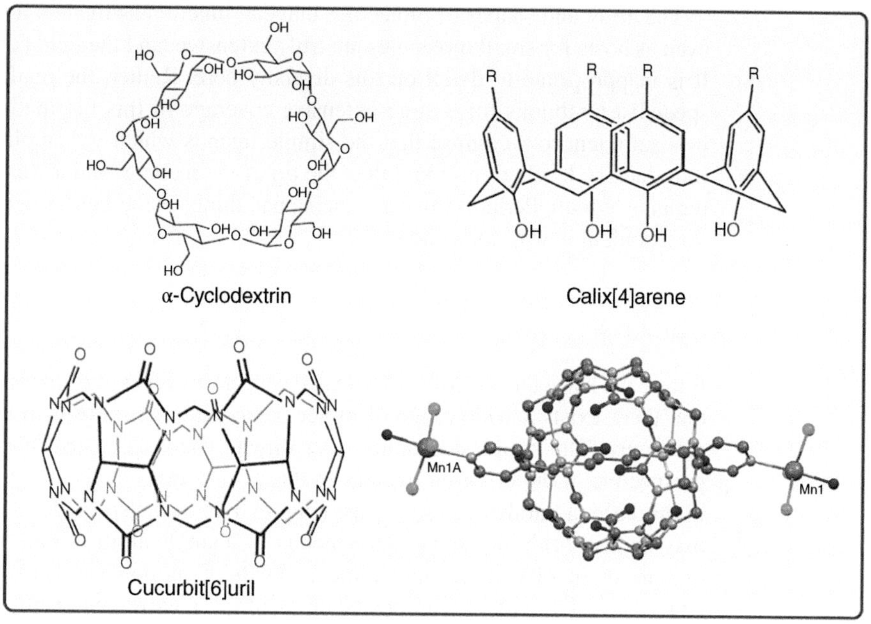

Figure 4.44

Structures of the macrocycles α-cyclodextrin, calix[4]arene and cucurbit[6]uril macropolycycles. The crystal structure of a cucurbit[6]uril rotaxane is shown where the long chain bridging between two Mn complexes threads the host.

cyclic molecules made up of typically four to eight phenol units linked via –CH_2– groups in the 2- and 6-positions. As their name suggests, they adopt bowl-like cavities of dimensions appropriate for sequestering a variety of small molecules, which is of great interest for purification, chromatography, storage and slow release of drugs *in vivo*. Hydrogen bonding involving alcohol groups plays a crucial role in the supramolecular chemistry of such assemblies. **Cucurbiturils** are a family of cyclic host molecules, akin to the cyclodextrins, featuring a hydrophobic cavity and polar carbonyl groups surrounding the two open ends (Figure 4.44). Better defined as cucurbit[n]urils (where n = the number of monomer units in the ring), they can be water-soluble, depending on substituent groups. When the typically hydrophobic guest enters the cavity, other non-covalent forces may also stabilise the complex, including H-bonds. An example is shown in Figure 4.44, where a protonated long chain polyamine with a Mn complex sited at each end is threaded through the cucurbit[6]uril host to give a host–guest assembly commonly known as a *rotaxane*. The thread can reversibly enter the host, and there are many other similar types of supramolecular assemblies now known.

For these different classes of hosts, the capacity to control the size of the rings and the introduction of various functional groups make it possible to 'tailor' them for a variety of chemical applications. Their sequestering properties can be exceptional, while appropriate substitution renders the cavities shape-selective and thus suitable for molecular recognition. Changes in a range of spectroscopic properties with the increasing formation of the host–guest adduct in water allow us to identify and quantify host–guest formation. The stability of host–guest assemblies in water can be high, even approaching that of some metal complexes. Aquated metal ions and other small metal complexes can also act as guests, depending on the ring cavity size.

The array and shapes of molecules that can function as ligands for metal ions and even as hosts for small molecules are truly extensive, and the field continues to grow. It is inappropriate to dwell on this diversity here. Rather, the reader is directed to specialist textbooks for a more extensive coverage of this fascinating field. For us, it is sufficient to recognise that the simple ligands which we employ largely at the introductory level are no more than the tip of the iceberg, and a wealth of chemistry remains hidden. Perhaps this isn't such a bad thing, as the field is demanding enough for a student new to its wonders.

4.5.4   Polymetallic Compounds

Before we leave the area of sophisticated shapes, however, it is important to recognise that there exists a wide range of molecules containing more than one metal ion in proximity. The simplest examples are dimers (introduced in Chapter 3, from the perspective of the bridging ligand). In these, only two metals are involved, but even in this case, a number of types exist (Figure 4.45; see also Figure 3.17). The dimers may involve bridging between the metals by a single atom or ion, by several atoms or ions or involve direct metal–metal bonds. Common bridging groups are Cl^-, HO^-, O^{2-}, S^{2-}, polyoxo anions (such as SO_3^{2-}), carboxylic acids and short-chain ligands with potentially coordinating groups at each end.

Complexes that contain at least three metals in proximity are often called *clusters*, typically when they are linked through single atom bridges or metal–metal bonds. Clusters have a venerable history, with metal-carbonyl species reported in the 1930s, although confirmation of structures had to await their characterisation by X-ray crystallography, which now forms the key method of identification in the solid state. The known shapes of clusters are extensive in number, and they may show some bond angle and bond length deformation due to the complexity of the structure and the associated demands of the ligands. Clusters grow complicated in structure as the number of metal centres increases and may even involve several layers of metal ions from a core outwards where, perhaps not surprisingly, metals in different environments (core or surface, for example) behave differently, even if inherently the same type and in the same oxidation state – in other words, 'environmental' effects

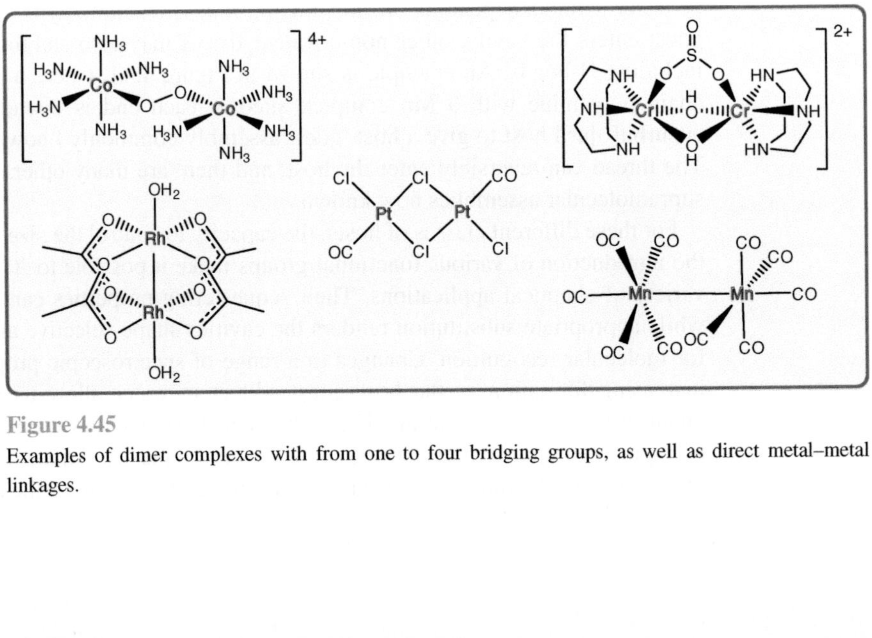

Figure 4.45

Examples of dimer complexes with from one to four bridging groups, as well as direct metal–metal linkages.

contribute to the metal's behaviour. Suffice to say that clusters are an enticing but difficult area of coordination chemistry. In clusters with more than one type of metal involved, an array of isomers may exist resulting from differing placements of the metals in the cluster framework, raising the complexity of this category.

In Figure 4.46, examples of some well-known 'simple' transition metal cluster families are shown. The mixed donor carbonyl/phosphine cluster compound $[Ru_3(CO)_9(PMe_3)_3]$ is but one of many examples that exhibit this triangular metal–metal bonded array unsupported by any bridging ligands. The oxido-bridged $[Mo_6O_{19}]^{2-}$ presents an interesting octahedral array of Mo atoms around a central six-coordinate O^{2-} ligand with an additional 12 bridging oxido ligands on the edges of the octahedron and 6 terminal oxido ligands (one on each metal); each individual Mo is in a distorted octahedral environment. There are a number of proteins with cluster cores, such as ferredoxins, which have a $\{Fe_4S_4\}$ core (similar to the hydrocarbon *cubane*, C_8H_8) which has inspired the synthesis of structural analogues. An example is shown in Figure 4.46, which bears terminal MeS^- ligands and bridging S^{2-} ligands. The Fe ions display a distorted tetrahedral geometry. This type of compound will be discussed in greater depth in Chapter 8.

Although metal–metal bonding is often present in organometallic clusters, more classic inorganic oxido and sulfido clusters do not exhibit such direct bonding between metals, although confinement in relatively close contact alone does clearly influence their electronic structure and properties of the assembly relative to monomers. Hybrid systems are common where the metal–metal bond is buttressed by additional bridging ligands; the carbonyl ligand is capable of doing this in addition to acting as a terminal ligand. Metal-carbonyl clusters are extensive in type, but one well-known type and interesting family is the series of clusters $[Pt_3(CO)_6)]_n^{2-}$ (where n varies from 1 to 10) developed by Paolo Chini, which can be described as containing stacks of triangular Pt_3 units comprising the flat $Pt_3(CO)_6$ core where both three terminal and three bridging carbonyl ligands are present. Three examples are shown in Figure 4.47.

Clusters formed from metal cations and oxide anions, the *metal oxido clusters*, exist in a vast array of sizes and shapes. Apart from the O^{2-} ion, clusters may form with some HO^- or H_2O molecules bound, which can be viewed as protonated forms of the oxido ligand. In general, the presence of several metal cations near a water oxygen alters the acidity sufficiently to promote deprotonation even to O^{2-}, with the dianion

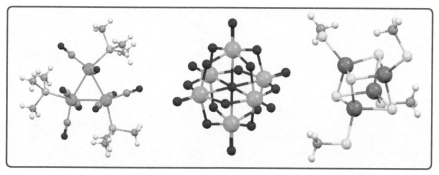

Figure 4.46

Examples of simple cluster compounds, $[Ru_3(CO)_9(PMe_3)_3]$ (left), $[Mo_6O_{19}]^{2-}$ (centre) and $[Fe_4S_4(SMe)_4]^{2-}$ (right).

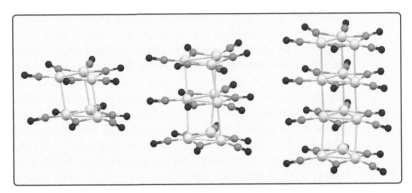

Figure 4.47
The triangular stacked cluster compounds $[Pt_3(CO)_6)]_n^{2-}$ (n = 2, 3 and 4).

stabilising the assembly formed with a set of closely-located metal cations. It is no accident that most metal ion clusters involve anions as bridging species. The O^{2-} ion, with four lone pairs, has the ability to be involved in a number of coordination modes, as a monodentate ligand to a single metal ion, or in bridging arrangement between two (termed μ^2-O) or three (termed μ^3-O) metal ions at once. A relatively simple example is $[Zr_6O_{36}H_{48}]$, where the cluster has six Zr(IV) metal ions (each eight-coordinate) arranged in an octahedral manner, each joined to four bridging O^{2-} and four terminal O-donors; protons are present (via H_2O or HO^- ligands) and balance the charges but are not structurally significant and, unlike other atoms, cannot be easily located by X-ray crystallography. The cluster can be viewed as a set of five approximate layers (Figure 4.48, left), with a single metal ion carrying four terminal O^{2-} at the top, a square of four O^{2-} below, a square of four metal ions (each carrying four terminal O^{2-}) below that layer, then another square of four O^{2-} further below prior to a single metal ion carrying four terminal O^{2-} at the base.

Although most examples developed above have involved simple ligands, it is notable that clusters incorporating larger organic molecules are common. One simple example has a $\{Fe_3O(CH_3COO^-)_6\}$ core where six carboxylates function as bridging groups (Figure 4.48, right). Elaborate metal-organic frameworks (called MOFs) have been developed, and their properties and potential applications continue to be explored, but these lie beyond an introductory level.

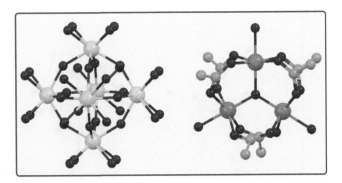

Figure 4.48
Structures of the metal-*oxido* cluster compounds $[Zr_6O_{32}H_{48}]$ (at left) and the carboxylate-bridged μ^3-O^{2-} centred $[Fe_3O(CH_3COO^-)_6(OH_2)_3]$ compound (at right). H atoms are not shown.

The examples in Figures 4.45–4.48 involve a single type of metal in the cluster, but this is not a requirement. Many examples of mixed-metal clusters exist, and, in addition, clusters may have a number of different ligands. In clusters, the bridging anions are less accessible to chemical reactions than those bound as terminal monodentate ligands to the periphery of the cluster, and it is here at the periphery that substitution by other molecules more readily occurs. Examples of clusters with mixed metals present, although often synthetically demanding, still abound in the chemical literature. Simple examples are complexes with *cubane* cores (see Figure 4.46) such as $\{Mo_3NiS_4\}$ and $\{Mo_3PdS_4\}$, where a Mo in the 'parent' $\{Mo_4S_4\}$ core has been replaced by a different metal ion. In organometallic compounds, there is a diversity of examples; a simple example is where the reaction together of structurally analogous $[Ni_6(CO)_{12}]^{2-}$ and $[Pt_6(CO)_{12}]^{2-}$ produces a series of $[Pt_{6-x}Ni_x(CO)_{12}]^{2-}$ mixed-metal clusters. The first bimetallic carbonyl compound, $[HFeCo_3(CO)_{12}]$, was reported decades ago in 1960 by the Paolo Chini research group.

The molecular assembly can be sufficiently large (such as $[Mo_{176}O_{528}(H_2O)_{80}]^{32-}$ and $[Au_{38}(SR)_{24}]$) that it may be defined as a *nanoparticle*, usually considered to be from 1 to 100 nm in size. These *nanomolecules* typically exhibit properties that differ from both simple complexes and related bulk solids. Nanomaterials in general may be inorganic-based (typically metals and metal oxides or silicates), carbon-based (either almost exclusively carbon, such as *fullerenes*, or based on organic molecules) or biomolecules (such as liposomes). They adopt a range of shapes, with roughly spherical shapes common, although tubular shapes (*nanotubes*) occur for carbon-based compounds.

Another well-known class of molecules that can occur in much larger sizes are *zeolites*, polymeric species of general formula $M_a[(AlO_2)_x(SiO_2)_y]\cdot zH_2O$, where M represents a metal ion or proton. In a simple sense, consider zeolites as based on pure SiO_2 where some Si(IV) ions are replaced by Al(III) ions, which then also require an addition of M^+ for charge balance. These crystalline solid zeolites, with well-defined networks of pores in their structures, are hydrated aluminosilicates of alkaline and alkaline earth metals where interlinked tetrahedra of $\{SiO_4\}$ and $\{AlO_4\}$ units define the structure. This requires the structure to have O^{2-} anions covalently bridging between Al(III) and Si(IV) centres. Around 40 naturally occurring zeolite minerals are known, but many more have been synthesised. With large volumes of free space in the structure, zeolites are of low density, exhibit shape selectivity due to the channel and chamber sizes, carry a large number of incorporated water molecules, and display capacities for ion sorption, ion exchange and catalysis. *Zeolite A* is one well-known compound of formula $[(Na_{12}(H_2O)_{27}]_8[Al_{12}Si_{12}O_{48}]_8$ as the sodium salt; the sodium ions can be exchanged with other metal ions, and it is this capacity for ion exchange that governs one of their applications. *Zeolite A* has a major internal cavity of about 0.4 nm; larger zeolites offer larger cavities, and it is the semi-porous nature of these compounds that direct their applicable chemistry. The X-ray crystal structures of two examples, complete with encapsulated molecules, appear in Figure 4.49.

Somewhat related are the *polyoxometallates*, a large class of generally anionic metal–oxygen clusters of particularly the early transition elements molybdenum, tungsten, vanadium and niobium. These exist in a very wide range of structures, sizes and compositions, although 10 basic types of clusters are routinely recognised, with each structural 'family' named after the key researcher associated with the development of each type. These discrete complexes are composed of an array of corner-sharing and edge-sharing approximately octahedral MO_6 units that form the ionic core (as distinct from the tetrahedral MO_4 units in zeolites). They exist

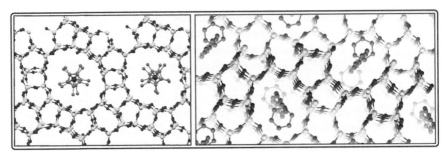

Figure 4.49

The structures of two zeolites: (left) zeolite UTD-1F with encapsulated [Co(Cp*)$_2$]$^+$ cations within its large cavities (Cp* = pentamethylcyclopentadienyl); and (right) silicalite-1 with encapsulated benzene molecules within smaller cavities. The Si and Al atoms (distributed randomly) are shown in light lime green, and H atoms are omitted; structures are not scaled identically.

as 'simple' polyanions {M$_x$O$_y$} such as the decavanadate [V$_{10}$O$_{28}$]$^{2-}$, and species incorporating other elements or small molecules, such as [SiW$_{10}$O$_{36}$]$^{8-}$. Oxides of d^0 metals form [MO$_4$]$^{n-}$ anions ($n = 2$ or 3) in strongly basic solution, which protonate and condense (eliminating water) forming M–O–M linkages as the pH is lowered to produce polyoxometallates, with known structures extending to extremely large units with over a hundred metal ions present. In the presence of anions such as phosphate, heteropolyoxo-metallates can form, incorporating additional elements, such as [PW$_{12}$O$_{40}$]$^{3-}$.

Clusters and nanomolecules are now a major research focus in chemistry, meaning that it is both appropriate and necessary to introduce them. Nevertheless, we have introduced just a few categories and examples above, as the complexity of structures and physical properties in this field makes further elaboration inappropriate for an introductory textbook. Thus, we shall avoid further discussion here and return to more basic aspects of coordination chemistry. However, we will return to discuss key synthetic aspects related to the formation of *polyoxometallates* and *zeolites* and their characterisation in Chapter 6.

4.6   Defining Shape

The above discussion of shape relies on our ability to actually identify and prove a shape. We have seen some examples of how this has been done in the past, using simple but revealing experiments that allow inferences to be drawn, on which base further experimental evidence has allowed structures to be well defined. Nowadays, we have moved into an era where we have available methods that allow us to define structure and shape with startling clarity and certainty in many cases. It is appropriate that we identify aspects of this methodology.

Three-dimensional structure can be probed by a wide variety of spectroscopic methods. Because most techniques tend to give limited or highly focussed outcomes, subject to some interpretation, they have sometimes been called 'sporting methods' – in other words, no certainty guaranteed! Nevertheless, specific techniques can be revealing; for example, the technique of mass spectrometry can provide information on the molecular mass of products in cluster formation reactions which can assist greatly in defining species formed. Where a crystalline solid can be isolated, the technique of single-crystal X-ray crystallography offers a highly accurate method of defining atom-by-atom location in three dimensions – a very unsporting, but popular and revealing physical method. These and other physical methods are discussed in some detail in Chapter 7.

It is also possible to employ pure computational methods (sometimes called *in silico* methods) to predict shape, or at least isomer preferences and relative stability, for complexes. The simplest approach to modelling employs *molecular mechanics*; this relies on a classic model that treats atoms as hard spheres with bonds as springs. This is introduced in Chapter 7. More sophisticated approaches are growing in popularity and capacity, such as density functional theory (DFT), a quantum-mechanical modelling method for calculating electronic structure developed by Nobel laureates Walter Kohn and Pierre Hohenberg in 1964, when they showed that the average density of electrons at all points in space is sufficient to uniquely determines the total energy and hence other properties of the system. Molecular modelling promises to provide excellent predictive capacity in the future without the need for laboratory synthesis, at least in the initial stages. However, laboratory-free chemistry is still far off, and synthesis and product identification remains the essence of chemistry.

Concept Keys

Coordination number and shape of complexes are influenced by the number of valence electrons on the metal ion, metal ion size and preferred coordinate bond lengths, inter-ligand repulsions and the shape and rigidity of the ligand.

Coordination numbers vary from 1 up to 9, and in some cases, reach into the 10s. For *d*-block metals, coordination numbers from 2 to 9 may be met, with 4–6 most common; for *f*-block metals coordination numbers above 6 and up to 14 may be observed, with 7–9 most common.

Shapes predicted by a simple amended VSEPR model are observed, but some others are found as well. Interconversion between shapes in a particular coordination number can occur, and many complexes are non-ideal in shape, showing distortion away from one limiting shape towards another.

Isomerism – the presence in a particular complex of a number of different spatial arrangements of atoms – is common in complexes.

The two key types of stereoisomers are positional or geometric isomers (such as *cis* and *trans*) and, where a complex is asymmetric or dissymmetric, optical isomers (such as Δ and Λ). The general class of isomers is diastereomers; where a complex is asymmetric or dissymmetric (optically active), a diastereomer will have a non-superimposable mirror image, called an enantiomer.

The absence of any plane of symmetry is a key requirement for a diastereomer having an enantiomer (and hence being chiral). Thus 'flat' molecules, such as those of square-planar shape, do not exist as optical isomers; octahedral complexes may be chiral, depending on the type and arrangement of ligands.

The possible existence of a set of geometric isomers does not mean that all are seen experimentally. They will differ in strain energy and thermodynamic stability, and as few as a single one may be isolable.

Ligand shape has an impact on the resultant complex shape, the number of possible isomers that could form, and the thermodynamic stability of complexes formed.

Apart from the formation of coordination complexes, some molecules bind other molecules or complexes by weaker non-covalent means, such as strong hydrogen bonding, forming outer-sphere (host–guest) complexes.

Some coordination complexes may contain more than one metal ion and employ bridging ligands between metal ions or even direct metal–metal bonding. Clusters are covalent molecular assemblies incorporating at least three metal ions, and these adopt a wide range of shapes and sizes. Large clusters may, as a result of their size, be classified as nanomolecules.

Further Reading

Atkins, P., Overton, T., Rourke, J., Weller, M. and Armstrong, F. (2006). *Shriver & Atkins Inorganic Chemistry*, 4th ed. Oxford: Oxford University Press.
    This popular but lengthy general text for advanced students contains some clearly-presented sections on shape and stereochemistry of coordination complexes.
Clare, B.W. and Kepert, D.L. (1994). *Coordination Numbers and Geometries*. In *Encyclopaedia of Inorganic Chemistry* edited by R.B. King. Vol. 2, p. 795, Chichester: Wiley.
    A detailed and readable early review, replete with examples.
Fergusson, J.E. (1974). *Stereochemistry and Bonding in Inorganic Chemistry*. New Jersey: Prentice-Hall.
    An old but valuable resource book for advanced students, but the depth and detail may concern others.
Gonzalez-Moraga, G. (1993). *Cluster Chemistry*. Springer-Verlag.
    A now venerable but still useful introduction for undergraduates to the chemistry of transition metal and main group molecular clusters, addressing synthesis, structure, bonding and reactivity.
Sharma, A. and Oza, G. (2023). *Nanochemistry: Synthesis, Characterization and Applications*. CRC Press.
    A recent book providing a useful overview of the diverse area of nanomaterials from a chemical perspective.
Steed, J.W. and Atwood, J.L. (2022). *Supramolecular Chemistry*, 3rd ed. Wiley.
    This advanced and lengthy but still readable text provides a comprehensive coverage of the field, for those who wish to explore the chemistry behind this class of compounds.
von Zelewsky, A. (1996). *Stereochemistry of Coordination Compounds*. Wiley.
    A mature but detailed coverage of topology in coordination chemistry, with a good many clear illustrations; a useful, though more advanced, resource book.

5 Stability

5.1 Making a Stable Assembly

Stability is something we all seek to achieve – molecules included. For molecules, stability is a relative term since it depends on the environment applying. Consider an aquated metal ion in aqueous solution under an inert atmosphere; in the absence of any added competing ligands, it is usually perfectly stable. The only reaction it can undergo is exchange between coordinated and free water, which, if it occurs, does not alter the coordination sphere at all. However, when ligands that bind strongly to the metal are added, a reaction may occur, leading to new complex species. How much of each complex species exists in solution depends on ligand preference and the stability of the various species. Two key aspects are involved in the above situation. The mixture changes to achieve the energetically most favourable products under the conditions applying, and this process does not occur instantaneously. The first aspect is governed by *thermodynamics*, the second by *kinetics*. These two aspects of stability are now discussed.

5.1.1 Thermodynamic Stability

A system in the process of achieving thermodynamic stability can be recognised readily in many cases if a change in observable properties occurs. For example, addition of ammonia to a solution containing Cu_{aq}^{2+} causes a rapid change in colour. This is a signal that a new complex species is forming, and it is apparent from the observation that new complexes between Cu(II) and ammonia form to a great extent, even for low concentrations of ammonia – evidently, the Cu(II)–ammonia complex is more stable than the Cu(II)–water complex. We have discussed some of the reasons for this type of outcome in Chapter 3. Now, we need to understand the processes involved so that we can quantify these observations. One important point arising from earlier discussion was that the amount of a complex species existing cannot be predicted simply from the ratio of the added ligand to solvent molecules because ligand preferences override purely statistical aspects. Establishing the actual composition in solution experimentally and expressing it in terms of some general parameter are thus important.

The process involved in the copper(II)–ammonia solution is a *ligand substitution* process, where one ligand is replacing another. In general, we can represent this, for reaction in an aqueous solution with a neutral monodentate ligand (L) at this stage, by the following equation:

$$M(OH_2)_x^{n+} + L \rightleftharpoons M(OH_2)_{(x-1)}L^{n+} + H_2O \qquad (5.1)$$

Introduction to Coordination Chemistry, Second Edition. Paul V. Bernhardt and Geoffrey A. Lawrance.
© 2025 John Wiley & Sons Ltd. Published 2025 by John Wiley & Sons Ltd.
Companion website: www.wiley.com/go/coordinationchemistry2e

Since we have written this reaction as an equilibrium, it is possible to write an equilibrium constant for this reaction, namely

$$K = \frac{[M(OH_2)_{(x-1)}L^{n+}][H_2O]}{[M(OH_2)_x{}^{n+}][L]} \quad (5.2)$$

in which the sets of square brackets in this case refer to the concentration of the species. In fact, thermodynamics tells us that the equilibrium constant above should be written in terms of activities (a), not concentrations. The relationship between these two is

$$a_S = [S]\gamma_S \quad (5.3)$$

where γ_S is the activity coefficient, which has a value $\gamma_S = 1$ in extremely dilute solutions but for practical solutions has a value of less than 1, caused by the influence of other solute species present on the behaviour of a particular solute molecule. Because activities are difficult to determine and vary with concentration and composition of the solution, it is convenient to work with concentrations by assuming that $\gamma_S = 1$ (pure water has this value itself). It is then important to quote the experimental conditions when reporting equilibrium constants, as the value will change with conditions. Fortunately, the size of the change with conditions usually met, where concentrations of complexes may vary between 0.001 and 0.1 M, is not large in the context of what we seek to determine. Of more concern is devising processes that permit accurate determination of the concentrations of species present, which is not a trivial task by any means.

Because water has a concentration of approximately 55 M, it varies by only trivial amounts for reactions of dilute species. As a result, it is convenient, and traditional, to leave out the solvent term and usually also (for simplicity in presentation) to ignore coordinated water molecules. We shall adopt this representation henceforth – but do not assume we are dealing with 'bare' metal ions! This reduces Eq. (5.1) to

$$M^{n+} + L \rightleftharpoons ML^{n+} \quad (5.4)$$

and thus

$$K = \frac{[ML^{n+}]}{[M^{n+}][L]} \quad (5.5)$$

The equilibrium constant K in this case is called the complex *formation constant* or *stability constant* since it is measuring formation of a metal complex and defining its thermodynamic stability. Experimentally, because we measure K values under non-ideal (in a thermodynamic sense) conditions, the term 'constant' here is not absolutely correct, as discussed above. Remember that it is defined only under the particular experimental conditions employed in reality, although it is fairly true to say that the value varies in only a limited way across the range of conditions that we are most likely to apply.

There is also a direct relationship between the stability constant and the change in Gibbs free energy ($\Delta G°$, in kJ mol^{-1}) of a reaction (developed by Josiah Willard Gibbs in the 1870s), expressed in terms of the relationship:

$$\Delta G° = -2.303 \cdot R \cdot T \cdot \log_{10}K \quad (5.6)$$

where R is the gas constant and T the temperature in Kelvin. This means that the higher is K, the more negative is the free energy of the reaction (and a negative value means a process is spontaneous). We feel this usually as a release of heat on complexation because of the direct relationship between $\Delta G°$ and reaction enthalpy change ($\Delta H°$), although the reaction entropy change ($\Delta S°$) also needs to be considered, as we shall do later. The fundamental relationship is

$$\Delta G° = \Delta H° - T \cdot \Delta S° \tag{5.7}$$

Examination of Eq. (5.5) shows that a large value of K means a high concentration of ML^{n+} relative to M^{n+} and L; in other words, a large K means a strong preference for complex formation. The size of K with metal complexes is usually so large that we tend to report $\log_{10} K$ values, for ease of use; obviously, it is simply easier to refer to, for example, a (log)K of 7.5 rather than a K of 3.16×10^7 (with its associated units; log values do not have units).

It is more common for a metal to bind several ligands of the same type, and this process occurs in succession. We can represent sequential substitution steps for formation of ML_x^{n+} by a series of equilibria (5.8):

$$M^{n+} + L \rightleftharpoons ML^{n+} \qquad\qquad (K_1)$$
$$ML^{n+} + L \rightleftharpoons ML_2^{n+} \qquad\qquad (K_2)$$
$$ML_2^{n+} + L \rightleftharpoons ML_3^{n+} \qquad\qquad (K_3)$$
$$\vdots \qquad \vdots \tag{5.8}$$
$$\vdots \qquad \vdots$$
$$ML_{(x-1)}^{n+} + L \rightleftharpoons ML_x^{n+} \qquad\qquad (K_x)$$

For *each* step, a *stepwise stability constant* can be written of the form

$$K_x = \frac{[ML_x^{n+}]}{[ML_{(x-1)}^{n+}][L]} \tag{5.9}$$

The *overall* reaction occurring through combination of the above steps can be written as

$$M^{n+} + xL \rightleftharpoons ML_x^{n+} \tag{5.10}$$

for which we can define an overall stability constant (β_x)

$$\beta_x = \frac{[ML_x^{n+}]}{[M^{n+}][L]^x} \tag{5.11}$$

This overall stability constant simply defines the formation of the overall or 'final' complex, where all of one ligand (here, water) can be considered replaced by another (L); it does not infer anything about the mechanism of the process.

It is a straight-forward task to show that the overall stability constant equals the product of all stepwise stability constants:

$$\beta_x = K_1 \cdot K_2 \cdot K_3 \cdot \ldots \cdot K_x \tag{5.12}$$

$$\log \beta_x = \log K_1 + \log K_2 + \log K_3 + \cdots + \log K_x \tag{5.13}$$

The overall stability constant is dealt with more fully in Section 5.1.3.

You may already be familiar with another form of equilibrium constant, the *acid dissociation constant* (K_a), which is so named because it numerically reports the dissociation of an acid (HX) into its ions (H^+ and X^-):

$$K_a = \frac{[H^+][X^-]}{[HX]} \qquad (5.14)$$

This is, in effect, an *instability* (dissociation) constant since it describes a break-up rather than a formation process; the equation for the constant is inverted relative to the form met for the stability constant of interest here (compare Eq. (5.14) with Eq. (5.5)). However, expressing the acid dissociation constant as $pK_a = -\log_{10}K_a$ makes it similar to a complex formation or stability constant expressed simply as $\log_{10}K$. We shall focus here on complex formation constants.

In the above discussion, we have tended to use a neutral ligand L throughout, as this avoids dealing with charge variation in equations. This does not imply that the outcome is in any real way different for a neutral or anionic ligand; the same forms of equations apply. While we can perhaps conceive more easily how an anionic ligand X^- may be attracted to and form a coordinate bond with a metal cation, recall that neutral ligands will often display polarity through having connected atoms of differing electronegativity. Therefore, an H–N bond, for example, may be considered a $^{\delta+}H–N^{\delta-}$ entity, and it is the negative end of the dipole (or, in effect, the lone pair) that attaches to the metal cation. Consequently, one might anticipate that neutral molecules that are more polar will make better ligands. Most heteroatoms in molecules that act as efficient ligands to metal ions carry substantial partial negative charge.

5.1.2 Factors Influencing Stability of Metal Complexes

We can identify several factors that contribute to the size of stability constants, and it is appropriate to summarise and review these effects. In doing so, we will also attempt to divide the effects into those more linked with the metal and those more associated with the ligand since both contribute to the partnership, and inevitably each contributes to the resulting metal–ligand assembly.

5.1.2.1 *Factors Dependent on the Metal Ion*

5.1.2.1.1 *Size and Charge*

Several factors based on pure electrostatic arguments contribute to a large stability constant. First, since we are bringing together a positively charged metal ion and either anionic ligands or polar neutral ligands carrying a partial negative charge on their donor atoms, there is certainly going to be a purely electrostatic contribution. We can see this experimentally, as shown in Table 5.1 for a series of hydroxido complexes. As the charge on cations, all of similar size, varies from +1 to +3 across the series, the size of $\log K_{OH}$ increases, remembering that we are looking at orders of magnitude differences in K with every unit increase in $\log K_{OH}$. The $\log K_{OH}$ value is the same as the pK_b (basicity constant) value of the $(MOH)^{(n-1)+}_{aq}$ species.

When the metal ion charge is fixed but the metal ionic radius is increased, the charge density (charge/volume) decreases. This means a less effective attractive force for the ligand applies, which leads to a decrease in $\log K_{OH}$ (Table 5.1) for the three Group 2 congeners. The relative importance of charge and ionic radius is

Table 5.1 The influence of metal ion charge, size and (for *d*-block elements) crystal field effects on experimentally determined stability constants of their complexes, illustrated with hydroxide and ammonia ligands.

Ionic charge effects: $M^{n+}_{aq} + OH^-_{aq} \rightleftharpoons M(OH)^{(n-1)+}_{aq}$

M(OH)	Li(OH)	Mg(OH)$^+$	Al(OH)$^{2+}$	
log K_{OH}	0.1	2.6	8.9	
charge (n)	+1	+2	+3	

Metal ion size effects: $M^{2+}_{aq} + OH^-_{aq} \rightleftharpoons M(OH)^+_{aq}$

M(OH)	Mg(OH)$^+$	Ca(OH)$^+$	Ba(OH)$^+$	
log K_{OH}	2.6	1.4	0.8	
r (Å)	0.72	1.00	1.35	

Crystal field effects: $M^{2+}_{aq} + 6NH_3 \rightleftharpoons [M(NH_3)_6]^{2+}$

M	Co^{2+}	Ni^{2+}	Cu^{2+}	Zn^{2+}
log β	5.0	7.8	13.0	9.0

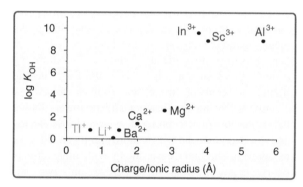

Figure 5.1

Variation of stability constant (log K_{OH}) with the charge/radius ratio for some M(OH)$^{(n-1)+}_{aq}$ complexes.

also seen if one plots this stability constant against the charge/ionic radius ratio for a range of M(OH)$^{(n-1)+}_{aq}$ complexes (Figure 5.1). Obviously, the effect of charge appears more important than that of ionic radius. The monopositive and dipositive ions form only weak complexes with OH$^-$ while the trications form complexes that are more stable by several orders of magnitude. Of course, our simple model of the cations as hard spheres applied in this analysis is also imperfect, even though the radii have been determined experimentally from interatomic dimensions of simple ionic solids by X-ray crystallography. There are other factors such as the degree of covalency (electron-sharing) in these bonds, which is not the same in this case, but the clustering of ions in Figure 5.1 according to their charge illustrates the overriding influence of charge on stability (strength) of the M–OH bond.

Although our focus currently is on the metal, it should be recognised that the size (or molar volume) of the ligand plays a role in the electrostatic effect on the stability of a complex, particularly for anionic ligands. This is sensible since the ligand can be assigned a charge density (when anionic) in the same way that we have done for the cation, and obviously an anion with a high charge density should form stronger complexes from an electrostatic perspective. This is best illustrated by examining halide monatomic anions, where a spherical volume has some meaning. Recall that cations display smaller radii than their parent element, whereas monatomic anions display larger radii than their parent element. This behaviour will impact on the charge density, which is typically lower for anions, though still important, as illustrated in the following example. With F^- (radius 1.33 Å, where 1 Å = 100 pm) and Cl^- (radius 1.81 Å), Fe^{3+} forms complexes with $\log K_X$ of 6.0 and 1.3, respectively, reflecting the greater charge density of the former. Fluorine and in fact all elements of the second row of the periodic table are distinctly different to their third and fourth row congeners (elements from the same group), a concept known as the *Uniqueness Principle*. The atomic/ionic radii of elements of the second row of the periodic table are especially small, which gives fluoride its exceptionally high charge density and strong ionic bonds. The concept of ionic radius becomes diffuse when we move to molecular anions, however, whose shape may not be anywhere near spherical: e.g. NO_2^- and SCN^-. However, at least for large reasonably symmetrically shaped and thus pseudo-spherical anions like ClO_4^-, the very low stability of its complexes is fairly consistent with an electrostatic model since this ion has a much greater radius than simple halide ions of the same charge.

5.1.2.1.2 *Metal Class and Ligand Preference*

We have examined the hard/soft acid/base 'like prefers like' concept as it applies to metal–ligand binding already in Chapter 3, so it is simply necessary to revise aspects here. Electropositive metals (lighter and or more highly charged ones from the *s*-, *d*- and *f*-block such as Mg^{2+}, Ti^{4+} and Eu^{3+}, belonging to *Class A*) tend to prefer lighter and highly electronegative *p*-block elements (such as O-donors and F^-). Less electropositive metals (heavier and or lower-charged ones such as Ag^+ and Pt^{2+} belonging to *Class B*) prefer heavier and less electronegative *p*-block donors (such as P- or S-donors and I^-). A more significant M–L covalent contribution is asserted to apply in the latter case, along with other effects such as π back-bonding. Of course, in any situation where there are only two categories, there is an intermediate or 'grey' area of metals and ligands who do not sit easily in either set. A summary is given in Table 5.2, in which a large number of the metal ions and simple ligands you are likely to meet are given.

Table 5.2 Examples for both ligands and metals of 'hard' and 'soft' character, with some less clearly defined intermediate cases also included.[a]

Hard	Intermediate	Soft
Ligands		
F^- O^{2-} ^-OH OH_2 HOR $RCOO^-$ NH_3 NR_3 RCN Cl^- NO_3^- CO_3^{2-} SO_4^{2-} PO_4^{3-}	Br^- $R\mathbf{S}^-$ $\mathbf{N}O_2^-$ $\mathbf{N}_3^-$ $SC\mathbf{N}^-$ $H_5C_5\mathbf{N}$	$\mathbf{P}R_3$ $\mathbf{S}R_2$ $\mathbf{S}eR_2$ $A\mathbf{s}R_3$ $C\mathbf{N}R$ $\mathbf{N}C^-$ $\mathbf{S}CN^-$ $\mathbf{C}O$ I^- H^- R^-
Metal Ions		
Mo^{5+} Ti^{4+} V^{4+} Sc^{3+} Cr^{3+} Fe^{3+} Co^{3+} Al^{3+} Eu^{3+} Cr^{2+} Mn^{2+} Ca^{2+} Mg^{2+} Be^{2+} K^+ Na^+ Li^+ H^+	Fe^{2+} Co^{2+} Ni^{2+} Cu^{2+} Zn^{2+} Pb^{2+}	Cu^+ Rh^+ Ag^+ Au^+ Pd^{2+} Pt^{2+} Hg^{2+} Cd^{2+}

[a]For ambidentate ligands, the donor atom is in bold type.

5.1.2.1.3   Crystal Field Effects

The above simple effects are not the sole metal-centric influences on complex stability, as is evident from examining trends in experimental K values with a wide range of ligands. This is particularly apparent when one examines a series of transition metal 2+ ions forming complexes with the same ligand; data with ammonia appear in Table 5.1. Overall, between Mn^{2+} and Zn^{2+}, there is a general trend in K values, essentially irrespective of ligand, of the order:

$$Mn^{2+} < Fe^{2+} < Co^{2+} < Ni^{2+} < Cu^{2+} > Zn^{2+}.$$

The relative sizes of the cations vary in a similar fashion but in the reverse order, with radii following the trend:

$$Mn^{2+} > Fe^{2+} > Co^{2+} > Ni^{2+} > Cu^{2+} < Zn^{2+},$$

so the change is in part (but only in part) assignable to charge/radius ratio ideas. Across this series of metal ions, the number of protons in the nucleus is increasing, but shielding by the electron clouds is imperfect so that as nuclear charge (atomic number, Z) increases, there is a progressively higher *effective nuclear charge* (Z_{eff}) 'felt' by the ligands, despite the common formal charge. This also contributes to a steady increase in the stability constants from left to right across the periodic block. However, there is an obvious discontinuity between Cu^{2+} and Zn^{2+}, with a drop that is consistently observed regardless of ligand type. The source of this effect is tied to the presence of an incomplete set of d-electrons, except for Zn^{2+}, and we shall expand on this below.

5.1.2.1.4   The Natural Order of Stabilities for Transition Metal Ions

For transition metal ions with incomplete sets of d-electrons, there is a contribution to stability from the *crystal field stabilization energy* (CFSE), whereas for d^{10} metal ions (such as Zn^{2+}), with a full set, there is no such stabilization energy. Crystal field stabilization energies of metal ions in complexes have been asserted to have a key influence on values of complex formation constants (K) for transition metals, reflected in the experimentally determined stabilities for the series of metal ions from Mn^{2+} across to Zn^{2+}. This effect is in addition to the general 'upward' linear trend from left to right across the row with increasing atomic number Z, associated with imperfect shielding of the nucleus, discussed earlier, and leads to what is often called the *natural order of stabilities* (also called the Irving–Williams Series since the trend was first reported by Harry Irving and Bob Williams in 1953). This experimentally based finding states that water-stable M^{2+} octahedral first-row transition complexes exhibit the 'natural' order of stabilities presented above regardless of the ligand, and this is exemplified in Figure 5.2.

This trend identifies copper(II) as the metal ion that forms the most stable complexes, irrespective of the ligand. The trend is preserved for the series of metal ions with didentate chelates even when the type of donor atom or the size of the chelate ring is varied (Figure 5.2). You may wonder why only the select range from Mn^{2+} to Zn^{2+} ions is included and not the entire transition metal series. The earlier transition metal M^{2+} (e.g. V^{2+} and Cr^{2+}) ions spontaneously oxidise to their M^{3+} or M^{4+} ions, making measurement of their complex formation constants difficult, so they are not included, but theoretically, the same concepts identified in the Irving–Williams series apply to all of the transition metals.

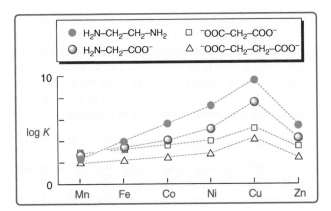

Figure 5.2

The *natural order of stabilities* in operation: consistent variation of the measured stability constant with different chelate ligands for the series of transition metal(II) ions from manganese across to zinc.

Table 5.3 Influences on and trends in the crystal field stabilisation effect (CFSE).

Order of stabilities (K)	$Mn^{2+} < Fe^{2+} < Co^{2+} < Ni^{2+} < Cu^{2+} > Zn^{2+}$					
Trend in CFSE	$Mn^{2+} < Fe^{2+} < Co^{2+} < Ni^{2+} > Cu^{2+} > Zn^{2+}$					
Electron configuration	d^5	d^6	d^7	d^8	d^9	d^{10}
High-spin CFSE value	0	$0.4\Delta_o$	$0.8\Delta_o$	$1.2\Delta_o$	$0.6\Delta_o$	0

In this series, we are dealing with octahedral systems for high-spin complexes. Only for Cu^{2+} do we have an especially distorted octahedral shape (from the Jahn Teller Effect discussed in Section 2.3.5). In general, electrons placed in the lower t_{2g} level are stabilised by $0.4\Delta_o$, whereas those placed in the upper e_g level are destabilised by $0.6\Delta_o$. Overall, the calculated CFSE follows the order shown in Table 5.3, where the trend is compared with the experimentally based order of stability constants.

The CFSE is highest for Ni^{2+}, and the ions at each end each have zero CFSE as a result of half-filled or filled d-shells. This behaviour closely follows the observed (experimental) behaviour of K with varying M^{2+} (Figure 5.2) *except* for Cu^{2+}. We should remember that the effective nuclear charge also contributes to the strength of the metal–ligand bond, and this continues to increase from Mn^{2+} to Zn^{2+} (independent of the number of d electrons). The net result is that the greater effective nuclear charge of Cu^{2+} over Ni^{2+} overrides the CFSE difference; in fact, both Ni and Cu form quite stable complexes, as can be seen in Figure 5.2. Thus, the influence of the d-electron configuration, through the CFSE, is important in understanding the natural order of stabilities for light transition metal ions; it links the basic crystal field theory to experimental observations effectively.

5.1.2.2   Factors Dependent on the Ligand

5.1.2.2.1   Base Strength

One aspect of complex formation that became apparent at an early date was the relationship between the Brønsted base strength of a ligand and its ability to form strong complexes. This is sensible, in the sense that the Brønsted base strength is a measure of proton affinity. Making a substitution of H^+ by M^{n+} (one Lewis acid for another) seems reasonable, permitting basicity to define likely complex strength. From this perspective, the more basic NH_3 should be a better ligand than PH_3 or H_2O, and F^- (a weak base) should be superior to all remaining halide ions that have no basic properties. This predicted behaviour holds quite well for *s*-block, lighter (first-row) *d*-block and *f*-block metal ions, which have been defined as 'hard' or *Class A* metal ions. We can say with some certainty that the greater is the base strength of a ligand (that is, its affinity for H^+), the greater is its affinity for (and hence stability of complexes of) at least *Class A* metal ions.

Unfortunately, when one examines heavier metal ions and those in low oxidation states, the behaviour is not the same (hence their definition as 'soft' or *Class B* metals). Obviously, there are electrostatic contributions still applying, but other influences are now important. A classic example is the reaction of the relatively large, low-charged Ag^+ ion in forming AgX, (where X^- is a halide ion) which follows the order ($\log_{10} K$ values in parentheses):

$$F^-(0.3) < Cl^-(3.3) < Br^-(4.5) < I^-(8.0).$$

Clearly, this does not follow the trend expected on electrostatic grounds, which should be opposite to that observed, as ion size increases from F^- to I^-. The experimental trend is thought instead to reflect increased covalent character in the Ag–X bond in transition from fluoride to iodide.

The steric bulk of ligands also introduces an influence that can counteract pure base strength effect; the ionic radius for halide anions increases down the column, with F^- (1.19 Å), Cl^- (1.67 Å), Br^- (1.82 Å) and I^- (2.06 Å). This steric bulk effect has been discussed earlier for NR_3 compounds also (Chapter 3.2.1). Substituted pyridines present another example of this influence; 2,6-dimethylpyridine is a poor ligand due to the location of the two methyl groups on either side of the pyridine N-donor atom, despite a similar base strength to unsubstituted pyridine.

5.1.2.2.2 *Chelate Effect*

From a purely thermodynamic viewpoint, the equilibrium constant is reporting the heat change (the enthalpy change, $\Delta H°$) in the reaction and the amount of disorder (called the entropy change, $\Delta S°$) resulting from the reaction, related to the reaction free energy change as defined in Eq. (5.7). The greater the amount of energy evolved in a reaction, the more stable are the reaction products; this heat change can sometimes be felt when holding a test tube in which a reaction has been initiated and is certainly experimentally measurable to a high level of accuracy through calorimetry. Further, the greater the amount of disorder resulting from a reaction, the greater is the entropy change and the greater the stability of the products. Disorder is a harder concept to grasp than some, but think of it in terms of particles involved in a reaction – if there are more particles present at the end of the reaction than at the start, or even if those present at the end are less structured or restricted in their locations, there is increased disorder and hence a positive entropy change. Again, this is experimentally measurable, but not as directly as a simple heat change associated with reaction enthalpy. In the coming together of two oppositely charged ions to form a complex, there is both a release of heat (negative enthalpy change, $\Delta H°$) and a release of solvent molecules both coordinated to the metal ion and more loosely associated in the second coordination sphere (positive entropy change, $\Delta S°$). Moreover, the magnitudes of both terms increase with charge on the metal ion.

When we employ molecules as ligands where they offer more than one donor group capable of binding to metal ions, there is the strong possibility that more than one donor group of such a ligand will coordinate to the one metal ion. We have entered the realm of polydentate ligands which were discussed in detail in Chapters 3 and 4. Just because a ligand offers two donor groups does not mean that both can coordinate to the one metal ion. For example, the *para* and *ortho* diaminobenzene isomers can both act as didentate ligands, but the former must attach to two different metal ions because of the direction in which the rigidly attached donors point. Only the *ortho* form has the two donors directed in such a way that they can occupy two adjacent coordination sites around a metal ion. This isomer alone has achieved *chelation*. The same behaviour is displayed by the linked pyridine molecules 2,2′-bipyridine and 4,4′-bipyridine (Figure 5.3).

In general, chelation is beneficial for complex stability, and chelating ligands form stronger complexes than comparable monodentate ligand sets. For example, consider the following:

$$[Ni(OH_2)_6]^{2+} + 6NH_3 \rightleftharpoons [Ni(NH_3)_6]^{2+} + 6H_2O \qquad (5.15)$$

$$\{(\textbf{7 \textit{molecules}}) \rightarrow (\textbf{7 \textit{molecules}})\}$$

$$\beta_6 = \frac{[Ni(NH_3)_6{}^{2+}]}{[Ni(OH_2)_6{}^{2+}][NH_3]^6} = 4.0 \times 10^8 \text{ M}^{-6}, \log \beta_6 = 8.6$$

$$[Ni(OH_2)_6]^{2+} + 3NH_2CH_2CH_2NH_2 \rightleftharpoons [Ni(NH_2CH_2CH_2NH_2)_3]^{2+} + 6H_2O$$
$$\qquad (5.16)$$
$$\{(\textbf{4 \textit{molecules}}) \rightarrow (\textbf{7 \textit{molecules}})\}$$

$$\beta_3 = \frac{[Ni(NH_2CH_2CH_2NH_2)_3{}^{2+}]}{[Ni(OH_2)_6{}^{2+}][NH_2CH_2CH_2NH_2]^3} = 2 \times 10^{18} \text{ M}^{-3}, \log \beta_3 = 18.3$$

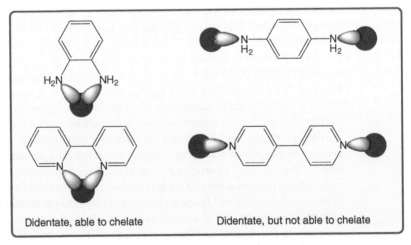

Didentate, able to chelate   Didentate, but not able to chelate

Figure 5.3
Lone pairs of *ortho*-diaminobenzene (top left) and 2,2′-bipyridine (bottom left) and *para*-diaminobenzene (top right) and 4,4′-bipyridine (bottom right) are arranged in space so that only the former pair can chelate to a single metal centre. The latter pair can only bind to two separate metal centres.

The three didentate chelate ethane-1,2-diamine ligands occupy six sites, as do the six monodentate ammonia molecules, and saturated nitrogen donors are involved in each case; however, it is the chelation of the former that causes the enhanced stability. However, direct comparison of the overall stability constants reported above is inappropriate since, as a result of the different terms in the equations for each β_n, the units are not equivalent, being M^{-3} for β_3 and M^{-6} for β_6. A better but still not perfect approach is to examine the experimental results for direct ligand exchange of two ammines by one ethane-1,2-diamine, as in reaction (5.17).

$$[Ni(NH_3)_6]^{2+} + H_2NCH_2CH_2NH_2 \rightleftharpoons$$

$$[Ni(H_2NCH_2CH_2NH_2)(NH_3)_4]^{2+} + 2NH_3$$

$$\{\log K = 2.4 \quad (\text{and } \Delta G^\circ = -13.7 \, kJ \, mol^{-1})\} \tag{5.17}$$

It is obvious that there is a strong driving force for this ligand replacement reaction, with formation being thermodynamically favourable, as seen by the negative ΔG° value. Introducing one $H_2NCH_2CH_2NH_2$ chelate in place of two NH_3 (ammine) ligands leads to a complex that is ~300-fold more stable than the analogue with two ammonia molecules bound; this is enhanced substantially once all sites around the octahedron are occupied by chelated amine donors, as exemplified by the overall reaction (5.18).

$$[Ni(NH_3)_6]^{2+} + 3H_2NCH_2CH_2NH_2 \rightleftharpoons [Ni(H_2NCH_2CH_2NH_2)_3]^{2+} + 6NH_3$$

$$\{\log \beta_3 = 9.3 \quad (\text{and } \Delta G^\circ = -53 \, kJ \, mol^{-1})\} \tag{5.18}$$

For this overall reaction broken down into its enthalpy and entropy terms (Eq. (5.7)), the components of ΔG° are $\Delta H^\circ = -16.8 \, kJ \, mol^{-1}$ and (at 298 K) $-T \cdot \Delta S^\circ = -36.2 \, kJ \, mol^{-1}$. Both contribute to the overall negative ΔG°, but the contribution from the entropy-based term $T \cdot \Delta S^\circ$ is the major component. The smaller ΔH° contribution is in part associated with slightly increased crystal field stabilisation (stronger Ni –N bonds) upon reaction, seen experimentally as a change in the maximum in the visible region of the electronic spectrum. We can interpret this reaction as dominated by a favourable entropy change associated with chelation (i.e. an increase in disorder and 'particles' or participating molecules), as exemplified in (5.15) and (5.16).

Another way of viewing the process, based on incoming donor group locations in sequential substitution steps, leads to an equivalent interpretation. Adding the first amine of an ethane-1,2-diamine molecule or adding an ammonia molecule in place of an existing ligand should occur with similar facility, as they are both strong basic N-donor ligands, both approaching from undefined initial locations. However, the next substitution step in each molecular assembly is distinctly different, as there is no special advantage for a second ammonia entering the coordination sphere (as its initial location, as was the case with the first, is not defined or fixed), whereas the second amine group of the partly bound ethane-1,2-diamine molecule is required to be closely located to a second substitution site as a result of the anchoring effect of the first substitution. Less translational energy is required, and the coordination step is more probable.

Some reactions appear to be driven almost entirely by a positive entropy effect. For the two related reactions (5.19) and (5.20), each leading to the formation of four Cd–N bonds, the ΔH° values are equivalent (–57 kJ mol^{-1} for the monodentate and –56 kJ mol^{-1} for the didentate chelate), and only the $-T \cdot \Delta S^\circ$ terms differ

(+20 kJ mol^{-1} for the monodentate and –4 kJ mol^{-1} for the didentate chelate), favouring the latter.

Monodentate : $Cd^{2+}_{aq} + 4(H_2NCH_3) \rightleftharpoons [Cd(H_2NCH_3)_4]^{2+}$ (5.19)

$(\log \beta_4 = 6.5)$

Chelate : $Cd^{2+}_{aq} + 2(H_2NCH_2CH_2NH_2) \rightleftharpoons [Cd(H_2NCH_2CH_2NH_2)_2]^{2+}$ (5.20)

$(\log \beta_2 = 10.5)$

However, it is not correct to assume that a favourable entropy change always drives ligand substitution reactions, as there are limited examples where the overall entropy change opposes the reaction, which is driven rather by a more substantial negative enthalpy change. Factors that can contribute to the latter include crystal field stabilisation energy variation, reduction in electrostatic repulsion terms on reaction, and solvation and hydrogen-bonding changes that favour the reaction. However, entropy-based effects are both easier to visualise and comprehend and more often the dominant contributor and so tend to attract greater attention.

As a general rule, the higher the denticity (or number of bound donor atoms) of a ligand, the higher will be its stability constant with a particular metal ion. Invariably, for occupancy of the same number of coordination sites around a metal ion, the outcome is greater stability (larger stability constants) for the chelate system over a set of monodentates [i.e. β(chelate) > β(monodentates)]. As the number of coordinated donor atoms for a single ligand increases, almost invariably the size of the stability constant likewise increases. This effect is assigned in large part, as discussed above, to entropy considerations, associated with favourable entropy change (or disorder) in the system as a chelate ligand replaces a set of monodentate ligands. Examples of stability constants and reaction energies for monodentate and analogous chelate ligands appear in Table 5.4. Recall, however, that direct comparisons of log β_n data for monodentate/chelate ligand substitution of the aqua metal ion will be complicated by the different metal/ligand stoichiometries of the reactions.

A more sophisticated aspect of chelation relates to the ligand's 'readiness' for coordination. This is illustrated in Figure 5.4, where one didentate chelate (1,10-phenanthroline, or phen) is rigidly fixed into an orientation immediately appropriate for chelation (that is, it is '*preorganised*'), whereas the other (2,2'-bipyridine or bipy) requires a 180-degree rotation of one ring (which costs energy) from its preferred

Table 5.4   Comparison of formation constants and reaction energies from the reaction of the nickel(II) ion $[Ni(OH_2)_6]^{2+}$ with monodentate and related didentate chelate ligands.

Monodentate	log β_n	$\Delta G°$ (kJ mol^{-1})	Didentate	log β_n	$\Delta G°$ (kJ mol^{-1})
with pyridine (py)			*with 2,2'-bipyridine (bipy)*		
n = 2× py	3.5	−20	n = 1× bipy	6.9	−39
4× py	5.6	−32	2× bipy	13.6	−78
6× py	9.8	−56	3× bipy	19.3	−110
with ammonia (NH$_3$)			*with ethane-1,2-diamine (en)*		
2× NH$_3$	5.0	−28	1× en	7.5	−43
4× NH$_3$	7.9	−44	2× en	13.9	−79
6× NH$_3$	8.6	−49	3× en	18.3	−104

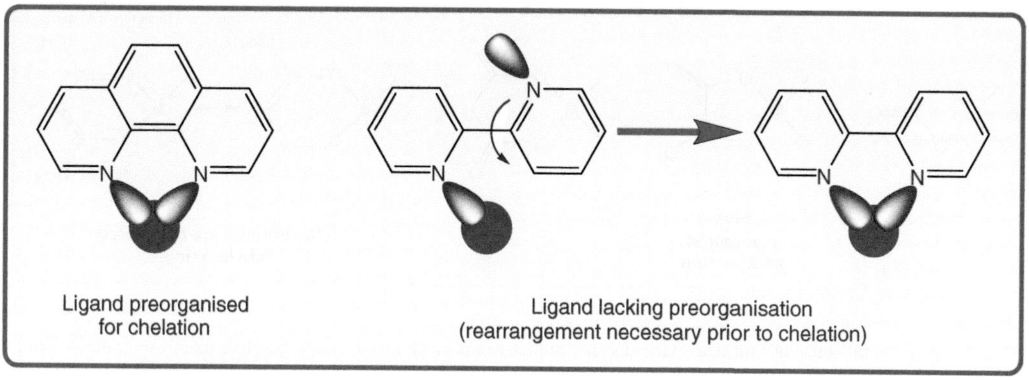

Figure 5.4

How a predefined appropriate ligand shape in 1,10-phenanthroline (phen) (*preorganisation*) provides chelation at less energy cost than in circumstances where energy-demanding ligand re-arrangement in 2,2′-bipyridine (bipy) must accompany chelation.

free ligand conformation prior to adopting a shape suitable for chelation. Usually, systems that are preorganised for binding metal ions exhibit larger stability constants than comparable systems where ligand rearrangement is required. This is the case for $[M(phen)_3]^{2+}$ complexes in comparison with their $[M(bipy)_3]^{2+}$ analogues, where the log β_3 values are consistently and significantly higher for the phen complexes: e.g. log $\beta_3([Co(phen)_3]^{2+}) = 19.9$, whereas log $\beta_3([Co(bipy)_3]^{2+}) = 16.0$.

5.1.2.2.3 Chelate Ring Size

Notwithstanding the above discussion, the *size* of the chelate ring also influences the size of the stability constant (an aspect we have already touched on earlier), being at its largest for five-membered rings and conjugated six-membered rings. Since we are constraining the metal by binding to two linked donor atoms, it can be hardly surprising that there is a relationship between the formed chelate ring size and stability. Chelate ring size may introduce a tension between the preferred metal–donor length and the preferred donor–donor 'bite' since perfect compatibility is rarely reached. Distortions of intraligand angles and bond distances have an influence on the stability of the assembly, as obviously a system under great strain is going to be unstable. Simply, the size of the stability constant depends on the number of atoms or bonds in the chelate ring. For saturated chelating ligands, five-membered rings are preferred for the lighter metal ions, with smaller or larger ring sizes being of lower stability (Figure 5.5). This can be understood to be a result of the assembly desiring the internal angles of the ligand and the bite angle subtended by the metal to be as close as possible to their preferred bond angles, e.g. 109.5° for an sp^3-hybridised C or N atom and 90° for an octahedral metal ion.

The exception to this observed preference for five-membered rings comes when unsaturated conjugated ligands are coordinated, where very stable complexes with six-membered chelate rings can exist with some light metal ions. This is associated with a shift from tetrahedral to trigonal planar geometry of ring carbons, with associated opening out of preferred angles around each carbon (from 109.5° to 120°) leading to a more appropriate ligand 'bite'. When the size of the chelate ring increases substantially, there is no particular stability arising from chelation. As a consequence, most examples of the chelate effect feature chelates with five or six

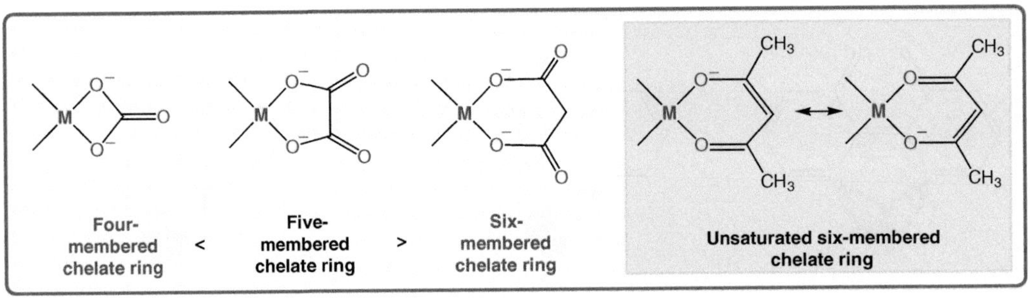

Figure 5.5

Influence of ring size on the stability of first-row *d*-block metal complexes; five-membered saturated rings are more stable. The exception is for chelation of unsaturated conjugated ligands, where six-membered rings (such as the acetylacetonate ligand illustrated at right) form complexes of enhanced stability.

atoms defining the ring; indeed, with lower or higher chelate ring size, the chelate effect rapidly decreases.

5.1.2.2.4   *Steric Strain*

As ligands can vary so much more in size and shape than metal cations, there must be other consequences, including simply size effects in terms of 'fitting' around the central atom. These effects of ligand bulk, resulting from molecules being required to occupy different regions of space to avoid 'bumping' against each other when confined around a central metal ion, tend to be termed *steric* effects; we have touched on these briefly already in Section 5.1.2.2.1. As a general rule, the bulkier a molecule, the weaker the complex formed when there is a set of ligands involved. Therefore, one would expect NH_3 to be a superior ligand to $N(CH_3)_3$ despite both being N-donors, and this behaviour is experimentally observed in solution. In the gas phase, where there is no metal ion solvation and usually only low coordination number complexes form due in part to a low probability of metal–ligand encounters, discrimination in complex stability based more on base strength than steric bulk may apply. Thus, $N(CH_3)_3$ (pK_a 9.8), a stronger Brønsted base than NH_3 (pK_a 9.2), forms the stronger complexes in the gas phase, contrary to behaviour in aqueous solution. However, gas phase coordination chemistry is rarely met. Nevertheless, even in solution, inherently sterically less demanding ligands can display stability constants that reflect basicity effects more; R–S–R ligands often form stronger complexes than R–S–H, for example.

Overall, we can assume that large, bulky groups which interact sterically (that is 'bump into' other ligands) when attempting to occupy coordination sites around a central metal ion usually lead to lower stability. The strain in such systems is seen in distortions of bond lengths and angles away from ideal for the particular stereochemistry applying. It is possible to model these distortions effectively even with molecular modelling that treats molecules as composed of atomic spheres joined by springs, based on Hooke's Law principles (see Chapter 7.3.10), although more sophisticated modelling based on quantum mechanics has been developed in recent years. Ligand bulk is particularly significant with regard to its introduction in the immediate region around the donor atom, as congestion will become greater as the closer bulky groups approach the central atom. Thus, coordination of $N(CH_3)_3$ will lead to greater steric interaction with other ligands than will occur with $^-OOC–C(CH_3)_3$, as the ligand bulk is displaced further away from the remaining donor atoms in the latter case.

As a consequence, the carboxylate is termed a more 'sterically efficient' ligand than the amine.

5.1.2.3 *Sophisticated Effects*

Although we understand complexation reasonably well and can make experimental measurements of stability constants using a number of different physical methods with high accuracy, the outcome of a suite of effects influencing the stability of metal complexes is that prediction is an approximate but not perfect art. There are other effects that arise as a result of molecular shape that add complications. For example, large cyclic ligands (introduced in Section 3.4.1) that carry at least three donor atoms (*macrocycles*) have a central cavity or 'hole', and the fit of a metal ion into this hole is an important consideration. In fact, fit (or misfit) of metal ions into ligands of predefined shape (or topology) is an important aspect of modern coordination chemistry, as we now tend to meet increasingly sophisticated and designed ligand systems. However, it is more appropriate to leave this aspect for an advanced book, and we shall restrict ourselves here to an introduction to one type.

5.1.2.3.1   *The Macrocyclic Effect*

We have already introduced the hole-fit concept for a macrocyclic ligand in Chapter 4 (see Figure 4.38), which effectively means that the most stable complexes form where the internal diameter of the ring matches the size of the entering cation (while allowing for achievement of appropriate bond distances on coordination in the ring). The effect can be significant; for example, the natural antibiotic valinomycin is a macrocycle that binds a potassium ion to form a complex $\sim 10^4$ times more stable than that formed with the smaller sodium ion, despite the chemical similarity of the cations. There is, however, an aspect of preorganisation that is important to macrocycle complexation also, and comparison of the stability constants for potassium ion binding to a linear polyether and a cyclic polyether with the same number of donors and linkage chain lengths (Figure 5.6) provides a clear illustration.

The flexible, long-chain linear molecule must undergo significant translational motion to 'stitch' itself onto the metal ion, which is not favoured. However, the cyclic

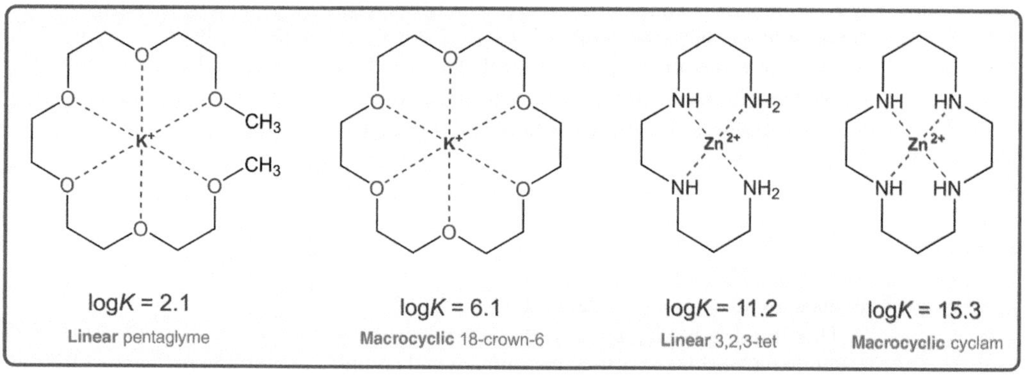

| $\log K = 2.1$ | $\log K = 6.1$ | $\log K = 11.2$ | $\log K = 15.3$ |
| **Linear** pentaglyme | **Macrocyclic** 18-crown-6 | **Linear** 3,2,3-tet | **Macrocyclic** cyclam |

Figure 5.6

Comparison of the stability constants for (at left) a linear and a cyclic polyether binding to potassium ions and (at right) a linear and cyclic polyamine binding to zinc ions, illustrating the macrocyclic effect in action.

molecule has the donors preorganised in more appropriate positions for binding to the metal ion, and its coordination is thus favoured. The higher stability achieved is mainly entropy-driven, although both enthalpy and entropy can contribute, as measured for the polyamines in Figure 5.6, for which changes in both ΔH° and ΔS° occur. In this case, the open-chain ligand must change from a highly disordered and freely moving conformation to a precisely ordered conformation when coordinated, which is opposed by entropy considerations. The macrocycle is preorganised in a conformation close to what is required for complexation, so entropy changes due to the ligand are negligible. This type of enhanced stability for complexation of macrocycles over open-chain analogues (illustrated in Figure 5.6) is shown by a wide range of macrocycles of different size, donor type and number and is well established.

5.1.3 Overall Stability Constants

Earlier, we introduced the overall stability constant β_n (Eq. (5.12)), where n is the number of ligands in the complex. The incorporation of several ligands within a complex occurs sequentially, not all at once. This is because ligand replacement is the result of molecular encounters, with the complex unit required to contact an incoming ligand with sufficient velocity and with an appropriate direction of approach so as to permit a ligand exchange to occur. The probability of all n ligands (L) replacing the entire set of water ligands at once (in a concerted reaction) is negligible. This is largely a matter of statistics. Most reactions take place with dilute concentrations of reagents much less than 1 M, so the chances of three, or even more, reactant molecules colliding simultaneously is so improbable that it may be dismissed. Hence, after formation of ML (from a collision between M and L), a series of other complexes ML_2, ML_3, ... , ML_x form in a sequence of *bimolecular* collisions between ML_x and L. So, although we refer to the stability constants for each of these processes as stepwise stability constants, there is no sudden 'step' from all of one species to all of another occurring in solution. Rather, and depending on the reaction conditions and relative sizes of the stepwise stability constants for each species, more than one complex will exist in solution at any one time, although one may be dominant. We can see this experimentally from analysis of accurate metal–ligand titrations, where complex formation curves (speciation diagrams) resulting from data analysis track the concentrations of each sequentially formed species; an example is shown in Figure 5.7.

From examinations of experimental results, it has been observed that, for sequential ligand substitution without change in coordination number during the reaction, there is a constant order for the sequential stability constants, whereby

$$K_1 > K_2 > K_3 > \cdots > K_x \tag{5.21}$$

Thus, as more and more of one type of ligand is introduced, the gain in stability in each step keeps diminishing. This occurs irrespective of whether neutral or anionic ligands are involved, as illustrated in Table 5.5.

This trend is explained to a modest level of satisfaction by a simple *statistical* argument. Consider a partly substituted metal complex containing m L and n OH_2 ligands, which we will define here simply as $ML_m(OH_2)_n$. If one reaches in and plucks out 'unseen' and at random any ligand from the coordination sphere, the probability of removing L (rather than OH_2) from $ML_m(OH_2)_n$ is greater than the probability of removing L from $ML_{(m-1)}(OH_2)_{(n+1)}$, simply because there is one more L to choose

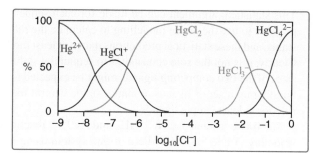

Figure 5.7

An example speciation diagram resulting from metal–ligand titrations, with the concentration profiles of the various $HgCl_x$ species throughout the process of successive addition of the ligand identified. Shown is the chloride ion concentration-dependent formation of the series of chloridomercury(II) species.

Table 5.5 Successive stability constants for several metal ion complexes with simple monodentate ligands.[a]

Reaction *(where m = 0 → 6)*	$\log K_1$	$\log K_2$	$\log K_3$	$\log K_4$	$\log K_5$	$\log K_6$
$Ni(NH_3)_m^{2+} + NH_3 \rightleftharpoons Ni(NH_3)_{(m+1)}^{2+}$	2.7	2.1	1.6	1.1	0.6	−0.1
$AlF_m^{(3-m)+} + F^- \rightleftharpoons AlF_{(m+1)}^{(2-m)+}$	6.1	5.0	3.9	2.7	1.6	0.5
$Cu(NH_3)_m^{2+} + NH_3 \rightleftharpoons Cu(NH_3)_{(m+1)}^{2+}$	4.0	3.2	2.7	2.0	<0	≪0
$CdBr_m^{(2-m)+} + Br^- \rightleftharpoons CdBr_{(m+1)}^{(1-m)+}$	2.3	0.8	−0.2	+0.1	Not formed	Not formed

[a]H_2O is the companion ligand in each case, but not shown.

from in the former case. At the same time, if you reach in and pluck out a ligand at random and replace it by a new L ligand (irrespective of whether you first remove an L or an OH_2), in some cases you will advance the number of L ligands (when you replace an OH_2 by an L), and in others you will make no change by inadvertently replacing one L by another L. From a statistical standpoint, the probability of adding L to $ML_{(m-2)}(OH_2)_{(n+2)}$ is greater than the probability of adding L to $ML_{(m-1)}(OH_2)_{(n+1)}$ because there is one extra OH_2 site to choose from in the former complex. Let us represent this in a 'real' case (Eq. (5.22)) below (where any charges on the complexes have been removed for viewing simplicity). Here, two sequential substitution steps, associated in turn with K_1 and K_2, are represented.

$$[M(OH_2)_6] \underset{-NH_3(+OH_2)}{\overset{+NH_3(-OH_2)}{\rightleftharpoons}} [M(NH_3)(OH_2)_5] \underset{-NH_3(+OH_2)}{\overset{+NH_3(-OH_2)}{\rightleftharpoons}} [M(NH_3)_2(OH_2)_4]$$

(5.22)

Six water substitution sites are available in $[M(OH_2)_6]$, with five water substitution sites available in $[M(NH_3)(OH_2)_5]$, so it is more probable for addition of ammonia to occur in $[M(OH_2)_6]$ than in $[M(NH_3)(OH_2)_5]$. For the reverse step of removal of ammonia, $[M(NH_3)_2(OH_2)_4]$ is more likely to lose one, as two are available versus only one in $[M(NH_3)(OH_2)_5]$. Consequently, there is a greater driving force for adding an NH_3 in the first step than in the second step. This means the concentration ratio of $[M(NH_3)(OH_2)_5]$ versus $[M(OH_2)_6]$ will be relatively greater than that of $[M(NH_3)_2(OH_2)_4]$ versus $[M(NH_3)(OH_2)_5]$, and hence K_1 will be larger than K_2.

This statistical argument can be made for each sequential substitution. Indeed, it is possible to use the above modelling to calculate the ratio of successive stability constants, and these statistical predictions match at least modestly to experimental values. Clearly, it is not the sole contributor to sequential stability (for example, the relative sizes of the two competing ligands might be expected to have some role to play when building up a set of ligands around a single central metal), but it plays a significant role for at least simple ligands.

Stability constants for Ni^{2+}/NH_3 and Al^{3+}/F^- reactions follow the predicted trend smoothly (Table 5.5), regardless of one system using a neutral ligand and the other an anionic ligand. For Cu^{2+}/NH_3, although the appropriate trend is followed, the far larger than usual drop from log K_4 to log K_5 (Table 5.5) is evidence for the operation in this example of the Jahn–Teller effect, which imposes elongation along one axis that does not favour strong complexation in the fifth and sixth axial positions. This was covered in Chapter 2.3.5.

There are also conditions that can lead to an exception to this general rule of size order of stability constants. For example, experimental log K_m values determined for $[Cd(OH_2)_6]^{2+}$ reacting with Br^- follow the usual trend from K_1 to K_3, but then there is an increase to K_4, and there are no further values measurable (Table 5.5); clearly something different is happening here. As it happens, we know from physical and structural studies that the complex resulting from introduction of four Br^- ions is the tetrahedral species $[CdBr_4]^{2-}$, whereas the immediately prior species with three Br^- ions coordinated is an octahedral species $[CdBr_3(OH_2)_3]^-$. Thus, it is clear that in the fourth substitution step, there is a change from six-coordinate octahedral to four-coordinate tetrahedral coordination. Yet sequential addition of Br^- still occurs, so why should it matter so much? The answer to this is associated with reaction entropy change, which we discussed earlier in explaining the added stability of complexes containing chelating ligands. The change in the K_4 step can be written as follows:

$$[CdBr_3(OH_2)_3]^- + Br^- \rightleftharpoons [CdBr_4]^{2-} + 3H_2O \qquad (5.23)$$

6-coordinate   4-coordinate

{(**2 molecules**)   →   (**4 molecules**)}

versus the usual situation met where there is no change in coordination number, as for K_3:

$$[CdBr_2(OH_2)_4] + Br^- \rightleftharpoons [CdBr_3(OH_2)_3]^- + H_2O \qquad (5.24)$$

6-coordinate   6-coordinate

{(**2 molecules**)   →   (**2 molecules**)}

The difference is the sudden release of two extra water molecules when the coordination number decreases from 6 to 4. This means there is an increase in 'disorder' in this first case (or a favourable entropy change) and not in the other, influencing K_4 but not K_3 and other prior steps. While this entropy-based effect has been illustrated for simple ligands through a reaction where change in coordination number is the key step, in effect any mechanism that leads to an unprecedented change in disorder (or a 'burst' of released molecules) will produce an enhanced stability constant.

5.1.4 Undergoing Change – Kinetic Stability

Reactions can be considered to have a start and an end; thermodynamics is concerned with the end, whereas kinetics is concerned with the processes leading from the start to that end. Thermodynamic stability is concerned with the extent of formation of a species under certain specified conditions when the system has reached equilibrium. The rate of formation of a species *leading to equilibrium* is a measure of what we can call **kinetic stability**. Reactions occur with a vast spread of half-lives (the time by which half the precursor molecules have reacted) or reaction rates. Complexes of metal ions which undergo reactions *rapidly* are termed **labile**. Complexes of metal ions which undergo reactions *slowly* are termed **inert**. Nobel laureate Henry Taube suggested a suitable definition of a labile metal ion as one that reacts with a ligand within the time needed to mix the reagents, typically a few seconds. Thermodynamic stability does not imply kinetic inertness, nor does thermodynamic instability imply lability. For example, in acidic solution, Co(III) amine complexes are thermodynamically unstable, but are inert towards dissociation, and as a result can be stored in acidic solution for extensive periods of time (even decades).

A key reaction in aqueous solution is *water exchange*, which is the process whereby a metal ion changes its coordination sphere of water molecules for other solvent water molecules. This process defines aqua metal ions as dynamic species that undergo a series of exchange reactions continuously (where H_2O and $\mathbf{H_2\underline{O}}$ represent the water molecules exchanged). A simple octahedral metal aqua ion is used here as an example:

$$[M(OH_2)_6]^{n+} + \mathbf{H_2\underline{O}} \rightleftharpoons [M(OH_2)_5(\mathbf{\underline{OH_2}})]^{n+} + H_2O \qquad (5.25)$$

The half-life for water exchange spans a broad range; experiment has shown that water exchange rates in transition metal complexes show differences of about 20 orders of magnitude and follows the pattern shown in Figure 5.8 (although the more inert metal ions are not included here). Whereas simple exchange of one ligand by another of exactly the same type is the simplest process inherently, the fact that such exchanges can occur suggests that, if other potential ligands of a different type are present, they may intercept the process and be inserted in the place of the original type – ligand exchange has become *ligand substitution*.

There is clearly a wide variation in rates of water exchange. Factors influencing how fast ligands are exchanged are metal ion size, metal ion charge and (to

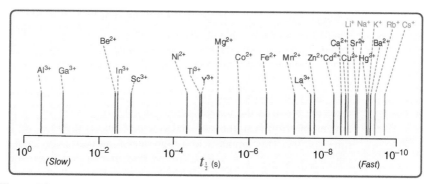

Figure 5.8

Half-lives for water exchange of some aquated metal ions.

some extent for transition elements) electronic configuration. For example, as the ionic radius increases from Mg^{2+} to Ca^{2+}, the exchange rate increases from $\sim10^5$ to $\sim10^8$ s^{-1} which equates to the reaction half-life decreasing from microseconds to nanoseconds. The left-hand side of Figure 5.8 equates with more inert (less labile) metal ions that have high charge density and half-lives on the timescales of seconds. The timescales shown on the x-axis of Figure 5.8 can extend much further to the left for classic inert metal ions such as Co^{3+}, Rh^{3+}, Ir^{3+} and Pt^{4+}, but measurement of water exchange reactions that are so slow becomes an exercise in patience as effectively nothing is happening over very long periods. The right-hand side of Figure 5.8 equates with labile compounds. There is obviously a 'grey' area, as will always be the case where we partition into two classes. However, it is clear that complexes need not be static species, and coordinated water molecules undergo exchange with their solvent environment, measurable where practicable by employing isotopically distinctive $^{18}OH_2$ or $^{17}OH_2$ as the solvent and following its introduction into the metal coordination sphere in place of most common $^{16}OH_2$ (such as by using ^{17}O-NMR spectroscopy). While we can measure these processes, understanding how they occur is another thing altogether.

Because we can initiate and observe reactions, we know that we start with certain compounds and end with others. We can usually separate and quantify the products of a reaction. However, we cannot really understand the reaction unless we can answer the question of how the reactants have been converted into the products. This process is the *mechanism* of the reaction, and it is difficult to tease out because the stages along the way involve only transient non-isolable species – at best, we can infer their nature from a range of experiments and by analogy to known stable species. At the core of reaction mechanisms is the concept of the transition (activated) state, the assembly present at the peak of the energy barrier prior to relaxation to form stable products. The activation barrier is like a mountain range – reacting molecules must expend energy to gain the peak of the barrier; if they falter, they fall back to their initial form, but if they succeed, they may cross over to form new products.

A simple macroscopic example may illustrate this process. Consider a thin flat piece of board painted white on one side and red on the other. Now arrange to turn this over on a tabletop with the board touching the table throughout. You can do this, of course, by turning the board onto its edge and then continuing the motion until it lies flat on its other side. In doing this transformation, the position where it is precariously standing on its edge can be considered a transition (or activated) state, and if you let go of it in that position, it may fall back so the white side is up (equivalent to no reaction) or fall forward so the red side is up (equivalent to a completed reaction). To get it from lying flat to on its side costs you effort, or energy – what we would call the activation energy at the molecular level. Molecules acquire this activation energy mainly through collision as a result of their motion; if the collision energy suffices to take the reactants to an arrangement where they can proceed to products without further energy, they have achieved the status of an activated or transition-state species. In our macroscopic example, your physical effort takes the board to the upright activated state. For a very heavy piece of board, the amount of effort is significant, and you may not get it into an upright position every time you try. This amounts to a high activation barrier or large activation energy, which equates with a slower rate of reaction after several failed attempts. If the board is small and light, the effort required is small and the task easy and rapidly completed on the first attempt; this amounts to a low activation barrier or small activation energy, consistent with a fast rate of reaction.

If you next consider your board starting on a different level to where it finishes (table to floor, for example), it is obvious that one is the lower level; the board could fall from the table to floor, but not easily the reverse way. On the molecular level, being on the 'floor' would be a thermodynamically more stable arrangement for the molecule than being on the 'table'. The difference in energy between the reactants and products is the reaction free energy change, and the stability of a complex depends on the size of this energy difference. This additional consideration (thermodynamics) does not change the process by which you turn the board over (kinetics), although there is a relationship between them at the molecular level, which we will not explore here. At the molecular level, however, raising temperature increases molecular velocities and increases the probability of a collision achieving the transition state, so the rate of a reaction always increases with increasing temperature.

At the crest of the activation barrier, reacting molecules are poised in a different configuration (the activated state) ready to proceed to product formation, a state structurally different from both the precursors and products. In coordination chemistry, the predicted activated state species are considered to be related to known geometries, but ones that are not inherently stable for the particular system under investigation; there is good supporting evidence based on reactivity and products that support this approach. There is also the expectation that gross changes in achieving an activated state would come with a higher energy cost, and thus be distinctly slower, than changes that are limited and have the activated species closer to both products and reactants in character.

To help illustrate the significance of these concepts, we can draw a reaction profile (Figure 5.9) that describes the free energy changes going from the reactants to the products (left to right along the horizontal axis) via the activated (transition) state in between. In the case shown, the products have a lower free energy than the reactants, so the reaction is spontaneous. However, the rate at which this reaction proceeds is determined by the height of the barrier. The activated state is inherently unstable as it sits atop the reaction potential energy profile and immediately converts to either the products or reverts to the reactants. It does not exist long enough to be observed by any physical method. On some occasions, however, an intermediate of high energy may form but which still has limited stability (sitting in a high energy potential energy trough) and sufficient lifetime to be detected and characterised. Such cases

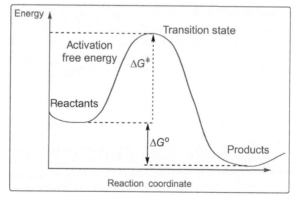

Figure 5.9

Defining the barrier to reaction (activation energy) in terms of a reaction coordinate. The activated state is an unstable form that immediately converts to products or reverts to the reactants.

provide clues about the form of the activated state, as the intermediate is assumed to structurally resemble the transition state. This will be discussed in more detail in Section 5.3.

5.2 Complexation – Will It Last?

You have probably heard the doomsayers, in commenting on a new partnership, predict that 'it will never last'. Of course, given the fragility of life, this is hardly a prophetic statement, but more of a comment on longevity. At the molecular level, we are also dealing with finite rather than infinite time periods, and the longevity of a molecular partnership is not constant for all assemblies. Coordination complexes are not uniform in their behaviour – in fact, each complex has exclusive physical properties, which include unique thermodynamic and kinetic stabilities. The fate of a complex relates to its environment; in solution in particular, it will need to deal with the presence of other competing ligands, which include the solvent itself. How it conducts itself, and whether it retains its integrity, depends on both its thermodynamic and kinetic stability. Let us look at a human-scale example: we recognise a bank as a (usually, at least) stable place that coordinates money, but if it parts with its money in too undisciplined a way, it can become, over time, insolvent and cease to exist. The money has not disappeared, merely moved elsewhere. Coordination compounds follow the same path; they may appear to have a certain form and stability yet may eventually under external influences undergo further change that leads to their demise – all a bit like life, really.

5.2.1 Thermodynamic and Kinetic Relationships

A large thermodynamic stability for a complex means that the complex is a greatly favoured product compared with its reactants and is possibly even the only significant product (see Figure 5.9), although one should not immediately assume that this is the case. This behaviour says very little about the manner by which, and pace at which, it undergoes its formation and any following reactions, however. Having a high stability constant may mean that a particular product forms essentially exclusively, but its accommodation of subsequent changes in its environment relates in part to the rate at which it can undergo reactions. There are relationships between thermodynamic and kinetic parameters, nevertheless. Consider a simple equilibrium involving M, L and ML. Let us define reactions that involve formation of ML (the 'forward reaction') and the reverse decomposition of ML (the 'back reaction') with rate constants for the forward (k_f) and back (k_b) reactions as follows:

$$M + L \rightarrow ML \quad k_f \tag{5.26}$$

$$ML \rightarrow M + L \quad k_b \tag{5.27}$$

We can then consider combining Eqs. (5.26) and (5.27) into one equation, namely,

$$M + L \underset{k_b}{\overset{k_f}{\rightleftharpoons}} ML \tag{5.28}$$

This expression, as written, also corresponds to the one we use to define the now familiar complex formation equilibrium constant, K. It can be shown that there is a simple relationship between these terms, namely,

$$K = \frac{k_f}{k_b} = \frac{[ML]}{[M][L]} \tag{5.29}$$

This holds provided no stable intermediate is involved in the reaction. If the rate at which ML dissociates into M and L is much slower than the rate at which M and L assemble into ML, then there will always, at equilibrium, be a larger amount of ML present than of M and L, which translates into a large thermodynamic stability constant ($K \gg 1$), which equates with the ratio of these two kinetic terms. In effect, for such simple steps, there is a thermodynamic–kinetic link. Returning to Figure 5.9, it can be seen qualitatively that in this case $k_f > k_b$ as the barrier height for the forward reaction is lower, and the transition state is closer in free energy to the reactants than the products. However, the barrier height can vary considerably depending on a number of factors, even though the end result (and overall free energy change) is the same.

5.2.2 Rate Coefficients

Reactions can occur in a number of ways, but we measure the processes underway in a common manner, by observing the change in the quantities of reactants and/or products over time. This can be done by employing a wide variety of physical methods (see Chapter 7); the key requirement is that the technique selected involves a measurable change over a given period of time and that the magnitude of the observed changes is proportional to the amounts of compound present.

There are different types of kinetic behaviour known for chemical reactions; defined by their so-called *reaction order* which is associated with a number. For example, in zeroth-order reactions, the rate of change in the amount of a reactant or product (the reaction rate) is constant over time. This is unusual, and we will not be covering this scenario here. However, it is often associated with reactions that are accelerated by small concentrations of catalysts, such as enzymes (see Chapter 8).

In first-order reactions, the rate of change slows down exponentially with time (Eq. (5.30)). First-order reactions are extremely common in coordination chemistry, and this is the type to which we shall restrict ourselves here. In its simplest form, the concentration of a reacting species varies over time as

$$[S]_t = [S]_0 \cdot e^{-k \cdot t} \tag{5.30}$$

where $[S]_t$ is the concentration of the reacting species at a particular time t during the reaction, $[S]_0$ is the original concentration (at time zero, before anything happens) and k is the rate coefficient, which is the characteristic measure of the rate of the reaction. This equation may also be expressed (integrated) as

$$\ln([S]_t) - \ln([S]_0) = -k \cdot t \tag{5.31}$$

so that a plot of $\ln([S]_t)$ versus t will be linear with a slope equal to $-k$. This rate coefficient, usually expressed in units of s^{-1}, is dependent on the experimental conditions operating (as it varies with temperature, solvent and added electrolyte). Typically, for reactions of coordination complexes, rate coefficients double for about every 5°C rise in temperature. Measurement of variations in rate coefficients with temperature allows determination of the activation parameters (activation enthalpy, $\Delta H^{\ddagger}$, and activation entropy, $\Delta S^{\ddagger}$) applying in the reaction, which assist in elucidating the mechanism. The relationship between these parameters and the activation free energy (of the transition state) is

$$\Delta G^{\ddagger} = \Delta H^{\ddagger} - T\Delta S^{\ddagger} \qquad (5.32)$$

but be aware that this is distinct from Eq. (5.7), which deals with standard free energy changes (ΔG°) at equilibrium, not activation free energies ($\Delta G^{\ddagger}$) during the reaction.

In some reactions, the rate coefficient is seen to vary with the concentration of one of the reactants (say molecule X). If a plot is made of the observed first-order rate coefficient (k_{obs}) versus [X] present where reaction rate increases linearly with [X], then the behaviour can be represented as

$$k_{obs} = k_{X} \cdot [X] \qquad (5.33)$$

and k_{X} is the second-order (bimolecular) rate constant and will have units of $M^{-1}s^{-1}$. The experimentally observed rate coefficient (k_{obs}, s^{-1}) is clearly *not a constant* as it is dependent on the concentration of X. This type of behaviour means that X is involved in the reaction mechanism in some way; it may appear as part of the product, although it may alternatively participate only in the transition state and not appear in the product. These are brief aspects of reactions that form a background to our discussion of the mechanisms of reactions below.

5.2.3   Lability and Inertness in Octahedral Complexes

Chemistry is full of 'opposites' – resulting from often setting two extremes as the limits for defining behaviour (the 'black or white' approach). Of course, these limits are not isolated options, but are usually the extremes of a continuum of behaviour (the chemical equivalent of saying that something is rarely ever black or white, but more a shade of grey). How rapidly metal ions take up or lose ligands is a good example of this characteristic. There are two extreme positions; very fast reactions of labile compounds or very slow reactions of inert compounds. Labile systems are the party animals of metal complexes – making and breaking relationships rapidly. We can measure the rate at which ligand exchange with the same type of ligand occurs, even though it appears that nothing changes, by using radioactive isotopes to allow the rate of the process to be monitored, provided the process is not too rapid. It is a bit like painting a white house with fluorescent white paint – nothing appears to have changed, until you see it at night. At the molecular level, we simply measure the uptake of the radioactive ligand into the coordination sphere of the metal over time as it replaces the non-radioactive ligand, which allows us to define the rate of ligand exchange.

An inert system is like some modern marriages – maybe not joined forever, but willing to give it a good try. These are complexes that, once formed, undergo any subsequent reaction very slowly. Inertness can be so great that it overcomes thermodynamic instability. This means that a complex may be pre-disposed towards decomposition, but this will happen so very, very slowly that to all intents and purposes, the complex is unreactive, or inert. Cobalt(III) amine complexes are the classic example of this; thermodynamically, they are unstable in aqueous solution, but they are so inert to ligand substitution reactions that they can exist in solution with negligible decomposition for years.

We have introduced the concept of rate of change for complexes, expressed in terms of the two extremes of *labile* (fast) and *inert* (slow), where charge and size are key factors. Experimental observations of reactivity allow us to define, at least approximately, which metal ions fall into which category. For transition metals, although size/charge effects are important, they do not fully explain experimental observations. It was Chemistry Nobel laureate Henry Taube who also showed that the *d*-electron configuration played an important role and that high-spin complexes are generally labile. The following groupings were identified from experimental observations of octahedral complexes:

labile – all complexes where the metal ion has electrons in antibonding (destabilising) e_g^* orbitals [e.g. Co^{2+} (d^7: $t_{2g}^5 e_g^{*2}$); high spin Fe^{3+} (d^5: $t_{2g}^3 e_g^{*2}$)], and all complexes with less than three *d*-electrons [e.g. Ti^{3+} (d^1)];

inert – octahedral d^3 complexes [e.g. Cr^{3+} (d^3: $t_{2g}^3 e_g^{*0}$)], and low spin d^4, d^5 and d^6 complexes [e.g. Co^{3+} (d^6: $t_{2g}^6 e_g^{*0}$)].

An analysis in terms of the relative energies of the precursor and reaction intermediate predicts lability for octahedral complexes with populated e_g^* levels, in line with the experimental observations.

There are, of course, 'grey' areas. Even within a traditionally inert system, reactivity may vary markedly with the type of ligand undergoing the reaction. For $[Co^{III}(NH_3)_5X]^{n+}$ reacting in water to replace X by H_2O, the rate constant varies by an order of $\sim 10^{10}$ from X = NH_3 ($k = 5.8 \times 10^{-12}$ s^{-1} at room temperature, a half-life of 3,800 years) to X = $OClO_3^-$ ($k = 1.0 \times 10^{-2}$ s^{-1}, a half-life of 7 seconds). Chemistry is astounding in its diversity, if nothing else. Such variations reflect the nature of the ligand rather than the metal since the ligand lability, or capability as a good leaving group, tends to extend across a wide range of metal ions.

In the context of the above discussion, it may be useful to reinforce earlier discussion – lability and inertness are *kinetic* terms, and all about the rate at which something reacts. A species that is inert is kinetically stable. This does not require that it be thermodynamically stable, however (although it may be). Thermodynamic stability is about being in a form which has no other readily accessible species lower in energy. The reason we can have kinetic stability in a system that is thermodynamically unstable is that the two differ. Remember, kinetics is about transition *to* equilibrium; thermodynamics is about the situation *at* equilibrium.

5.3 Reaction Mechanisms

We can categorise reactions in a number of ways. So far, we have done so in terms of rate (how fast or slow). This approach, however, tells us nothing about what is happening in a reaction or how a particular reaction occurs. We shall now define reactions in terms of what happens. We shall explore examples of the following reactions in Chapter 6, when describing examples of chemical synthesis in coordination chemistry.

Inorganic reactions involving metal complexes fall into three major categories:

(1) reactions involving the inner **coordination sphere** of the complex;
(2) reactions involving the **oxidation state** of the central metal; and
(3) reactions involving the **ligands** themselves.

It is possible also to include metal ion exchange as another category, but since this involves a change in the inner coordination sphere of each metal participating in this process, it is appropriate to consider this as a subset of (1). Indeed, within each category, there may be several further convenient subdivisions applicable. In particular, the sub-categories may be defined for *reactions involving the inner coordination sphere* as

(a) increase in coordination number
     (*addition*: $ML_x + nL \rightarrow ML_{(x+n)}$);
(b) decrease in coordination number
     (*elimination*: $ML_xY_n \rightarrow ML_x + nY$);
(c) partial or complete ligand replacement
     (*substitution*: $ML_x + nZ \rightarrow ML_{(x-n)}Z_n + nL$);
(d) change in geometry for a fixed coordination number
     (*geometric isomerisation*: e.g. tetrahedral geometry $\rightarrow$ square planar geometry);
(e) change of relative positions of atoms for a fixed coordination number
     (*stereoisomerisation*: e.g. $cis\text{-}ML_xZ_n \rightarrow trans\text{-}ML_xZ_n$);
(f) change of central metal for a fixed ligand set
     (*transmetallation*: e.g. $ML + M' \rightarrow M'L + M$).

These cover a vast array of reactions, and we shall restrict ourselves here to a limited selection of some important reactions, in order to further define their character. We shall do this in terms of not only the nature of the reaction but also the manner in which it occurs, or the *mechanism* of the reaction.

When we initiate and observe a reaction, we can of course clearly define the chosen reactants, and we are able also, through their separation and characterization, to identify the stable products. What we cannot define so readily is how the process of change from reactants to products has occurred. The process by which change occurs is what we call the *mechanism* of the reaction. As already described, transition state theory proposes that reactants pass through an *activated (transition) state*, a transient and unstable intermediate or activated state of higher energy. Energy is required to reach this transition state, with stable products of lower energy then forming from this transition state. The change in free energy of reactants in reaching the transition state is the *activation free energy* ($\Delta G^{\ddagger}$, Eq. (5.32)). The overall change in energy between reactants and products is the *reaction free energy* change (ΔG°, Eq. (5.7)). The former relates to the mechanism *en route* to the products and is associated with the kinetics; the latter relates to the relative stabilities of reactants and products and is a thermodynamic term and is independent of the mechanism. The key to a reaction mechanism lies in understanding the transition state. This is something we cannot 'see', so our identification of it must rely on inference from secondary observations. For example, the variation of reaction rates with concentrations of various reagents, the measured activation parameters, the range and various amounts of products and incorporation of isotopically labelled components of the reaction mixture in products are devices that can assist the chemist in understanding the way a reaction proceeds.

5.3.1 Substitution

We shall examine in some detail here one of the most important, simplest and best understood classes of reactions in coordination chemistry, namely, *ligand substitution*. This involves one (or several) ligand(s) departing the coordination sphere to be replaced by one (or several) others, without a change in coordination number or coordination geometry (shape) between the reactant complex and its product. It is usually a clearly observable reaction, at least where the nature of the ligands involved in the substitution are distinctively different and can often be followed by simple approaches such as following the change of colour over time as reaction proceeds.

In considering how a substitution can occur, there are really a very limited number of sensible options, which are as follows:

(a) the departing ligand can leave first (leaving a temporary 'vacancy' in the coordination sphere), and then the incoming ligand enters in its place;

(b) the entering ligand can add first (causing a short-lived increase in the number of ligands around the metal ion), and then the departing ligand leaves or

(c) one ligand can leave as the other enters, in a concerted manner (where a 'swapping' of these ligands occurs with neither tightly bound in the transition state).

In all cases, regeneration of the preferred coordination number and geometry in forming the stable product occurs.

In the above options, (a) involves ligand loss (or *dissociation*) as the key initiation step, whereas (b) involves ligand gain (or *association*) as the key initiation step. The third option (c) involves concerted replacement (or *interchange*) as the key step. As a consequence, they tend to be referred to as *dissociative* (abbreviated as **D**), *associative* (**A**) and *interchange* (**I**) mechanisms in turn.

The concepts rely also on our understanding of preferred coordination numbers and geometries. For example, for many metal ions, six-coordination is preferred, but stable examples of both five- and seven-coordination are known. Therefore, it is not too much of a jump to consider a short-lived five- or seven-coordinate species existing for a complex most stable in a six-coordinate ligand environment. As an example, (six-coordinate) octahedral is the overwhelmingly dominant coordination mode for stable cobalt(III) complexes. Yet, in recent years, rare examples of isolable but usually very reactive compounds with five-coordination and seven-coordination have been prepared. It does not take a great leap of faith to assert that such geometries form as short-lived intermediates in substitution reactions of this whole family of complexes.

5.3.1.1 *Octahedral Substitution Mechanisms*

Since octahedral complexes dominate coordination chemistry of lighter transition elements, the mechanisms of these reactions are both important and well-studied. We can commence by asserting that there are two limiting mechanisms, *dissociative* and *associative*; they are distinguished by the order in which the process of substitution of the ligand to be replaced (the leaving group) by a new ligand (the entering group) occurs, as outlined above, and in Figure 5.10. The outcome is in each case identical, of course – one ligand has been replaced by another. As discussed generally earlier, the mechanism that operates and the intermediates and transition states cannot be observed directly but must be inferred from various experiments. Moreover, as stated, these are limiting mechanisms, that is, they lie at the extremes of potentially a continuum of behaviour, which infers that reactions may not be mechanistically ideal.

The above discussion could suggest that a successful forward step always occurs. However, if energy acquired in an encounter or collision process between reacting molecules is insufficient to allow the reactant to reach the energy level of the activated state, then no reaction can occur; this is a key aspect of transition state theory. Moreover, the transition state, when reached, does offer a 'forward or back' option in both mechanisms. With reference to Figure 5.10, if 'B' re-enters the coordination

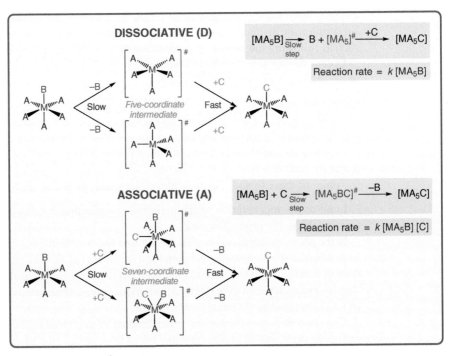

Figure 5.10

The operations involved in the dissociative and associative reaction mechanisms. Two different shapes for the intermediates are represented, based on known shapes for these coordination numbers. (All reactions are represented here as proceeding in one direction.)

sphere (in the **D** mechanism) or 'C' departs before 'B' (in the **A** mechanism), then the initial reactant is regenerated. Thus, there is a potential reversibility in the reactions, although this has been neglected here for the sake of simplification. The experimental observation of change in a reaction tells us that there is a driving force towards the products.

5.3.1.1.1 Dissociative Mechanism

In the reaction of a six-coordinate complex, slow and therefore rate-determining loss of one ligand to produce a *five-coordinate intermediate* is the key to this process. This intermediate is short-lived and of higher energy than the reactants. The process occurring is illustrated in Figure 5.11 in the form of a reaction coordinate diagram. To reach this high energy 'metastable' state, the activation energy barrier must be overcome; this is the slow (or rate determining) step in the overall reaction process, which can be expressed as

$$[MA_5B] \underset{slow}{\overset{-B}{\rightleftharpoons}} \{[MA_5]\} \underset{fast}{\overset{+C}{\rightarrow}} [MA_5C] \qquad (5.34)$$

Once reached, the intermediate $\{[MA_5]\}$ can 'fall back' via a now shallow activation barrier to reform the reactants (so nothing is achieved) or else can 'fall forward' over another small activation barrier to yield the products (and a reaction is observed to occur).

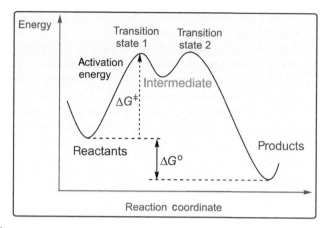

Figure 5.11
Defining the barrier to reaction (activation energy) in terms of a reaction coordinate, which represents the progress of a reaction. Shown here is the process where there is an intermediate of limited stability and lifetime formed.

The *dissociative*, or **D**, mechanism is also sometimes called an S_N1 reaction, a form of nomenclature used in organic reaction mechanisms. In the S_N1 representation, **S** refers to the process (substitution), subscript **N** to the character of the leaving and entering group (a nucleophile, equated in this case to a Lewis base with a lone pair of electrons) and **1** to the number of molecules involved in the rate-determining step (unimolecular). For the **D** mechanism, the reaction rate can be expressed by

$$\text{rate} = k[MA_5B] \tag{5.35}$$

It is first-order in reactant only since the first step of ligand loss from the precursor complex, which involves no other molecule, is rate-determining.

5.3.1.1.2 Associative Mechanism

Likewise, for a six-coordinate complex reacting, slow rate-determining gain of one ligand to produce a *seven-coordinate intermediate* is the key to this process. We can represent this reaction as a sequence of a slow and therefore rate-determining formation of a seven-coordinate intermediate, followed by the fast loss of a ligand which is only significant if the ligand lost differs from the one added (5.36).

$$[MA_5B] \underset{\text{slow}}{\overset{+C}{\rightleftharpoons}} \{[MA_5BC]\} \underset{\text{fast}}{\overset{-B}{\rightarrow}} [MA_5C] \tag{5.36}$$

The associative, or **A**, mechanism is also sometimes called an S_N2 mechanism (nomenclature favoured for reactions in organic chemistry). In the S_N2 representation, like before, **S** refers to the process (substitution), subscript **N** to the character of the entering group (a nucleophile) and **2** to the number of molecules involved in the rate-determining step (bimolecular, as the precursor complex and the entering group together are involved in forming the activated state). For the **A** mechanism, the reaction rate can be expressed by

$$\text{rate} = k[MA_5B][C] \tag{5.37}$$

Since the rate-determining step involves two species, two concentration terms appear in the rate expression.

For the **D** mechanism, *bond-breaking* is important in the rate-determining step, and this should not be influenced by the entering group, which only appears after the leaving group has departed. Thus, we expect the rate constant to be largely independent of the entering group. For the **A** mechanism, *bond-making* is important in the rate-determining step, and clearly this process will be influenced by the character of the group entering to form the new bond. As a result, we expect k to be dependent on the entering group. As a consequence, examining how the rate constant changes with the type of entering ligand may provide evidence for a particular mechanism. Unfortunately, 'pure' or limiting **A** or **D** mechanisms, as mentioned earlier, are uncommon; as a result, the distinction is not always clear-cut.

5.3.1.1.3   *Interchange Mechanism*

The interchange (**I**) mechanism is offered as a 'half-way house' between the limiting **A** and **D** mechanisms, involving concomitant entry and departure of the two participating ligands (Figure 5.12).

As a result of the process proposed, you could view the transition state as a five-coordinate species with two tightly bound outer-sphere exchanging ligands or else as a seven-coordinate species with two very long bonds to the exchanging ligands. The descriptions amount to the same thing, in effect.

5.3.1.1.4   *Probing the Transition State*

The problem with the transition or activated state is its extremely short-lived nature, which means we cannot 'see' it easily by any physical method. However, this does not prevent experiments being devised that probe the transition state structure through inspecting the products. There are several long-standing experiments that do this neatly. One is to make use of the opportunity for stereochemical change in passing through particular transition states, experiments most accessible using inert complexes for which both the precursors and products can be easily isolated and characterised. The sole example we shall use to illustrate this aspect is the reaction of *cis*- or *trans*-$[Co(en)_2Cl(OH)]^+$ to form (*cis*- or *trans*-) $[Co(en)_2(OH_2)(OH)]^{2+}$, which is a simple ligand substitution (hydrolysis) reaction proceeding (as for the vast majority of cobalt(III) complexes) via a dissociative mechanism. In the **D** mechanism in Figure 5.10, two different five-coordinate intermediates were suggested – trigonal bipyramidal and square-based pyramidal (in this case, of $\{[Co(en)_2(OH)]^{2+}\}$). This reaction offers us the opportunity to both support this mechanistic assertion and identify the likely geometry of the transition state. We can do this by examining the stereochemistry of the products. For *cis*-$[Co(en)_2Cl(OH)]^+$ reacting, there is 100%

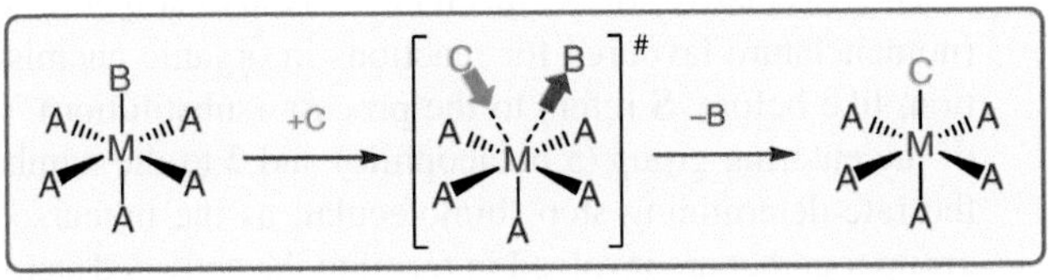

Figure 5.12

A schematic representation of the interchange mechanism for octahedral substitution. The key to this mechanism is the initial bound ligand (B) departing as the new potential ligand (C) enters the coordination sphere.

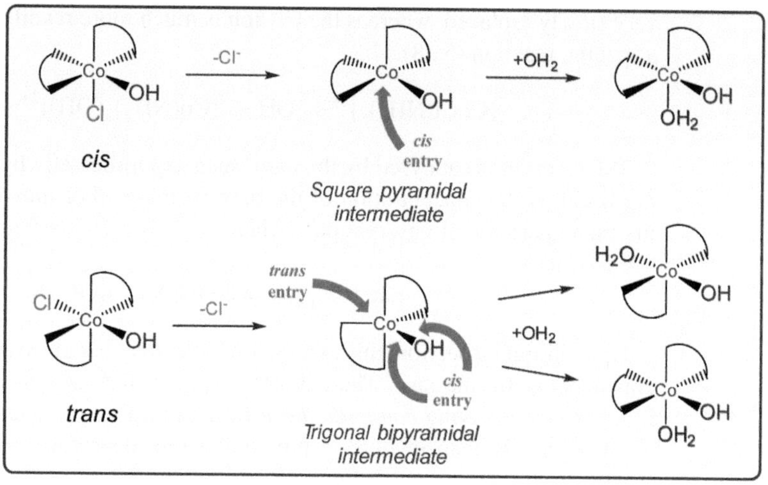

Figure 5.13

Stereochemical control of the substitution reaction in *cis*- and *trans*-[Co(en)$_2$Cl(OH)]$^+$ via different transition-state geometries. The square pyramidal intermediate offers only one choice to the entering group and 100% retention of geometry, and the trigonal bipyramidal intermediate offers three entry points around the triangular central plane, two leading to *cis* entry and one to *trans* entry, for a 2:1 ratio. The bidentate en ligands are shown as curved lines for simplicity.

conversion to *cis*-[Co(en)$_2$(OH$_2$)(OH)]$^{2+}$, whereas for *trans*-[Co(en)$_2$Cl(OH)]$^+$ reacting, there is ~70% *cis* and ~30% *trans* isomers of the product formed. Given the similarity of the precursor complexes, a common gross dissociative mechanism is presumed – so why the change in the isomer ratio of the products? The answer lies in the form of the transition state.

For a square-based pyramidal intermediate, full retention of stereochemistry for the *cis* isomer is predicted (Figure 5.13, top) as only one site of attachment for the incoming water group to regenerate the original stereochemistry is offered, whereas an intermediate with the trigonal bipyramidal shape for the *trans* isomer allows choice because water entry may occur from any of three different positions around the triangular plane, some of which lead to different geometric isomers (in fact, a theoretical ratio of 2:1 *cis*:*trans* is predicted, not too distant from the experimental results). Thus, simple experiments with geometric isomers, involving separating and identifying product isomers, allow the reaction mechanism to be probed in some detail.

5.3.1.1.5   *Base Hydrolysis*

Although most reactions we meet occur in aqueous solution, we have not yet looked at the effect of one obvious variable in aqueous solution, which is pH. Reactions may occur in neutral, acidic or basic solution. What has been observed is that some reactions are accelerated by protons and others by hydroxide ions. Such observations infer a role for these ions in the mechanism. We shall examine the origin of such effects in one case only here, with hydroxide ions.

Reactions in aqueous solution in the presence of added base (hydroxide ion) may lead to substitution of one donor group by hydroxide, a process termed *base hydrolysis*. As some ligands are not readily replaced by hydroxide ions, the leaving group is not selected arbitrarily. Some ligands are more susceptible to reactions than others. A simple example is the complex ion [CoCl(NH$_3$)$_5$]$^{2+}$, where the NH$_3$ groups are only very slowly replaced, whereas the Cl⁻ ion is much more readily removed, leading to a specific reaction (5.38).

$$[CoCl(NH_3)_5]^{2+} + {}^-OH \rightarrow [Co(NH_3)_5(OH)]^{2+} + Cl^- \qquad (5.38)$$

The reaction is catalysed by the base, seen experimentally by the reaction becoming faster as the concentration of the base increases. For this behaviour, Eq. (5.39) fits the experimental observations (where k_{obsd} is the experimentally observed rate coefficient).

$$\text{rate} = k_{obsd}[CoCl(NH_3)_5]^{2+}[OH^-] \qquad (5.39)$$

Upon initial inspection, this expression looks like that given earlier for an **A** mechanism, but is this the case? There is strong support for a separate, distinct mechanism in this case, involving *conjugate base* formation. From a range of studies, the now accepted mechanism involves a pre-equilibrium deprotonation and a following **D** mechanism, as illustrated in Figure 5.14.

The reaction commences with formation of the conjugate base of the complex by proton loss from an ammine ligand to generate an amido ligand (⁻NH₂), from which complex the leaving group Cl⁻ dissociates (slowly), generating a five-coordinate intermediate. Entry of a water molecule fills the vacant coordination site, with rapid proton transfer (from coordinated H₂O to ⁻NH₂ to form ⁻OH and NH₃) completing the reaction (Figure 5.14). A somewhat complicated concentration dependence of the reaction rate for the mechanism, shown as an inset in Figure 5.14, simplifies to fit the experimentally observed linear dependence on complex and hydroxide

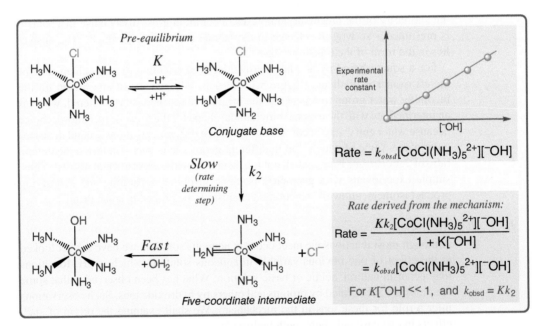

Figure 5.14

Mechanism for base hydrolysis of cobalt(III) complexes. A pre-equilibrium ammine ligand deprotonation step is followed by slow loss of the leaving group (Cl⁻ here) to form a short-lived five-coordinate intermediate that reacts readily with solvent water to form the hydrolysed product. A first-order dependence on [OH⁻] is observed experimentally (top inset), with the mechanism yielding the equivalent expression (lower inset).

ion concentration under conditions when the hydroxide ion concentration is high. The intermediate five-coordinate species (as in a standard **D** mechanism) is stabilized further in this case by the special $^-NH_2$ bonding. This can be visualised as having a double-bond character through the π-type interaction with the metal ion of the additional lone pair created upon deprotonation, a process that is unavailable to an ammonia molecule, which has only the one lone pair which is required for coordinate bond formation. As a result, the activation energy for forming this intermediate is lowered compared with the usual **D** mechanism, and therefore *catalysis* has occurred. This is related to the intermediate being more stable, though still not sufficiently so as to be isolable. This mechanism obviously relies on the capacity of an ammine ligand to deprotonate, and the general observation that N–H bonds are very rapidly replaced by N–D bonds in alkaline D_2O solution supports facile deprotonation/protonation reactions for ammine ligands. Moreover, if all ammine ligands are replaced by the heterocyclic N-donor pyridine, which has no N–H bond and is thus unable to react by the same mechanism, the reaction is much slower, despite the bulkier ligands possibly promoting a dissociative process.

We have introduced *catalysis* above, and you should be aware that this is common behaviour in chemistry and biology. Catalysts are important because they play key roles in several industrial processes (see Chapter 9). A catalyst accelerates a reaction through a lowering of the activation barrier. Although the catalyst influences the reaction, it is not itself consumed in the process. The hydrolysis reaction discussed here, because a hydroxide ion ends up coordinated to the cobalt ion, would appear to break the rule of the catalyst not being consumed. However, here, the mechanism involves hydroxide performing the role only of promoting deprotonation of an ammine ligand, and it is a water molecule that is the reactant that 'splits' to insert a hydroxide ion as the ligand and adds a proton to the $^-NH_2$ ligand to regenerate the ammine ligand. Thus, we can legitimately describe this base hydrolysis reaction as a catalysed process.

5.3.1.1.6 *Ion Pair Formation*

Another reaction that poses similar problems in interpretation is where an anionic group other than hydroxide replaces another group in a cationic complex. For these types of reactions, as the cation and anion are oppositely charged, they can undergo strong electrostatic attraction to form what is called an *ion pair* in an equilibrium reaction. This outer-sphere process, where X^{m-} is not yet part of the coordination sphere but attracted to and in proximity of the cationic complex, precedes ligand substitution, and therefore K_{IP} represents a *pre-equilibrium*.

$$[MA_5B]^{n+} + X^{m-} \underset{}{\overset{K_{IP}}{\rightleftharpoons}} \{[MA_5B]^{n+} : X^{m-}\}^{\#} \xrightarrow{k'} [MA_5X]^{p+} + B^{q-} \qquad (5.40)$$

A complication is that either the free complex cation or the ion pair can proceed forward to form the final products. The unpaired cation will react differently to the ion-paired cation (which is, apart from the ion-pair formation, otherwise identical). If the reaction is accelerated due to ion-pair formation, we can derive

$$\text{rate} = k'K_{IP}[MA_5B][X] = k[MA_5B][X] \text{ (where } k = k'K_{IP}) \qquad (5.41)$$

which cannot be distinguished from the direct **A** mechanism, which has the same apparent rate law. Fortunately, ion-pair strength for complexes in aqueous solution is usually low, so it is not usually a mechanistic issue in water but is important in organic solvents where ion pairing is stronger.

5.3.1.1.7   Substitution by Solvent

Other types of reactions may also not present themselves as mechanistically obvious due to complicating factors. For example, in a reaction where a solvent molecule is replacing another ligand, namely,

$$[MA_5X] + solvent \rightarrow [MA_5(solvent)] + X \tag{5.42}$$

we cannot distinguish between a dissociative mechanism, where

$$rate = k[MA_5X] \tag{5.43}$$

and an associative mechanism, where

$$rate = k[MA_5X][solvent] = k'[MA_5X] \text{ (where } k' = k[solvent]) \tag{5.44}$$

because the concentration of the solvent is high and cannot be varied. Changing solvent concentration by dilution through addition of another solvent changes the resultant solvent properties since no solvent is 'innocent' and without effects, invalidating that process. This means that an associative mechanism would appear experimentally as a simple first-order expression. Thus, the **D** and **A** mechanisms present the same apparent kinetic behaviour.

These examples are presented to show that defining a reaction mechanism is not a simple field of study since we are dealing with transient species that cannot be isolated or often even sensed directly by any method. Nevertheless, the field has progressed due to a suite of sophisticated experiments to the point where basic concepts, as discussed above, are fairly well understood and we have a true understanding of the mechanism at a molecular level.

5.3.1.2   The Square Planar Substitution Mechanism

Since the square planar geometry has a coordination number of only 4, and there are no ligands bound above and below the plane of the metal and its surrounding ligands, substitution where the entering ligand attaches first in an axial position prior to departure of the leaving group seems very reasonable. Although no ligands exist above and below the metal and ML_4 plane, this region is occupied by electrons in the d_{z^2} orbital, as explained in Chapter 2. Upon entry of a ligand, electronic rearrangement is necessary. This process, with bond-making dominant, is an associative mechanism (Figure 5.15). There is good experimental evidence to support it as the dominant mechanism for square planar complexes. A dissociative mechanism would

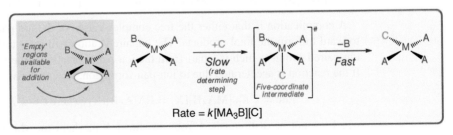

Rate = $k[MA_3B][C]$

Figure 5.15

Associative mechanism for substitution reactions in square planar complexes. Sites above or below the plane are inherently readily accessible to incoming ligands.

generate a three-coordinate intermediate, a rare geometry reserved exclusively for d^{10} complexes (e.g. Cu(I) and Pd(0)) which suggests it may be disfavoured compared with a five-coordinate intermediate as formed in the associative mechanism.

Evidence favouring the **A** mechanism can be exemplified for the well-known square planar d^8 Pt(II) complexes, with three pieces of evidence relevant:

(1) The rates of *aquation* (the process where a ligand, in this example chloride ion, is replaced by water) of $[PtCl_4]^{2-}$, $[PtCl_3(NH_3)]^-$, $[PtCl_2(NH_3)_2]$ and $[PtCl(NH_3)_3]^+$ are all similar; that is, the rate is independent of complex charge. This is consistent with rate-determining attack by neutral H_2O in an **A** mechanism. For a **D** mechanism, rate-determining release of the anion Cl^- should be affected significantly by the original complex charge, as one would expect that it would be easier for an anion to be repelled from an anionic complex than a neutral or cationic one, and this is clearly *not* the case supported by the experiment.

(2) Reactions show a dependence on the entering group concentration. For the halide substitution reaction in Eq. (5.45), the rate = $k[PtCl(NH_3)_3][Br^-]$, which is behaviour consistent with an **A** mechanism.

$$[PtCl(NH_3)_3]^+ + Br^- \rightarrow [PtBr(NH_3)_3]^+ + Cl^- \qquad (5.45)$$

(3) The effect of changing the leaving group (here the anion X^-) on the rate is not as large as usually observed with dissociative mechanisms. For the series of complexes $[Pt(dien)X]^+$, that include the tridentate amine ligand dien, $(H_2NCH_2CH_2NHCH_2CH_2NH_2)$, reacting with pyridine (py) to form $[Pt(dien)(py)]^{2+}$ and release X^-, the observed rate constant varies only modestly across a wide range of X^- ligands; with 'softer' leaving groups with a higher affinity for Pt^{2+} leading to a slower reaction.

Discussion of square planar complexes in terms of mechanistic concepts that were developed in more detail for octahedral complexes serves to illustrate that the concepts devised initially for octahedral geometry can be transferred in large part to other coordination numbers and stereochemistries. In doing so, of course, it is necessary to consider the most likely way a reaction may occur, as exemplified above for square planar geometry.

5.3.2 Stereochemical Change

As outlined at the commencement of Section 5.3, there are other types of reactions apart from substitution reactions. One of these of key importance involves what we can describe as a change in the stereochemistry. This may involve a change in gross shape or symmetry of the molecule, such as a switch from square planar to tetrahedral, or, more often observed, a change in the relative position or three-dimensional arrangement of ligands in a particularly shaped complex.

5.3.2.1 Change in Coordination Geometry

A complex prepared and isolated usually exists in a thermodynamically stable form. As such, there seems to be little reason for the complex to undergo a change in shape or stereochemistry since this would cost energy and not usually lead to a more stable complex. For a complex that exists in a particular geometry, to undergo a change in its overall shape without any change in the ligand set, there must be some change in its environment. This may involve a change of temperature or else dissolution in a different solvent that may favour a different shape. An example of this type of behaviour is transition from square planar to tetrahedral geometry, for which the activation barrier can be low. This has been exemplified earlier (Chapter 4.2). Another example involves the reaction in Eq. (5.46) (X = halide). This system displays a relatively small activation barrier of $\sim$45 kJ mol^{-1}, with a rate constant of $\sim$10^5 s^{-1} at 298 K.

$$(5.46)$$

Molecules that exhibit stereochemical non-rigidity are said to display *fluxional character*. All molecules undergo vibrations about an equilibrium position that does not alter their average spatial location; for a limited number, however, rearrangement that changes the configuration can occur. One of the most common ligands, ammonia, is a simple example of the concept (Eq. (5.47)) since, as a pyramidal molecule, it can invert.

Transition state

$$(5.47)$$

This has a low energy barrier ($\sim$25 kJ mol^{-1}), and inversion occurs many times every nanosecond ($k \sim 2 \times 10^{10}$ s^{-1}). However, for pyramidal PR$_3$ compounds, the activation barrier has increased to over 100 kJ mol^{-1}, so inversion is very slow (timescale of hours), sufficiently slow to even allow enantiomer (optical isomer) separation in the case of chiral phosphines where the three R groups are different. The related organic arsines (AsR$_3$) only invert on timescales of years, by comparison.

In all of the Group 15 congeners (ER$_3$, E = N, P, As) though, coordination of the E lone pair 'freezes' the configuration indefinitely as there is no pathway to inversion without breaking the M$-$ER$_3$ bond. Examples of fluxional complexes tend to be found mostly, but not exclusively, amidst organometallic compounds. Spectroscopic methods, particularly nuclear magnetic resonance spectroscopy (Chapter 7.3.5), assist in defining the fluxional behaviour.

5.3.2.2   *Change in the Mode of Ligand Coordination – Linkage Isomerisation*

Whereas many ligands offer a single lone pair on only one atom or at least a single donor and are thus able to coordinate at one site and in only one way, there are a range of *ambidentate* ligands introduced in Chapter 4 that offer two or more potential donors carrying lone pairs of electrons. Put simply, they have a choice of donors by which they may be attached to a particular site. While isolated species usually are the thermodynamically most stable form, it may be possible to prepare a complex where the ligand is bound in its less stable form. For example, the thermodynamically unstable [Co(NH$_3$)$_5$($\underline{S}$CN)]$^{2+}$ (S-coordinated form of thiocyanate) is formed preferentially by capture in basic solution of the [Co(NH$_3$)$_4$($^-$NH$_2$)]$^{2+}$ intermediate by the less solvated S nucleophile of thiocyanate; it is sufficiently stable to be isolable in the solid state if precipitated rapidly. However, in solution, this complex spontaneously undergoes ligand donor group rearrangement, forming the thermodynamically more stable $[Co(NH_3)_5(\underline{N}CS)]^{2+}$ (N-coordinated form). This occurs via an intramolecular mechanism, proceeding through an intimate ion-paired intermediate. This process of changing from one type of coordinated donor group to another for an ambidentate ligand is termed *linkage isomerisation*, the name referring to a change in the way the ligand is linked to the metal centre that generates *linkage isomers*.

The process can be represented in a general form as

$$[L_nM(\mathbf{X}-\mathbf{Y})]^{m+} \rightarrow [L_nM(\mathbf{Y}-\mathbf{X})]^{m+} \tag{5.48}$$

where there is no change in the chemical composition, but the coordination sphere is altered when one donor of an ambidentate ligand swaps for another. Because the two alternate donors are usually different elements, a change in the colour of the complex may occur in this reaction.

Another example of an ambidentate donor is the nitrite ion (NO_2^-), which can coordinate via either the N or an O atom. One of the earliest known examples is the pair of linkage isomers $[Co(NH_3)_5(\underline{O}NO)]^{2+}$ and $[Co(NH_3)_5(\underline{N}O_2)]^{2+}$ (the latter known in the 1800s as xanthocobalt) which exhibit contrasting colours due to the differing ligand field strengths of the O- and N-bound nitrite anions (Figure 5.16). Coordination modes for *nitrito* (O-bound; nitrito-*O*) and *nitro* (N-bound; nitro-*N*) linkage isomers are also shown in Eq. (5.49), along with the proposed transition state that may be involved in the process of linkage isomerisation.

$$\tag{5.49}$$

Because isotopic labelling studies have shown that the reaction above occurs without the coordinated anion departing the coordination sphere, the mechanism of isomerisation for the nitrito–nitro reaction has been proposed to involve a symmetrical transition state where both oxygen atoms are disposed equidistant from the metal. Relaxation from this activated state occurs by either recapture of an oxygen atom to

Figure 5.16
The linkage isomers $[Co(NH_3)_5(\underline{O}NO)]^{2+}$ (left) and $[Co(NH_3)_5(\underline{N}O_2)]^{2+}$ (right) both isolated as their chloride salts.

give the nitrito isomer (which means no effective reaction occurs, apart from possibly oxygen 'scrambling') or capture of the nitrogen alone (leading to the more stable nitro isomer). Although this isomerisation typically occurs in aqueous solution, the same reaction also takes place in the solid state by heating the red/orange salt $[Co(NH_3)_5(\underline{O}NO)]Cl_2$ (Figure 5.16) in an oven to generate the more stable yellow $[Co(NH_3)_5(\underline{N}O_2)]Cl_2$ quantitatively, which further supports the mechanism shown in (5.49).

5.3.2.3 *Change in the Relative Position of Ligands – Geometrical Isomerisation*

Ligand rearrangement of the type mentioned in the previous section involves a change in the way a particular ligand is bound. Another process focussed on ligand arrangement involves changing the way a set of ligands are assembled around a central metal. We have already established in Chapter 4 that ligands can in some cases be organised around the central metal with a number of outcomes, forming *geometric isomers*. Apart from separating and observing geometric isomers, it became apparent early on to researchers that interconversion between geometric isomers was possible. Many of these interconversions are accompanied by a colour change, while analysis shows clearly that the set of ligands originally present have been maintained in the reaction. Two key conclusions were made: the way ligands are arranged around a central metal ion influences physical properties such as colour; and there must be a sufficiently low energy barrier to reaching the activated state so as to permit the observed interconversions.

In effect, there are two types of isomerisation reactions (Figure 5.17), and we shall examine each in turn. However, it is appropriate initially to identify the two and the differences. In one form, called *optical isomerisation*, change in optical properties alone (such as the direction of rotation of plane polarized light interacting with a solution of the complex) occurs. Because no other physical change is observed, this type of reaction is not readily 'seen', as it requires the use of devices that measure specific optical properties. In the other and more familiar form, *geometrical isomerisation*, change in most physical properties (such as colour) occurs, so this type is readily observed, often even by the naked eye. The reason for the stark difference between observable behaviour lies in the way the set of donors in the coordination sphere is affected. In optical isomerisation, one optical isomer is converted into the other. Optical isomers are identical in nearly every way, except they are mirror images of each other; like our left and right hands (or feet). The conversion of one optical isomer into the other occurs without any change in the relative arrangement of the set of donors themselves and thus has no influence on most physical properties. In geometrical isomerisation, although the set of donors remains unchanged, the relative arrangement of donors in space around the central metal changes. This is accompanied by a change in molecular symmetry, with subsequent effects on physical properties, even stability.

The mechanism by which one geometric isomer converts to another can employ the type of intermediates used in substitution reactions, discussed earlier. The clear difference here is that no new ligand can replace one already bound; any additional molecule present in the coordination sphere in the activated state must be non-competitive as a ligand, allowing the original coordination sphere to remain intact. While both **A** and **D** mechanisms can account for isomerisation reactions represented in Figure 5.17, another prospect is for the molecule to undergo a bond angle distortion and twisting (without any bond-breaking or participation of any

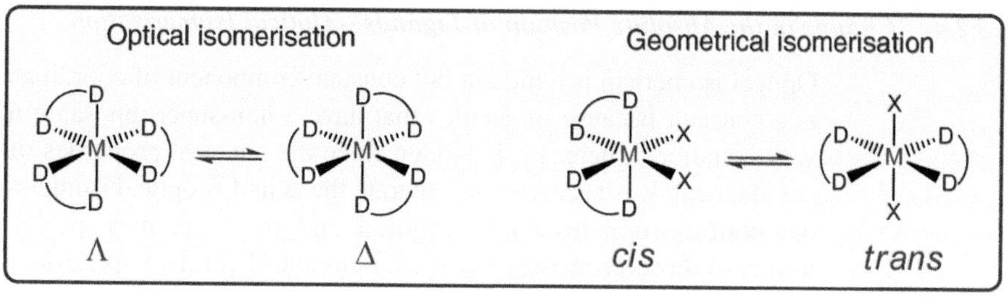

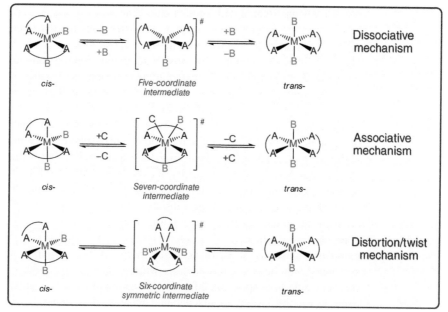

Figure 5.17

Optical and geometrical isomerisation represent the two classes of reaction where change of the relative position of ligands is involved.

Figure 5.18

Mechanisms for geometrical isomerisation: dissociative, associative and distortion/twist mechanisms.

entering group) that takes it to a more symmetrical six-coordinate trigonal prismatic geometry in the activated state from which it can revert to the original isomer or continue to form the other optical isomer (Figure 5.18, bottom).

While geometric isomerisation without any ligand substitution processes interfering is well known, it is not unusual to observe substitution reactions *also* occurring along with geometrical isomerisation. An example is given in Eq. (5.50).

$$trans\text{-}[CoCl_2(en)_2]^+ + OH_2 \rightarrow cis\text{-}[CoCl(en)_2(OH_2)]^{2+} + Cl^- \qquad (5.50)$$

Introduction of a different ligand alters the relative stabilities of *cis*- and *trans*-isomers, so there may be a driving force for isomerisation, leading to this occurring concomitant with substitution. In circumstances where two geometric isomers are energetically similar, both may exist in solution once thermodynamic equilibrium is reached, with the amount of each form present defined by the relative stability of the isomers.

5.3.2.4   *Change in the Absolute Position of Ligands – Optical Isomerisation*

Optical isomerism is a hidden but constant component of coordination chemistry. It is a constant because molecules that have a non-superimposable mirror image are very common, whereas it is hidden since the physical properties of optical isomers are identical. It is necessary to separate the Δ and Λ optical isomers before their hidden optical properties can be exposed, and then it requires special instrumentation to record those properties (such as a circular dichroism spectrometer, described in Chapter 7.3.3). If this is the case, one might ask why the interest in optical isomerism anyway? One answer lies in biological systems, where optical isomers do behave differently as a result of the way they match with a specific optically active (chiral) environment. This is merely an extension of the way optical isomers are separated in chemistry; while the Δ and Λ forms of a complex behave identically as isolated species, when partnered with another chiral entity Δ', the $\{\Delta\cdot\Delta'\}$ and $\{\Lambda\cdot\Delta'\}$ assemblies are no longer equivalent in their properties (they are diastereomers), allowing their separation. Presented with a naturally chiral protein, the Δ and Λ forms of a complex may likewise interact differently. Natural coordination complexes may exist as one particular optical isomer, as exemplified in Chapter 8. Another observation is that one of the spectroscopic methods for reporting optical activity, circular dichroism, is important in providing information on the splitting of energy levels in complexes under reduced symmetry and also is very sensitive to the surrounding environment of a complex, providing a method for probing outer-sphere interactions of complexes with other molecules or ions. Apart from these aspects, it is simply important to understand the full breadth of the chemistry of complexes, of which chirality is an important component. *Chiral* molecules are those that cannot be superimposed on their mirror image through any rotational or translational processes, whereas *achiral* molecules are identical to their mirror image. For complexes, in terms of molecular symmetry concepts, chiral molecules do not have an S_n (rotation–reflection) axis so that chiral complexes exist only in C_n or D_n point groups. A point group describes molecular symmetry, and the single term defines what symmetry operations can be performed on a molecule (see Appendix 2 for an explanation of these concepts; fortunately, you can get by at this level without a detailed understanding of point groups). Complexes *without* reflection or inversion symmetry (i.e. no plane or point of symmetry and termed *dissymmetric*), are chiral, but may have some elements of symmetry (rotation axes). An *asymmetric* compound has no symmetry elements at all (apart from trivial C_1) and is, of course, also chiral.

It is known that an optically pure complex (100% of either the Δ or the Λ form) can undergo a process called *racemisation*, whereby it returns to an optically inactive racemic (50:50 Δ:Λ) mixture of its two optical isomers. For this to occur, it is assumed that it must proceed through a symmetrical transition state, from which there is no particular preference regarding whether it reverts to the original or the opposite optical isomer, thus providing a process that will lead to a racemate. For most complexes, it is possible to devise an appropriate transition state by invoking traditional **D** or **A** mechanisms. In addition, the *twist mechanism* developed above (Figure 5.18) for geometrical isomerism can be usefully applied to optical isomerism; unlike the other two mechanisms, this involves no coordinate bond-breaking or bond-making but merely bond angle distortion in moving to and from the activated state. These mechanisms are illustrated in Figure 5.19 for one of the classical examples of optical activity in coordination complexes, octahedral compounds with three didentate chelate ligands.

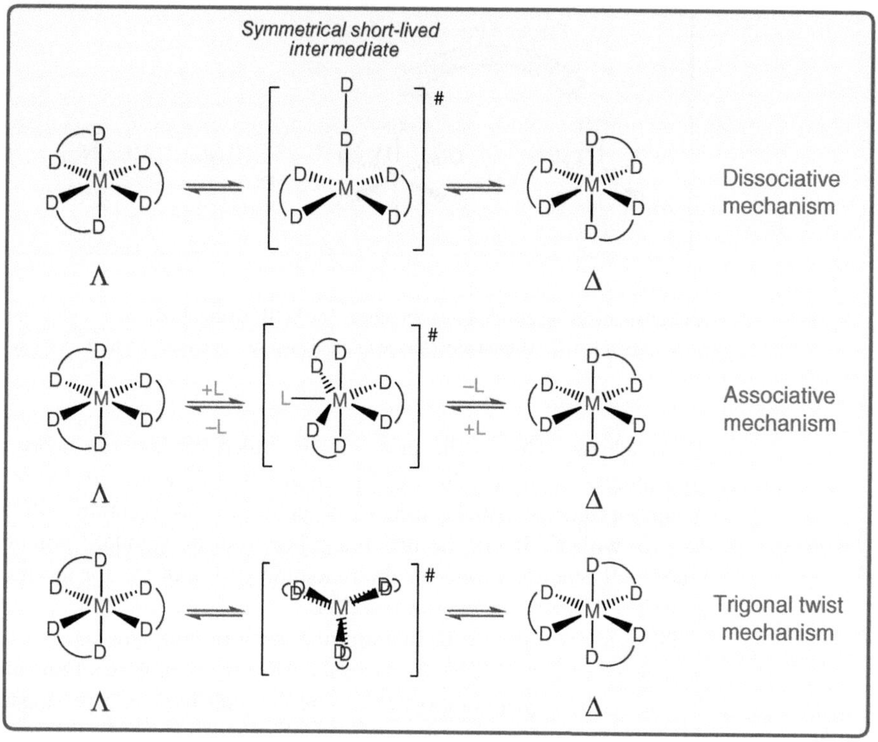

Figure 5.19

Mechanisms for isomerisation of an octahedral complex with three didentate chelates, proceeding through a symmetrical short-lived intermediate in each possible mechanism.

All three mechanisms involve a transition state, so there is an activation barrier to overcome. For some complexes, the isomerisation is very slow (as for the inert cobalt(III) complex [Co(en)$_3$]$^{3+}$), whereas for others, the process is sufficiently fast (as for the nickel(II) analogue [Ni(en)$_3$]$^{3+}$) that the complex cannot be separated into its optically pure forms in solution and can only be obtained in an optically pure form in the solid state through conventional crystallisation methods. The isomerisation reaction can be readily followed by observing the loss of optical rotation at a selected wavelength over time; this is usually a simple first-order exponential decay process.

5.3.3 Reduction–Oxidation Reactions

Reduction–oxidation (electron transfer) reactions are important in chemistry and biology. When a chemical oxidation of $\underline{A}$ by $\underline{B}$ occurs, $\underline{B}$ itself is reduced – an *electron transfer* process has occurred. For such chemical processes, there is always a partnership between an oxidant (which is reduced in carrying out its task) and a reductant (which is oxidised in the reaction); thus, we frequently talk of reduction–oxidation, or (for ease of use) *redox*, reactions for what are essentially electron transfer processes. Of course, an electron can be supplied or accepted electrochemically via an electrode rather than through using a chemical reducing or oxidising agent, and this is another way to initiate change in the oxidation state

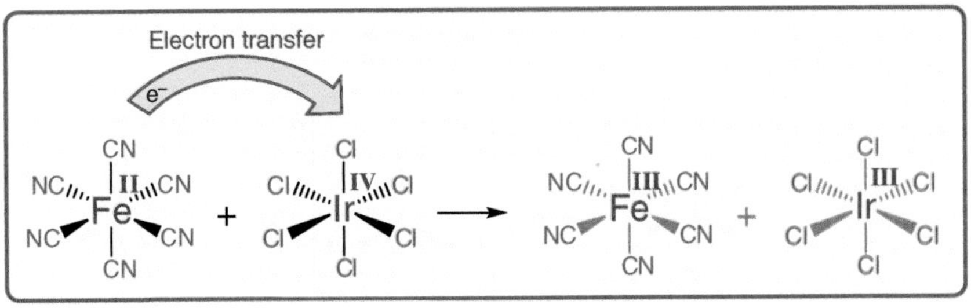

Figure 5.20

A reaction involving electron transfer alone. The Fe(II) centre is oxidised to Fe(III), and the Ir(IV) centre is reduced to Ir(III), without any change in the ligand set for either complex. (Charges are not shown.)

of complexes. This is the principle by which all batteries work. Here, we shall concentrate mainly on chemically based systems, where an oxidant and reductant of appropriate potentials need to be combined. A reaction will be favourable if $E^o > 0$, with E^o being the difference between the standard potentials for the two redox half-reactions (one for the oxidation part and one for the reduction) that are combined to form the overall reaction.

There are two processes that are important in inorganic redox reactions. The key process is, of course, *electron transfer*. However, some reactions also involve *atom transfer*, whereby a component of one reacting molecule is transferred to another during the reaction. It is not essential for both to occur, and many reactions are purely electron transfer reactions. A classic example of a pure electron transfer alone is the reaction shown in Figure 5.20.

This reaction of the two octahedral complexes occurs without any change in the coordination spheres, or ligand sets, of either metal. However, if you inspect the two metal centres, you will note that the iron complex (the reductant) is oxidised from Fe(II) to Fe(III), and at the same time, the iridium complex (the oxidant) is reduced from Ir(IV) to Ir(III) – an electron has been transferred from one metal to the other. This reaction can be conveniently followed since the colour of each species changes as it is converted from one oxidation state to another.

Atom transfer alone can also occur, although it is a more difficult concept to come to grips with. Think of it as an atom, ion or molecule moving with its normal complement of valence electrons, and no others, to or from a metal. *Oxidative addition reactions*, common in organometallic chemistry, can be considered a form of atom transfer. The classical reaction involves introduction of two anionic ligands (X$^-$ and Y$^-$) that originate from the neutral molecule X–Y to a metal centre, which leads to an equal increase in coordination number and metal oxidation state. The formal negative charges on the ligands X$^-$ and Y$^-$ are both provided by electrons from the metal whose oxidation state increases by two in this case. The reverse process is called *reductive elimination* where the coordination number and the metal oxidation state both decrease by the same amount. A typical example of oxidative addition is given in Eq. (5.51), where the complex also steps from four-coordinate Ir(I) to six-coordinate Ir(III):

$$[Ir^ICl(CO)(PR_3)_2] + Cl_2 \rightarrow [Ir^{III}Cl_3(CO)(PR_3)_2] \qquad (5.51)$$

A variation on this electron/atom transfer mechanism, of considerable historic importance, is the reaction of the two octahedral complexes shown in Figure 5.21.

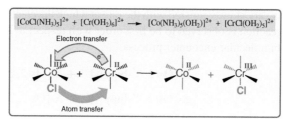

Figure 5.21
An example of electron transfer with associated atom transfer.

In the above example, an electron is formally transferred from the Cr(II) to the Co(III), and a chloride ion is transferred from the cobalt to the chromium ion as well. The Cr(III) product is inert, allowing it to be isolated and identified, thereby defining the presence of the chloride ion in its coordination sphere. Colour changes in this reaction allow the reaction to be readily monitored spectrophotometrically.

Clearly, the different processes of electron transfer alone, and electron transfer along with atom transfer illustrated above need not, and most likely do not, occur by a common mechanism – in fact, there are two general mechanisms that apply to all chemical compounds (metal complexes and organic compounds included). It was as recent as the 1950s that this distinction was first recognised and demonstrated through experiment by Chemistry Nobel laureate Henry Taube who defined these two pathways as the *outer sphere mechanism* and the *inner sphere mechanism*, and we shall examine each in turn to learn about their characteristics, relying on classical rather than quantum-mechanical concepts for simplicity.

5.3.3.1 The Outer Sphere Electron Transfer Mechanism

The key to the outer sphere mechanism is that electron transfer from reductant to oxidant occurs with the coordination spheres of each reactant staying intact at least up to the point where the electron 'jumps' from one metal to the other. Since the coordination (or inner) sphere, that is, the set of bound ligands, is not changed during the reactions, it appears that the key to the process lies beyond these, in the outer sphere around the reactants. We have seen an example earlier (Figure 5.20) involving two different metal centres.

Another curious example of pure electron transfer alone, involving two oxidation states of the one metal ion, is Eq. (5.52), a so-called *self-exchange* electron transfer reaction where the reactants and products are essentially identical. As you can imagine, it is not easy to measure a reaction when nothing appears to change, so some special experimental design is necessary. In this case, the Ru^{II} ion is bound to six water molecules with the most abundant ^{16}O isotope present, while the Ru^{III} ion has the less common ^{18}O isotope at the start. This 50:50 mixture of $[Ru^{II}(^{16}OH_2)_6]^{2+}$ and $[Ru^{III}(^{18}OH_2)_6]^{3+}$ converts to an equal (25%) mixture of $[Ru^{II}(^{16}OH_2)_6]^{2+}$, $[Ru^{III}(^{18}OH_2)_6]^{3+}$, $[Ru^{II}(^{18}OH_2)_6]^{2+}$ and $[Ru^{III}(^{16}OH_2)_6]^{3+}$ at equilibrium.

$$[Ru^{II}(^{16}OH_2)_6]^{2+} + [Ru^{III}(^{18}OH_2)_6]^{3+} \rightleftharpoons [Ru^{III}(^{16}OH_2)_6]^{3+} + [Ru^{II}(^{18}OH_2)_6]^{2+}$$
$$(5.52)$$

There is no scrambling (swapping) of the isotopically labelled water between metal ions; all water molecules remain completely with the one metal centre throughout the reaction because the rate at which Ru-coordinated water is exchanged with bulk water is very slow compared with the rate at which the electron transfer reaction occurs. The reaction is observed to be first order with respect to both reactants, consistent with a bimolecular encounter process.

$$\Delta\text{-}[\mathbf{Os}(2, 2'\text{-bipy})_3]^{3+} + \Lambda\text{-}[\underline{\mathbf{Os}}(2, 2'\text{-bipy})_3]^{2+} \rightleftharpoons$$

$$\Delta\text{-}[\mathbf{Os}(2, 2'\text{-bipy})_3]^{2+} + \Lambda\text{-}[\underline{\mathbf{Os}}(2, 2'\text{-bipy})_3]^{3+} \tag{5.53}$$

The above reaction (Eq. (5.53)) of the octahedral Os complexes of didentate 2,2'-bipyridyl occurs rapidly ($k = 5 \times 10^4$ M^{-1} s^{-1}), but, as in the reaction in Eq. (5.52), there appears to be nothing happening at first glance. However, if you inspect the two Os complex ions (differentiated above by one being underlined as well as in bold type), you will note that one is Os(III) on the left and Os(II) on the right, and vice versa – an electron has been transferred from one complex ion to another, although there is no colour change in the reaction as the products equate with the reactants exactly. The only way this reaction can be conveniently followed, and indeed shown to occur, is by using an optically active form of the complex in one of the two oxidation states in the initial mixture since its optical spectroscopic properties (see Chapter 7.3.3) change as it is converted from its original oxidation state to the other, even though the overall percentage of Os(II) and Os(III) complex remains constant.

There is a sequence of elementary steps that must occur for an outer sphere reaction to proceed. Chemistry Nobel laureate Rudoph A. Marcus is credited with providing the theoretical foundation that allowed accurate predictions of the rates of electron transfer reactions. We will not dive deep into the equations that Marcus derived, but instead provide a qualitative overview of the important factors that either accelerate or slow electron transfer. Firstly, the oxidant and reductant complexes must come together sufficiently close and in an arrangement in space appropriate for electron transfer to proceed, through collision and rotational orientation processes. Because outer sphere electron transfer over 'long' distances (>20 Å or 2 nm) is a high-energy process and is so slow as to be negligible, a closer approach of reactants is necessary. To achieve optimal contact, the complexes usually shed some of their immediate outer sphere of non-covalently bound solvent molecules to make a close and specifically oriented approach appropriate for orbitals involved in the electron transfer to interact sufficiently, sometimes termed forming an *encounter* or *precursor* complex. It is important to understand that no coordinate bonds are made or broken up to this point and the two reactants have come to a compromise and their metal–ligand bond lengths are the same, halfway between their oxidised and reduced form. Next, this non-covalent encounter complex assembly undergoes electron transfer, with subsequent adjustment of the successor complexes to their new oxidation states (sometimes termed *relaxation*).

Since metal–ligand bond lengths typically vary for many metal ions in different oxidation states (Figure 5.22), this becomes a key part of the adjustment process. For example, Co(III)–N distances are typically 196 pm, whereas Co(II)–N distances are near 217 pm. For cobalt complexes of N-donor ligands, there is consequently a significant change in Co–N bond lengths as a result of electron transfer, associated with the shrinking or expanding first coordination sphere tied to these bond length variations. If we return to ligand field theory, recall that the *d*-electrons for an octahedral complex lie in the t_{2g} and e_g^* levels (alternatively designated as π and σ^*, respectively, to equate with the character of bonding in which they participate in a complex exhibiting both σ-donor and π-donor/acceptor bonding character;

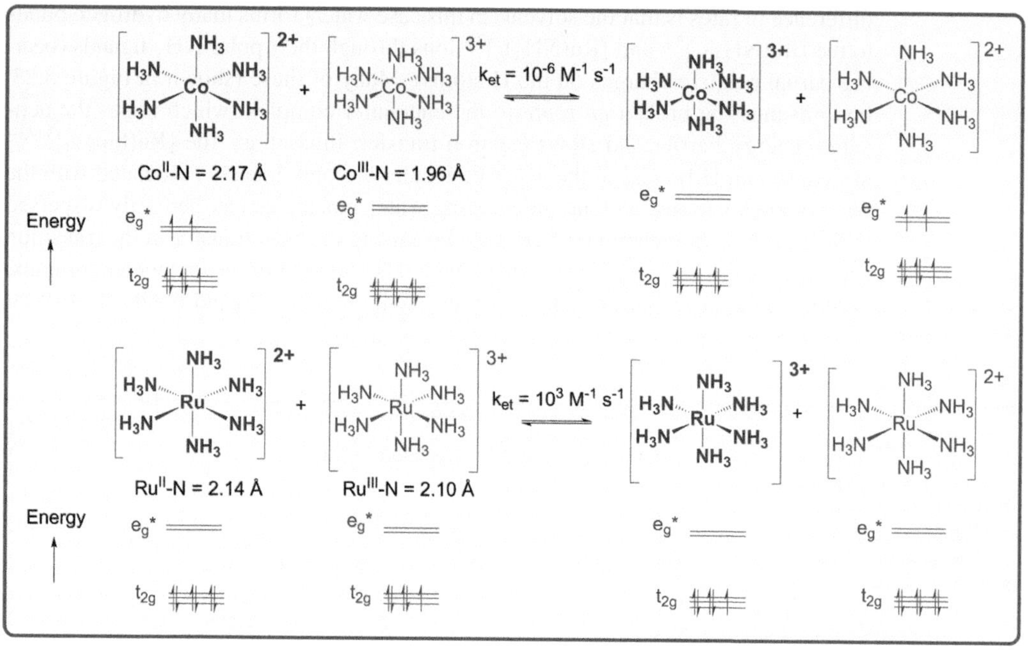

Figure 5.22

Self-exchange electron transfer reactions of $[Co(NH_3)_6]^{3+/2+}$ and $[Ru(NH_3)_6]^{3+/2+}$ with their preferred d-electronic configurations.

for consistency in this book, we will use the former nomenclature). It is these electrons that are exchanged in reactions of transition metals, and it is this feature that sets their chemistry apart from all other elements in the periodic table. For electron transfer, a metal-based d-electron needs to move from a location in a t_{2g} or e_g^* orbital on one metal ion to a t_{2g} or e_g^* orbital on the other metal ion. Change in spin state from low spin (Co(III)) to high spin (Co(II)) plays a key part in the above changes, as high-spin arrangements populate antibonding states more, affecting bond lengths.

For some ions, however, the change in bond distance with metal oxidation state alteration is small; ruthenium is such an ion, with Ru(III)–N distances of 210 pm and Ru(II)–N distances of 214 pm. In both oxidation states of this $4d$ metal, the complexes are low spin, as large Δ_o for $4d$ and $5d$ metals force low-spin behaviour, so the influences seen above with cobalt do not apply. All of these changes in metal–ligand bond length (large or small) can be associated with what we call the *inner sphere reorganizational energy* upon electron transfer; the smaller this is, the faster the reaction will occur, which explains why the self-exchange rate constant for $[Ru(NH_3)_6]^{3+/2+}$ is nine orders of magnitude larger than $[Co(NH_3)_6]^{3+/2+}$.

Another separate but important feature is how the solvent, which may interact with the complex via non-covalent forces such as hydrogen bonding and electrostatic (dipolar) forces, is affected by electron transfer. To continue with the system $[Ru(NH_3)_6]^{3+/2+}$ discussed above, where inner sphere reorganisation is small due to small changes in the Ru–N bond lengths, a three-orders of magnitude increase in the self-exchange electron transfer rate is found going to the $[Ru(bipy)_3]^{3+/2+}$ analogue (Figure 5.23). The changes in the Ru–N bond lengths are the same, so the inner sphere reorganisational energies do not differentiate the two systems. The reason for the large difference in rates is that the solvent (in this case water) forms many hydrogen bonds to the $[Ru(NH_3)_6]^{3+}$ and $[Ru(NH_3)_6]^{2+}$ ions through their polar NH_3 ligands (bearing partial positive charges on the H atoms). Many of these (shown in Figure 5.23, bottom) must be broken *en route* to the encounter complex, which raises the activation energy barrier and slows electron transfer. In contrast, the $[Ru(bipy)_3]^{3+/2+}$ system has no H-bonds, so the water molecules are only loosely associated with the two complex cations, and any intervening water molecules can be easily dispersed at little cost, so the reaction is very fast. Following electron transfer in the transition state, the *successor complex* is formed, which is identical to the encounter complex, except the metal oxidation states have reversed; the metal–ligand bonds are still far

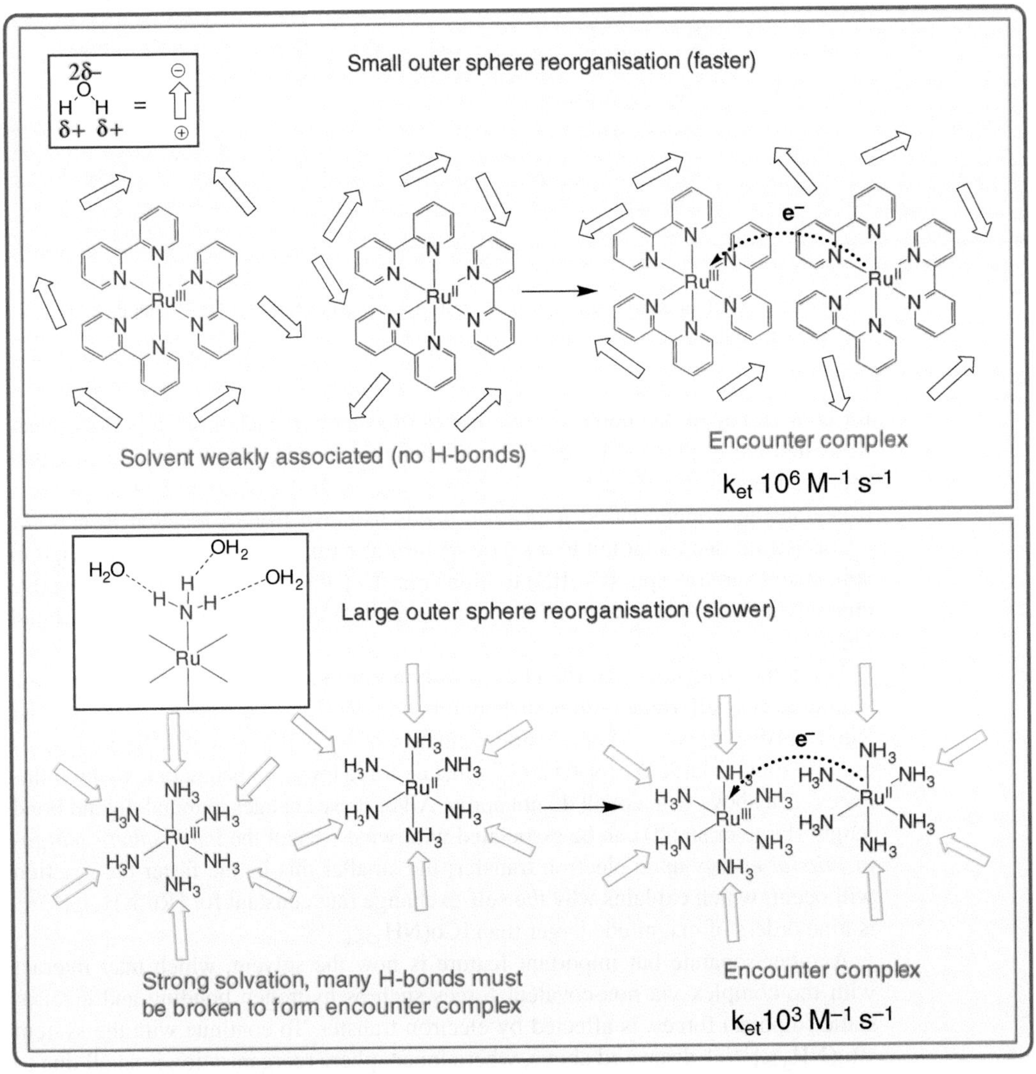

Figure 5.23

Outer sphere (solvent) reorganisation upon electron transfer where there is (top) weak solvation and (bottom) strong solvation: the green arrows represent strongly H-bonded water molecules (dipoles) that must be displaced before electron transfer can occur.

from their ideal values, and relaxation occurs rapidly to give the solvent separated complex ions comprising the products, each in their lowest energy states.

The same concepts extend to the more common situation involving reactions between complexes where the metals and ligands are different, so-called electron transfer *cross-reactions*. Here, unlike a self-exchange reaction, there is a change in the free energy of the system and a thermodynamic driving force that usually favours the products over the reactants. The reaction of cobalt(III) complexes in mixed-metal outer-sphere electron transfer reactions is usually slow also, as a result of the influence of effects discussed above. The reaction in Eq. (5.54) of the hexaamminecobalt(III) ion with the hexaaquachromium(II) complex occurs slowly, with a rate constant of $\sim 10^{-3}$ M^{-1} s^{-1}.

$$[Co^{III}(NH_3)_6]^{3+} + [Cr^{II}(OH_2)_6]^{2+} \xrightarrow[slow]{} [Co^{II}(NH_3)_6]^{2+} + [Cr^{III}(OH_2)_6]^{3+} \quad (5.54)$$

$$[Co^{II}(NH_3)_6]^{2+} \xrightarrow[fast]{H^+} [Co^{II}(OH_2)_6]^{2+} + 6NH_4^+ \quad (5.55)$$

Decomposition of the initially formed Co(II) ammine complex product in aqueous acidic reaction conditions to release ammonium ion and form the Co_{aq}^{2+} ion is characteristic of the outcome for Co(III) systems upon forming a labile Co(II) product (Eq. (5.55)). It is important to note that the occurrence of the reaction in Eq. (5.55) does not affect the electron transfer mechanism, which is still an outer sphere process, as no bonds are made or broken up to the electron transfer event.

5.3.3.2   *The Inner-Sphere Electron Transfer Mechanism*

In 1953, subsequent Nobel laureate Henry Taube and co-workers studied the series of reactions (Eq. (5.56)) where X = NH_3, H_2O, Cl^- and Br^-.

$$[Co^{III}(NH_3)_5X]^{n+} + [Cr^{II}(OH_2)_6]^{2+} \rightarrow [Co^{II}(NH_3)_5(OH_2)]^{2+} + [Cr^{III}(OH_2)_5X]^{n+} \quad (5.56)$$

Notably, the reaction where $[Co(NH_3)_6]^{3+}$ (X = NH_3 in (5.56)) is replaced by $[Co(NH_3)_5Cl]^{2+}$ (X = Cl^- in (5.56)) is much faster ($\sim 10^9$-fold). Such a substantial acceleration in reaction rate for an apparently minor change could not be explained if the same mechanism was operating.

This study proved to be a turning point in our understanding of coordination chemistry and electron transfer reactions in general. Here, Taube first proposed the formation of a chloride-bridged intermediate which then underwent electron transfer (Eq. (5.57)) before dissociation to Co(II) and Cr(III) products (Eq. (5.58)), and the *inner sphere electron transfer mechanism* was born.

$$[(H_3N)_5Co^{III} - Cl - Cr^{II}(OH_2)_5]^{4+} \xrightarrow[fast]{} [(H_3N)_5Co^{II} - Cl - Cr^{III}(OH_2)_5]^{4+} \quad (5.57)$$

$$[(H_3N)_5Co^{II} - Cl - Cr^{III}(OH_2)_5]^{4+} \rightarrow [Co^{II}(OH_2)_6]^{2+}$$
$$+ [ClCr^{III}(OH_2)_5]^{2+} + 5NH_4^+ \quad (5.58)$$

When X = NH_3 in Eq. (5.56) an intermediate of the type proposed in Eq. (5.57) cannot form as this ammine ligand has but one lone pair which is needed to bind to its parent Co(III) centre. When lone pair-rich X = Cl^- or Br^- are involved, then the bridged intermediate $[(NH_3)_5Co^{III}–X–Cr^{II}(OH_2)_5]^{4+}$ may form. The identification by Taube of $[Cr^{III}(OH_2)_5Cl]^{2+}$ as the chromium(III) product of the reaction in Eq. (5.58), which involves a necessary handover of chloride from inert Co(III) to labile Cr(II) concomitant with electron transfer, was the key observation. The alternative source for formation of $[Cr^{III}(OH_2)_5Cl]^{2+}$ from ligand substitution by chloride on inert $[Cr^{III}(OH_2)_6]^{3+}$ was a known slow reaction, whereas in this case, $[Cr^{III}(OH_2)_5Cl]^{2+}$ is formed within fractions of a second. Indeed, the chlorido ligand of $[Co^{III}(NH_3)_5Cl]^{2+}$ was transferred quantitatively to $[Cr^{III}(OH_2)_5Cl]^{2+}$ even in the presence of radioactively labelled chloride ions in solution without any radioactivity introduced into the product.

The key to the inner sphere mechanism is that electron transfer from the reductant to oxidant occurs across a *bridging group*, which is a ligand shared between the reductant and the oxidant at the point of electron transfer. This bridging group has one obvious requirement – it must be able to bind to two metals simultaneously, which means that it must have two lone pairs or else several donors orientated in such a way as to accommodate this shared arrangement. Of course, there are many examples of stable compounds where ligands bridge between two metal centres, in dimer and higher oligomer complexes, so invoking such a short-lived intermediate in a reaction profile does not introduce a large leap away from conventional chemistry. The chloride ion is an example of a single-atom ligand capable of bridging, being replete with lone pairs of electrons; many examples of M–Cl–M′ coordination exist.

Another type commonly met is an ambidentate ligand with two different donor atoms available, such as thiocyanate (SCN^-), which offers both an N- and an S-donor and can form M–SCN–M′ linkages. As an example, we have the reaction between the cobalt(III) complex $[Co(NH_3)_5(\underline{N}CS)]^{2+}$ (the oxidant) with $[Cr(OH_2)_6]^{2+}$ (the reductant), where thiocyanate acts as the bridging ligand. The overall reaction in aqueous acidic solution is shown in Eq. (5.59).

$$[Co^{III}(NH_3)_5(\underline{N}CS)]^{2+} + [Cr^{II}(OH_2)_6]^{2+}(+5H^+) \rightarrow \quad [Cr^{III}(OH_2)_5(\underline{S}CN)]^{2+}$$
$$+ Co^{2+}{}_{aq} + 5NH_4{}^+ \quad (5.59)$$

The original inert Co(III) complex with N-bound thiocyanate has been reduced to a labile Co(II) complex. In the acidic conditions present in this reaction, ammonia ligands on the labile Co(II) complex first formed have subsequently rapidly dissociated and protonated into free ammonium ions, leaving a Co(II) aqua ion as already met before. The labile Cr(II) centre is not only oxidised to an inert Cr(III) complex, but an S-bonded thiocyanate group has been introduced into its coordination sphere. The only source of this ligand is the original Co(III) complex. While this anion may have been captured from solution following dissociation of the reduced Co(II) species, the observation that it is bonded entirely by the S-donor group suggests that it was transferred in a 'handshake' operation where it is shared between the cobalt and chromium centres; an intermediate of the type $\{[(H_3N)_5Co–NCS–Cr(OH_2)_5]^{4+}\}$ is presumed to form, where one water ligand in the labile Cr(II) complex has been displaced and replaced by the thiocyanate S atom exclusively, as the N atom is already attached to the Co centre. The overall mechanism that has been proposed is shown in Figure 5.24, for a general bridging ligand represented as X–Y.

The process of bridge formation leads to a marked acceleration of the electron transfer process compared with outer sphere reactions of related compounds. In general, the key requirements for an inner sphere mechanism to operate can be summarised as follows:

(a) one reactant must possess at least one ligand capable of binding simultaneously to two metal ions in a bridging arrangement (this reactant is usually the oxidant) and

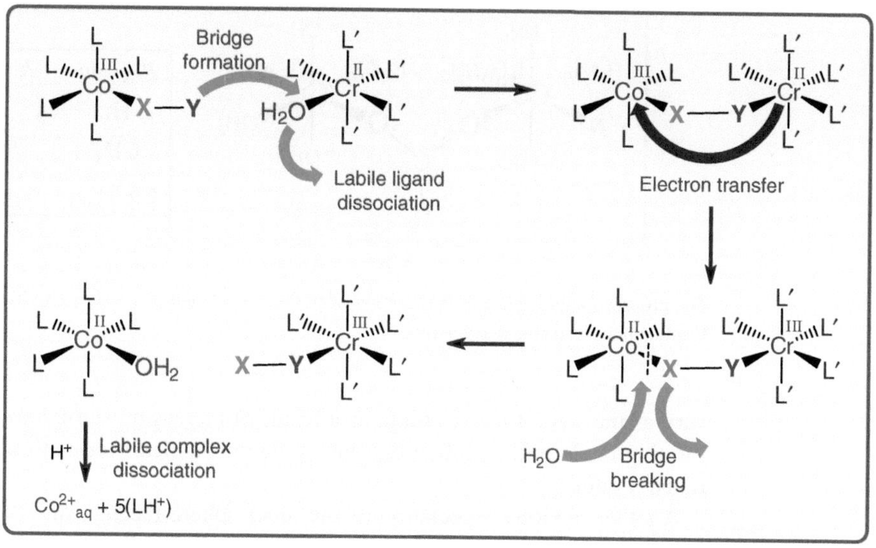

Figure 5.24

Electron transfer between Co(III) and Cr(II) complexes by the *inner sphere* mechanism, involving bridge formation in the intermediate. Transfer of the bridging group from the oxidant to the reductant is not required, but is characteristic of the mechanism where it does occur.

(b) at least one ligand of one reactant must be capable of being replaced by a bridging ligand in a facile substitution process (this replaced ligand in many examples is found on the reductant).

The latter requirement means a relatively labile site must be available and often involves substitution of a coordinated water. Note that atom transfer is *not* a requirement in this mechanism. However, it can occur, and its occurrence is usually good evidence for the mechanism. For example, when atom transfer occurs with an ambidentate ligand (like SCN^-), the donor preferred and bound by the precursor metal ion is often not the one attached to the product, as described earlier; this helps define the mechanism. Nevertheless, the bridging ligand *may* remain with its parent metal ion. The mechanism is then implied by the rate acceleration compared with analogous outer sphere electron transfer reactions.

The type of the bridging ligand can affect the observed electron transfer rate markedly, not surprisingly, given that its role is to bring the metal ions together and mediate the transfer through itself. For the reaction described by Eq. (5.56), the rate increases with increasing size of the halide ion in the order $X = F^- < Cl^- < Br^- < I^-$. One interpretation of this trend relates it to increasing polarizability with increasing halide ion size and a tied tendency towards ready distortion of the electron charge cloud. For the bridged binuclear intermediate, this tendency towards distortion has potential consequences. The higher-charged Co(III) attracts more halide electron density towards its side of the bridging unit, depleting the side near the lower-charged Cr(II) from where the electron departs; this effect increases with halide ion size. The process facilitates attraction of the transferring electron initially on the Cr(II) centre to the bridging ligand, from where it moves on to the Co(III).

The effect of altering the bridging group can be seen more starkly in the reaction of Cr_{aq}^{2+} with the carboxylate complex $[(NH_3)_5Co(OOCR)]^{2+}$, where it arises from a different cause. Attack by the Cr(II) ion at the carboxylate carbonyl of the Co(III) complex leads to the proposed intermediate in Figure 5.25, and rate constants clearly

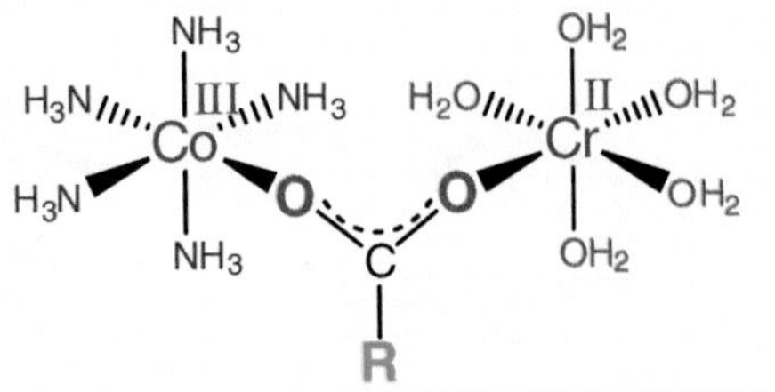

Figure 5.25
The proposed intermediate in the reaction of $[Cr(OH_2)_6]^{2+}$ and $[Co(NH_3)_5(OOCR)]^{2+}$; the size of the R-group impacts on the reaction rate.

vary as the size of the R group, as a result of increasing steric bulk of the R-group limiting facile bridge formation; that is, there is a clear steric effect on the electron transfer rate.

Another obvious expectation of the inner sphere mechanism is that the electron is transferred via the ligand; think of it like a ball rolling across a bridge. If this is the case, then for some time, albeit very briefly, the electron will reside on the bridging ligand, leading to a ligand radical. This has also been probed, using as a bridging ligand one that includes an aromatic heterocyclic group. Such groups help stabilise ligand radicals, and the presence of the radical intermediate has then been detected by electron paramagnetic resonance (EPR) spectroscopy, which is a highly sensitive method for detecting radicals (see Chapter 7.3.6). Thus, overall, experimental evidence for the inner sphere mechanism is strong.

The two mechanisms of electron transfer, or reduction–oxidation reactions, for metal complexes have some common aspects, with which we shall finish this section. These are the common keys to electron transfer reactions. To occur efficiently, the molecular orbital in which the donated electron originates, and that to which it moves, should be of the same type. The most efficient *outer sphere* electron transfer requires transfer between t_{2g} *d*-orbitals. Effective *inner sphere* electron transfer occurs with transfer between either both t_{2g} or both e_g^* orbitals. The chemical activation process prior to electron transfer will be much greater if the above does not apply, meaning that under such circumstances, reactions may not occur or occur only very slowly. Further, structural deformation and or electron configuration changes may be necessary (costing energy, and thus raising the activation barrier), and reactions under these conditions will, in general, be slower than those requiring little solvent reorganisation and or no electron configuration change. Overall, only where the activation barrier leading to the intermediate is sufficiently low will reactions proceed at measurable rates.

5.3.3.3 *Electrochemical and Radiolytic Electron Transfer Reactions*

It is useful to note that there are other methods for supplying electrons to complexes that will lead to a change in oxidation state. Whereas reduction–oxidation reactions are partnerships between two compounds, an alternative is to offer a direct source of or sink for electrons; this is achieved by electrodes in an electrochemical cell. At the appropriate potential, a complex may be reduced by transfer of an electron from an electrode to the complex; oxidation can occur by transfer of an electron to an electrode. What is perhaps apparent is that, for a solid electrode and a solution of a complex, it is only near the surface of the electrode that the process can occur efficiently, so both the rate of transport to an electrode surface as well as the rate at which an electron is transferred between the solid surface and dissolved complex ion are of relevance. In general, we can write a simple expression for the process (represented here for reduction).

$$[ML_m]^{n+} + e^- \rightleftharpoons [ML_m]^{(n-1)+} \tag{5.60}$$

The reduction–oxidation potentials of complexes are usually probed by the experimental methods of *voltammetry*, which is a sensing rather than a complete conversion technique; the full reduction or oxidation of a solution of a complex is achieved by *electrolysis*, using large surface area electrodes and stirring to enhance mass transport and thus speed up the process. These techniques and their application are developed in Chapter 7.3.7.

Yet another, and perhaps more exotic, method of supplying an electron in solution is to use **radiolysis**, where a very high energy source is directed into an aqueous solution to generate a range of products from reaction with the abundant water molecules, with the aquated electron (e^-_{aq}) and the hydroxyl radical ($^\bullet OH$) being dominant species, and of particular importance. The former is a powerful (one-electron) reductant, the latter a powerful one-electron oxidant (subsequently producing the ^-OH ion). The technique requires the use of high-energy devices like a van der Graaf generator or synchrotron and so is not widely available. However, the powerful radical species formed can initiate definitive one-electron reduction or oxidation of complexes, either at the metal centre (direct reduction or oxidation) or at the ligand (with metal reduction or oxidation via a following intramolecular electron transfer possible). The former process is a form of outer sphere reduction, with e^-_{aq} as the reductant here. The latter process is followed by electron transfer from the ligand radical to the metal ion, which is, in effect, 'half' of the reaction in the electron transfer through a bridge in a chemical inner sphere process. When the high-energy source is pulsed (the technique is called *pulse radiolysis*), a reaction in solution can be initiated rapidly and the reaction kinetics of the outcome followed. Electron transfer reactions between two metal complexes can also be initiated by pulse radiolysis under certain circumstances. For example, if a relatively high concentration of Zn(II) and a low concentration of a Cr(III) complex are both present in solution, creation of e^-_{aq} causes it almost exclusively to react rapidly with the Zn(II), because it is in such very large excess, to form the extremely rare and unstable Zn(I) ion. Then two reactions can occur with the highly reducing Zn(I), in competition: *intermolecular electron transfer* between Zn(I) and the Cr(III) complex; and *disproportionation* of Zn(I) to form equal amounts of Zn(II) and Zn(0). By examining colour change associated with the chromium centre, the intermolecular reaction can be examined. However, this technique is not one you are likely to meet often.

5.3.4 Ligand-centred Reactions

Changing ligands is a common feature of reactions that we have discussed in above sections. However, it is possible to chemically alter a ligand while it remains attached to the metal centre. Coordination need not prohibit chemistry going on with the bound ligand; in fact, it may promote reaction as a result of electronic and positional influences resulting from coordination. Because reactions of coordinated ligands is a topic intimately tied up with synthesis, since new stable complex molecules are formed and from which new ligands can be isolated, it is addressed in detail in Chapter 6.

Concept Keys

The thermodynamic stability of a complex can be expressed in terms of a stability constant, which reports the ratio of the complexed ligand to free metal and ligand in an equilibrium situation.

The stability of metal complexes is influenced by an array of effects associated with the metal or ligand: the size of and charge on the metal ion, hard–soft character of metal and ligand, crystal field effects, base strength of the ligand, ligand chelation where it is possible, chelate ring size where applicable and ligand shape influences.

For complexation of a set of ligands, each successive addition has a stability constant (K_n) associated with it, with each successive stability constant progressively smaller for a fixed stereochemistry. An overall stability constant $(\beta_n = K_1 \cdot K_2 \cdot ... \cdot K_n)$ expresses the stability of the completed assembly.

Whereas thermodynamic stability is concerned with complex stability at equilibrium, kinetic stability is concerned with the rate of formation of a complex, leading to equilibrium. Complexes undergoing reactions rapidly are labile, and those reacting slowly are inert.

Reactions of complexes may involve changes to the coordination sphere, the metal oxidation state and or the ligands themselves. Partial or complete ligand replacement (substitution) is the most common reaction met in metal complexes.

Octahedral substitution reactions may occur through a dissociative mechanism featuring a five-coordinate intermediate (ligand loss rate-determining), an associative mechanism featuring a seven-coordinate intermediate (ligand addition rate-determining) or else by an interchange mechanism where ligand exchange happens in a concerted manner. Similar concepts can be applied to other coordination numbers and shapes.

Mechanisms for optical and geometrical isomerisation reactions similar to those employed for substitution reactions can be envisaged. Additionally possible is a twist mechanism, involving distortion of the polyhedral framework in the activated state but in which no ligands depart or join the coordination sphere.

Oxidation–reduction (or electron transfer) reactions involving two metal complexes may occur by one of two mechanisms: outer sphere (no change in the coordination spheres of the reactants) or inner sphere (where one ligand on one complex forms a bridge to the other metal in the transition state).

Electron addition or extraction from a complex to cause reduction or oxidation may also be achieved electrochemically (at an electrode in an electrochemical cell) or radiolytically (using an aquated electron or hydroxyl radical).

Further Reading

Atwood, J.D. (1997). *Inorganic and Organometallic Reaction Mechanisms*, 2nd ed. New York: Wiley.

Classical examples are covered clearly for the student in good depth; reactions, procedures for their examination, and mechanisms by which they react are explained – with the bonus of coordination complexes and organometallic systems presented in a single text.

Kragten, J. (1978). *Atlas of Metal–Ligand Equilibria in Aqueous Solution*. Chichester: Ellis Horwood.

A dated but still useful resource book for stability constants, should you seek numerical data.

Martell, A.E. and Motekaitis, R.J. (1992). *Determination and Use of Stability Constants*, 2nd ed. New York: Wiley.

Provides a description of the potentiometric method commonly used to determine stability constants along with how complex speciation is determined, with examples and software.

Wilkins, R.G. (1991). *Kinetics and Mechanisms of Reactions of Transition Metal Complexes*. New York: VCH.

A somewhat older text, but one of the classics in the field, that gives a fine student-accessible coverage of the basic concepts, which have remained fairly constant over time.

6 Synthesis

6.1 Molecular Creation—Ways to Make Complexes

Synthesis, the process of making compounds, lies at the heart of chemistry, with the preparation of new and possibly useful compounds occupying a great number of chemists. Every new compound presents challenges not only in its preparation and isolation but also there are the challenges of characterisation and defining its reactivity and possible uses. Whereas synthesis directed towards formation of a specific product is a mature field in organic chemistry, the challenges to the synthetic chemist when a metal ion is present are more profound, so coordination chemistry remains at a relatively more developmental stage, despite over a century of research. Nevertheless, a number of clear strategies exist, which we shall explore. Knowledge of thermodynamic and kinetic stabilities of complexes, developed in Chapter 5, is invaluable as an aid in devising methods for synthesis of coordination compounds and or understanding synthetic reactions. In this chapter, we will draw on categories of reactions developed in Section 5.3 for discussion of synthetic approaches. In addition, we will look beyond classic coordination complexes to explore examples of inorganic polymers or clusters. Prior to this detailed look at synthesis, it would seem appropriate to provide a brief overview of the chemical and physical properties of metals that may be expected based on their position in the periodic able.

6.2 Core Metal Chemistry—Periodic Table Influences

The periodic 'membership' of metallic elements (see Figure 1.4) and location in the periodic table influences their chemical behaviour and reactivity. Prior to an examination of synthetic strategies, it is useful to introduce the various groups of metallic elements that function as the central atom in most coordination complexes, to identify aspects of their chemistry that impact on complexes that are accessible and the subsequent development of synthetic procedures. In general, the periodic block origins of a metal will play a part in the type of complexes that can be easily and conveniently synthesised and the types of ligands that the metal will prefer.

6.2.1 *s*-Block: Alkali and Alkaline Earth Metals

The metals of the *s*-block of the periodic table have properties that can be interpreted in terms of the trend in their ionic radii down each periodic column. There is a very strong tendency towards formation of M^+ for Group 1 (ns^1 or alkali) and M^{2+} for Group 2 (ns^2 or alkaline earth) metal ions, whereas other oxidation states are rarely

Introduction to Coordination Chemistry, Second Edition. Paul V. Bernhardt and Geoffrey A. Lawrance.
© 2025 John Wiley & Sons Ltd. Published 2025 by John Wiley & Sons Ltd.
Companion website: www.wiley.com/go/coordinationchemistry2e

encountered. With relatively high charge density, the alkali and alkaline earth metal ions are 'hard' Lewis acids and prefer small 'hard' Lewis bases. They particularly like O-donors, but can also accommodate N-donors, especially when present as part of a molecule offering a mixed O,N-donor set. The number of metal–donor bonding interactions varies a great deal, depending in one respect on the shape of the ligand and orientation of potential donors. For example, Mg^{2+} forms a complex with the polyaminoacid anion $EDTA^{4-}$ where all of the six heteroatom groups in the molecule are bound as a N_2O_4-donor set, in addition to an extra water molecule. Some of the strongest complexes they form are with large cyclic or polycyclic polyether molecules (see Figure 4.42). In these situations, the size of the ligand cavity relative to the size of the metal cation is an important factor in defining the 'strength' of the complex formed (concepts developed in Figure 4.38 earlier). Aquated cations where all ligands are H_2O (M_{aq}^+ and M_{aq}^{2+}) aside, few stable complexes with other simple ligands that offer just one donor atom occur, apart from some reported for beryllium, which forms complexes (mostly four-coordinate) with hard ligands such as halide ions to give $[BeX_4]^{2-}$ (X = F or Cl). However, organolithium and organomagnesium compounds such as $Li_n(CH_3)_n$ and the famous Grignard reagents RMgX (R = alkyl or aryl, X = Cl, Br) form a major part of the organic synthesis toolbox where they supply highly reactive carbanions (R^-), stabilised by metal–carbon bonds, in new C–C bond forming reactions. The Na^+ and Mg^{2+} cations in particular have important biological roles.

6.2.2   *p*-Block: Metals and Non-metals

Chemistry of the *p*-block elements is a vast area of research, best addressed through specialist textbooks on the topic. In the context of this book, non-metallic elements from the *p*-block are most likely to provide the best donor atoms for metal coordination, e.g. C, its neighbouring heteroatoms N, O, S and P and the halogens. It also houses the metalloids, which display mixed non-metal–metal character, and engage in coordination chemistry sometimes as electron pair donors and sometimes as acceptors. However, the *p*-block is a source of a number of important metals as well, including the most common metal in the Earth's crust, aluminium, which is in Group 13. The metallic elements of the *p*-block in Group 13 mostly form trications that are generally hard Lewis acids. As in the *s*-block, metals of the *p*-block tend to form stable ions through complete loss of their valence shell of electrons to leave the inert gas core electrons only:

$$\textbf{Al} \; [Ne]3s^23p^1 \; \rightarrow \; \textbf{Al}^{3+} \; [Ne]$$
$$\textbf{Sn} \; [Kr]5s^25p^2 \; \rightarrow \; \textbf{Sn}^{4+} \; [Kr]$$

However, the heavier *p*-block elements (most notably those in the sixth row of the periodic table) form cations in two different oxidation states that differ by the presence or absence of the so-called inert pair of $6s^2$ electrons. Examples include Tl^+/Tl^{3+}, Pb^{2+}/Pb^{4+} and Bi^{3+}/Bi^{5+} where the lower oxidation state (retaining the $6s^2$ electrons) is more stable. This contrasts with their fifth row congeners In^{3+}, Sn^{4+} and Sb^{5+} which prefer the inert gas valence electronic configuration although all can form cations where the $5s^2$ pair is retained. We can interpret this simply by seeing it as loss of at least sufficient electrons to leave a stable subshell:

$$\textbf{In} \; [Kr]5s^25p^1 \; \rightarrow \; \textbf{In}^{3+} \; [Kr]$$
$$\textbf{Tl} \; [Xe]6s^26p^1 \; \rightarrow \; \textbf{Tl}^+ \; [Xe]6s^2$$

However, it is only the heaviest metals of the *p*-block that tend to display this trait.

Hence, aluminium exists in its ionic form almost exclusively as Al^{3+}, and, as a consequence of being a light element with a high charge (high charge density), it is a 'hard' Lewis acid. It has, like the *s*-block elements, a preference for O-donor ligands, but binds efficiently to chelating ligands that also carry other types of donors. It forms an octahedral $[Al(OH_2)_6]^{3+}$ complex, although this readily undergoes proton loss (hydrolysis) from some coordinated water groups to form bound ^-OH as the pH rises, due to the influence of the highly charged metal ion on the acidity of the bound water groups. Chelators comprising particularly O-donor or mixed O, N-donor ligands form relatively strong complexes with Al(III).

6.2.3   *d*-Block: Transition Metals

The *d*-block transition metals, which form a group of elements ten-wide and four-deep in the periodic table associated with filling of the five *d*-orbitals, represent the classic metals of coordination chemistry and the ones on which there is significant and continuing focus. In particular, the lighter and usually more abundant or accessible elements of the first row of the *d*-block are the centre of most attention. Whereas stable oxidation states of *p*-block elements correspond dominantly to empty or filled valence shells, the *d*-block elements characteristically exhibit stable oxidation states where the n*d* shell remains partly filled; it is this behaviour that plays an overarching role in the chemical and physical properties of this family of elements, as covered in earlier chapters.

The ability of complexes of *d*-block metal ions to readily undergo oxidation or reduction involving one or several electrons is a unique feature of their chemistry. Because many transition elements have this capacity to exist in a range of oxidation states, they offer different chemistry even for the one element as a result of the differing *d*-electrons present in the different oxidation states; each oxidation state needs to be considered separately. Lighter elements of the *d*-block (in the third row of the periodic table) tend to prefer O-, N- or halide ion donor groups, whereas heavier elements in the fourth and fifth rows lean towards coordination by ligands featuring heavier *p*-block elements such as S- and P-donors. The diversity of oxidation states, shapes, and donor groups met in transition metal chemistry makes it both fascinating and puzzling – it is not simple chemistry, and one size does not fit all.

The *d*-block elements display perhaps the greatest variation of the various groups, with the first series row (with a 4*s*3*d* valence electron shell) in particular distinct from the second (5*s*4*d*) and third (6*s*5*d*) series rows. The fully synthetic fourth 'super-heavy' (7*s*6*d*) row has little chemistry reported yet. Given the miniscule quantities of these elements that have been made by nuclear synthesis and their extremely short half-lives (at best a few minutes), it is doubtful that we will ever be able to conduct any meaningful coordination chemistry with these radioactive and inherently unstable 6*d* transition metals so they will not be included in discussions of *d*-block chemistry, as there is really nothing to say.

Radii of the heavier *d*-block elements and ions are larger than those of the first series, although the so-called 'lanthanide contraction' of the 4*f* subshell (Section 6.2.4) determines that radii of the 5*d* series differ little from radii of the 4*d* series, despite increased atomic number. Predicting 'common' oxidation states in transition metal chemistry is fraught with exceptions and qualifications since, as we have already shown, the metal is only half of the partnership and ligand substitution can transform the redox reactivity of a metal. However, there are some periodic trends that emerge. Elements on the far-right-hand side of each transition series, replete with nd electrons, prefer lower oxidation states (+1 and +2). Elements on the left-hand side of the transition series are commonly found in an inert gas electronic configuration (e.g. Ti^{4+}, Nb^{5+}). Both of these trends stem from the concept of *effective nuclear change*. Addition of a proton to the nucleus and an electron (to a d orbital) to give the next element in a row leads to this newly introduced d electron being more strongly attracted to its nucleus than the preceding element. This happens because electrons in the d orbitals do not shield the nuclear charge as effectively as the s and p orbitals, so when a proton is added to the nucleus, there is no shielding provided by the accompanying d electrons and all valence electrons feel the electrostatic force of the higher charged nucleus. The later transition elements (Groups 10–12), where the effective nuclear change is high, will only relinquish one or two electrons going from their elemental state. Conversely, the transition elements at the left-hand side of the transition series (Groups 5 and 6) have only a modest effective nuclear charge to deal with and readily give up all their $(n+1)s$ and nd electrons to attain the nearest inert gas electronic configuration.

Elements in the middle of the transition series are capable of many oxidation states (see Figure 1.5). *Higher oxidation states* (>+III) are much more common for $4d$ and $5d$ than for $3d$ elements; the +II and +III states are dominant in the $3d$ series. The $4d$ and $5d$ metals are much more prone to *metal–metal bonding* than the $3d$ row, and multiple metal–metal bonds are common (Figure 6.1). For $3d$, M–M bonds are only common in metal carbonyls. *Polynuclear (and cluster) compounds* are more common for $4d$, $5d$ than for $3d$ species. *Magnetic properties* differ, with heavier elements tending to form dominantly low-spin compounds due to larger ligand field splitting energies (see Section 2.3). *Higher coordination numbers* (>6) are also more common for $4d$, $5d$ complexes.

If we examine a few of the columns within the d-block selectively, the chemistry behind these generalisations becomes a little clearer. It is in Group 9 (cobalt, rhodium, iridium) where there exists the greatest similarities, so it is instructive to explore it first. Cobalt is a classic $3d$ element, with Co(II) and Co(III) the sole

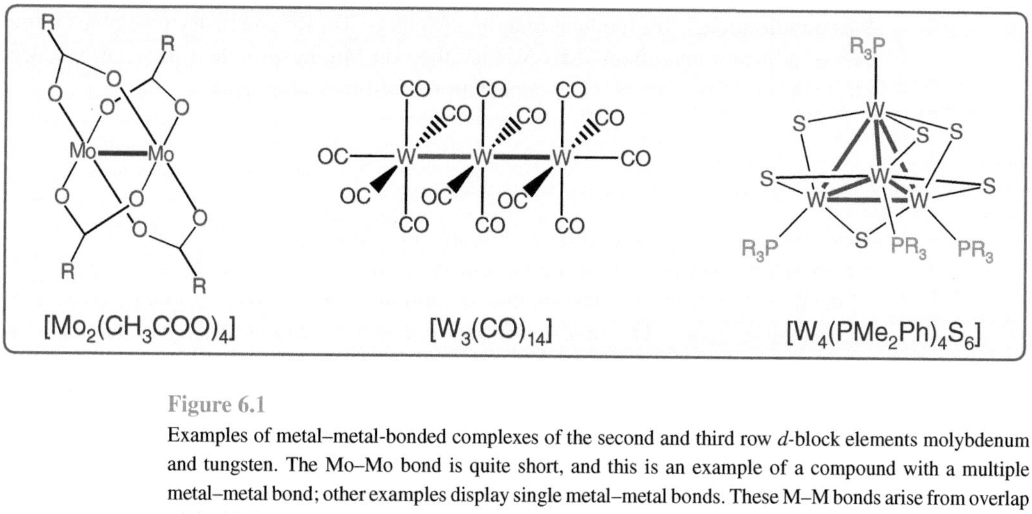

$[Mo_2(CH_3COO)_4]$ $[W_3(CO)_{14}]$ $[W_4(PMe_2Ph)_4S_6]$

Figure 6.1

Examples of metal–metal-bonded complexes of the second and third row d-block elements molybdenum and tungsten. The Mo–Mo bond is quite short, and this is an example of a compound with a multiple metal–metal bond; other examples display single metal–metal bonds. These M–M bonds arise from overlap of d-orbitals.

significant oxidation states (although Co(IV), Co(I) and Co(0) are known but rarer), with a preference for N-, O- and halide donor atoms. Complexes dominantly feature octahedral stereochemistry, particularly for Co(III) where virtually all structurally characterised Co(III) complexes are six-coordinate, while four- and five-coordinate structures make up the remainder. By contrast, six-coordinate Co(II) compounds are not as prevalent (~65%) compared with their four- and five-coordinate analogues (each 15–20%) . While Co(IV) is extremely rare, Ir(IV) is more common with compounds such as $K_2[IrCl_6]$ commercially available, albeit a powerful oxidant. The M(III) oxidation state is common for all of Co, Rh and Ir, but the inertness of complexes increases down the column; typical reactivity trends for the complexes Co:Rh:Ir are about 1,000:50:1. This reactivity trend is characteristic of the *d*-block in general. It is for the M(II) state that differences are starkest; Co(II) is common and stable, whereas Rh(II) and Ir(II) form few stable monomeric complexes. Despite some common behaviour, there are clear differences across their chemistries: for example, M(III) polyhalide complexes such as $[IrCl_6]^{3-}$ and $[RhCl_5(OH_2)]^{2-}$ are stable, whereas no more than three halide ions, as in $[CoCl_3(NH_3)_3]$, can be present for stability in Co(III). Hydrido (H^-) complexes of Rh(III) and Ir(III) are common, with even the simple species $[RhH(NH_3)_5]^{2+}$ formed, whereas Co(III) hydrido species have rarely been reported, even though an example, $[CoH(tmen)_2(OH_2)]^{2+}$ (tmen being a substituted chelating diamine) was identified two decades ago.

Group 7 (manganese, technetium, rhenium) is one with clearer differences between the chemistry of the lightest member and the heavier members. Manganese is known in oxidation states all the way from Mn(I) to Mn(VII), with a changeover from preferred six- to four-coordinate complexes occurring around Mn(V). The Mn(II) oxidation state is common, with higher oxidation states progressively less common, although some particularly important compounds exist at higher oxidation states, namely the simple Mn(IV) oxide MnO_2 and the powerful and popular Mn(VII) oxidant $[MnO_4]^-$. Technetium and rhenium have no analogue of Mn_{aq}^{2+} and form very few M(II) species; indeed, they have little cationic chemistry in any oxidation state. Unlike Mn, they have an extensive chemistry in the M(IV) and M(V) oxidation states; the latter is the least common for Mn. The formation of clusters and M–M bonds is much more common for Tc and Re than with Mn and is a feature of the (II) and (IV) oxidation states for Tc and Re. The chemistry of radioactive Tc features in Chapter 9 as a radionuclide for medical imaging. Much of Tc, Re chemistry resembles more that of adjacent neighbours Mo, W than they do Mn, despite their different valence electron sets. Diversity is a key expectation of *d*-block chemistry.

6.2.4 *f*-Block: Inner Transition Metals (Lanthanoids and Actinoids)

However, it is not entirely time to despair, since at least one family of metals shows a remarkable consistency in their chemistry. This occurs in the oft-ignored *f*-block of the periodic table, yet they are just as valid participants in coordination chemistry as transition metals. The first row of that block is the *lanthanoids*. The lanthanoids (also called the lanthanides) were once called the 'rare earths', but that is a misnomer as they are not particularly rare. They exist as the first row of the 14-wide 4*f*-block elements, where filling of the seven *f*-orbitals is the key to their chemistry. All but radioactive promethium, the rarest of rare earth elements whose most stable isotope ^{145}Pm has a half-life of only 17.7 y, occur naturally. The scarcest naturally occurring lanthanoid (thulium) is as common as bismuth and more common than arsenic, cadmium, mercury and selenium. As their name suggests, they are named after lanthanum (La), electronic configuration $[Xe]6s^2 5d^1$, a *d*-block element that sits in Group 3. Immediately after this element, the $4f$ orbitals lie slightly lower in energy than the $5d$, and so fill first (with 14 electrons) up to Lu ($[Xe]4f^{14}6s^2 5d^1$) before returning to filling of the $5d$-shell. This means the *f*-block breaks up the *d*-block of the periodic table after the first column, but it is more traditional (and convenient, in the context of the shape of most printed pages) to represent the *f*-block essentially as a subscript to the main periodic table in what is sometimes referred to as the 'short form' of the Table. This means that the *f*-block elements sit virtually between Groups 3 (Sc, Y, La, Ac) and 4 (Ti, Zr, Hf and Rf), like the famous 'Platform 9¾' known to Potterheads. Because *f*-orbitals are expansive and separated, the shielding of one *f*-electron by others from the effect of the nuclear charge is weak. Thus, with increasing atomic number (and nuclear charge), the effective nuclear charge experienced by each *f*-electron increases markedly with atomic number, causing a significant shrinkage in the radii of the atoms or ions from La (1.06 Å) across to Lu (0.85 Å), commonly called the *lanthanide* (or *lanthanoid*) *contraction*.

As splitting of the degenerate *f*-orbital set in crystal fields is small, so crystal field stabilisation issues are only of minor importance. Preferences between different coordination numbers and geometries are controlled dominantly by metal ion size and ligand steric effects. *Coordination numbers* greater than six are usual (Figure 6.2), with seven, eight and nine all important. Examples with coordination numbers up to 14 exist. Coordination numbers of greater than nine are restricted to the *f*-block and never found in the *d*-block (see Section 2.3). Some low coordination numbers (<6) are known but are rare, and only exist when stabilised with bulky ligands, such as aryloxy ($^-$OR) or amido ($^-$NR$_2$) ligands. Dominantly, the lanthanoids exist in only one stable oxidation state, M(III), through loss of nominally the $6s^2 5d^1$ outer electrons. Certain elements form M(II) (Eu(II)) or M(IV) (Ce(IV)) ions, but these are readily oxidised or reduced to the M(III) ion. Because the $4f$ electrons are essentially inner electrons, due to effective shielding, their spectroscopic properties are little affected by the surrounding ligands. Their coordination chemistry tends to be remarkably similar for the whole family of elements, and as hard Lewis acids, they have a common preference for O-donors, though they can accommodate other donors, even including forming some metal–carbon bonded species. All form $[M(OH_2)_n]^{3+}$ (where $n > 6$), such as $[Nd(OH_2)_9]^{3+}$, although these are readily deprotonated (hydrolyse); this tendency increases from La to Lu as the ionic radius and pK_a decreases. Chelating ligands

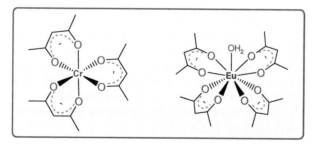

Figure 6.2

High coordination numbers are common in the *f*-block. The didentate chelate ligand acetylacetonato (acac$^-$) forms six-coordinate octahedral complexes with first-row *d*-block metal ions (like Cr(III), shown at left), but eight- or nine-coordinate complexes with *f*-block lanthanide ions (like Eu(III), shown at right).

give the most stable complexes, in line with expectations developed for *d*-block metals in Chapter 5; basically, they follow the normal rules of complexation developed in detail for *d*-block elements. However, no significant electronic stability is offered by adopting a particular coordination geometry, which contrasts with the significant crystal field stabilisation energies attained in octahedral and tetrahedral *d*-block ions (see Section 2.3).

The row below the lanthanoids comprises the *actinoids* (also called the actinides), which are mainly synthetic elements. Those that are found on Earth naturally are only long-lived thorium and uranium isotopes, but all are radioactive. ^{232}Th boasts an impressive half-life of 14 billion years so it has certainly not become extinct yet (the current dating of the Earth is 4.5 billion years). ^{238}U is not far behind with a half-life of, coincidentally, 4.5 billion years. Beyond element 92 are the so-called trans-uranium elements which are the product of chemistry requiring nuclear reactors which has limited their availability and investigation, in addition to their inherent nuclear instability. After the *d*-block parent actinium, in principle the *f* orbitals are then filled sequentially for the following elements. However, energies of the 5*f* and 6*d* orbitals are so close that elements immediately following Ac (and their ions) may have electrons in both 5*f* and 6*d* orbitals, at least until 4 or 5 electrons have been entered, when 5*f* alone seems to be more stable. This means that the early actinoid elements tend to show more *d*-block character (variable oxidation states and associated chemistry). Consequently, resemblance of the series to the parent is less marked than with the lanthanoids, at least until americium. Only after americium (about half-way along the series) are the elements similar to the corresponding lanthanoid in their chemistry, with only the M(III) oxidation state stable. Earlier elements, such as uranium, display oxidation states of up to M(VI); in fact, U(VI) is the most common oxidation state for that element. High coordination numbers (up to 14) are characteristic; for example, $[Th(NO_3)_6]^{2-}$ is 12-coordinate, as each nitrate ion acts as an O,O-didentate ligand. Their solution chemistry is often complicated; hydrolysis in water (even to oxido species) is common. Elements above fermium are short-lived, and isolable only in trace amounts; others can be prepared in gram or even kilogram amounts. Being radioactive and in most cases rare synthetic elements, they are not met by most scientists let alone in everyday life, although they have found applications. You might be surprised to learn that radioactive ^{241}AmO$_2$ (with a half-life of about 400 y) is the essential ingredient of one class of domestic smoke detectors. Its radioactive decay to relatively stable ^{237}Np is achieved by release of an alpha particle (effectively ^{4}He^{2+}) which is accelerated towards an electrode but slowed down in the presence of smoke particles, triggering the alarm.

6.2.5 Beyond Natural Elements

The periodic table has been extended from U (Z = 92) to currently Z = 118 (Og) since 1940 by synthetic methods. Some synthetic elements have long half-lives (e.g. ^{248}Cm, $t_{1/2}$ 3.5 × 10^5 y), others moderate half-lives (e.g. ^{249}Bk, $t_{1/2}$ 300 d) or short half-lives (e.g. ^{261}Rf, $t_{1/2}$ 65 s). Their syntheses have involved high-energy fusion and bombardment reactions, e.g.

$$^{249}_{97}\text{Bk} + ^{18}_{8}\text{O} \rightarrow ^{260}_{103}\text{Lr} + ^{4}_{2}\text{He} + 3\,^{1}_{0}n$$

and, in general, the more unequal the mass of the two nuclei, the greater the possibility that a fusion reaction proceeds. Separation of a new element is a key problem.

Separations involve methods such as volatilisation, electrodeposition, ion-exchange, solvent extraction and precipitation/adsorption. Separation relies on unique chemistry for each element; although not heavy elements, but useful as an illustration, $^{64}Zn/^{64}Cu$ (the latter a synthetic radioactive isotope) are separated by dissolution in dilute HNO_3 followed by selective electrodeposition of metallic Cu (a very simple task, as the $Cu^{II/0}$ and $Zn^{II/0}$ redox potentials differ by ~ 1 V).

The totally synthetic fourth row of the *d*-block and seventh row of the *p*-block have now been created fully, with all member elements prepared, albeit in tiny amounts, and characterised. Lifetimes of these new radioactive elements are not long, so their coordination chemistry has not been explored in any detail and probably never will be. However, it is highly likely they would behave like their third-row analogues. They are more chemical curiosities than applicable species at this time; however, they do stand as monuments to the human inventive spirit and technological capacity.

What may be apparent from this very brief overview is that the periodic table location of metals plays a role in determining their coordination chemistry—yet each has truly unique coordination chemistry. However, certain 'global' traits exist to guide the synthetic chemist. The above notes may serve to support the following specific discussion of synthetic methods.

6.3 Reactions Involving the Coordination Shell

Often, simple commercially available hydrated salts of metal ions (halides, sulfates or nitrates in particular) are the starting point of much synthetic chemistry involving the coordination shell. Oxides are occasionally used but tend to be highly insoluble and their use can involve dissolution with acid to form simple salts as a first step in any case, the 'rare earth' 4*f*-block elements being an example. Furthermore, anhydrous simple salts (dominantly halides) are employed, particularly where reaction is to proceed in non-aqueous solvent; however, some of these are polymeric and as a result are often converted first to monomeric compounds that include coordinated solvent, which are more soluble and reactive. For example, polymeric $TiCl_3$ is reacted in tetrahydrofuran (THF) to form the monomer $[TiCl_3(THF)_3]$, and polymeric $\{MoCl_4\}_n$, with pairs of bridging Cl^- ligands, is reacted in acetonitrile (MeCN) to form the monomer $[MoCl_4(MeCN)_2]$. Ligand substitution reactions using precursors where a didentate chelate ligand is replaced by stronger chelators is another approach; typical complexes for such reactions are $[M(CO_3)_3]^{x-}$ and neutral $[M(acac)_n]$, featuring didentate carbonato and acac$^-$ (acetylacetonato or butane-2,4-dionato, Figure 6.2) ligands, respectively. Much organometallic chemistry relies on metal carbonyl compounds, such as $[Cr(CO)_6]$, as starting points for the substitution chemistry. However, ligand substitution can occur with an almost infinite number of starting materials, providing thermodynamics and kinetics for the proposed reaction are appropriate.

6.3.1 Ligand Substitution Reactions in Aqueous Solution

The most common synthetic method used in coordination chemistry involves ligand substitution in an aqueous environment. It is a relatively cheap approach, employs the commonest and safest solvent and has the advantage that, as many reagents and complexes are ionic, the reactants often have good solubility in water. Moreover, most metal salts are supplied commercially as hydrated forms (such as $Cu(SO_4)\cdot 5H_2O$, see Figure 3.2) since they are prepared and isolated from aqueous solution during manufacture. As a consequence, this class of reaction often involves aqua ligand replacement, although substitution of other ligands is well known.

As an example, consider reaction (6.1). Here, the Cu_{aq}^{2+} ion initially has water as the ligand in its coordination sphere that is substituted by a stronger ligand, in this case ammonia, to form $[Cu(NH_3)_4]^{2+}$. Multiple ligand substitution is assisted by the use of an excess of the incoming ligand to drive equilibria towards the fully substituted compound.

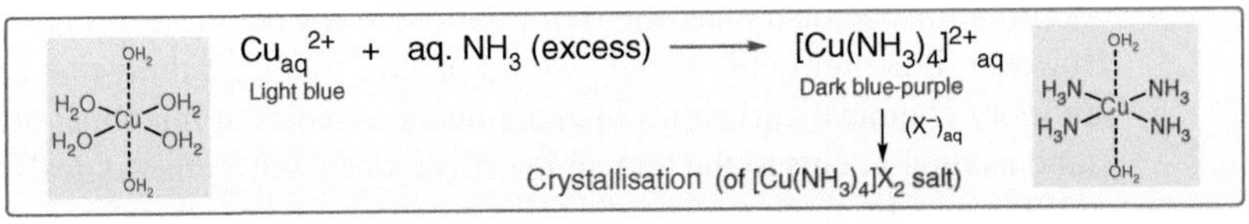

$$Cu_{aq}^{2+} + \text{aq. } NH_3 \text{ (excess)} \longrightarrow [Cu(NH_3)_4]^{2+}_{aq}$$

Light blue   Dark blue-purple

$\downarrow (X^-)_{aq}$

Crystallisation (of $[Cu(NH_3)_4]X_2$ salt)

(6.1)

There is no change in oxidation state in this substitution chemistry, and usually (though not always) the coordination number is also preserved. What is readily seen is a very rapid colour change, associated with the replacement of coordinated water by ammonia. Colour change is a common characteristic signal of change in the coordination sphere for metal complexes that absorb light in the visible region. The rapidity of the reaction is an indication that we are dealing with a *labile* metal ion, one that exchanges its ligands rapidly. We might more properly represent Cu_{aq}^{2+} as the distorted octahedral complex ion $[Cu(OH_2)_6]^{2+}$, but due to significant elongation of the axial bonds due to the Jahn–Teller Effect (Section 2.3.5), the axial aqua ligands exchange much faster than equatorial groups, indicative of poor ligand binding in these sites. Consequently, stability constants for the $[Cu(NH_3)_5(OH_2)]^{2+}$ and $[Cu(NH_3)_6]^{2+}$ species are very low ($\beta_4 \gg \beta_5 > \beta_6$), so in practice the substitution in aqueous solution effectively halts after four ammonia ligands are added around the copper ion, making the tetraammine complex the dominant species in strong aqueous ammonia solution. On crystallisation, it is this species that is usually isolated. Knowledge of spectroscopic behaviour and stability constants supports an understanding of the outcome.

Often, substitution is an uncomplicated process of one type of ligand being replaced fully by another type of ligand, such as for reaction of $[V(OH_2)_6](SO_4)$ in reaction (6.2).

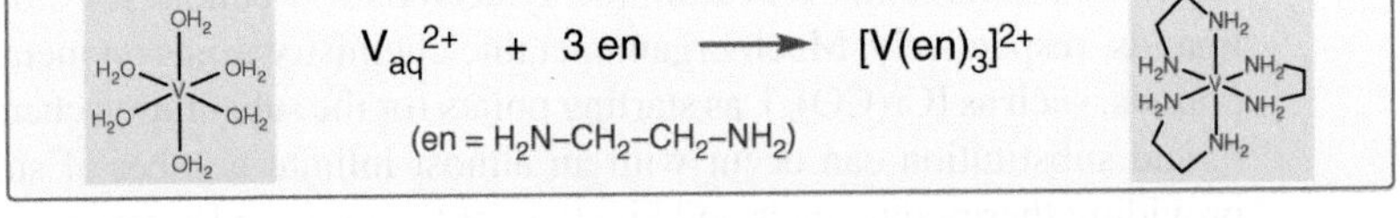

$$V_{aq}^{2+} + 3 \text{ en} \longrightarrow [V(en)_3]^{2+}$$

$(en = H_2N-CH_2-CH_2-NH_2)$

(6.2)

However, the outcome of such reactions does depend on entering ligand, as the above reaction performed with the aromatic monodentate amine pyridine instead of ethane-1,2-diamine (en) leads to only four pyridines binding, with *trans*-$[V(py)_4Cl_2]$ the product.

Substitution with change in coordination number may also occur in some cases. A clear example of this occurs for Ni_{aq}^{2+} reacting with pyridine, which undergoes the process (reaction (6.3)), moving from an octahedral (six-coordinate) to a square planar (four-coordinate) complex.

$$Ni_{aq}^{2+} + aq.\ pyridine \longrightarrow [Ni(py)_4]^{2+}$$

Light green (Excess) Brown

$(X^-) \downarrow$

Crystallisation (of $[Ni(py)_4]X_2$ salt)

(6.3)

The strong-field ligand pyridine supports the spin-paired d^8 diamagnetic square planar geometry for the product (see Chapter 2.3.5), whereas the relatively weaker ligand water supports the high-spin paramagnetic octahedral geometry in the precursor. Polyamines and other strong-field ligands (e.g. CN^-) generally tend to yield square planar complexes with this metal ion.

Complex formation is but the first stage of synthesis, as it is usually true that we require the product in a solid form, isolated as a salt or neutral compound. There are a number of approaches to isolation of a solid:

- cooling, usually in an ice-bath, which may lower solubility sufficiently to permit crystallisation of the product from solution, with addition of a 'seed' crystal of the product from an earlier synthesis an option for assisting the process;
- concentration to a reduced volume, usually using a rotary evaporator that operates at reduced pressure which lowers the solvent boiling point, followed by cooling, may be necessary if the solubility of the product is too high to permit precipitation from a dilute reaction mixture;
- slow addition to the reaction mixture, with stirring, of a water-miscible non-aqueous solvent (such as ethanol) in which the product is only sparingly soluble, until the commencement of precipitation;
- addition of an excess of a different simple salt of an anion (often in the soluble Na^+ or K^+ form) that provides a high concentration of counter anion that forms a less soluble complex salt, allowing its ready crystallisation;
- evaporation to dryness, appropriate where the product is sufficiently pure and stable and unreacted species (such as excess ammonia, for example) are removed by evaporation, with subsequent recrystallisation from a non-aqueous solvent in which the product is sparingly soluble possibly employed;
- sublimation or distillation of the product following removal of solvent, which is not usually applicable with ionic products, but finds limited use with some neutral organometallic complexes.

As an example of the value of anion exchange reactions, the product from reaction of $NiSO_4$ with pyridine in water, the complex cation $[Ni(py)_4](SO_4)$, is highly soluble. However, the addition of excess sodium nitrite leads to ready precipitation of the much less soluble $[Ni(py)_4](NO_2)_2$ complex; the change of the counter ion alone is sufficient (reaction (6.4)).

$$[Ni(py)_4](SO_4) + 2\ Na(NO_2) \longrightarrow [Ni(py)_4](NO_2)_2 + Na_2(SO_4)$$

High solubility in water   Excess,   Low solubility in water
dissolved in water

$\downarrow$

Ready crystallisation

(6.4)

Where the product of a reaction is neutral, as occurs when red anionic $[PtCl_4]^{2-}$ has two chloride anions replaced by two neutral ammonia ligands to form yellow neutral *cis*-$[PtCl_2(NH_3)_2]$, the solubility of this neutral product will be lower than the ionic starting material or other possible ionic products, leading to its ready and selective precipitation from aqueous solution (reaction (6.5)).

$$
\begin{array}{c}
K_2[PtCl_4] + 2\,NH_3 \longrightarrow [PtCl_2(NH_3)_2] + 2\,KCl \\
\text{Water soluble} \qquad\qquad \text{Low solubility in water} \\
\downarrow \\
\text{Ready crystallisation}
\end{array}
$$

(6.5)

The reaction above is also an example where substitution of a ligand other than coordinated water is occurring, reminding us that ligand substitution is not inherently governed by the nature of the leaving and entering groups. In the absence of any other options, the metal will take what is left to satisfy its desired coordination number.

Indeed, there is no requirement that only one type of ligand can be replaced in the one reaction; for example, purple $[CoCl_3(NH_3)_3]$ reacted with ethane-1,2-diamine (en) forms on heating yellow $[Co(en)_3]^{3+}$, where both chloride and ammonia ligands are substituted, the reaction (6.6) driven by the higher stability of the chelate complex formed (Chapter 3.1.3):

$$
[CoCl_3(NH_3)_3] + 3\,(en) \longrightarrow [Co(en)_3]Cl_3 + 3\,NH_3
$$

(6.6)

While reactions of labile metals like Cu(II), Co(II), Mn(II) and Zn(II) are fast, metals ions like Ni(II) react slower, and Pd(II) much slower. Metals like Co(III), Rh(III), Cr(III) and Pt(II) are *inert*, and to make their reactions occur in a convenient timeframe of hours or a day, instead of weeks or months, it is often necessary to heat their reaction mixtures. This is helpful because reactions typically double in rate for approximately every five degrees rise in temperature. An example is the inert red-purple $[RhCl_6]^{3-}$ ion, which when boiled in aqueous solution with the chelating ligand oxalate (ox^{2-}, $C_2O_4^{2-}$) for approximately 2 hours forms yellow $[Rh(C_2O_4)_3]^{3-}$ (reaction (6.7)), whereas reaction at room temperature takes days.

$$
K_3[RhCl_6] + 3\,Na_2(ox^{2-}) \xrightarrow{\text{Heat; hours}} K_3[Rh(ox)_3] + 6\,NaCl
$$

(6.7)

Most examples above have involved total ligand exchange to introduce just a single new type of ligand, driven usually by use of an excess of incoming ligand. This does not mean that partial substitution cannot occur. Often, partial substitution is driven by significant differences in the rate of substitution of different ligands on the reacting complex along with differing thermodynamic stability in products. For example, reaction (6.8) effectively stops following substitution of the coordinated chloride ions even with excess oxalate ion, because the two ethane-1,2-diamine (en) chelate ligands are more strongly bound and not readily substituted even by another chelating ligand.

$$\text{cis-[CoCl}_2\text{(en)}_2\text{]Cl} \longrightarrow \text{[Co(en)}_2\text{(ox)]Cl} \qquad (6.8)$$
$$+ \text{Na}_2\text{(ox}^{2-}) \qquad\qquad + 2\,\text{NaCl}$$

We have already seen an example earlier (reaction (6.5)), where the formation of *cis*-[PtCl$_2$(NH$_3$)$_2$] is driven in this case by low solubility of this neutral species allowing it to crystallise out of the reaction mixture rather than continue reaction to form ionic [Pt(NH$_3$)$_4$]$^{2+}$. Another way to achieve partial substitution is through use of a stoichiometric amount of a reagent. For example, reaction (6.9) may occur, where only one of two available coordinated chloride ions are substituted predominantly because of the availability of only one molar equivalent of added cyanide anion.

$$\text{cis-[CoCl}_2\text{(en)}_2\text{]Cl} + \text{Na(CN)} \longrightarrow \text{cis-[CoCl(CN)(en)}_2\text{]Cl} \qquad (6.9)$$
$$\text{One molar} \qquad\qquad\qquad\qquad + \text{NaCl}$$
$$\text{equivalent only}$$

However, partial substitution is compromised in many cases by the formation of lesser amounts of species with both greater and lower levels of substitution than the target complex. This is a particularly common outcome with labile systems; for example, addition of two molar equivalents of ammonia to Cu$_{aq}$$^{2+}$ ([Cu(OH$_2$)$_6$]$^{2+}$) will not lead to only [Cu(OH$_2$)$_4$(NH$_3$)$_2$]$^{2+}$ but also to some [Cu(OH$_2$)$_5$(NH$_3$)]$^{2+}$, [Cu(OH$_2$)$_3$(NH$_3$)$_3$]$^{2+}$ and possibly even [Cu(OH$_2$)$_2$(NH$_3$)$_4$]$^{2+}$, the outcome depending on the relative stepwise stability constants (K_1, K_2, K_3, etc., Section 5.1.1) of the various species.

Where a mixture of products results from a reaction, it may be possible to separate these by selective crystallisation. However, this is not always successful, and difficult where small amounts of a particular product are present. An option in these circumstances is to employ ion exchange chromatography to separate the various complexes present in the reaction mixture as long as they are substitution inert. For ionic and neutral complexes, it is possible to separate mixtures in aqueous solution reliably and readily using cation or anion exchange chromatography. Separation of cations, for example, can be conducted successfully using either an acid-stable resin such as Dowex® 50W-X2 or a neutral pH resin such as SP-Sephadex® C-25, typically employing acids (0.5–5 M) for the former and neutral salts (0.1–0.5 M) for the latter as eluents (solutions that mobilise and separate the mixture components on the column). These resins separate first by charge, with separation also influenced by molecular mass and shape. The order of elution of complexes of comparable molecular masses from a column will be charge-neutral $\gg$ 1+ complex >2+ complex >3+ complex. For example, a mixture containing [CrCl$_2$(OH$_2$)$_4$]$^+$, [CrCl(OH$_2$)$_5$]$^{2+}$ and [Cr(OH$_2$)$_6$]$^{3+}$ will exit the column (elute) in this order. These resin columns can be re-used many times, requiring no more than 'washing' with a simple acid or salt solution and then water prior to use.

An example of separation with a SP-Sephadex® C-25 column is shown (Figure 6.3) for a mixture of the Co(III) complexes [Co(NH$_3$)$_5$(OH)]$^{2+}$ (pink) and [Co(NH$_3$)$_6$]$^{3+}$ (yellow). A solution of the mixture was loaded onto the column where it was absorbed at the top of the resin bed, the column was washed with water, then eluted with a salt solution. The dication [Co(NH$_3$)$_5$)(OH)]$^{2+}$ separates clearly and elutes well before the trication [Co(NH$_3$)$_6$]$^{3+}$. The total amount of a mixture of

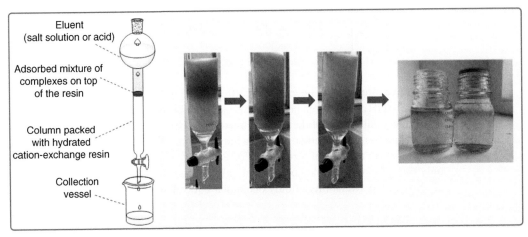

Figure 6.3

Simple chromatographic separation of differently-charged cobalt(III) complexes $[Co(NH_3)_5(OH)]^{2+}$ (pink) and $[Co(NH_3)_6]^{3+}$ (yellow) by cation exchange column chromatography. The components are shown at left. A solution of the complex mixture is introduced and sorbed onto the top of the resin, then eluted with an appropriate electrolyte (salt) solution to separate the complexes. The centre images show the separation on the column at different time points and eventual collection as pure complexes (at right), the pink sample taken first. (The complexes separated here were used for recording the NMR spectra shown in Figure 7.14.)

complexes and volume of resin in the column influence the separation; these factors are important in defining the capacity of the column, a consideration in all forms of chromatography.

One observation from a vast collection of experimental results is that ligands not undergoing substitution themselves ('spectator' ligands) can influence substitution at sites directly opposite them (*trans*) and, to a lesser extent, at adjacent sites (*cis*). The best-known examples lie with Pt(II) chemistry, where some groups exhibit a strong *trans effect*, causing groups opposite them to be more readily substituted (a kinetic effect) than those in *cis* dispositions. Groups opposite a chloride ion, for example, are substituted more readily than those opposite an ammine group. Extensive studies have produced an order of *trans* effects for various ligands bound to Pt(II), being

$$CO \approx CN^- > PH_3 > NO_2^- > I^- > Br^- > Cl^- > NH_3 > {}^-OH > OH_2$$

The *trans* effect means that reaction of $[PtBr_4]^{2-}$ with two equivalents of ammonia will lead preferentially to the *cis*-$[PtBr_2(NH_3)_2]$ isomer being formed, as the second substitution step occurs selectively opposite a bromide ion, which has a stronger *trans* influence than ammonia (reaction (6.10)).

(6.10)

Taking account of the influence of different *trans* groups allows for construction of different isomers through stepwise substitution processes from different precursors. The reactions are displaying what is called *stereospecificity*—favouring a specific stereoisomer. The influence of 'spectator' ligands on substitution processes relates to their capacity to withdraw electron density from or donate it to the metal centre, which influences their opposite partner; we see these influences in the variation in metal–donor distance for a particular group as the group opposite is varied.

Although all the examples above relate to the synthesis of mononuclear complexes, it should be appreciated that polymetallic compounds are accessible, and sometimes form preferentially, depending on the ligands involved. A simple example involves the reaction of copper(II) carbonate with a carboxylic acid, which can produce a dinuclear compound where the carboxylate acts as a bridging didentate ligand (reaction (6.11)).

$$2\ CuCO_3 + 4\ \text{R-COOH} \longrightarrow [Cu_2(OOCR)_4(OH_2)_2] + 2H_2CO_3 \qquad (6.11)$$
$$\text{Aqueous solution}$$

The metals are in proximity to each other which begs the question of whether they are bonded, or not. This can be controversial as, although interatomic distances can be measured now to great exactitude, the question of whether there is an attractive force between the metals required a consideration of molecular orbital interactions. The $Cu^{II} \ldots Cu^{II}$ distance is certainly close enough that the unpaired electrons on each formally d^9 centre feel each other's presence. In other related transition metal complexes (e.g. Cr, Mo, Tc, Ru, Rh, W, Re, Os), the metal–metal distance is so short that there is little doubt that there is a bonding interaction between them.

As an aside, these metal–metal bonding attractions, not surprisingly, require the involvement of d-orbitals. Historically, the importance and nature of metal–metal bonding was highlighted in the 1960s by characterisation of the dimeric $[Re_2Cl_8]^{2-}$ ion (formally Re^{III}) which, unlike the carboxylate-bridged dimers above, is only held together by a Re–Re bond (Figure 6.4). Although we will not explore this in the same depth as in Chapter 2, a rather beautiful set of molecular orbitals arise from symmetry-adapted combinations of d-orbitals oriented either along or perpendicular to the metal–metal bond. In the case of $[Re_2Cl_8]^{2-}$ it boasts a quadruple bond as shown by different combinations of $5d$ atomic orbitals. The sigma (σ) and pi (π) bonding interactions are similar to those discussed in Chapter 2 except atomic orbitals from each metal overlap. However, the delta (δ) bond is new and unique to transition metals. The only d-orbitals not involved in metal–metal bonding are the $5d_{x^2-y^2}$ orbitals on each Re^{III} ion which overlap with the four chlorido lone pairs. Only the bonding MOs are shown but of course the corresponding antibonding MOs will exist. In $[Re_2Cl_8]^{2-}$, both metals provide four $5d$ electrons which perfectly fill the four bonding MOs in Figure 6.4. In other M–M systems, the bond order is naturally determined by the number of d electrons present. In higher total d-electron configurations (above $2 \times d^4 = 8$) populating, the corresponding antibonding orbitals weakens the M–M bond.

Limited examples of other dinuclear and higher polynuclear complexes will appear herein. However, the focus in this section will remain on mononuclear systems.

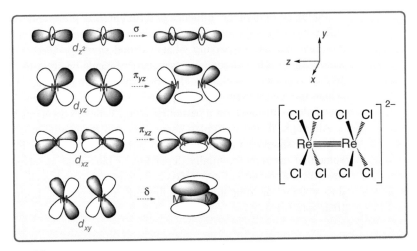

Figure 6.4
Metal–metal bonding molecular orbitals resulting from overlap of *d*-atomic orbitals; sigma (σ), pi (π) and delta (δ) bonding MOs are shown. Multiple metal–metal bonds occur in $[Re_2Cl_8]^{2-}$, shown at right.

6.3.2   Substitution Reactions in Non-aqueous Solvents

Water may not be a suitable solvent in all cases due to: reactant insolubility of the precursor complex and or the ligand, but usually the latter; extreme inertness that demands use of a higher boiling point solvent for reaction; or high stability of undesired hydroxido or oxido species that prevent or interfere with formation of desired products. This is usually solved by using another solvent, either a conventional molecular organic solvent or a low melting point (room temperature) ionic liquid.

The aqueous chemistry of particularly Al(III), Fe(III) and Cr(III) involves the formation of strong M–O bonds, and in basic aqueous solution, insoluble hydroxide species (or unreactive oligomers) usually precipitate preferentially. The behaviour of Fe(III) in a weakly basic aqueous solution is a good example and can be represented simplistically by reaction (6.12).

$$\left[\begin{array}{c} OH_2 \\ H_2O \diagdown | \diagup OH_2 \\ Fe \\ H_2O \diagup | \diagdown OH_2 \\ OH_2 \end{array}\right]^{3+} \xrightarrow[-3H^+]{OH^-} \left[\begin{array}{c} OH_2 \\ H_2O \diagdown | \diagup OH \\ Fe \\ H_2O \diagup | \diagdown OH \\ OH \end{array}\right] \xrightarrow[-3H_2O]{\text{Rapid dehydration and precipitation}} Fe(OH)_3 \downarrow$$

Deprotonation to produce
coordinated hydroxides

(6.12)

Hydroxide is both an effective *bridging* ligand and a strong base, so it promotes the formation of polymeric hydroxido complexes, which are rarely the desired target. This can be avoided simply by working in the absence of water, using instead an anhydrous aprotic solvent. For example, using a hydrated salt of chromium(III) in water (reaction (6.13); upper eq.) or an anhydrous salt in anhydrous diethyl ether (reaction (6.14); lower eq.) with ethane-1,2-diamine (en) as introduced ligand has distinctly different outcomes, due to the absence of water in the latter reaction.

$$
\text{Cr}_{aq}{}^{3+} + 3(\text{en}) \xrightarrow{\text{water}} \text{Cr(OH)}_3\downarrow + 3(\text{enH}^+)
$$

Anhydrous chromium(III) chloride

$$
\text{CrCl}_3 + 3(\text{en}) \xrightarrow{\text{ether}} [\text{Cr(en)}_3]\text{Cl}_3
$$

Purple → Yellow

(6.13–6.14)

In the first case, the basic diamine effectively extracts protons from coordinated water molecules, leaving an insoluble metal hydroxide, rather than initiating any ligand substitution. In the aprotic solvent where no water is present, thus preventing formation of any hydroxido species, the diamine achieves coordination. It is also possible to prepare solvated salts other than hydrated ones for use. For example, $[\text{Ni(OH}_2)_6](\text{ClO}_4)_2$ may be replaced by $[\text{Ni(DMF)}_6](\text{ClO}_4)_2$ in dimethylformamide (DMF, O-bound) as solvent, providing a way of ensuring that the initial complex coordination sphere as well as the solvent is a non-aqueous species.

Where solubility alone is the issue, simply changing solvent to permit all species to be dissolved allows the chemistry to proceed essentially as it would in aqueous solution were the species soluble. Typical molecular organic solvents used in place of water include other protic solvents such as alcohols (e.g. ethanol), and aprotic solvents such as ketones (e.g. acetone), amides (e.g. dimethylformamide), nitriles (e.g. acetonitrile) and sulfoxides (e.g. dimethyl sulfoxide). Solvents termed ionic liquids, which are merely salts with very low melting points such that they are liquid at or near room temperature, have been employed for synthesis. Typically, these consist of a large organic cation and an inorganic anion (e.g. *N,N'*-butyl(methyl)imidazolium nitrate) and their ionic nature supports dissolution of, particularly, ionic complexes. Their ionic nature can promote reaction pathways or product ratios different from those found with traditional solvents.

In some cases, where hydrolysis reactions are not of concern, mixtures of organic solvents and water can be employed. For example, it is possible to mix an aqueous solution of the metal salt with a miscible organic solvent containing the water-insoluble organic ligand, with the two reactants sufficiently soluble in the mixed solvent to remain in solution and react. Furthermore, it is possible sometimes to react an insoluble compound with a dissolved one with sufficient stirring and heating, since 'insoluble' usually does not mean absolute insolubility (there is always an equilibrium present between solid and solute), with sufficient dissolving to permit reaction, and its 'removal' from solution by complexation allowing further compound to dissolve and react.

In certain cases, complexes may not form, or else are not stable, in water because they are not thermodynamically stable in that solvent. Their formation in the total absence of water can be achieved, however, as exemplified by reaction (6.15), where the neutral tetrahedral complex can be isolated readily and is stable in the absence of water. However, dissolution in water leads to very rapid formation of the octahedral $\text{Co}_{aq}{}^{2+}$ cation and dissociation of the initial ligands.

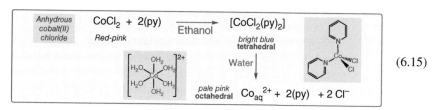

$$
\text{CoCl}_2 + 2(\text{py}) \xrightarrow{\text{Ethanol}} [\text{CoCl}_2(\text{py})_2]
$$

Anhydrous cobalt(II) chloride — Red-pink → bright blue tetrahedral

pale pink octahedral $\xrightarrow{\text{Water}}$ $\text{Co}_{aq}{}^{2+} + 2(\text{py}) + 2\,\text{Cl}^-$

(6.15)

Neutral complexes of the type [M(acac)$_2$] and [M(acac)$_3$], where acac$^-$ is the butane-2,4-dionate (or acetylacetonate) anion, usually show good solubility in organic solvents and serve as useful synthons, since they undergo ligand substitution reactions by other chelates fairly readily. Moreover, they tend to stabilise otherwise relatively unstable oxidations states; for example, the Mn(III) complex [Mn(acac)$_3$] (reaction (6.16); upper eq.) is very stable, whereas the hydrated ion Mn$_{aq}^{3+}$ readily undergoes disproportionation (reaction (6.17); lower eq.) in water (to Mn(II) and Mn(IV)). The former complex is itself readily prepared from reaction of MnCl$_2 \cdot$4H$_2$O and butane-2,4-dione (acacH) in basic solution with a strong oxidant (reaction (6.16)). Subsequent reaction of [Mn(acac)$_3$] in an organic solvent with potential polydentate ligands offers a route to a wide range of complexes.

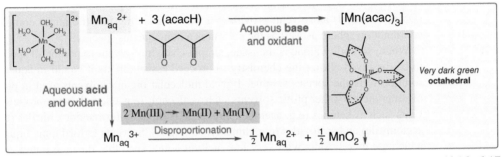

$$(6.16\text{–}6.17)$$

6.3.3   Substitution Reactions Without Using a Solvent

In several reactions, the ligand to be introduced is a liquid, and where there are no issues relating to a need for stoichiometric amounts of reagents it is possible to use the ligand in excess as effectively its own solvent in the reaction. This is particularly appropriate where the ligand can be removed readily by evaporation or distillation or where the complex product precipitates from the reaction mixture and can be separated directly by filtration. The best example of this type of reaction is with ammonia, which has a boiling point of only −33°C. For example, reaction (6.18) proceeds readily and avoids the hydrolysis problems that lead to preferential formation of Cr(OH)$_3$ in aqueous ammonia.

$$
\begin{array}{llll}
\text{Anhydrous}\\
\text{chromium(III)} & \text{CrCl}_3 \;+\; 6(\text{NH}_3) & \xrightarrow{\;\;\text{Liquid ammonia}\;\;} & [\text{Cr}(\text{NH}_3)_6]\text{Cl}_3 \\
\text{chloride} & \quad\;\; \text{Purple} & & \qquad\quad \text{Yellow}
\end{array}
\qquad (6.18)
$$

While this reaction works well with most anhydrous metal ion salts, it is not necessarily convenient because of the handling difficulties involving liquid ammonia, and some salts form isolable ammine complexes sufficiently well using aqueous ammonia, as is the case with Ni(II) and Cu(II), for example.

Higher boiling point and also chelating amines can be reacted effectively with anhydrous metal salts in the same manner as described for ammonia. These reactions do generate heat, as expected in an acid/base (metal cation/ligand) reaction, so careful mixing of reagents is required. In principle, this technique also offers a route to a wide range of [M(solvent)$_6$]$^{n+}$ salts through reaction of an anhydrous salt with traditional molecular solvents, provided the solvent is an effective ligand. Nitriles, amides, sulfoxides and alcohols may be introduced in this manner in many cases.

Another approach, which permits introduction of anionic ligands of strong acids, is to react a complex containing coordinated chloride ion directly with an anhydrous strong acid, such as trifluoromethanesulfonic acid, in the total absence of any other solvent. One example, where HCl is released as a diatomic gas and leaves the anhydrous reaction mixture and the $CF_3SO_3^-$ anion enters the coordination sphere as an O-bound monodentate ligand, is reaction with chloropentaamminecobalt(III) chloride (reaction (6.19)).

$$[CoCl(NH_3)_5]Cl_2 + 3(CF_3SO_3H) \xrightarrow[\substack{\text{heat in excess} \\ 100\% \text{ CF}_3\text{SO}_3\text{H}}]{} [Co(NH_3)_5(OSO_2CF_3)](CF_3SO_3)_2 + 3HCl \uparrow$$

Purple   Red

precipitated after cooling
by slow addition of
cold diethyl ether;
caution – heat generated!

(6.19)

The complex product can be isolated as a solid by very slow and careful addition of cold diethyl ether to the stirring cooled reaction mixture; a great deal of heat is generated in this process, so great caution is required. However, the product in this example is especially useful as a synthetic reagent. The coordinated trifluoromethanesulfonate anion is an extremely reactive ligand even when bound to an inert metal ion like Co(III) and is thus readily replaced. Consequently, these complexes find use as reagents for the introduction of other ligands through simple substitution reactions in a poorly or non-coordinating solvent, as exemplified in reaction (6.20).

(6.20)

Another approach is to employ the effect of heat on a solid complex. It has been known for a long time that metal complexes, if heated strongly, undergo decomposition reactions that eventually take them through to usually simple salts or oxides. Generally, ligands are lost in a series of steps, related in part to their volatility, and this can be probed using a technique called *thermal gravimetric analysis* (TGA), which effectively amounts to following weight change with a sensitive balance during heating of a small sample. Simple neutral ligands often depart the coordination sphere as molecular species over a reasonably small and well-defined characteristic temperature range so that heating to a defined temperature can allow controlled conversion to occur. The simplest examples are the hydrated salts of metal ions, such as $[M(OH_2)_n](SO_4)$, which on heating lose water to form anhydrous $M(SO_4)$, usually with a distinctive colour change (such as seen for the copper(II) salt in Figure 3.2). For complexes containing water as one of several ligands, its loss tends to occur ahead of other ligands such as organic amines, allowing partial change of the coordination sphere. The metal centre involved still seeks to retain its original coordination geometry so that loss of water is usually associated with replacement in the coordination sphere by an involatile anion of the original salt, such as in reaction (6.21) which is similar to the hydrate isomerism described in Chapter 4.4.2.1.

$$[Co(NH_3)_5(OH_2)]Cl_3 \xrightarrow[\substack{Heat\ the\ solid;\\110°C,\ hours}]{} [CoCl(NH_3)_5]Cl_2 + H_2O \qquad (6.21)$$

Red-pink   Purple

This approach permits the insertion of a wide range of stable anions apart from chloride into the coordination sphere, simply by commencing with them present as the counter-ion. At higher temperature, ammine ligands can be lost in the same manner that water is lost and may occur in a stepwise process that permits isolation of useful intermediate complexes. Eventually, at sufficiently high temperature, all but the chlorido ligands are lost, as in reaction (6.22). Charge balance must be maintained, so the chlorides remain tightly bound.

$$[Pt(NH_3)_4]Cl_2 \xrightarrow[\substack{heating;\\250°C,\ hrs}]{} [PtCl_2(NH_3)_2] + 2\ NH_3\uparrow \xrightarrow[\substack{extend\ heating;\\500°C,\ hrs}]{} \{PtCl_2\}_n + 2\ NH_3\uparrow \qquad (6.22)$$

colourless   yellow   anhydrous platinum(II) chloride

Reaction (6.22) is general for a range of Pt(II)-coordinated amines and appears to yield exclusively the *trans* geometric isomer (which is called, because of this exclusivity, a *stereospecific* reaction). Even chelating diamines can be substituted, as they are inevitably more volatile than any anions present; thus, ethane-1,2-diamine (en) can be replaced by two chloride ions in [Co(en)$_3$]Cl$_3$ to form *cis*-[CoCl$_2$(en)$_2$]Cl and by two thiocyanate ions in [Cr(en)$_3$](NCS)$_3$ to form *trans*-[Cr(en)$_2$(NCS)$_2$](NCS) on heating. Anions such as thiocyanate are ambidentate ligands that can coordinate through either of two donors, in this case, the S or N atoms. Because heating is employed in these syntheses, the thermodynamically stable form will always be isolated, which in the case of thiocyanate with chromium(III) is the N-bound form. This behaviour can be easily demonstrated by commencing with the less stable isomer, and observing behaviour on heating; for example, pink [Co(NH$_3$)$_5$(ONO)]Cl$_2$ on mild heating changes to the thermodynamically stable yellow [Co(NH$_3$)$_5$(NO$_2$)]Cl$_2$, where an O- to N-donor isomerisation has occurred, as also occurs in solution and discussed in Chapter 5.

6.3.4 Chiral Complexes

Many ligands are chiral as a result of asymmetric centres, which mean they exist as two different optical isomers. These ligands may be present as a racemate (a 50:50 mixture of the two optical forms) or else as a chiral form (that is, as one optical isomer). The presence of chirality in a ligand is 'felt' by the complex and seen in its chiroptical properties (see Chapter 7.3.3), but there are rarely any particular specific requirements for synthesis involving introduction of a chiral ligand as opposed to its racemic form. However, in certain circumstances, the complex itself can be chiral as a result of the disposition of ligands. As a general rule, the presence of at least two chelate rings is sufficient for some octahedral complexes to be resolvable into their enantiomeric forms. The classic chiral metal complex is the octahedral

[Co(en)$_3$]$^{3+}$ ion, which can be separated into Δ and Λ optical forms through successive recrystallisation of its salt formed with an optically active anion such as *d*-tartrate (reaction (6.23)). This approach is based on the Δ-[Co(en)$_3$]$^{3+}$/(*d*-tartrate) combination producing a salt that is less soluble than the Λ-[Co(en)$_3$]$^{3+}$/(*d*-tartrate) combination, allowing the former diastereomer to be preferentially precipitated.

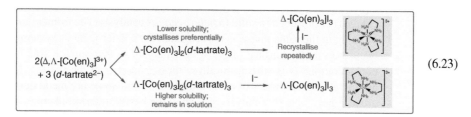

(6.23)

The basis of this differentiation is different H-bonding patterns between the chiral complex cations and the chiral tartrate anion (Figure 6.5). On the left is the Λ-[Co(en)$_3$]$^{3+}$ (with a δδδ conformation of its three five-membered chelate rings) paired with *d*-tartrate, whereas on the right is the Δ-[Co(en)$_3$]$^{3+}$/*d*-tartrate combination. Although the complex cations are mirror images, the tartrate anions have the same chirality so the H-bonding patterns of each pairing must be different, leading to different solubilities.

It is also possible to separate the optical isomers through chromatography on a cation-exchange resin (such as SP-Sephadex® C-25 resin, Figure 6.3) by using a chiral eluent such as a *d*-tartrate salt solution. Differential ion pairing of the kind seen in the solid state (Figure 6.5) means the complex separates on a sufficiently long column into two bands comprising the two optical forms of the complex. This is a facile method for the separation of optically pure isomers.

Chirality is often met with polydentate ligands at a number of different centres in the complex, and separation of all optical isomers is either impractical or, in effect, essentially impossible. In many cases, working with a racemate has no significant influence on the chemistry, and optical resolution of complexes is attempted on only very limited occasions, such as where researchers wish to record the chiroptical properties (see Section 7.3.3), or where a chiral complex is required to assist in achieving a chiral reaction, such as use of a chiral complex as a catalyst in synthesis of organic molecules where a particular optical isomer is sought (exemplified in Chapter 9.4.2).

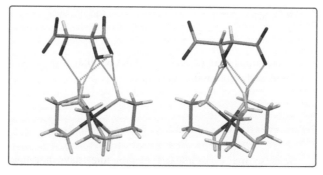

Figure 6.5

Crystal structures of (left) Λ-[Co(en)$_3$]$^{3+}$ and (right) Δ-[Co(en)$_3$]$^{3+}$ as their *d*-tartrate salts, showing the different H-bonding/ion pairing patterns.

6.3.5   Catalysed Reactions

Catalysis is a general chemical term dealing with an acceleration of extremely slow reactions without resorting to an increase in temperature or pressure. The concept of catalysis is that the catalyst lowers the activation barrier for a reaction, thus allowing it to proceed at a faster rate; it is not usually considered to influence the position of an equilibrium. The catalyst may be dissolved in the reaction solution (a *homogenous* catalyst) or present as a solid (a *heterogenous* catalyst). The former has the advantage of being in the same phase as the reagents; for a heterogenous system, reaction must occur only at the surface of the solid catalysts so that surface area is an important consideration, as is stirring to allow mass transport in the system. Many catalysts are metals, metal oxides or other simple salts, but some are metal complexes. There are also non-metallic catalysts, of which the best-known simple ones are the various forms of activated carbon.

Formation of platinum(IV) complexes involving ligand substitution is an extremely slow process, due to the kinetic inertness of this oxidation state. However, the addition of small amounts of a platinum(II) complex to the reaction mixture leads to excellent catalysis of the reaction, assigned to mixed oxidation state bridged intermediates that promote ligand transfer. Isomerisation of inert, chiral cobalt(III) complexes is accelerated significantly by the presence of activated carbon (effectively charcoal); this is assigned to reduction of cobalt(III) on the surface of the carbon to labile cobalt(II), allowing rapid ligand rearrangement, with air subsequently rapidly re-oxidising the cobalt(II) to cobalt(III) before ligand dissociative processes can become involved.

Complexes of some metal ions show a capacity towards photo-activation, which means that they undergo different chemistry in the presence of light. This is because light can cause electron transfer from metal orbitals to ligand orbitals, leading to a reactive excited state that can undergo different chemistry; ruthenium(II) is one metal that exhibits this capacity. Usually, simple complexes exhibit only light (*hv*) catalysed aquation reactions, but redox chemistry can result with other reagents. Both outcomes are illustrated in the following simple example, where both aquation (reaction (6.24); upper eq.) and oxidation (reaction (6.25); lower eq.) can occur.

$$[Ru(L)(NH_3)_5]^{2+} + H_2O \xrightarrow[\text{Aquation}]{hv} \begin{array}{l} [Ru(NH_3)_5(OH_2)]^{2+} + L \\ [Ru(L)(NH_3)_4(OH_2)]^{2+} + NH_3 \end{array}$$

$$[Ru^{II}(L)(NH_3)_5]^{2+} + H^+ \xrightarrow[\text{Oxidation}]{hv} [Ru^{III}(L)(NH_3)_5]^{3+} + \tfrac{1}{2}H_2 \qquad (6.24-6.25)$$

Examples of reactions that undergo change in oxidation state are covered in more detail in the following section.

6.4   Reactions Involving the Metal Oxidation State

Because many complexed metals ions have a range of oxidation states accessible in usual solvents, it is not surprising to find that syntheses may involve reduction–oxidation reactions. The classic example is cobalt, which is usually supplied commercially as hydrated Co(II) salts, but whose complexes are best known as Co(III) compounds. This is because Co(III) compounds are inert and usually robust, readily isolable compounds, whereas Co(II) compounds are labile and prone to rapid reactions. This lability is put to beneficial use in synthesis, since it is convenient to use the Co(II) form to rapidly coordinate ligands initially, and then oxidise the mixture to the Co(III) form. This oxidation can often be achieved by oxygen in the air alone, depending on the ligand environment and the redox potential (E^o) of the complex. For example, in an aqueous ammonia/ammonium chloride buffered solution, Co(II) reacts in a sequential manner as in reaction (6.26).

$$
\left[\begin{smallmatrix} & OH_2 \\ H_2O_{\text{\tiny III}} & | & _{\text{\tiny III}}OH_2 \\ & Co & \\ H_2O & | & OH_2 \\ & OH_2 \end{smallmatrix}\right]^{2+} \quad Co_{aq}{}^{2+} \; + \; NH_3/NH_4Cl \longrightarrow [Co(NH_3)_6]^{2+} \xrightarrow{\;O_2\;} [Co(NH_3)_6]^{3+} \; \left[\begin{smallmatrix} & NH_3 \\ H_3N_{\text{\tiny III}} & | & _{\text{\tiny III}}NH_3 \\ & Co & \\ H_3N & | & NH_3 \\ & NH_3 \end{smallmatrix}\right]^{3+} \tag{6.26}
$$

$$\text{Cobalt(II)} \qquad \text{Excess} \qquad\qquad\qquad\qquad \text{Cobalt(III)}$$
$$\text{Pale pink} \qquad\qquad\qquad\qquad\qquad\qquad \text{Yellow}$$

Substitution reactions without redox chemistry being involved are available for Co(III) but not commonly met. One example is the use of $[Co^{III}(CO_3)_3]^{3-}$ as a synthon, since the chelated carbonate ion is readily displaced by other better chelating ligands such as polyamines. An example of this type of reaction, .where the entering polyamine (cyclam) is a saturated and flexible macrocyclic tetraamine, appears in reaction (6.27).

$$
[Co(CO_3)_3]^{3-} \quad + \quad \text{cyclam} \quad \longrightarrow \quad [Co(CO_3)(\text{cyclam})]^+ \; + \; 2\,CO_3{}^{2-} \tag{6.27}
$$

Where stronger oxidising agents than air are required to take the Co(II) form to Co(III), which usually applies where fewer N-donor and more O-donor groups are bound, hydrogen peroxide is particularly useful, because it leaves no problematical products to separate from the desired complex product. One of the problems with oxidation of Co(II) to Co(III) compounds is that in some cases a bridged peroxido complex, featuring a $Co^{III}-O_2{}^{2-}-Co^{III}$ linkage, forms as a stable intermediate. One can envisage its formation through a redox reaction whereby an oxygen molecule is reduced to peroxide ion, and two Co(II) ions are oxidised to Co(III). The $O_2{}^{2-}$ ion in the bridge can be readily displaced by reaction with strong acid and heating, with the use of hydrochloric acid leading to monomers with $Co^{III}-Cl^-$ components replacing the bridging group (reaction (6.28)).

$$
Co_{aq}{}^{2+} + 5\,L \;\rightleftharpoons\; [L_5Co(OH_2)]^{2+} \xrightarrow{\;O_2\;} \left[L_5Co^{III}{-}O^-{-}O^-{-}Co^{III}L_5\right]^{4+} \tag{6.28}
$$

$$\xrightarrow{HCl_{aq}}$$

$$\left[\begin{smallmatrix} & L \\ L_{\text{\tiny III}} & | & _{\text{\tiny III}}L \\ & Co & \\ L & | & Cl \\ & L \end{smallmatrix}\right]^{2+} \quad 2\,[Co^{III}(Cl)L_5]^{2+}$$
$$(+\,H_2O + \tfrac{1}{2}O_2)$$

Many other metal complexes can be chemically oxidised to higher oxidation states with an appropriate oxidising agent. This can occur with complete preservation of the coordination sphere (reaction (6.29)), which means the oxidation–reduction reaction is reversible if a suitable reducing agent is then employed for reduction of the oxidised form. Alternatively, the coordination number and or the ligand set can change substantially in an irreversible oxidation reaction. In this case, any reduction reaction of the product will not regenerate the original complex.

$$[IrCl_6]^{3-} \underset{\substack{+e \\ \text{Reduction}}}{\overset{\substack{\text{Oxidation} \\ -e}}{\rightleftharpoons}} [IrCl_6]^{2-}$$

(6.29)

Not only are oxidation reactions fairly common but also one may employ reduction reactions in simple synthetic paths, provided the reduced form is also in a stable oxidation state. Two examples of reduction reactions, with the metal oxidation states included, are given in reactions (6.30; upper eq.) and (6.31; lower eq.).

$$[Pt^{IV}(en)_3]Cl_4 \xrightarrow{+2e^-} [Pt^{II}Cl_2(en)] \quad (+ 2(en) + 2Cl^-)$$

$$[Mo^{V}Cl_5O]^{2-} \xrightarrow{Zn/HCl} [Mo^{III}Cl_6]^{3-} \quad (+ OH^- + Zn^{2+})$$

(6.30–6.31)

Reduction reactions occurring along with substitution chemistry are also well known, and the examples above are such cases. Another simple example involves the $[IrCl_6]^{2-}$ ion, which undergoes both reduction (with hypophosphorous acid) and substitution in reaction (6.32). Because of the slow substitution chemistry of iridium, the sequential reaction steps are well separated in terms of reaction time so that the initial product of the redox-substitution reaction undergoes further simple substitution only with prolonged heating.

$$[Ir^{IV}Cl_6]^{2-} \xrightarrow[\substack{H_2PO_2^- \\ \text{pyridine} \\ \text{Boil, 30 min}}]{} fac\text{-}[Ir^{III}Cl_3py_3] \xrightarrow[\text{Boil, 6 hours}]{} cis\text{-}[Ir^{III}Cl_2py_4]^-$$

(6.32)

Where a reactive lower oxidation state results, a key concern is the necessary protection of the reduced complex from air or other potential oxidants, as the complexes are often readily re-oxidised. Usually, this requires their handling in special apparatus such as inert-atmosphere boxes or sealed glassware in the absence of oxygen. Where active metal reducing agents (such as potassium) are employed, particular care with choice of solvent is also necessary. The nickel reduction reaction (6.33) can be performed in liquid ammonia as solvent, since the strongly bound cyanide ions are not substituted by this potential ligand. While the Ni(II) is reduced to Ni(0), the potassium metal is oxidised to K(I); it is a standard redox reaction. The tetrahedral Ni(0) complex formed exemplifies the necessity for careful handling of many low-valent complexes; it is readily oxidised by air, and also reacts with water in a redox reaction that liberates hydrogen gas.

$$[Ni^{II}(CN)_4]^{2-} + 2K \xrightarrow[\substack{liquid \\ NH_3}]{} [Ni^{0}(CN)_4]^{4-} + 2K^+$$

(6.33)

Lower-valent metal complexes may be prepared in reduction reactions with full substitution of the coordination sphere. One example results from reduction of the vanadium(III) complex $[VCl_3(THF)_3]$ with a suitable reactive metal in the presence of carbon monoxide under pressure (reaction (6.34)). This is an example of synthesis of an organometallic compound; more examples occur in Section 6.6.

$$(6.34)$$

On some occasions, a reducing agent may not be able to reduce the metal centre in a complex but may be sufficient to initiate chemical change in a ligand. This is, in effect, a reaction of a coordinated ligand (see Section 6.5.2), rather than a metal-centred redox reaction. A simple example involves the reduction of coordinated N_2O to N_2 and H_2O, using Cr(II) ion as reducing agent (reaction (6.35)); the metal ion retains its original oxidation state throughout. The reaction is promoted through coordination lowering the N–O bond strength.

$$(6.35)$$

Many complexes can be prepared by electrochemical reduction or oxidation under an inert atmosphere. Exhaustive reduction (*electrolysis*) using a large working electrode can lead to clean formation of the reduced form, if it is sufficiently stable, since electrons alone are used in the reduction process. As an example, the octahedral vanadium(III) complex $[V(phen)_3]^{3+}$ undergoes a series of sequential and reversible reduction steps (reaction (6.36)) with retention of the three chelate ligands down to oxidation state of formally vanadium(−I) but the phen ligands are not 'innocent' and the last three reductions are almost certainly ligand centred to give formally $[V^{II}(phen^-)_3]^-$.

$$(6.36)$$

With unsaturated ligands like phen, the location of introduced electrons is not easily assigned, and they may reside on the ligand (leading to a ligand radical) rather than on the metal ion (leading to a lower oxidation state). This does not invalidate the chemistry but does bring into question the nature of the product. Overall, electrochemistry provides a useful way of performing not only reduction reactions but also oxidation reactions. The sole concern is that exhaustive electrolysis to convert all of one form to another oxidation state takes some minutes to perform, so the kinetic stability of the product does influence the validity of the method.

When undertaking electrochemical reactions, it is important to note that the E^o for a complex is dependent on the set of ligands coordinated and can change substantially with donor set. This is represented in reaction (6.37) for the simple substitution of a M_{aq}^{n+} complex, with the potential required to reduce the precursor complex invariably differing from that for the product complex ($E^o_1 \neq E^o_2$).

$$[M(OH_2)_y]^{n+} + y\,L \;\rightleftharpoons\; [M(L)_y]^{n+} + y\,H_2O$$
$$+e \downarrow E^o_1 \qquad\qquad\qquad +e \downarrow E^o_2 \qquad\qquad (6.37)$$
$$[M(OH_2)_y]^{(n-1)+} \qquad\qquad [M(L)_y]^{(n-1)+}$$

The shift in redox potential with ligand substitution is particularly obvious for cobalt, as the E^o for $Co^{III/II}_{aq}$ is near +1.8 V, making the Co(III) aqua complex inaccessible from Co_{aq}^{2+} as water is oxidised before Co_{aq}^{2+}, whereas introduction of other donor groups lowers the potential substantially, and in some cases sufficiently to make even air (oxygen) an effective oxidant of the Co(II) complex.

6.5   Reactions Involving Coordinated Ligands

6.5.1   Metal-directed Reactions

There exist both metal-catalysed and metal-directed reactions, which require definition to distinguish the two classes. *Metal-catalysed* reactions are those in which the metal-containing species in the reaction is regenerated in each reaction cycle, so stoichiometric amounts are not required. It is the transition state of the catalysed reaction, rather than the product, which is most strongly complexed by the metal. Thus, it is the rate of establishment of equilibrium (where k_f and k_b, the forward and back reaction rates, are important) rather than the position of the equilibrium ($K = k_f/k_b$) that is altered (reaction (6.38)).

$$\text{Reactants} + M_{catalyst} \;\underset{k_b}{\overset{k_f}{\rightleftharpoons}}\; \text{Products} + M_{catalyst} \qquad (6.38)$$

A homogeneous transition metal catalyst usually employs its coordination sphere as the site of the chemistry it promotes. One example is the rhodium catalyst used for promoting ethene hydrogenation. The keys to the process are an addition reaction of dihydrogen to the rhodium centre and a substitution reaction that also introduces ethene to the coordination sphere in place of a solvent molecule. It is in the intermediate produced where the former adds to the latter to produce ethane; this then, as an exceedingly poor ligand, departs the coordination sphere and leaves vacancies for the process to occur again (Figure 6.6). The catalyst is reused many times. This type of chemistry is explored further in Chapter 9.

For a *metal-directed* reaction, the product is formed as a metal complex, and stoichiometric amounts of metal are consumed in the process (reaction (6.39)). In virtually all cases, the driving force for metal-directed reactions is the stabilisation associated with the formation of a chelating ligand from monodentate ligands or the

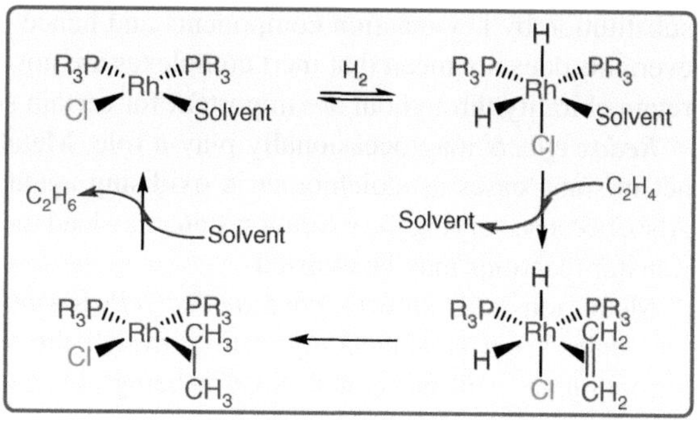

Figure 6.6

A simplified catalytic cycle for the hydrogenation of ethene (C_2H_4) to ethane (C_2H_6), which employs a rhodium(I) catalyst and involves reaction of coordinated ligands.

conversion of a weakly chelating ligand to a stronger chelator. Often, the major product in the presence of the metal ion is not even detected from the same reaction in the absence of the metal ion. The metal has either caused an extreme displacement of an equilibrium or promoted a new and rapid reaction pathway by complexation and stabilisation of an otherwise inaccessible transition state.

$$\text{Reactants} + M \longrightarrow [\text{Products} - M] \tag{6.39}$$

Normally, metal-directed reactions involve the synthesis of a larger organic molecule from smaller components, with the product being an effective ligand. In fact, these reactions result in the organic product being bound to the metal. There are several general principles considered of importance in governing metal-directed reactions, the most important of which are as follows:

Chelation: This is probably the most important factor. In nearly all cases, it is the formation of a (more) stable metal chelate as the primary reaction product that drives the equilibrium to favour that product.

Ligand Polarisation: Nucleophilic and electrophilic reactions of organic molecules (such as condensation, hydrolysis, alkylation and solvolysis, amongst others) can be greatly enhanced by their coordination to metal ions as ligands. The metal ion can act variously as a Lewis acid, π-acid or π-donor to alter electron density or distribution on the bound organic molecule (or ligand), thus altering the character of the ligand as a nucleophile or electrophile, and hence its reactivity.

Template Effects: The metal ion can function as an 'organiser' or 'collector' of ligands into arrangements around it that are most suitable for the desired reactions.

There are, in addition, several other contributing effects:

Enantiomer discrimination relates to the fact that ligand binding and reactivity can be affected by other ligands not actually participating in the reaction (the so-called spectator ligands—like spectators at a football game can 'lift' their team's performance without actually playing themselves, spectator ligand will influence what happens at reaction sites by their presence). If these spectator ligands are bulky and optically active, or if the complex is made chiral by a chiral and rigid spectator ligand, differential binding of another chiral ligand, or stereospecificity of reactions, can be introduced.

Metal ion lability is the ability of a metal ion to exchange its ligands rapidly, which is of importance in template reactions. Very slow exchange will effectively prevent substitution by key reaction components and hence limit reactions occurring. However, this does not mean that inert complexes are not relevant, as inert metal ions that retain chirality throughout are important for certain stereospecific syntheses.

Redox effects may occasionally play a role. Metals in high oxidation states may act in some cases as stoichiometric oxidising agents of a ligand functional group. Also, because varying the oxidation state may lead to more stable complexes, electron transfer reactions may be assisted.

Metal ions can direct *spontaneous self-assembly* of larger and often cyclic molecules from smaller components through the above effects. Nature does this exceptionally well but is not alone in being able to build cyclic molecules readily. Simple, one-step, comparable syntheses can be achieved in an open beaker, directed by a metal ion, as exemplified in Figure 6.7. This condensation reaction (generating 4 molecules of water) occurs spontaneously in high yield when appropriate amounts of copper(II) nitrate, ethane-1,2-diamine, aqueous formaldehyde and nitroethane are mixed in methanol in a beaker, warmed for a brief period, and then stood overnight to allow crystallisation. The metal ion acts as a 'collector' of ligands as well as a promoter of ligand reactivity by means of chelation and polarisation effects.

A reaction in the absence of a metal ion may differ completely from what occurs in the presence of a metal ion, as exemplified in Figure 6.8. Here, the linear organic molecule formed in the presence of stoichiometric amounts of a metal ion is absent in the metal-free chemistry, where small heterocyclic ring formation occurs.

Metal-directed reactions have been used to prepare a wide range of cyclic and acyclic ligand systems. Often, they involve reaction of a coordinated amine with an aldehyde or ketone. Reaction of a carbanion with an electron-deficient site is also

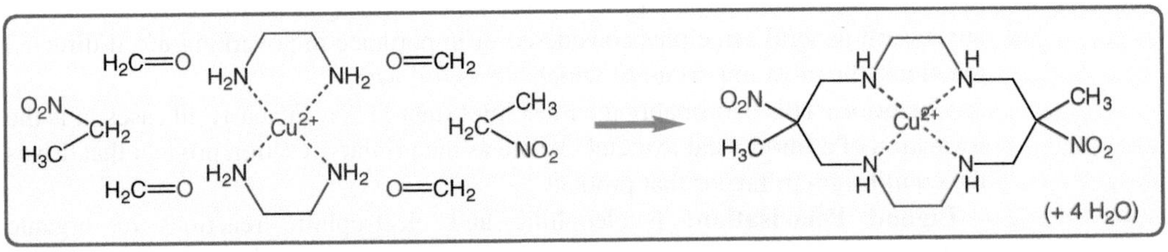

Figure 6.7

An example of a spontaneous self-assembling template reaction, in which small organic components are organised by the metal ion and undergo reaction to form a large cyclic organic product that includes the metal ion.

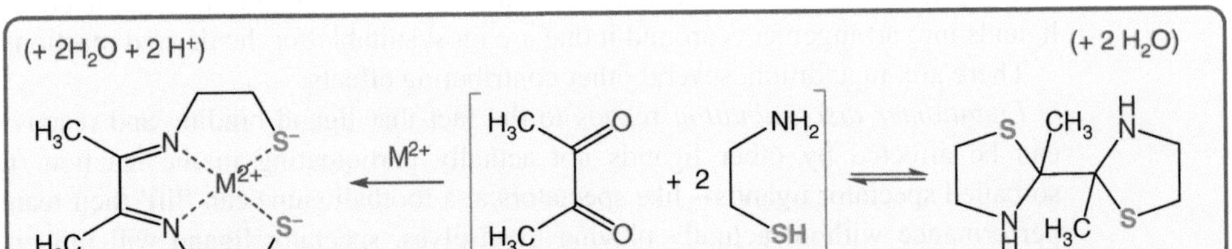

Figure 6.8

An example of the normal organic reaction route compared with the template reaction in the presence of a metal ion; distinctly different paths are followed.

commonly featured. Zinc(II)-directed condensation between an aldehyde (RCHO) and an aromatic nitrogen heterocycle (pyrrole) has been used since the 1960s to prepare substituted porphyrin rings (aromatic tetraaza macrocycles, analogous to hemes found as iron complexes in blood). Once formed, the new ligands can be removed from the templating metal ion and different metal ions bound to it. This usually involves one of the following: treatment with acid to protonate the ligand and cause the complex to dissociate; reduction or oxidation of the metal to an oxidation state which will not bind the ligand effectively, allowing it to be removed; treatment with a strongly binding anion (such as CN^-) that removes the metal ion competitively, leaving the free ligand; addition of a competing metal ion to a solution of the templated product to which the ligand binds preferentially, causing ligand exchange with the added metal ion.

Polynuclear complexes form through self-assembly also, where both the ligand and precursor metal complex geometry play a role in the outcome—the pieces tend to fit together in a particular way like children's building blocks. An early example involves the self-assembly of 4,4′-bipyridyl and [Pd(en)(ONO$_2$)$_2$]. The two *cis*-disposed O-bound nitrate ions are readily substituted by the preferred pyridine nitrogen donors, but they impose an L-shape when including the palladium, while the ligand imposes a rod-like shape; thus, an assembly of four Pd 'corner' L-shapes and four ligand 'rod' shapes creates a square 'picture frame' shape, which is now a large cyclic molecule (Figure 6.9) also known as a catenane.

The product in the above reaction is stable and of lower solubility than the precursor monomer complex; precipitation from the reaction solution may drive product formation, which occurs in high yield. However, studies of analogues in solution using NMR attest to the tetranuclear complex being dominant and formed essentially irreversibly even prior to isolation. The reaction is, in effect, a sequence of substitution steps, with coordinated nitrate ions each replaced in turn by the pyridine nitrogen donors.

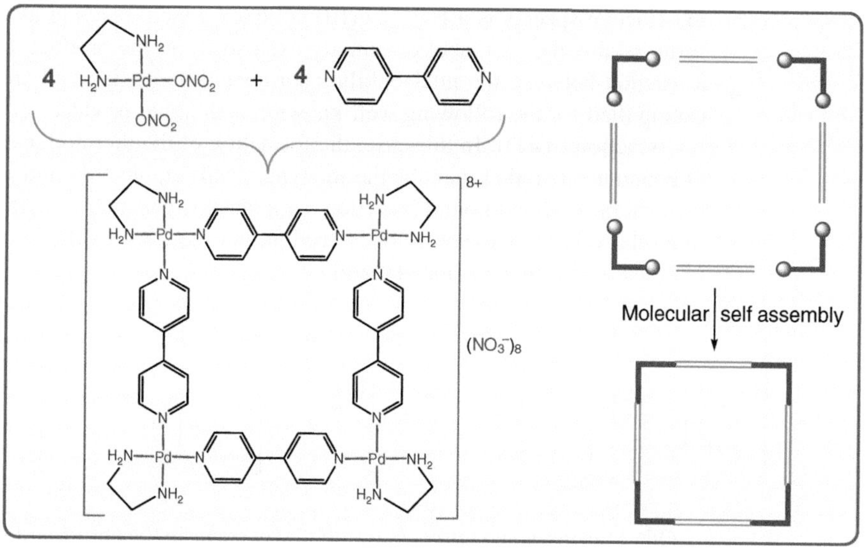

Figure 6.9

An example of a self-assembly reaction of a monomeric complex and a ligand, forming a macrocyclic tetranuclear complex.

6.5.2   Reactions of Coordinated Ligands

It is well known that there occur a range of reactions that involve chemistry of the ligands and in which metal–ligand bond cleavage is not involved. We can regard these as reactions of coordinated ligands. These early and deceptively simple studies provide fine examples of chemical detective work. One of the earliest studies probed the preparation of an H_2O–Co(III) species from a O_2CO–Co(III) precursor, a reaction which was seen to release CO_2 gas. Two quite different options for this reaction are: dissociation of carbonate ion from the complex followed by release of CO_2 from decomposition of the released anion, with a solvent water molecule entering the coordination sphere in its place; or cleavage of a C–O bond on the coordinated ligand to release CO_2 while leaving the residual O atom bound to cobalt, with reprotonation of the residual bound oxygen dianion to form coordinated water. In the former case, a completely different ligand is inserted, making it a traditional substitution reaction; in the latter case, part of the original ligand remains behind with the metal–donor atom bond staying intact, making it a different class of reaction which we now define as a reaction of a coordinated ligand. The key to distinguishing these reactions was to use isotopically-labelled $^{18}OH_2$ as solvent, which showed that the product contained just the normal dominantly $^{16}OH_2$ coordinated with <u>no</u> entry of $^{18}OH_2$ into the coordination sphere — therefore, the CO_2 must depart from the coordinated carbonate, as shown in reaction (6.40).

$$(6.40)$$

Not only can this type of reaction occur for this and a range of other coordinated oxyanions but also what is effectively the reverse reaction can occur, where the reactive species is a HO^-–Co(III) complex which, reacting as the ^{18}O-labelled form, retains the ^{18}O label essentially exclusively in the product, proving that it is a reaction between the nucleophilic coordinated hydroxide and an electrophile, as exemplified for the following well-known reaction that produces coordinated nitrite ion (reaction (6.41)). In this case, the formation of the thermodynamically unstable O-bound nitrite rather than the thermodynamically stable N-bound nitrite is supporting evidence. The O-bound form does spontaneously isomerise to the N-bound form in a relatively slow isomerisation reaction, but the rate is sufficiently slow that the O-bound form can be isolated readily.

$$(6.41)$$

Other reactions of simple anions that may occur in the absence of coordination can also be observed to occur for the complexed form, with the rate of reaction usually changed significantly as a result of complexation. This is anticipated, since the coordinated anion is bonded directly to a highly-charged metal ion, which must influence the electron distribution in the bound molecule and hence its reactivity. Two well-known examples where the product is ammonia occur through either reduction of nitrite with zinc/acid or oxidation of thiocyanate with peroxide. The former example is exemplified in reaction (6.42).

$$\begin{bmatrix} & NH_3 & \\ & | & O^- \\ H_3N\!-\!\overset{II}{Pt}\!-\!N & \\ & | & \diagdown \\ & NH_3 & O \end{bmatrix}^{1+} \xrightarrow{\;Zn\,/\,H^+_{aq}\;} \begin{bmatrix} & NH_3 & \\ & | & \\ H_3N\!-\!\overset{II}{Pt}\!-\!NH_3 \\ & | & \\ & NH_3 & \end{bmatrix}^{2+} + 2\,H_2O + Zn^{2+}_{aq} \qquad (6.42)$$

Transition metal complexes can promote reactions by organising and binding substrates. We have already seen this in terms of metal-directed reactions. Another important function is the supply of a *coordinated nucleophile* for the reaction, which is incorporated in the product. We have already seen a coordinated nucleophile at work in reaction (6.41) discussed above of Co–$^-$OH with NO$^+$; nucleophiles, which are electron-rich entities, are best represented in coordination chemistry by the important coordinated hydroxide ion, formed by proton loss from a water molecule; this is a common ligand in metal complexes.

Normally, water dissociates only to an extremely limited extent, via

$$H_2O \rightleftharpoons H^+ + {}^-OH$$

for which we define $K_w = [H^+][^-OH]/[H_2O] \approx 10^{-14}$ and consequently for which $pK_w \approx 14$ (approximately the pK_a in this case). However, when bound to a highly-charged metal ion, its acidity is significantly enhanced, to the extent that, at neutral pH, a coordination complex will have a significant part of its M–OH$_2$ present as M–$^-$OH, via

$$M^{n+}\text{–}OH_2 \rightleftharpoons H^+ + M^{n+}\text{–}^-OH \qquad (pK_a \approx 5\text{–}9)$$

Although the coordinated hydroxide is an inferior nucleophile compared with free hydroxide, due to electronic effects of the bonded metal cation, its substantially higher concentration in the bound form at any pH more than compensates. A coordinated water molecule with a pK_a of 7 will be 50% in the hydroxide form at neutral pH, for example. Importantly, because it can often be placed adjacent to a bound substrate (thus *preorganised* for reaction with it), it is highly effective, and marked catalysis is commonly observed.

Although the most important, hydroxide is not the sole example of a coordinated nucleophile met in coordination chemistry. The next most important, as a result of the prevalence of ammonia as a ligand, is the amide ion. Ammonia is usually thought of simply as a base, but it has the capacity to lose protons and thus function as an acid,

$$NH_3 \rightleftharpoons H^+ + {}^-NH_2 \qquad (pK_a \approx 35).$$

This reaction has such a high pK_a that it is not of significance for free ammonia in water as water will be deprotonated first. However, as for coordinated water, acidity of ammonia is significantly enhanced through coordination

$$M^{n+} - NH_3 \rightleftharpoons H^+ + M^{n+} - {}^-NH_2 \qquad (pK_a \approx 10\text{–}14)$$

so that sufficient concentrations of the bound amido ligand can form to permit reaction. Overall, the coordinated amido ligand is a more abundant nucleophile than the free amide ion. Alkylamines can show the same activity as nucleophiles, via

$$M^{n+}\!\!-\!\!NHR_2 \rightleftharpoons H^+ + M^{n+}\!\!-\!\!^-NR_2$$

as long as they have at least one amine hydrogen atom (as above) to release as a proton.

Reactions of coordinated ligands with organic substrates occur usually where the organic molecule enters the coordination sphere in a position adjacent to the nucleophile, with the following reaction involving attack of the coordinated nucleophile at a relatively electron-deficient site on the organic substrate. These reactions lead to a new organic molecule that is usually chelated to the metal ion. This product may depart from labile complex centres through substitution by other ligands (providing a mechanism for repeating the reaction or catalysis) or else may occur with inert metal centres as a single stoichiometric reaction. These reactions can also induce a particular stereochemistry and may be defined as *stereospecific* (producing exclusively one stereoisomer) or *stereoselective* (producing an excess of one isomer). *Selectivity* can be introduced simply by preference for a particular conformation in a chelate ring equilibrium, as illustrated in Figure 6.10. Here, the λ conformation (left hand side) is preferred and not the δ conformation (right hand side) of the chelate ring, as steric clashing (of the ring methyl substituent with other axial ligands on the complex) is minimised in the former. Any subsequent reaction will 'carry forward' this selectivity into the reaction outcome, leading to selectivity in the product.

Either a coordinated ^-OH or $^-NH_2$ group is able to initiate chemistry with appropriate ligands present in an adjacent (*cis*) site. This reactivity was probed in detail for several decades commencing in the 1960s by the groups of New Zealander David A. Buckingham and Australian Alan M. Sargeson. The coordinated hydroxide ion in particular has been the subject of extensive studies. Buckingham studied peptide cleavage by an inert cobalt(III) complex, employing isotopic labelling to probe the origin and fate of oxygen atoms in the product. The alternate processes proposed are illustrated in Figure 6.11. This work clearly showed that, while pathways for both intramolecular (Path A) and intermolecular (Path B) hydroxide attack exist, the coordinated hydroxide is the significant player, consistent with its enhanced concentration and preorganised location.

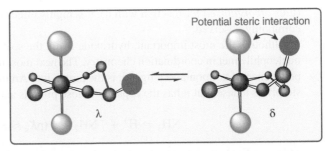

Figure 6.10

Two conformations of a propane-1,2-diamine chelate ring. The left-hand conformer has the methyl substituent on the chelate ring displaced away from the rest of the molecule (equatorial), whereas the right-hand conformer places it in an axial disposition where it may clash with another axial ligand, which is thus unfavourable.

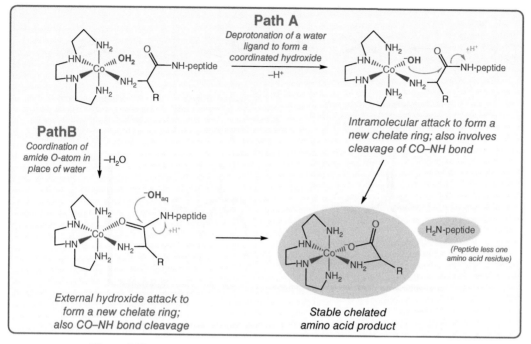

Figure 6.11

Proposed competing mechanisms for reaction of coordinated and free hydroxide with a coordinated peptide to form a new chelated amino acid ligand; the coordinated hydroxide route (path A) is more efficient.

For a coordinated $^-NH_2$ group in aqueous solution, there is no capability in the presence of free ammonia for competition that would lead to the same product due to the exceedingly high pK_a of free ammonia, and hence the intramolecular nature of the reaction is quite clearly defined. The reactivity of this group is illustrated in the simple reaction with $[Co(NH_3)_5(OPO_3R)]^+$ in aqueous base (Figure 6.12). Here, the coordinated nucleophile can attack at the adjacent and relatively electron-deficient P atom, leading to a new N–P bond forming. The phosphorane intermediate formed has in effect one too many bonds around the P, relieved by P–O bond breaking to release an alkoxide ion and form a stable chelated aminophosphate.

A more complicated example of a reaction featuring a coordinated $^-NR_2$ nucleophile is the reaction of N-bound 2-aminoacetaldehyde with a coordinated polyamine

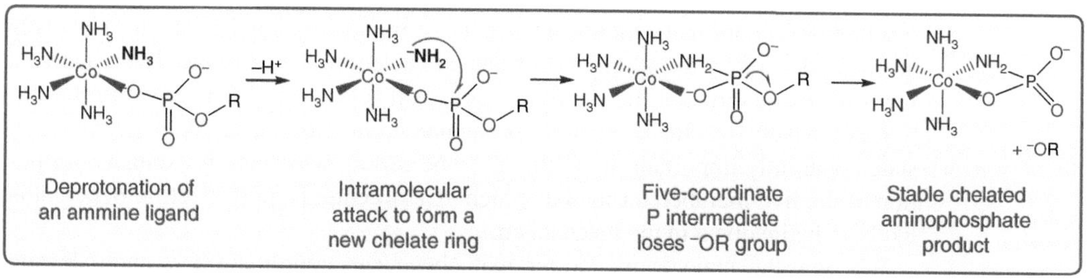

Figure 6.12

Proposed mechanism for reaction of a deprotonated ammine ligand with an adjacent coordinated phosphate monoester, leading to formation of a new chelated aminophosphate ligand.

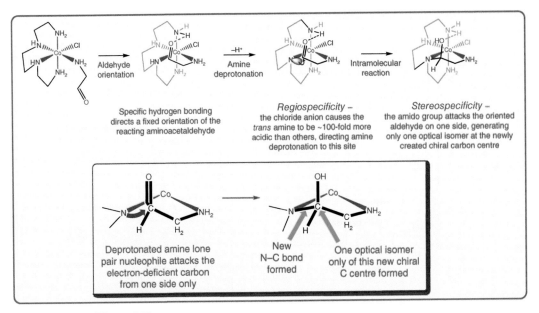

Figure 6.13

Proposed mechanism for reaction of a particular amine group (an example of *regiospecificity*) with a pendant aldehyde of coordinated 2-aminoacetaldehyde, leading to only one optical isomer of an aminol (an example of *stereospecificity*). The core reaction appears in the inside box.

(Figure 6.13). In this molecule, there are three amine groups sufficiently close (in positions adjacent or *cis* to the aldehyde) to participate in reaction, and yet the reaction occurs at only one of these three—it is therefore *regiospecific*. We discussed regiospecificity with respect to substitution reactions earlier, and it is the influence of a *trans* ligand that was used as the example. In this case, the regiospecificity arises in the same manner, since the *trans* chlorido ligand makes the secondary amine group opposite significantly more acidic than those in the other two sites so that it is deprotonated much more readily to produce a reactive $^-NR_2$ group. Furthermore, this reaction is *stereospecific* since a particular chirality is introduced at the newly created tetrahedral carbon centre. This is ascribed to specific hydrogen bonding interactions in the transition state holding the carbonyl oxygen in a particular orientation that carries through into the product once the amide ion attacks the carbon of the carbonyl to form a new N–C bond. Intramolecular hydrogen-bonding may often play an important role in directing stereoselectivity.

Hydration of olefins is another reaction involving a coordinated hydroxide. This reaction with coordinated monodentate maleate monoester (Figure 6.14, top) involves alkene hydration by the adjacent coordinated nucleophile, hydroxide. The reaction is *regiospecific* in terms of the *site* of reaction, as it selects five-membered ring formation exclusively over the alternative six-membered ring formation. With the transition state apparently important, it is difficult to be certain why there is exclusive formation of the five-membered ring, but which –CH= centre is more electron-deficient is likely to be involved in the mechanism.

Whereas the maleate monoester met above can coordinate as a monodentate ligand through the one free carboxylic acid group, the diacid can employ both acid groups to bind as a didentate chelate, forming a seven-membered chelate ring (Figure 6.14, bottom left). This chelated maleate dianion bound to Co(III) can also

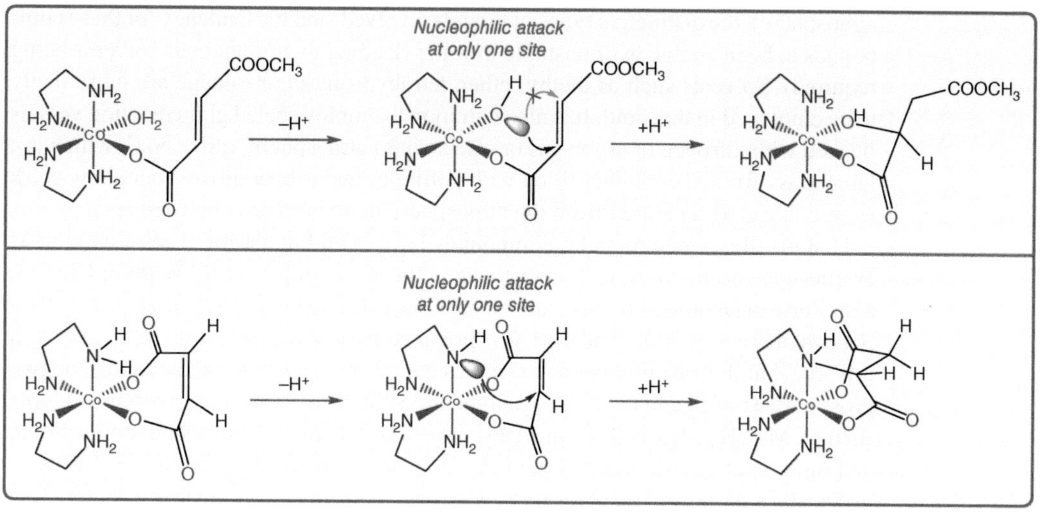

Figure 6.14

Hydration of a pendant alkene by coordinated hydroxide (top), displaying specificity in site of attack, with the five-membered chelate ring only formed, and regiospecific attack of a deprotonated amine nucleophile (bottom) at the alkene of a chelate maleate ion.

undergo intramolecular attack, but in this case, the nucleophile is a deprotonated amine group of an adjacent chelated ethane-1,2-diamine (Figure 6.14, bottom centre), as no coordinated hydroxide is present. The reaction has a choice of two sites for attack (at either end of the C=C bond), but again there is stereospecificity observed. This specificity in ring formation may arise if in the transition state we require one carboxylate to be coplanar and thus conjugated with the diene. This leads to two discrete conformations, depending on which of the two carboxylate groups is coplanar. From examining models, it appears substantially more favourable for nucleophilic addition to occur at the 'front' –CH= group, as a result of the spatial orientation of the lone pair and the preferred conformation adopted by the chelate ring leading to a closer and appropriately directed approach at that site, depicted in Figure 6.14 (bottom centre and right).

It is notable that reaction does not occur between free maleate ion and free ethane-1,2-diamine, although it does occur (but only very slowly) with maleate diester and ethane-1,2-diamine. Acceleration of the reaction rate resulting from coordination to a metal ion is significant and lies in the range 10^6–10^{10}. Accelerations of this size are common for reactions of coordinated nucleophiles. The observation of stereospecificity and large accelerations suggests that these types of reactions may be relevant to the modes of action of certain metalloenzymes, where reactions are very rapid, specific and stereoselective. Such reactions are met in Chapter 8.

6.6   Organometallic Synthesis

Compounds with metal–carbon bonds usually require specialised approaches to their synthesis that differ from those discussed above for traditional Werner-type coordination complexes. This relates to the typically low oxidation state of the target compounds that makes many of them air-sensitive (requiring the use of an inert atmosphere), the distinctive types of ligands involved, and the tendency for these compounds to be insoluble in or unstable in water (leading to non-aqueous solvents being required). Solvents such as diethyl ether, tetrahydrofuran or toluene are more likely to be employed in this field. It is also common to employ sealed glass reaction vessels flushed with nitrogen or argon gas or else an inert atmosphere 'glove box', which is a large glass-fronted container fitted with portholes and rubber gloves that allow work to be conducted separated from the atmosphere in an inert gas environment.

Historically, organometallic compounds have been known for at least as long as Werner-type complexes, with coordinated ethene first reported by William Christopher Zeise in Denmark in 1827, and metal carbonyls like $PtCl_2(CO)_2$ prepared by Paul Schutzenberger in 1868, and $Ni(CO)_4$ prepared by Ludwig Mond in 1890, although it is true that the vast number of examples date from the 1950s and beyond. The two gross classes of ligands met in organometallic chemistry are: simple σ-bonded type, such as $M–CH_3$, that behave in many ways like a conventional metal–donor bond; and multi-centred π-bonded systems such as occurs with an alkene that binds symmetrically side-on and involves its π-electrons in the linkage to the metal centre. This has been discussed earlier in Chapter 2.4.

As an aid to understanding the outcomes of organometallic reactions and in synthesis, there is a convenient way in which the stability of a compound can be predicted, called the 18-electron rule. In the light elements of the *p*-block, we traditionally invoke an octet (or 8-electron) rule to probe stability. This assumes that an *s* and three *p* valence orbitals are used in bonding and allows us to understand why CH_4 is stable, whereas CH_5 is not. In *d*-block chemistry, it is possible to use what is in that case an 18-electron rule (using an *s*, three *p* and five *d* orbitals) to help predict stability of a complex. Nevil Sidgwick in 1927 noted that in metal complexes, the metal tends to bind sufficient ligands so that its effective atomic number replicates that of the noble gas in the same periodic table row. This rule holds for and thus is used mostly for organometallic complexes and has value since it limits the number of combinations of metal and ligand that can lead to the desired electron count. This concept was explained in Chapter 2.4.4 and will not be developed further here but is invariably met in more specialised and advanced courses. It works best for low-valent metals with small neutral high-field ligands like carbonyls but is less effective for high-valent metal ions involving weak-field ligands, and thus is not usually invoked for traditional Werner-type coordination complexes.

Metal carbonyls represent a key class of organometallic compounds and are often a starting point for other chemistry. They tend to be monomers, dimers or small oligomers, such as $Ni(CO)_4$ and $Mn_2(CO)_{10}$, the latter involving a formal metal–metal bond. Most metal carbonyls are made by reduction of simple salts or oxides in the presence of CO and a reducing agent (reductive carbonylation) or by the direct reaction of CO with finely-divided metals at elevated pressure. Examples appear in reactions (6.43) and (6.44).

$$Ni + 4\,CO \xrightarrow[\text{1 atmosphere}]{25°C} [Ni(CO)_4]$$

$$Re_2O_7 + 17\,CO \xrightarrow[\text{100 atmospheres}]{200°C} [Re_2(CO)_{10}] \quad + 7\,CO_2$$

(6.43–6.44)

The carbonyl ligands are able to be substituted by some other ligands or else the coordination sphere can be expanded by addition reactions with other compounds. It is also straightforward to prepare compounds that incorporate carbonyl and other ligands such as amine ligands. This can sometimes be achieved directly, such as in reaction (6.45). The product with en (ethane-1,2-diamine) chelated is able to be isolated readily. The transition metal carbonyls are usually far more robust than those formed by main group elements; for example, $H_3B(CO)$ decomposes below room temperature, whereas $Cr(CO)_6$ can be sublimed without decomposition.

$$Cu^ICl + en + CO \longrightarrow [CuCl(CO)(en)] \qquad (6.45)$$

An early example of organometallic synthesis was the reaction of $[PtCl_4]^{2-}$ with ethylene in dilute aqueous hydrochloric acid solution, which yields the π-bonded orange complex $[Pt(C_2H_4)Cl_3]^-$, where the Pt and two carbon atoms form an equilateral triangle (or in other words, the ethene is bound symmetrically side-on to the platinum(II) centre). A vast range of alkene complexes have been reported subsequently, including the tris(η^2-ethylene)nickel(0) which features three side-on ethylene molecules arranged in a trigonal pattern around the metal atom. Bonding in these complexes requires molecular orbital theory for a satisfactory explanation. This assigns σ-bonded character through a filled π molecular orbital of the alkene overlapping symmetrically with an empty metal orbital, and π-bonded character through overlap of an empty π^* molecular orbital on the alkene with a filled metal d-orbital.

Another important class of compounds are the so-called sandwich compounds introduced in Chapter 3.6, featuring a metal bound (or 'sandwiched') between two flat aromatic anions, the best known of which is the cyclopentadienyl ion ($C_5H_5^-$). These compounds can be prepared as in reaction (6.46) and analogues, performed in a non-aqueous solvent such as diethyl ether.

$$FeCl_2 + 2\,Na(C_5H_5) \longrightarrow Fe(C_5H_5)_2 \qquad + 2\,NaCl \qquad (6.46)$$

This compound is robust – it is air stable, able to be sublimed without decomposition, and resistant to strong acids and bases. It can undergo reversible one-electron oxidation chemically or electrochemically, with a significant change in colour from yellow to blue. An array of other related compounds are known, including those featuring larger aromatic ring systems, such as $[Cr(C_6H_6)_2]$. Examples with metals from most parts of the periodic table, even f-block elements, have been prepared bound to arene ligands such as benzene and the cyclopentadienyl anion.

Metal-alkyl σ-bonded compounds can be formed conveniently by reactions employing alkyl halides or alkyl magnesium bromides, such as reaction (6.47), performed in diethyl ether. These reactions are facilitated by the presence of spectator ligands such as phosphines and carbonyls. Phosphines, common co-ligands in organometallic compounds, are usually conveniently introduced through direct reaction with metal halides.

$$[PtBr_2(PR_3)_2] + 2\ CH_3MgBr \longrightarrow [Pt(CH_3)_2(PR_3)_2] \quad + 2\ MgBr_2 \qquad (6.47)$$

Organometallic compounds also undergo reactions of coordinated ligands readily. A simple example involves the susceptibility of coordinated carbon monoxide towards nucleophilic attack. The $[Mo(CO)_6]$ complex reacts with methyllithium [reaction (6.48)], with the new ligand produced at one site also able to undergo additional reactions, not described here.

$$[Mo(CO)_6] + LiCH_3 \longrightarrow Li[Mo(CH_3CO)(CO)_5] \qquad (6.48)$$

A reaction with a similar outcome involves migration of one ligand to attack another adjacent ligand (a 1,1-migratory insertion), promoted by the availability of another ligand to occupy the site vacated by the first movement. In the example in reaction (6.49), an initial R–M–C=O component of the molecule is converted to an –M–C(R)=O component (equivalent to that formed in reaction (6.48) by a different process), leaving a vacant coordination site, filled by an added phosphine ligand in this case.

$$[Mn(CH_3)(CO)_5] + PR_3 \longrightarrow [Mn(CH_3CO)(CO)_4(PR_3)] \qquad (6.49)$$

The hydride ion (H^-) is an efficient small ligand in organometallic chemistry. The first transition metal hydrides were prepared using the Hieber base reaction, exemplified in reaction (6.50). The hydroxide adds to the carbon of one CO ligand to produce an intermediate that rapidly loses carbon dioxide, leaving the hydride ion to occupy the coordination site.

$$[Fe(CO)_5] + {}^-OH \longrightarrow [Fe(CO)_4(COOH)]^{\#} \longrightarrow [Fe(CO)_4(H)] + CO_2$$
$$\text{\textit{Intermediate}} \qquad (6.50)$$

These are but a few examples of an array of reactions available to organometallic systems. Clearly, the level of difficulty is greater when performing many organometallic reactions compared with Werner-type coordination chemistry reactions as special equipment must be employed because air and or protic solvents can lead to unwanted reactions, although the reactions themselves do not necessarily impose any greater inherent complexity. The storage and handling of products may pose problems, however, due to their reactivity, particularly in redox reactions. A detailed examination of their chemistry may be pursued through advanced and or specialist texts.

6.7 Nanomolecular Synthesis

Nanomolecules are appreciably larger than simple coordination complexes or even those involving several metals. A nanoparticle is usually defined as an entity of from 1 to 100 nm in size. They may display different chemical and physical properties to both small molecules and larger materials, which is a key basis for their synthesis and study. They may be inorganic (such as silicon and titanium oxides), organic (such as carbon nanotubes) or biomolecular (such as liposomes and proteins). Although many examples incorporate metals in their structure, they are too large to be considered classic coordination compounds. Nanoparticles are important as a result of their unique properties, which reflect the combined outcome of their chemical composition, size, shape and any surface coating. Although inorganic nanomolecules are often structurally and chemically sophisticated, and thus tend to lie beyond the scope of this book, their existing and expanding importance suggests that it is appropriate to present selected examples, without attempting to cover the field broadly and at depth.

6.7.1 Traditional Coordination Complexes and Nanomolecules

Coordination compounds that we have dealt with to date are obviously far from nanomolecular in size. However, they may relate to this evolving field in three ways: first, as components of a much larger molecular entity involving covalent linkages (such as metallocycles); second, as low molecular weight complexes encapsulated and non-covalently bound in the cavities of large molecular entities (such as cyclodextrins); or third, as compounds that direct the formation of nanomolecules in some manner. Furthermore, so-called surface coordination chemistry deals with how small molecules with potential as ligands are covalently 'coated' on the surface of solids and typically metallic nanoparticles, and the way these influence the properties of the nanoparticles. Although simple physical properties such as shape and size are significant in determining the properties of a nanoparticle, surface ligand choice is important for certain properties such as formation of stable colloids (fine suspensions) through control of surface energy and for the chemical function of the nanoparticles.

The types of small molecules covalently linked to the surface of nanoparticles are diverse, and known examples range from organic acids and small polymers such as glycols, to biological species such as peptides, proteins and oligonucleotides. Small coordination complexes that introduce catalytic properties can be bound in a number of ways. One fairly simple example appears in reaction (6.51), where an iron oxide nanoparticle is linked by surface Fe–O–Si covalent bonds with the silicon-containing head of a pendant group on a palladium(II) complex. Several such linkages may be present on the surface of a nanoparticle.

$$\text{(6.51)}$$

In almost the reverse of the above approach, it is possible to commence with simple metal complexes and, under controlled conditions, initiate chemistry that yields new metal nanoparticles with ligands from the dissociated precursor complex bound to the surface. Clusters of metals with small organic molecules covalently bound to the surface represent one well-studied class. Reaction of a simple gold(I) thiolate with a powerful reducing agent in an inert solvent is one route to nanoscale gold clusters with thiolates on the surface, involving internal Au−Au and surface Au−S covalent bonds (Figure 6.15). These reactions yield atomically precise nanomolecules with defined numbers of gold atoms and thiolates; the thiolates act effectively as protecting groups on the metal surface. In a period of little more than a decade from the turn of the Millenium, examples from $Au_{10}(SR)_{10}$ to $Au_{187}(SR)_{68}$ were characterised, and the field continues to grow. The nanomolecules exhibit size-dependent electrochemical and spectroscopic properties. Not only may species with only one type of metal present be prepared but also mixed-metal compounds may be prepared; the alloy series $[Au_{38-n}Ag_n(SPh)_{24}]$ is one example reported. This series is based on the well-studied gold-based nanomolecular parent $[Au_{38}(SAr)_{24}]$ (Figure 6.15). Mass spectrometry techniques (see Chapter 7.3.9) are often employed in characterisation. Moreover, success in employing X-ray crystallography, first applied in 2007–2008 to crystals of $[Au_{102}(SC_6H_4COOH)_{44}]$ and $[Au_{25}(SCH_2CH_2Ph)_{18}]^-$, has revolutionised the field as this technique provides minute details of the structures of gold nanomolecules including surface protection modes provided by the thiolates. Notably, not every possible $Au_m(SR)_n$ compound can be readily synthesised, with stability of the assembly variable; as yet, exactly what properties govern stability remains unclear. Similar nanomolecules arise from reaction of mixtures of gold(III) salts and thiols.

Whereas the above examples are relatively simple, sophisticated chemistry involving coordination complexes may be involved in some nanomolecule syntheses, but it is well beyond the remit of this book to explore this aspect here. We simply note

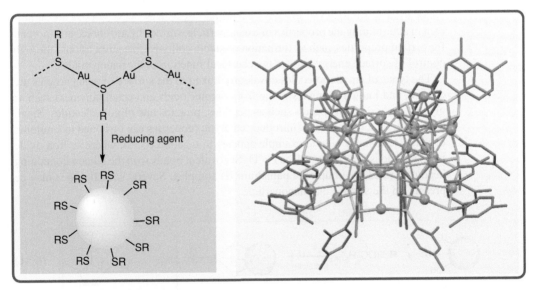

Figure 6.15

A synthetic route to surface-coated gold clusters from $\{Au(SR)\}_n$. To illustrate this class of metal-containing nanomolecules, the X-ray structure of $[Au_{38}(SAr)_{24}]$ is shown. This compound has a 23-atom Au(0) core layered with three monomeric (ArS–Au–SAr) and six dimeric (ArS–Au–SAr–Au–SAr) units containing Au(I), with the external thiolates providing protection.

that the growing diversity of types of and routes to nanoparticles reflects the extensive research in this continually evolving field. A key driving force is the use of the products in potentially commercial applications, such as catalysis. Without exploring this field deeply, it is important to note that the basic rules of coordination chemistry that we have explored for traditional low molecular weight complexes apply equally for much larger molecules that incorporate metal centres.

6.7.2   Polyoxometalates and Other Clusters

Cluster complexes are molecular assemblies that involve several metal ions covalently linked through bridging groups in an ordered fashion. The metal centres display individual coordination shells that are structurally like those already met in simple complexes, but it is the extensive array of bridging ions or molecules present that distinguish this class. They can vary in size from those containing as few as three metal centres to compounds containing a large number of centres that are of nanomolecular size. Perhaps the best known and most studied are the *polyoxometalates*: polynuclear metal-oxido clusters, which are mostly anionic, and contain three or more metal centres. They form in usually aqueous acidic solution through self-assembly from simple monomeric precursors. Typically, they comprise an assembly of approximately octahedral MO_6 or tetrahedral MO_4 units that involve some corner- and or edge-sharing of oxido ligands (Figure 6.16) with elimination of H_2O (condensation) at each new linkage formed

Polyoxometalates tend to involve early transition elements such as vanadium, molybdenum and tungsten in high oxidation states, and, although dominated by the presence of oxido ligands, may undergo reactions that also incorporate a limited number of a very wide range of other metals (such as Na^+, Co^{2+}) or *p*-block elements, (Si and P). Reactions leading to incorporation of organometallic fragments have also been reported. The high charge and relatively small ionic radii of metal ions such as Mo(VI) and W(VI) supports the essential multiple bridging via O^{2-} ions and also promotes, where appropriate, M–O π-bonding. Polyoxometalates are

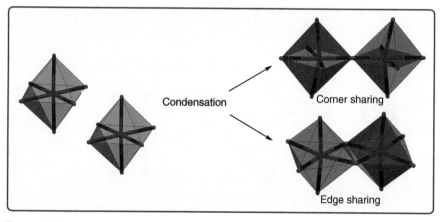

Condensation   Corner sharing

Edge sharing

Figure 6.16

Corner and edge sharing of oxido ligands in polyoxometalates. Large clusters with both types of sharing present are common. Note that the metal centres above retain six-coordination in this process, with the corner- and edge-sharing ions being essentially bridging ions between metal centres.

considered some of the most promising building blocks for developing functional nanomaterials.

Their often simple formulation (as, e.g. $[M_pO_q]^{n-}$) belies the structural diversity now established. Extensive studies have led to classification of polyoxometalates into a number of structurally related families, such as isopolyanions (composed of metals and oxido ligands only, such as $[Mo_6O_{19}]^{2-}$ (Figure 6.17)), and heteropolyanions ($[XM_pO_q]^{n-}$) that include additional *p*-block elements. The heteropolyanions include 1:12 X:M species (Keggin type, based on the $\{XM_{12}O_{40}\}$ parent), 1:9 X:M species (Dawson type, such as $\{X_2M_{18}O_{62}\}$), and 1:6 X:M species (Anderson-Evans type, such as $\{XM_6O_{24}\}$); in each case, M is a transition metal and X is a hetero atom. The Keggin and Dawson structures are considered the basic building units, since other more complex structures can be viewed as assemblies of several of their fragments. The Keggin structure was structurally characterised by X-ray crystallography by J.F. Keggin in 1933, the Anderson–Evans structure by H.T. Evans in 1948, and the Dawson structure by B. Dawson in 1953. Within another 40 years, about 200 additional structures had been characterised by X-ray crystallography, and this number has ballooned substantially since. Although polyoxometalates are synthetic, in 2017, the first example of a naturally occurring mineral with an albeit complicated polyoxometal (Mg/Cu) cation was recognised: greenish-blue *ramazzoite*.

The synthesis of many polyoxometalates occurs from reaction of simple salts such as Na_2MoO_4, which polymerise under acidic conditions to yield a range of species. These salts are accessible from the d^0 metal oxides, such as V_2O_5 and MoO_3, which dissolve at high pH to give their orthometalates, such as VO_4^{3-} and MoO_4^{2-}. Lowering the pH leads to protonation to, e.g. $MoO_3(OH)^-$, and it is these species that polymerise via a process called *olation*, involving the loss of water and formation of M–O–M bonds. For two monomers alone reacting, we can describe the reaction simply via:

$$2\,MoO_3(OH)^- \quad \rightarrow \quad [O_3Mo\text{—}O\text{—}MoO_3]^{2-} + H_2O \qquad (6.52)$$

although the dimer above actually exists as $[Mo_2O_5(OH)(H_2O)_5]^+$ in strongly acid solution, with each Mo in an octahedral environment. This olation process then continues to yield higher clusters, such as hexamolybdate:

$$6\,MoO_3(OH)^- + 5H^+ \quad \rightarrow \quad [Mo_6O_{19}]^{2-} + 5\,H_2O \qquad (6.53)$$

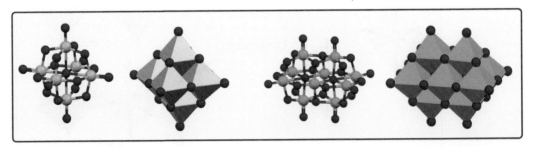

Figure 6.17
Example polyoxometalate isopolyanions $[Mo_6O_{19}]^{2-}$ (left) and $[V_{10}O_{28}]^{6-}$ (right). Two representations are given for each, being the standard ball-and-stick view as well as the filled octahedra view, the latter emphasising the retention of six-coordinate octahedral geometry for each metal centre as well as the linking in the clusters.

Molybdate and analogues undergoes olation readily, driven by a favourable enthalpy effect for the condensation process leading to polyoxometalate formation as a result of the tetrahedrally coordinated Mo(VI) in the simple MoO_4^{2-} ion expanding its coordination spheres from four to six during cluster formation.

In aqueous acidic solutions of molybdate, even from concentrations as low as 10^{-4} M, a number of polymeric species form, the outcome depending on the pH and ratio of acid to monomer reacting. In highly alkaline solution, the parent molybdate monomer anion MoO_4^{2-} is dominant, whereas in strongly acidic solution, the dominant species are the monomer $[MoO_2(OH)(H_2O)_3]^+$ and the dimer $[Mo_2O_5(OH)(H_2O)_5]^+$. At pH 5.5, the molybdate system is dominantly composed of $[Mo_7O_{24}]^{6-}$. As many as 16 other species are known, particularly $[Mo_8O_{26}]^{4-}$, although even the large cluster $[Mo_{36}O_{112}(OH_2)_{16}]^{8-}$ has been isolated in the solid state and also identified in solution. Simply ageing a solution of $[Mo_7O_{24}]^{6-}$ for differing periods, days or weeks, leads to transformation into other clusters. A range of physical methods have been employed since the 1960s to detect species in solution, including potentiometry, nuclear magnetic resonance (NMR) and Raman spectroscopies, and electrospray ionisation mass spectrometry (ESI–MS); these techniques are discussed in Chapter 7. Apart from species in solution, many isopolyoxometalates have been defined in the solid state by X-ray crystallography, as control of reagent concentrations and pH promotes the formation of target species in sufficient concentrations to permit their crystallisation.

For tungsten(VI) reacting as WO_4^{2-} in aqueous acid, 13 polyoxo species have been defined. Overall, polyoxotungstates are more stable than their molybdenum analogues, but equilibria in solution tend to establish more slowly. Aged, equilibrated tungstate solutions close to neutrality dominantly form paratungstate, $[W_{12}O_{40}(OH)_2]^{10-}$. With vanadium(V), both tetrahedrally coordinated vanadium compounds such as $[V_5O_{14}]^{3-}$ and octahedrally coordinated compounds such as $[V_{10}O_{28}]^{6-}$ (see Figure 6.17) have been defined. Studies of polyoxoniobates and polyoxotantalates are limited compared to polyoxovanadates, and they are stable only in basic solution, with the $[M_6O_{19}]^{8-}$ compound dominant, although other species have been defined in recent times.

Heteropolymetalates are the large, well-known family of compounds that include additional elements other than simply M and O. The range of elements that can be incorporated in the cluster framework species $\{[X_pM_qO_r]^{n-}\}$ is extensive, with almost all elements now reported. For example, other transition metal ions can be easily inserted, as in the case of nickel(II) with molybdate:

$$Ni^{2+} + 6\,[MoO_4]^{2-} + 6\,H^+ \rightarrow [Ni(OH)_6Mo_6O_{18}]^{4-} \qquad (6.54)$$

For molybdenum(VI), however, molybdophosphate $[PMo_{12}O_{40}]^{3-}$ and the analogous molybdosilicate, $[SiMo_{12}O_{40}]^{4-}$ have attracted the most attention, but known compounds are nevertheless extensive. For example, 14 molybdophosphate species have been reported and form in acidic solutions of molybdate ion with simple phosphate ion present; molybdoarsenate species are also readily formed. Only for tungsten has a comparable range been reported. The simple reaction in aqueous acid of ammonium molybdate and sodium phosphate forming canary yellow molybdophosphate was reported by Jöns Jacob Berzelius in 1826:

$$12\,(NH_4)_2MoO_4 + Na_3PO_4 + 3\,HNO_3 \rightarrow (NH_4)_3[\mathbf{PMo_{12}O_{40}}]$$

$$+ 12\,H_2O + 3\,NaNO_3 + 21\,NH_3 \qquad (6.55)$$

This polyoxometalate, about 1 nm in size, is one of the Keggin structure family. The yellow compound may be reduced to produce an intense blue complex mixture of species, observed in 1778 by the prolific chemist Carl Scheele (who, incidentally, was also credited with the isolation and identification of the elements oxygen and chlorine). Scheele did not, at that time, know the composition of this blue mixture. The blue colour of this complex provides an effective visual method of detecting and determining the concentration of phosphate. This venerable field of chemistry was revitalised with the discovery and structural characterisation of a spectacular 4 nm diameter wheel-shaped polyoxomolybdate, $[Mo_{154}(NO)_{14}O_{448}H_{14}(H_2O)_{70}]^{28-}$, by Achim Müller and his team in 1995, with even larger examples reported since that study.

Given the now vast number of polyoxometalates reported (even by the first decade of the 20th century, about 750 were known), it is not appropriate here to present an exhaustive review and or exemplify their synthesis exhaustively. In all cases, only with the careful control of conditions can a particular target species be prepared and isolated. Thus, although the reactions leading to polyoxometalates are typically initiated by just the mixing of simple salts in solution, careful control of pH, reagent concentrations and reaction times is essential for success.

6.8 Silicate and Aluminosilicate Synthesis

6.8.1 Silicates

The majority of minerals in the Earth's crust are silicate-based. Apart from natural materials, synthetic examples of silicate clusters and analogues involving replacement of some Si(IV) centres by other cations (notably Al(III), forming aluminosilicates or zeolites) exist, with products of nanomolecular scale common, although their typically polymeric nature suggest they are best treated separately.

We meet the simplest silicate as *sodium metasilicate* (Na_2SiO_3), the main component of commercial sodium silicate solutions, in which the metasilicate anion is a linear polymer $(SiO_3)_n^{2n-}$; it is stable in basic and neutral solution. Sodium metasilicate is made as a solution by treating a mixture of silica (SiO_2, usually as quartz sand), caustic soda, and water with hot steam; think of it as a solution of silica in sodium hydroxide. A related compound is sodium orthosilicate (Na_4SiO_4), made by fusing a mixture of sodium hydroxide and quartz silica at about 500°C in an electric furnace.

Polymerisation of monomer orthosilicate can be considered to occur by the *olation* reaction that was described for polyoxometalates, effectively involving condensation, or removal of H_2O, which we can represent simplistically (ignoring protonation) for steps leading to the two-centre species disilicate (see reaction (6.56) and Figure 6.18) and subsequently the trisilicate (see reaction (6.57)).

$$2\,[SiO_4]^{4-} + 2H^+ \rightarrow [Si_2O_7]^{6-} + H_2O \tag{6.56}$$

$$[Si_2O_7]^{6-} + [SiO_4]^{4-} + 4H^+ \rightarrow [Si_3O_9]^{6-} + 2H_2O \tag{6.57}$$

Formation of larger polymers involve exactly the same processes to those outlined above.

The mechanism thought to apply for dimerisation involves attack of an oxygen anion at another silicon centre, producing a five-coordinate intermediate that rapidly loses water (Figure 6.18). The actual species present in the precursor solution is

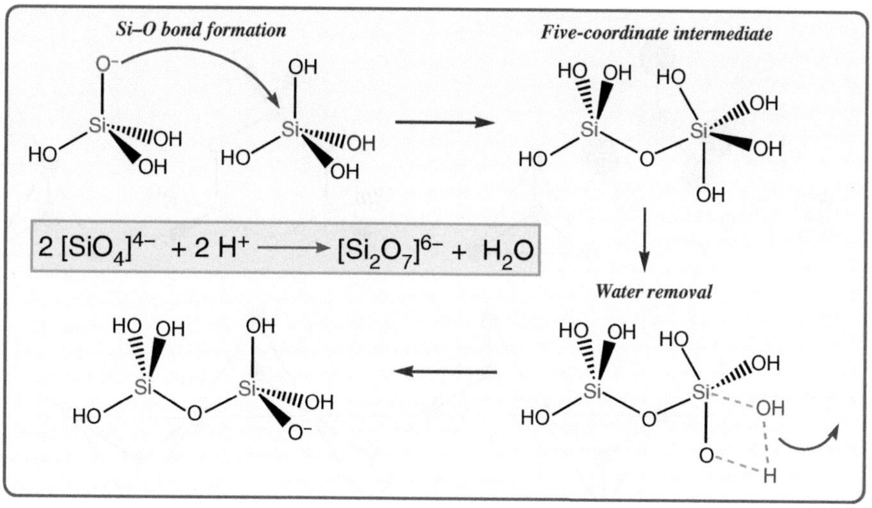

Figure 6.18

Representations of the process leading to dimerisation in silicate formation. This is the first step in a polymerisation process that can continue to produce, next, a trimer, then continue to yield higher polymers.

both concentration and pH dependent, but in the monomeric domain key species are $Si(OH)_4$, $SiO(OH)_3^-$ and $SiO_2(OH)_2^{2-}$, with the first dominant up to about pH 9. In Figure 6.18, it is clear that the two silicon centres retain their tetrahedral form after reaction and are linked by a corner-shared bridging oxygen ion. Continuation of the polymerisation process can lead to a row of linked silicates (chain silicates), linked chains of silicates (leading to ribbon silicates) or rings of silicates (cyclic silicates). The simplest example of a cyclic silicate is shown in Figure 6.19 as well as part of a chain silicate.

6.8.2 Aluminosilicates

Aluminosilicates are closely related to the silicates discussed above, but with some Si(IV) ions replaced by Al(III) ions (with each replacement also requiring an addition cation to maintain the charge balance); both natural and synthetic examples abound. *Erionite*, $Na_2K_2CaMg[(AlO_2)_2(SiO_2)_2]\cdot 6H_2O$, is an example of a natural aluminosilicate. The commercially relevant form are the microporous *zeolites*, which are hydrated alkaline and alkaline-earth metal salts of aluminosilicates, which occur naturally but can also be readily prepared. Aluminosilicates find extensive use as catalysts and adsorbents. They are synthesised by several methods, but the principal one involves reaction of $NaAlO_2$, $NaSiO_3$ (or else alumina and silica) and NaOH in water, which on ageing overs hours to days produces a reactive gel that is collected by filtration and then heated at between 80 and 300°C; this hydrothermal process yields a zeolite that is then washed and dried. A surprising range of what would otherwise be waste products are used commercially as silica sources; these include waste glass, rice husk ash and paper sludge. Purity, particle size and structural form control achieved with synthetic zeolites means they are preferred over natural materials in most industrial applications. The formation reaction involves olation analogous to that described for silicate formation, represented for dimer and trimer (with Al:Si = 1:2, and without

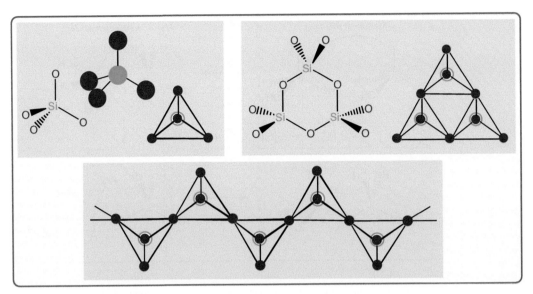

Figure 6.19
Views (*top left*) of the silicate tetrahedron including a 'triangular' view of the tetrahedron (looking down an O–Si bond, a view with the tetrahedral frame edges not involved in bonding also included) that is commonly used in representations of polymers. This is followed by (*top right*) two drawings of the cyclic silicate $[Si_3O_9]^{6-}$ (the covalent bonding and tetrahedral frame representations), and then (*at the base*) a representation of part of a chain silicate. (Charges have been left off throughout.)

including monomer/dimer protonation) as:

$$[SiO_4]^{4-} + [AlO_4]^{5-} + 2H^+ \rightarrow [AlSiO_7]^{7-} + H_2O \tag{6.58}$$

$$[AlSiO_7]^{7-} + [SiO_4]^{4-} + 4H^+ \rightarrow [AlSi_2O_9]^{7-} + 2H_2O \tag{6.59}$$

with reaction following the process shown in Figure 6.18 but with one silicon centre replaced by an aluminium.

The isolated zeolites have three-dimensional structures that can be viewed as consisting of networks of $[SiO_4]^{4-}$ and $[AlO_4]^{5-}$ tetrahedra linked to each other with bridging oxygen anions, generally by corner sharing. The general formula is $M_a[(AlO_2)_x(SiO_2)_y] \cdot zH_2O$, with M an alkali or alkaline earth metal cation. Silicon is predominant; the Si:Al ratio represented as y/x usually ranges from 1.0 to 5.0, but may be as large as 100. Other metallosilicates (e.g. with Ti(IV)) are known, but far less studied. Zeolites, of which over 250 are known, are structurally complex; e.g. a synthetic zeolite of formula $Ca_{40}Na_6[(AlO_2)_{86}(SiO_2)_{106}]$ is used commercially. The zeolites are microporous, meaning they contain cavities with a typical diameter of 0.3–0.8 nm; within these, they can accommodate water molecules and loosely bound cations (held by non-covalent forces) that can be readily exchanged with others, as indicated schematically in Figure 6.20.

Although zeolites appear structurally complex, universally they can be considered composed simply of linked tetrahedra (Figure 6.20). Their syntheses are straight-forward, to the point that they are made on a very large scale industrially in high yield using quite simple equipment. In general, aluminosilicates, silicates and polyoxometalates are all examples of stable inorganic polymers, where the primary key to their formation and structure is simply the presence of bridging

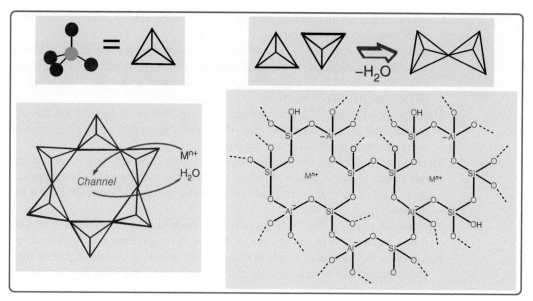

Figure 6.20

Corner-sharing (illustrated at top right) arising from olation reactions is the key to zeolite type polymers. A small ring polymer (lower left) clearly displays only corner-sharing and demonstrates simplistically the type of cavities that exist in these structures in which exchangeable cations and water molecules may reside and exchange. A drawing (lower right) shows the type of linkages met in large zeolites; note that each Si and Al displays four-coordination. Metal ions are needed to balance the negative charge on each $Al^{III}O_4$ centre, whereas there is charge neutrality for $Si^{IV}O_4$ centres.

oxido ligands. The secondary key is that, upon formation, the metal centres reside in common repeating tetrahedral or octahedral units joined by M–O–M linkages. Their accessibility via straightforward reactions in aqueous solution without any need to exclude oxygen promotes the interest in these vast families of compounds that often achieve nanomolecular size, and they have found application in a diversity of processes.

Concept Keys

The position of a metal in the periodic table has an impact on the type of chemistry that element will undergo, and the complexes that can be readily synthesised.

The majority of complexes are prepared through reactions involving ligand substitution in the coordination shell.

Syntheses typically involve chemistry performed in aqueous or non-aqueous solution; however, solvent-free reactions can sometimes be employed.

Reactions may involve a deliberate change in the oxidation state of the central atom; this is particularly the case where the target complex is in a very inert oxidation state, for which prior ligand substitution in a more labile oxidation state facilitates synthesis.

Substitution reactions of low-valent complexes (such as organometallic complexes), in which the central metal is often sensitive to oxidation by air, must be performed in apparatus that ensures oxygen-free conditions.

Coordinated ligands themselves may in some cases be able to undergo chemistry while bound to a central metal.

Reactivity of a coordinated ligand can change significantly due to polarisation effects that alter its acidity; this is exemplified by the acidity of coordinated water rising markedly on binding to a metal ion, the deprotonated hydroxide form being an efficient nucleophile in reactions.

Metal-catalysed reactions are accelerated processes that occur by a ligand binding and reacting to form a product that subsequently departs, leaving a regenerated complex which is able to repeat the process.

Metal-directed reactions involve the metal promoting chemistry in its coordination sphere that produces a new molecule to which the metal remains bound; hence the process occurs only once, and stoichiometric amounts of the central metal are involved.

Spontaneous self-assembly of larger molecules from smaller components can result from a metal acting as an organiser or director of components leading to reaction.

Reactions in the coordination sphere of a complex involving chelate formation may yield a sole isomer (a *stereospecific* reaction) or at least a preferred isomer (a *stereoselective* reaction).

Whereas traditional coordination complexes are usually prepared in aqueous solution, organometallic complexes (with M–C bonds, and typically involving metals in low oxidation states) are usual prepared in non-aqueous solution and reactions may need to be conducted in an inert atmosphere.

Metal carbonyls (with carbon monoxide acting as a ligand) are readily prepared compounds and are common starting points for syntheses of a wide range of more elaborate organometallic complexes.

Polynuclear coordination compounds, or clusters, can in some cases be prepared sufficiently large that they become nanomolecular in size.

Spontaneous olation reactions of high oxidation state early transition metal ions in aqueous acid lead to stable polyoxometalate clusters and likewise polymeric silicates and aluminosilicates, which feature M–O–M bonds; other cations may also be inserted in such clusters, which may even be of nanometre size.

Further Reading

Constable, E.C. (1996). *Metals and Ligand Reactivity*. Weinheim: VCH.
     This is a clearly written and well-illustrated text for the more advanced student, which contains some good descriptions of metal-directed reactions.
Cotton, S. (2006). *Lanthanide and Actinide Chemistry*, Wiley.
     This is one of few introductory student-focused texts on the *f*-block elements, covering the field from isolation of elements through to physical properties of complexes.
Davies, J.A., Hockensmith, C.M. and Kukushin, V.Yu. (1996). *Synthetic Coordination Chemistry: Principles and Practice*, World Scientific.
     This is a fairly comprehensive text, which a motivated student may find valuable for elucidating practical techniques and their context in experiments.
Elschenbroich, C. (2006). *Organometallics*, 3rd ed. Wiley-VCH.
     Well-organised, popular, and comprehensive in nature, with a good coverage of transition metal compounds – but more for the advanced student.
Garnovskii, A.D. and Kharisov, B.I. (2003) *Synthetic Coordination and Organometallic Chemistry*, CRC Press.
     A deep coverage of methods across the field, with examples; best for advanced students.
Girolami, G.S., Rauchfuss, T.B. and Angelici, R.J. (1999). *Synthesis and Technique in Inorganic Chemistry: A Laboratory Manual*, 3rd ed., University Science Books.

A popular coverage of not just experiments but experimental methods, and the student may find the latter part in particular worth a read.

Hanzlik, R.P. (1976) *Inorganic Aspects of Biological and Organic Chemistry*. New York: Academic Press.

An aged but still readable account of the role of metals in organic and biochemical systems; a clear account of fundamentals of metal-directed organic reactions is a useful section.

Housecroft, C.E. (1999). *The Heavier d-Block Metals: Aspects of Inorganic and Coordination Chemistry*. Oxford, UK: Oxford University Press.

One of few textbooks devoted to exemplifying differences between the coordination chemistry of different rows of the *d*-block.

Komiva, S. (Ed.) (1997). *Synthesis of Organometallic Compounds: A Practical Guide*, Wiley.

Apart from surveying metals and their organometallic compounds and introducing important synthetic approaches in this field, detailed synthetic protocols for key reactions are given. More for the graduate student, but undergraduates may find some parts valuable.

Marusak, R.A., Doan, K., and Cummings, S.D. (2007). *Integrated Approach to Coordination Chemistry: An Inorganic Laboratory Guide*, Wiley-Interscience.

An accessible and useful set of experiments in coordination chemistry; more for the instructor, but may assist students with understanding reactions.

Pope, M.T. (1983). *Heteropoly and Isopoly Oxometalates*, Springer.

Though advanced, this remains a key resource book for polyoxometalate chemistry, despite being published some decades ago.

Woollins, J.D. (Ed.) (2003). *Inorganic Experiments*, 2nd ed., Wiley.

Details of over 80 tested laboratory experiments in inorganic chemistry, including examples of coordination and organometallic chemistry; more for the teacher than student, but a useful resource of examples of reactions.

7 Properties

7.1 Investigative Methods

Molecules are not inherently endowed with the capacity to communicate their characteristics. To find out about their structures and properties, we must develop ways to interrogate them efficiently and effectively – we must make them talk or reveal their character. Working at the atomic and molecular level, this involves the application of an array of chemical and physical methods which in concert can tell us about the compound we have selected for examination. Our first task is straightforward – just what do we want to know? This may seem a simple question, but is not, since the answer will vary with the scientist asking the question. One person may simply want to know what elements a compound contains, which relates to the chemical composition; another may want to know what shape the compound takes, which is a far more complex question involving defining what components are involved and how they are assembled overall. More sophisticated questions usually require access to more sophisticated instrumentation to provide the answers. Overall, general tasks involve the *separation* of mixtures and/or the isolation of individual compounds from solution, the *identification* of an isolated compound, and the *quantification* of individual species in a solution or mixture.

It has been suggested that it is easier to make a new molecule than to discover exactly what it is you have made, and what are its properties. In synthesis, the difficulty usually lies not in the execution, but in the separation, isolation and identification of products. This is particularly so with metal complexes, where the often large array of options for products and their capacity to sometimes undergo further reactions in the process of separation makes life difficult for coordination chemists. Further, species that exist as dominant components in solution may not be the same as the dominant species isolated in the solid state. All this means that defining molecular structure in both solution and the solid state requires a call on a wide range of physical methods for characterisation. Of course, beyond mere characterisation lies the elucidation of chemical and physical properties, which are likely to be extensive and also may prove difficult to ascertain.

The presence of a metal ion in a coordination complex provides a centre of attention for probing the properties of the assembly and in due course for making use of special properties in diverse applications. Unfortunately, the metal may also introduce complications that require the use of a range of sophisticated approaches to interrogate the complex and thus determine aspects of both the structure and properties of the compound. Our focus in this chapter will be restricted to the more commonly employed

Introduction to Coordination Chemistry, Second Edition. Paul V. Bernhardt and Geoffrey A. Lawrance.
© 2025 John Wiley & Sons Ltd. Published 2025 by John Wiley & Sons Ltd.
Companion website: www.wiley.com/go/coordinationchemistry2e

techniques in use. In particular, techniques (for example, electronic spectroscopy and magnetic properties) that intersect with our models of bonding that have been developed will be probed. We shall briefly examine other selected key techniques, while leaving some methodologies untouched. However, it is inappropriate to explore selected methods in ignorance of others, so we will begin by at least identifying key methods that may be of use to the coordination chemist. This is not an exhaustive list, but there are texts and other resources devoted to reporting the details of instrumental methods that can be employed if you need to go beyond this introductory overview.

7.2 Physical Methods and Outcomes

A surprisingly large number of physical methods have been applied to metal complexes to provide information on molecular structure, stereochemistry, properties and reactivity. While characterisation may be tackled by first isolating a complex in the solid state, as mentioned earlier, this approach is complicated by there being sometimes differences between the character of a species in solution and in the solid state, as well as its stability and reactivity. In fact, it may be inappropriate to assume that the isolated solid-state species even persists in solution. Thus, it may be necessary to define species in different environments. Fortunately, the extensive array of chemical and instrumental methods now available makes the task of defining structure achievable, but what to choose and how a method assists the definition is not always obvious.

It is not the task of an introductory text to pursue this issue to exhaustion, but it seems essential to at least acquaint you with the basic and advanced techniques and what they can achieve. This has been produced in summary form in Tables 7.1–7.3. There are other ways of classifying the techniques that you may meet. For example, sometimes you will meet physical methods classified as sub-classes related to the core mode of examination, such as *resonance techniques* (nuclear magnetic resonance [NMR], electron paramagnetic resonance [EPR] and Mössbauer spectroscopy), *absorption spectroscopy* (UV–visible, circular dichroism [CD] and infrared [IR]), *ionisation-driven techniques* (mass spectrometry [MS] and photoelectron spectroscopy) and *diffraction methods* (X-ray diffraction and neutron diffraction). The list of available techniques is extensive, but in effect most coordination chemists employ a limited number of key techniques initially. While each chemist's key list will vary depending on the type of compounds worked with, it is generally true to list elemental analysis, UV–visible spectrophotometry, IR spectroscopy, NMR spectroscopy, single crystal X-ray diffraction, electrochemical methods and mass spectrometry as commonly employed techniques. These widely available methods are augmented by several of the vast array of others, depending on the task and target of the study. The cost and availability of instruments is an issue; for example, a modern high-field NMR spectrometer can cost 100 times the price of a good quality UV–visible spectrophotometer. Due to both purchase and ongoing maintenance costs, the number of some types of specialist instruments worldwide may be quite small (such as is the case with synchrotrons), so users may need to travel long distances to access them to perform experiments.

Of course, the first task in any study is the acquisition of a pure compound. In coordination chemistry, this is often a non-trivial task, as several stable compounds may form during a synthesis. *Separation*, by chromatography and/or crystallisation, is often required (Table 7.1). Chromatography columns filled with gel-form or bead

Table 7.1   Key physical methods available for qualitative identification and separation of complexes and for determining basic information about the isolated compound.

Method	Sample and *device* requirements	Outcome expected
Basic Analytical Procedures		
Traditional methods	Samples as solids or solutions, form used depending on the method. Simple methods include chemical tests, flame tests, titrations and gravimetric analysis *Basic laboratory glassware/equipment*	Determination of the presence (or absence) of an element or compound and certain components (e.g. water content by weight change on heating, precipitation)
Conductivity	Solution of a compound of known concentration *Conductivity meter and probe*	Identification of neutral or ionic character and the charge of an ionic compound
Thermal analysis	Solids for both thermal analysis (weight change on heating) and calorimetry (heat transfer with change of state) *Thermogravimetric analyser or differential scanning calorimeter*	Definition of the dehydration and decomposition processes, and change of state, on heating
Magnetic moment measurement	Pure complexes, usually as solids *Accurate balance with switchable (on/off) magnetic field*	Determination of diamagnetic or paramagnetic character and the number of unpaired electrons in a complex
Elemental analysis	Pure compound as liquid, solid or solution of known concentration. Instrumentation developed for the task: *Elemental autoanalyser* (for C, H, N, S, O usually) using sample combustion/chromatographic analysis, *or else* (applicable for most elements) *atomic absorption spectrometer (AAS) or* *inductively coupled plasma optical emission spectrometer (ICP-OES)*	Percentage composition of elements determined, allowing component elemental ratios to be determined and an empirical chemical formula to be defined
Separation Techniques		
Crystallisation	A solution of either the pure complex or a mixture of complex species *Standard laboratory glassware*	Selective crystallisation of a pure complex (this may, but need not, follow chromatographic separation of species in solution)

Table 7.1 *(Continued)*

Method	Sample and *device* requirements	Outcome expected
Ion exchange chromatography	Solutions of soluble charged complexes *Ion chromatography using columns packed with appropriate cationic or anionic polymer resin*	Separation of ions according to mainly ionic charge (Examination of separated bands of compounds directly as they exit the column by tandem instrumental methods is also possible)
High-performance liquid chromatography (HPLC)	Liquids or solutions *Commercial HPLC instrument with appropriate packed columns*	Separation of neutral and/or charged species
Gas chromatography	Gaseous or volatile liquid samples *Gas chromatograph instrument with capillary or packed columns. Detection by mass spectrometry (GC–MS)*	Separation of volatile samples (usually applicable only to thermally stable and neutral low molecular weight complexes). Possible to identify species by comparison with known standards or from the characteristic mass spectrum

resins, which are anionic or cationic in character and carry counter-ions that can be replaced readily, allow their use as ion-exchangers. For coordination complexes introduced to the resin column as dilute solutions of low ionic strength, the exchange process effectively traps cationic complexes present at the top of the column, following which elution with acid (for strongly acidic resins like Dowex® 50W) or salt solution (for resins like SP Sephadex® C-25, first developed for protein purification) removes complexes as bands which move down the column at rates determined by the charge on the complex and molecular size (see Figure 6.3). From the separated solutions of pure complexes, pure solids usually need to be recovered. Crystal growth through slow solvent evaporation, dilution with another solvent, lowering temperature or addition of other counter-ions are common approaches in coordination chemistry, whether or not prior chromatographic separation has been performed, although it is now possible to define compounds reasonably well in solution and this may be sufficient for some purposes.

Scientific instruments have been developed to exploit the electromagnetic spectrum, which ranges from gamma rays to radio waves (Figure 7.1). The interaction

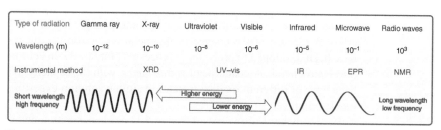

Figure 7.1

The electromagnetic spectrum and the regions employed for some of the common instrumental methods.

with matter varies depending on the energy associated with photons. Towards the high energy end of the spectrum, the change in electronic state is important; this applies to X-ray absorption and UV–visible absorption. Nearer the centre of the spectrum, changes in molecular vibrational states are important; this relates to infrared and Raman techniques. At the low energy end of the spectrum (microwave), changes in rotational states and of electron and nuclear spin are important, the last two applying to EPR and NMR methods. Not everything is spectroscopy. Of course, there are also key instrumental techniques that do not rely on the electromagnetic spectrum, such as mass spectrometry and electrochemistry.

Spectrometers, as a broad class, provide enhanced levels of information about coordination complexes that are not accessible using basic techniques. Spectroscopic and spectrometric techniques are summarised in Table 7.2; this is not a complete list, but contains methods discussed in this chapter and/or mentioned in other chapters. As an example, for the simple complex $[Co(NH_3)_5(OOCCH_3)]Cl_2$, *conductivity* experiments may allow us to identify the complex cation, specifically a 2+ cation, but little else. *Elemental analysis* may define the elemental ratio as $C_2H_{18}N_5O_2Cl_2Co$ but says essentially nothing about the actual structure of the complex or its ligands. However, a technique like *IR spectroscopy* will produce vibrations consistent with the presence of ammine and acetato ligands, *UV–visible spectroscopy* will produce a spectrum consistent with Co(III) and its presence in an octahedral CoN_5O donor environment, and *NMR spectroscopy* will not only confirm that we are dealing with a diamagnetic complex but also provide resonances assignable to five ammonia molecules and one acetate anion. It is often the ***combination of a set of methods*** like the above that leads to a firm definition of the complex form. Ultimately, if suitable single crystals can be obtained, the technique of *X-ray crystallography* can provide a detailed and accurate three-dimensional image of the complex in the solid state. There is one note of caution: the structure in solution may mirror that in the solid state, but this need not be the case.

Where a coordination complex is isolable as a solid, and particularly as a crystalline solid, several additional methods are available to assign structure, up to a detailed picture of the three-dimensional structure including accurate bond distances and angles obtained by diffraction methods (final entry in Table 7.2). Where non-crystalline samples or solids deposited on surfaces are obtained, alternate methods (including various types of advanced microscopy) can probe structure, but these are not central to introductory coordination chemistry and are not discussed here. Deliberate formation of extended solids of particular shapes defined by the way metal ions and selected ligand components self-assemble is the realm of materials science, an area of exceptional growth and promise that is taking coordination chemistry into new frontiers. An array of different physical methods are now available for the investigation of such species, but lay beyond the scope of this textbook.

The fact that metal ions in complexes often have the ability to undergo oxidation and/or reduction, with the resultant complexes having distinctly different properties as a result of the change in the metal *d*-electron set, means that techniques have been developed to probe these processes (Table 7.3). In coordination chemistry, the electrochemical technique of *voltammetry,* with the capacity to rapidly probe the behaviour of different oxidation states in simple solution experiments, is now commonly employed.

Thermodynamic properties (Chapter 5.1), such as reaction enthalpy and complex stability constants, can be routinely determined. Kinetic properties (Chapter 5.3),

Table 7.2 Selected spectroscopic and spectrometric methods available for providing information on complex formation, speciation and molecular structure.

Method	Sample and *device* requirements	Outcome expected
Nuclear magnetic resonance (NMR) spectroscopy (^{1}H, ^{13}C and many other nuclei)	Liquid or a solution, often in a usually deuterated solvent; solids can be examined. Usually for diamagnetic compounds, but paramagnetic ones can be examined *Fourier-transform multinuclear NMR spectrometer*	Ligand and complex environment and symmetry. Molecular structure in solution; highly detailed three-dimensional shape in solution sometimes obtainable for complex systems through sophisticated 2D NMR experiments
Electron paramagnetic resonance (EPR) spectroscopy (also called electron spin resonance [ESR] spectroscopy)	Liquids, solids or solutions. Paramagnetic compounds (unpaired electrons) *EPR spectrometer (usually with a variable-temperature facility)*	Information on electronic ground state, complex geometry, metal–donor environment and symmetry
Electronic spectroscopy: ultraviolet–visible–near-infrared (UV–vis–NIR) spectrophotometry	In absorption mode, uses solutions of known concentrations or thin single crystals, while powders can be examined by the diffuse reflectance technique *UV–vis(–NIR) spectrophotometer*	Approximate complex geometry, possible ligand environment and symmetry, and information on electronic structure
Circular dichroism (CD) spectroscopy and optical rotary dispersion (ORD) spectroscopy	Solution of optically resolved chiral compound that absorbs in the visible and/or UV region *CD or ORD spectrophotometer*	Identification of the optical isomer and information on symmetry, stereochemistry and conformation
Emission spectroscopy	Typically, solutions or solids *Fluorescence spectrophotometer*	Fluorescence or phosphorescence behaviour of complexes; information on electronic structure, especially excited states
Infrared (IR) spectroscopy and Raman spectroscopy	A gaseous, liquid or solid sample; solutions can be handled by subtraction of the solvent spectrum, with an aqueous solution acceptable if using a Raman instrument *Fourier-transform infrared or Raman spectrometer*	For simple compounds, molecular shape based on symmetry about the central atom Functional group and ligand identification

(Continued)

Table 7.2   *(Continued)*

Method	Sample and *device* requirements	Outcome expected
Mass spectrometry (MS)	A liquid or solid of reasonable volatility, or a gaseous sample *Quadrupole or time-of-flight (TOF) mass spectrometer*	Molecular mass. Compound characterisation from fragmentation pattern
Tandem techniques (general)	Where two (or even more) instruments or methodologies are joined to allow for expansion of the analysis and collected information (such as ESI–MS)	Enhanced information through the combination of methods. Detection of short-lived species can be achieved
X-ray absorption spectroscopy (XAS) and related advanced techniques	Usually non-crystalline solids *X-ray spectrometer*	Information on the molecular environment of the metal
Diffraction (X-ray [XRD], neutron, electron)	Crystalline sample; single crystals for full structure information or powders for bulk sample analysis *X-ray Diffractometer*	Atomic positional coordinates of all atoms allowing measurement of bond lengths and angles to high precision (bond lengths $\pm 0.1\,pm$ and bond angles $\pm 0.01°$)

such as rates of reactions that transform a complex from one species into another, and the mechanisms through which this occurs, can be sought employing modern equipment that measures change in a physical property such as colour or conductivity. As distinct from the early days of coordination chemistry, what is obvious nowadays is that it is more the inventiveness and choice selection of chemists that limits our achieving clear structural assignment rather than a lack of methods to use as probes.

7.3   Probing Complexes Using Physical Methods

The behaviour of life on Earth that we can see with our eyes can be probed by observation and recorded using cameras. Move down from the macroscopic level through the microscopic level to the molecular level, and we eventually lose our ability to 'see' the species directly under most circumstances. Thus, to probe the 'life' of complexes or their behaviour as molecular species, we must resort to far more elaborate instrument-based techniques, outlined in the previous section. Two of the most accessible spectroscopic techniques for coordination chemists, because the instrumentation is relatively cheap and usually accessible in most modern laboratories, are *ultraviolet–visible* (UV–vis) or *electronic spectrophotometry* and *infrared* (IR) or *vibrational spectroscopy*. Of the more expensive instrumentation available, *NMR*

Table 7.3 Electrochemical and thermodynamic/kinetic analysis techniques.

Method	Sample and *device* requirements	Outcome expected
Electrochemical methods		
Voltammetry (particularly cyclic voltammetry)	Solution (aqueous or non-aqueous) of a compound *Potentiostat, electrodes and electrochemical cell*	Metal-centred and ligand redox behaviour. Potentials and reversibility of reduction and oxidation processes. Information on the kinetics of electron transfer and decomposition processes
Coulometry	Solution (aqueous or non-aqueous) of a compound *Potentiostat/galvanostat and coulometry cell*	Number of electrons in a particular redox process
Kinetic and thermodynamic analysis techniques		
Stopped-flow and conventional kinetics	Solutions of a known complex *Stopped-flow instrument with UV–visible detector (using rapid mixing for initiating fast reactions), or a standard UV–vis spectrophotometer (for slow reactions); temperature-controlled reaction cell compartment*	Information on mechanisms of reactions via measurement of reaction kinetics; rates of particular reactions and associated activation parameters
Potentiometric and spectrophotometric titration	Solutions of a metal ion and a pure ligand *Auto-titration assembly and detector system (pH meter and/or UV–vis spectrophotometer)*	Metal–ligand speciation in solution and thermodynamic formation constants for complexes
Calorimetry	Solutions of a metal ion and a pure ligand *Calorimetry cell with temperature change detection electronics*	Heat of reaction. Information on complexation thermodynamics

spectroscopy is perhaps the one most commonly employed. In both electronic and vibrational spectroscopy, the observation of a molecule happens extremely rapidly ($<10^{-11}$ s), so we get an 'instant' view, and any chemical changes are too slow to be seen or influence outcomes. With NMR, the timescales are on the order of 10^{-2} to 10^{-5} s, so that rearrangements where the activation energy is small (as in some fluxional molecules) may be probed, such as by varying the temperature. We call this aspect the spectroscopic timeframe of the techniques; each method will have one, which influences the tasks it can perform somewhat. It is now possible using highly sophisticated instruments to probe dynamic processes in molecules that occur on extremely short timescales (down to 10^{-16} s), called *time-resolved spectroscopy*, but the discussion of this technique is not addressed here.

Whereas IR spectroscopy is often employed in coordination chemistry to identify functional groups or ligand types, and thus differs little from its application in organic chemistry, UV–vis spectroscopy for coordination complexes relies heavily on the special theories (crystal and ligand field theory) developed for these complexes. Thus, as we have introduced these theories in Chapter 2, it will be a target for particular attention here, as well as the related but less commonly met technique employed for chiral complexes of *CD spectroscopy*. Another physical method closely tied up with these theories is magnetochemistry, which is again experimentally simple and also the subject of attention below.

There are a number of techniques that rely on applying specific protocols that lead to excitation and a resonance condition, reported experimentally as the appearance of a peak(s) in a spectrum when using appropriate instrumentation. It is beyond the scope of this textbook to develop the detailed theory behind these many methods, but it is appropriate to illustrate two of the most common techniques already mentioned – *IR spectroscopy* and *NMR spectroscopy*. Further, the less common but important technique of *EPR spectroscopy* will also be addressed.

The examples in Sections 7.3.1–7.3.10 are designed to be merely illustrative of some key techniques. Details of these and other techniques may be found in advanced textbooks on physical inorganic chemistry and/or analytical chemistry. What should shine through the above, at least, is how molecular size, shape and symmetry relate to spectroscopic analysis.

7.3.1   Infrared (IR) Spectroscopy

IR spectroscopy is based on the absorption of infrared radiation associated with molecular vibrations caused by the oscillation of molecular dipoles involving at least two bonded atoms, such as bond stretching, bending and bond angle deformation. Deformations can be exemplified simply using models for the simple case of three linked atoms (Figure 7.2). These specific processes lead to resonances and

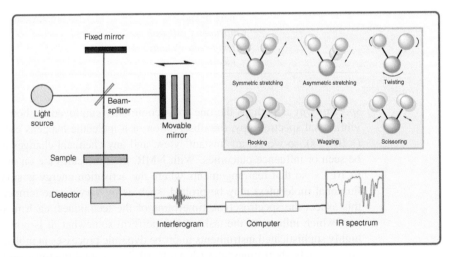

Figure 7.2

A schematic drawing of an FT-IR instrument and (inset) representations of vibrations probed by the IR technique.

tied absorbance peaks in specific positions characteristic of the molecule or its components. Vibrations associated with the metal–donor atom bonds reflect the overall molecular symmetry and usually occur in a different region to vibrations associated with coordinated molecules.

IR spectra in modern Fourier-transform (FT-IR) instruments may be reported as absorbance spectra or the inverse, transmittance spectra. In FT-IR, all frequencies of light are measured simultaneously with the IR spectrum obtained through a mathematic conversion process (Fourier transformation) operating on the initially obtained interferogram, rather than by scanning through the wavelengths. The interferogram is in the time domain and the transformation converts it to the frequency domain displayed as a spectrum. The IR technique can accept samples as solutions, powdered or dispersed solids, as liquids and thin films, and as gases; the instrument usually operates in transmission mode (where light passes through the sample), but may also operate in the reflectance mode (where light is reflected off the sample surface).

FTIR instruments are single-beam devices and contain an IR light source, an interferometer whose key components are a beam-splitter that directs light to both a fixed mirror and a moving mirror that oscillates at a constant velocity, and a detector that produces the interferogram, which is Fourier-transformed in the inbuilt computer to the desired spectrum, reporting signal intensity versus frequency (Figure 7.2). Apart from FT-IR, the Fourier-transform technique is also met in Raman spectrometers and NMR spectroscopy.

For a simple diatomic molecule XY, the vibrational frequency of the X−Y bond (typically expressed as its wavenumber in units of cm^{-1}) is dependent on both the strength of the bond (given by its force constant, k) and the masses of the atoms involved in the vibrational mode (m_X and m_Y) according to Eq. (7.1), where c is the velocity of light in units of $cm\,s^{-1}$.

$$\tilde{v} = \frac{1}{2\pi c}\sqrt{k\left(\frac{m_X + m_Y}{m_X m_Y}\right)} \qquad (7.1)$$

The term that includes the masses (known as the *reduced mass*) becomes smaller as the masses increase as the denominator is the product of the masses, while the numerator is just the sum. Therefore, *heavier* atoms joined in a bond (for a similar bond order and k value) will vibrate at *lower* frequency. Isotopic substitution (deuterium for hydrogen) can be used and is very clearly seen by IR spectroscopy, as the reduced mass will be essentially halved by replacing the atomic mass of ^1H with ^2H while k is unaffected. This in turn lowers the frequency by a factor of approximately $1/\sqrt{2}$ due to a doubling of the denominator of the reduced mass term in Eq. (7.1). For example, a typical vibrational frequency v_{CH} ~3,000 cm^{-1} drops to v_{CD} ~2,100 cm^{-1} upon deuteration solely due to this mass effect. However, when a different atom is introduced, there will also be a change in the bond strength (and force constant k) which may even override the influence of the mass term. A good example is to compare the X–H (X = O, N, C) IR vibrations which follow the series v_{OH} (3,600 cm^{-1}) > v_{NH} (3,300 cm^{-1}) > v_{CH} (3,000 cm^{-1}). This trend goes against the reduced mass prediction because the k_{OH} term is significantly larger than k_{NH} and k_{CH} due to the higher polarity of the O–H bond, which has an ionic as well as a covalent component, strengthening its bond.

For simple inorganic complexes, the molecular symmetry (Appendix 2) allows the definition of expected vibrations based on the whole molecule. For more complex

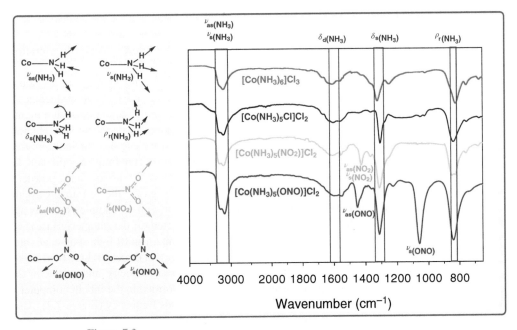

Figure 7.3

IR spectra of a selection of cobalt(III) ammine complexes. Representations of the vibrational modes are also shown (at left). All vibrations of the ammine ligands occur at essentially the same frequency in the various complexes, and vibrations of the O-bound nitrito and N-bound nitro ligands are clearly distinguished.

molecules, it is the parts of the whole, particularly functional groups, that are more likely to be detected and identified. For example, $[Co(NH_3)_6]^{3+}$, of approximately O_h symmetry, and $[CoCl(NH_3)_5]^{2+}$, of C_{4v} symmetry (see Appendix 2), will produce inherently different spectra, but it is really the peaks associated with motions of the ligand molecules themselves (asymmetric stretching (v_{as}) and symmetric stretching (v_s), twisting (τ), scissoring or bending (δ), rocking (ρ) and wagging (ω), illustrated in Figure 7.2) that dominate and are reported in the region accessible in most commercial IR spectrometers ($4{,}000$–$350\,cm^{-1}$). Vibrations associated with the metal and its donors tend, due to the large mass of the metal (see Eq. (7.1)), to be of low frequency (energy) and lie at the edge of or below the accessible region. The simple complex $[Co(NH_3)_6]Cl_3$, for example (Figure 7.3, top), displays IR bands associated with the N–H bonds of the ammine ligands at around $3{,}240$ (v_{as}(N–H)), $3{,}120$ (v_s(N–H)), $1{,}580$ (δ_d(N–H)), $1{,}320$ (δ_s(N–H)) and 820 (ρ_r(N–H)) cm^{-1}; bands associated with Co–N bonds are at $360\,cm^{-1}$ and below. The chlorido analogue $[Co(NH_3)_5Cl]Cl_2$ exhibits an IR spectrum that is essentially the same, since again the Co–Cl vibration (with an even smaller reduced mass compared to the ammine N-donor) is too low to be seen here. When nitrite ion is introduced as a ligand, it can be O-bound or N-bound, and the Co–O–N=O (nitrito) isomer is different in bonding mode and symmetry to the Co–NO$_2$ (nitro) linkage isomer, and thus produces a different IR spectrum (Figure 7.3). It is the molecular components present as ligands that are commonly probed in this technique, as also occurs in NMR spectroscopy. Since the technique relies on vibrations, single atoms or ions (like Cl$^-$) cannot themselves give an IR band; only vibrations involving the M–Cl bond are possible but typically occur too low to be observed.

Table 7.4 The variation in the IR stretching vibration of CO as a result of complexation.[a]

Species	IR resonance (cm^{-1})	Species	IR resonance (cm^{-1})
Free CO *(not coordinated)*	2,143	Free CO *(not coordinated)*	2,143
M(CO) *(monodentate-bound to a single metal)*	2,120–1,850	[Ni(CO)$_4$] *(bound to a M(0) centre)*	2,060
M$_2$(μ^2-CO) *(bridging to two metal centres at once)*	1,850–1,750	[Co(CO)$_4$]$^-$ *(bound to a M(-I) centre)*	1,890
M$_3$(μ^3-CO) *(bridging to three metal centres at once)*	1,730–1,620	[Fe(CO)$_4$]$^{2-}$ *(bound to a M(-II) centre)*	1,790

[a]The left-hand column reports trends for differing coordination modes, whereas the right-hand column compares the resonances of a series of structurally related complexes with different metal centres in different oxidation states.

IR spectroscopy can also provide information about ligand binding character. This is readily illustrated with carbon monoxide as a ligand, since upon coordination to a transition metal the CO bond is weakened, and this is seen as a shift in the position of the stretching vibration (Table 7.4). Two sets of data illustrate the effect: on the left, the effect of increasing the number of metals of one type to which the CO is bound is seen, with the bond becoming progressively weaker as the number of metals bound rises (k decreases in Eq. (7.1)); on the right, the effect of increasing the formal charge on the central metal is shown, with again a relationship between metal oxidation state and coordinated ligand bond strength apparent. Here π-backbonding from the more electron-rich metals weakens the CO bond strength.

Closely related to IR spectroscopy is *Raman spectroscopy*; they are complementary techniques, as both techniques result from changes in molecular vibrational modes. Whereas a signal in an infrared spectrum arises from a change in the dipole moment during vibration, a signal in Raman spectroscopy arises in molecules that undergo a change in polarizability during the vibration. IR spectra arise simply through energy absorption. Raman spectra arise from scattering of light: when light causes excitation, relaxation from an excited state to a different rotational or vibrational state causes a photon of a different energy to be emitted. As we are observing a shift from the incident light energy, we refer to it as a Raman shift. Since different symmetry rules govern activity in the two techniques, some bands may be IR active only, Raman active only, or else active in both techniques. Molecules with a centre of symmetry, for example, do not have vibrations that are active in both techniques, meaning aspects of symmetry may be probed by employing both techniques. Advantages of Raman spectroscopy are that, since both the incident and Raman-scattered light are in the visible region, glass can be employed in solution cells, and water can be used as a solvent. The Raman instrument is more expensive, however, due to the necessity of measuring weak signals amidst other unwanted light scattering.

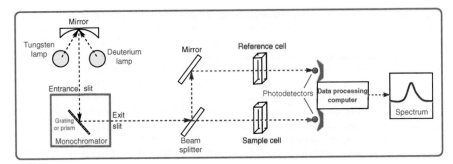

Figure 7.4

A schematic drawing of a two-beam UV–visible spectrophotometer. The tungsten lamp provides the visible (and near-IR) light, and the deuterium lamp provides the ultraviolet light; the monochromator permits selection of specific wavelengths and allows scanning across a region of the spectrum.

7.3.2 Electronic Spectroscopy

Basic ultraviolet–visible (UV–vis) spectrophotometers (also called dispersive spectrometers) illuminate a sample with a bright light source covering the UV to visible wavelength region (usually 190–800 nm), employing a monochromator to disperse light into its components and select narrow wavelengths of light employed during scanning. Subsequently, the instrument measures the light transmitted or reflected by the sample at each wavelength across a selected range. The intensity of light that leaves the sample (I) compared with its initial intensity (I_o) is converted to absorption by a simple formula ($A = \log_{10}(I_o/I)$) and it is the absorption at each wavelength that constitutes the *absorption spectrum*. Some more expensive spectrophotometers contain a detector that offers an extended wavelength range including the near-infrared (NIR) (800–3,200 nm), called UV–vis–NIR spectrophotometers. The basic transmission/absorption mode of operation of the spectrophotometer is illustrated in Figure 7.4. Of modern instruments, it is one of the simplest, reflected in its price. Most instruments are the double-beam type illustrated, allowing interference to be minimised. Combination of both a deuterium lamp source (for 190–350 nm) and a tungsten–halogen filament lamp source (for 330–3,200 nm) provides full access to the UV, visible and even near-IR regions with appropriate detectors. The measured absorbance (A) is related to the molar absorptivity (ε), concentration (c) and the path length (l, usually 1 cm) according to the Beer–Lambert law (Eq. (7.2))

$$A = \varepsilon \, c \, l \tag{7.2}$$

Absorbance varies linearly with concentration, with ε (in units of $L\,mol^{-1}\,cm^{-1}$) being a constant for a particular compound at a given wavelength. The variation of ε as a function of wavelength constitutes the unique UV-vis-NIR spectrum of the complex.

Many, but by no means all, coordination complexes absorb light in the visible region of the spectrum, leading to their distinctive colour. The central metal ion, its oxidation state, ligand environment and shape all play a role in the form of the spectrum observed. The crystal field and ligand field theories that we have developed to some extent earlier in this textbook provide a reasonable interpretation of colour. The ligand set leads to, for octahedral geometry, a stabilisation of diagonal (t_{2g}) orbitals by $-0.4\Delta_o$ and destabilisation of axial ($e_g{}^*$) orbitals by $+0.6\Delta_o$, giving a total separation of Δ_o. For most transition metal complexes, Δ_o lies in the range from $\sim$10,000 to $\sim$30,000 cm^{-1} (1,000–330 nm) which spans the near-infrared, visible and near-ultraviolet regions, accessible by UV–vis–NIR instruments. Energy from the absorption of (visible) light may promote an electron from a lower energy to a higher energy orbital if, and only if, its wavelength (energy) is aligned with the $t_{2g}-e_g{}^*$ d-orbital splitting energy (Δ_o). In this case, part of the visible spectrum is absorbed, and the remaining transmitted light (complementary to that absorbed) gives the complex its characteristic hue. If we examine the octahedral splitting diagram for all of the first-row transition metal ions in an octahedral field (Chapter 2, Figure 2.20), we can appreciate the concept, and even understand why some compounds are colourless.

It should be said that all d–d electronic transitions are theoretically orbitally (Laporte) forbidden, as the orbitals involved in the transition are of the same type. A Laporte-allowed transition is one that occurs between different orbital types, such as $s \rightarrow p$, $p \rightarrow d$ or $d \rightarrow f$. This does not mean that transitions cannot occur for same types (such as $d \rightarrow d$) but rather that their intensities (molar absorptivities) are low ($\varepsilon < 10^2$ M^{-1} cm^{-1}) compared with organic chromophores where transitions between orbitals of different types are typical (when $\varepsilon > 10^4$ M^{-1} cm^{-1}). That said, the colours apparent in Figure 7.5 are attributed to d–d electronic transitions. The solutions comprise the divalent hexaaqua ions of the first transition series spanning the electronic configurations d^3–d^{10}. For examples of d^0, d^1 and d^2 ions, see Figure 2.17 in Chapter 2.

Consider the $Co_{aq}{}^{2+}$ ion (d^7, $t_{2g}{}^5 e_g{}^{*2}$) in Figure 7.5. Its wavelength of maximum absorbance across the visible region (400–700 nm) is 520 nm (Figure 7.6, right), which is in the green region of the spectrum. As green is complementary to red, the solution appears red. The energy associated with the light of 520 nm promotes one of its five t_{2g} electrons to a singly unoccupied $e_g{}^*$ level, to produce an excited state ($t_{2g}{}^4 e_g{}^{*3}$). The ion $Zn_{aq}{}^{2+}$, however, has a full complement of ten 3d-electrons ($t_{2g}{}^6 e_g{}^{*4}$), so d–d transitions are impossible, and the compound is predicted and observed to be colourless (Figure 7.5).

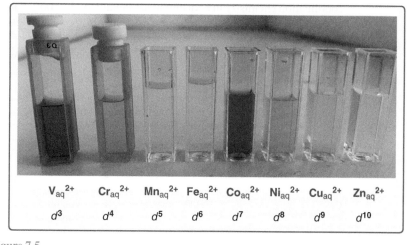

$V_{aq}{}^{2+}$	$Cr_{aq}{}^{2+}$	$Mn_{aq}{}^{2+}$	$Fe_{aq}{}^{2+}$	$Co_{aq}{}^{2+}$	$Ni_{aq}{}^{2+}$	$Cu_{aq}{}^{2+}$	$Zn_{aq}{}^{2+}$
d^3	d^4	d^5	d^6	d^7	d^8	d^9	d^{10}

Figure 7.5

The colours of all accessible $M_{aq}{}^{2+}$ ions of the first transition series. Note that $V_{aq}{}^{2+}$ and $Cr_{aq}{}^{2+}$ are both highly air sensitive so their cuvettes are sealed from the atmosphere. The absent d^1 Sc^{2+} and d^2 Ti^{2+} ions are unstable in water.

The *spin selection rule* says that the number of unpaired electrons does not change going from the ground to the excited electronic state. Accordingly, Mn_{aq}^{2+} is essentially colourless in Figure 7.5 (at best pale pink) as its unique half-filled set of five $3d$-orbitals with all electron spins the same ($t_{2g}^{3}e_{g}^{*2}$) does not allow electron promotion ($t_{2g} \rightarrow e_{g}^{*}$) without a theoretically forbidden change in electron spin, and hence the electronic transition intensity is very small ($\varepsilon < 10^{-1}$ M^{-1} cm^{-1}). A rule that is *never* broken under any circumstances is that no two electrons can occupy the same orbital without having opposite spins (Pauli exclusion principle). The pale colour of the Fe_{aq}^{2+} ion in Figure 7.5 is due to its rather broad d–d electronic maximum appearing in the near-IR region (~900 nm) where our eyes are insensitive to spectral changes.

The capacity of the crystal and ligand field models to respond to and accommodate ligand-directed influences is notable. This is best exemplified by the respective ligand and metal spectrochemical series discussed in Chapter 2.3. The influences of ligands and metal ions on each contribute to the values of Δ_o for various metal–ligand combinations. To appreciate the influence of each partner we must keep one of them constant.

We can see the influence of ligands on the splitting by examining a straightforward example, the d^3 system. Most Cr(III) complexes (like isoelectronic V(II)) absorb in the visible region of the electromagnetic spectrum. The size of Δ_o equals that of the lowest energy absorption maximum shown in Table 7.5 as the wavenumber ($\bar{v}$ (cm^{-1}) $= 10^7/\lambda$ (nm)) and follows the trend shown in the vertical columns of the table, increasing with ligand field strength of the donor. The metal's influence is shown in the horizontal rows in Table 7.5 with the data for the $[MCl_6]^{n-}$ and $[M(OH_2)_6]^{n+}$ complexes as examples; in both cases, the increase in charge correlates with an increase in Δ_o. As mentioned in Chapter 3, the magnitude of Δ_o also increases down any group (triad) with the same set of ligands ($5d > 4d > 3d$), principally due to greater covalency of the coordinate bonds. This is exemplified for the $[M(NH_3)_6]^{3+}$ series where Δ_o varies from Co (22,900 cm^{-1}) to Rh (34,000 cm^{-1}) to Ir (41,200 cm^{-1}).

Cobalt(III) ammine complexes abound and hold a special place in the history of coordination chemistry in both our understanding of the coordinate bond (see Chapter 1) and reaction mechanisms (Chapter 5). They also provide an excellent platform for studying ligand field effects on stable and inert chromophores. The series shown in Figure 7.6 (left) comprises three Co(III) pentaamine moieties with the sixth ligand

Table 7.5 Comparison of ligand field splitting energy (Δ_o) for different ligands bound to d^3 electronic configuration octahedral metal complexes $[ML_6]^{n+}$.

Ligand	Δ_o (cm^{-1})		
	V^{2+}	Cr^{3+}	Mn^{4+}
Br^-	—	12,700	—
Cl^-	7,200	13,200	17,900
F^-	—	14,900	21,800
OH_2	11,700	17,400	~24,000
NH_3	—	21,600	—
CN^-	—	26,700	—

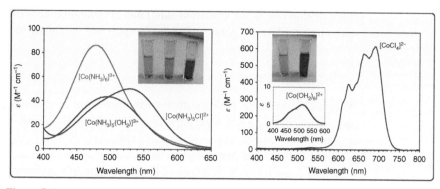

Figure 7.6

The electronic spectra of (at left) the octahedral complexes $[Co(NH_3)_6]^{3+}$, $[Co(NH_3)_5(OH_2)]^{3+}$ and $[Co(NH_3)_5Cl]^{2+}$ (inset shows solutions of each complex), and (at right) the octahedral $[Co(OH_2)_6]^{2+}$ complex (red line) and the tetrahedral $[CoCl_4]^{2-}$ (blue line). The weak absorption maximum of $[Co(OH_2)_6]^{2+}$ is amplified in the inset, and a picture of the two solutions of the same concentrations is shown above the inset spectrum.

the only variable (NH_3, H_2O or Cl^-). When the strongest field ligand is present (the N-donor ammonia), yielding $[Co(NH_3)_6]^{3+}$, the lowest energy absorption maximum wavelength appears at 478 nm; this is followed by $[Co(NH_3)_5(OH_2)]^{3+}$ (495 nm) and $[Co(NH_3)_5Cl]^{2+}$ (528 nm). While these wavelength shifts may not seem numerically remarkable, the resultant differences in their colours certainly are.

Apart from the horizontal position of absorption bands in the spectrum (represented by wavelength in nm, or frequency/wavenumber in cm^{-1}), it is notable that the intensity (molar absorptivity ε, Eq. (7.2)) of bands varies with coordination geometry and symmetry. The orbital (Laporte) selection rule has already been discussed, but distortions away from ideal octahedral symmetry also increase the probability of the transition and its intensity. For the two most common shapes, octahedral and tetrahedral, it is noted that ε values for tetrahedral complexes are invariably greater (more than 100-fold) than those found for octahedral complexes. With tetrahedral symmetry, the intensely (cobalt) blue colour of $[CoCl_4]^{2-}$ ($\varepsilon_{max} = 600\,M^{-1}\,cm^{-1}$) compared with the pale pink of octahedral Co_{aq}^{2+} ($\varepsilon_{max} = 5\,M^{-1}\,cm^{-1}$) is due to the former lacking a centre of inversion, which relaxes the Laporte selection rules and greatly enhances the absorption intensity (Figure 7.6, right).

It is also noted that the absorbance bands we see in the electronic spectra of d-block complexes are broad. This arises because complexes are constantly undergoing an array of molecular vibrations and rotations that, for example, are changing bond lengths slightly and thus influencing the size of Δ_o in the process. Because absorption of a photon of light is an extremely fast process compared with these minor internal structural changes in the complex, the form of the complex at the particular instant of photon capture is itself 'captured', leading to a range of energies associated with different vibrational and rotational states, so that we see an averaged outcome, and a broad peak. It is notable that f-block elements display sharp but weak (Lapore forbidden) $f \rightarrow f$ electronic absorption bands, as the f-orbitals involved are more 'buried' and, overall, there is little influence of metal–ligand vibrational motion in that block of the periodic table.

In summary, electronic (UV–vis–NIR) spectroscopy is central to much transition metal coordination chemistry as it readily provides information on not only electronic configuration (including oxidation state) and coordination geometry but also more subtle features like the symmetry of the complex and the ligands that are present.

7.3.3   Circular Dichroism (CD)

Whereas colour is an obvious distinguishing feature of many complexes, chiral compounds, having a non-superimposable mirror image (your two hands are a good example) can only be spectroscopically differentiated with techniques involving polarised light. Given that chirality, or so-called optical activity, is a natural and extremely common property of compounds, it is appropriate that we briefly mention and illustrate the instrumental method most often employed in its study. Whereas UV–vis–NIR spectroscopy involves the absorbance of standard isotropic light, CD spectroscopy measures the difference in absorbance of right- and left-circularly polarised light by a compound in solution. Circular dichroism is a useful method as it has the capability of displaying the presence of multiple components under a usually broad absorption envelope and displays even subtle changes in different environments. Because biomolecules are almost invariably chiral (through their chiral amino acid and sugar components), CD finds ready application in probing their structure (especially of proteins and nucleic acids), but also is employed in the study of metal complexes that have been resolved.

Electromagnetic radiation consists of oscillating electric and magnetic fields that are perpendicular to each other and to the direction of propagation. Polarised light can be linearly or circularly polarised, and in circularly polarised light the electric field vector rotates around the propagation axis with the magnitude staying constant but the direction varying (or oscillating). It appears to trace a circle when looking along the propagation axis, with a full rotation occurring at each wavelength. As propagation occurs, a helical shape is traced out by the vector head, the origin of 'circular' in the name circular dichroism; the 'dichroism' aspect relates to the capacity to absorb one of two components more strongly, the standard definition of a dichroic material. Counterclockwise rotation of the vector defines left-circularly polarised light, whereas clockwise rotation defines right circularly polarised light. In the presence of an optically active molecule, there will be preferential absorption of one direction of circularly polarised light. The different amplitudes of left and right circularly polarised light resulting leads to elliptical polarisation (the vector head traces out an ellipse), with the molar ellipticity (θ degrees) used as one form of measurement in CD spectroscopy. The wavelength-dependent molar absorptivity (defined as $\Delta\varepsilon = \varepsilon_l - \varepsilon_r$, the difference in absorption of the left and right circularly polarised light) provides the other way CD spectra are typically reported, particularly in coordination chemistry; $\Delta\varepsilon$ and θ are related, and interconversion is practicable.

Commercial CD instruments, which became readily available in the 1960s, operate usually in the visible–UV region, and pass polarised light through a monochromator (that yields a variable single wavelength output), then through a modulating device (called a photoelastic modulator, invented in the 1960s by Jacques P. Badoz) which transforms the linearly polarised light to circular polarised light before it is passed through the sample and reaches the detector (Figure 7.7). As the wavelength is varied, the difference in absorptivity at every wavelength is observed and the CD spectrum calculated and reported.

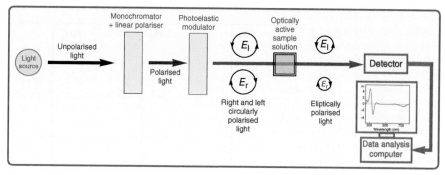

Figure 7.7

Schematic diagram of a circular dichroism instrument. Linear polarised light is passed through a photoe-lastic modulator and switched continuously to produce right- and left-circularly polarised light. Unequal absorption of circularly polarised light yields elliptical polarisation, the basis of the circular dichroism spectrum.

Importantly for coordination compounds, because the electronic transitions that are detected in a standard UV–vis spectrum are the same ones detected in a CD spectrum, CD occurs in the same spectral region as the electronic absorption bands in the UV–vis spectrum. Moreover, bandwidths of an absorption band and a corresponding CD band are about the same, allowing for a robust assignment of the CD band to the corresponding absorption band. A key difference, however, is that a component band in the CD spectrum can have a positive or negative sign on the vertical axis, and each optical isomer displays identical, but mirror image, spectra reflected about the wavelength axis. For a CD spectrum to be observed, the complex must itself be chiral (due to inherent dissymmetry or due to the arrangement of two or more chelate rings) and resolved into one of its optical isomers. The CD of a racemic mixture of two enantiomers will be a flat line. The metal centre may also come under the influence of the asymmetry of one or more chiral ligands, including the effect of chiral con-formations of chelates, leading to a CD signal. Further, since the observed spectrum is the sum of component electronic transitions, it may display several bands under a single absorption envelope, which is clear evidence that the absorption envelope is not the result of just a single electronic transition.

For an octahedral cobalt(III) complex, the lowest energy absorption envelope seen in the visible region is assigned to a transition from an orbitally non-degenerate ground state to a triply degenerate excited state. If symmetry of the complex is low-ered due to the presence of several ligand types and their coordination locations, the threefold degeneracy can split into different components (a doubly degenerate and non-degenerate state, for example) which leads to two separate transitions. Often undetected in the broad envelope of the visible absorption spectrum, the CD spectrum may, however, as a result of its component transitions having the potential to display inherently opposite signs, clearly resolve two or more transitions. This is one role for CD spectroscopy. However, CD is particularly sensitive to environment (a role put to good use with proteins and nucleic acids), and significant variation in the spec-trum can occur upon change in ligand substituents even when distant from the metal ion, upon addition of small amounts of electrolytes, or change of solvent; so-called 'outer sphere' interactions that do not involve changes in covalent bonding. CD sig-nals may emerge when a complex is placed in an asymmetric environment through non-covalent interactions with a chiral solute. The further the source of asymmetry

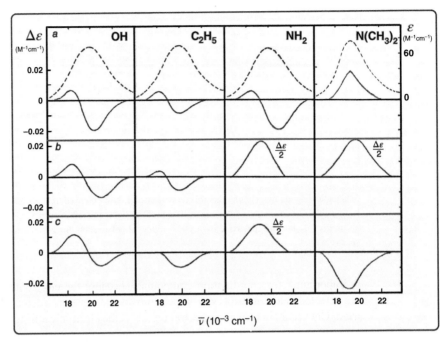

Figure 7.8

The lowest energy visible absorption band (broken line) and associated circular dichroism (solid line) spectra of four $[Co(NH_3)_5(OOC-CH(X)-CH_3)]^{2+}$ complexes, with each X-group identified on the figure. Spectra are shown in (a) water only, (b) 0.05 M aq. Na_2SO_4 and (c) 0.05 M aq. K_2SeO_3.

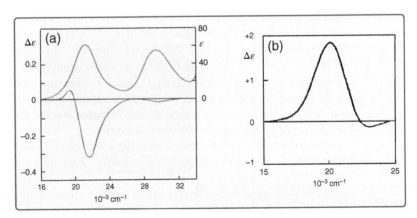

Figure 7.9

(a) The visible absorption (top) and circular dichroism (bottom) spectrum of $[Co(NH_3)_4(S\text{-pn})]^{3+}$ (pn = $H_2N-CH_2-CH(CH_3)-NH_2$) in water and (b) the larger circular dichroism spectrum under the lowest energy absorption band of Λ-$[Co(en)_3]^{3+}$ (en = $H_2N-CH_2-CH_2-NH_2$) in water. (The units of the vertical axes are M^{-1} cm^{-1}.)

lies from the metal centre, the smaller is the observed CD spectrum, as illustrated in Figures 7.8 and 7.9.

The simple series of complexes $[Co(NH_3)_5(OOC-CH(X)-CH_3)]^{n+}$ contain a single chiral centre located at the 2-position of the propionate ligand. Although there is no inherent chirality at the metal centre, a weak induced CD (Figure 7.8) is observed due to the asymmetric field of the optically pure ligand bound to the metal centre (termed a *vicinal effect*). Moreover, more than one CD band may be observed, consistent with several component transitions under the broad absorption envelope, as discussed earlier. Changing the substituent X-group changes the CD spectrum significantly, despite its distance from the metal chromophore and there being no other variation in the complex. The solution CD spectrum is even influenced by low concentrations of simple salts added to the solution, with added anions being involved in specific ion-pairing interactions with the cationic complex. Sensitivity to the environment, illustrated in Figure 7.8, is a common feature of circular dichroism spectra.

When a chiral ligand is coordinated as a chelate rather than a monodentate, there are two contributions to the CD – a *vicinal effect* (introduced earlier) and a *conformational effect* from the presence of a chelate ring, of which the latter effect is dominant. An example is $[Co(NH_3)_4(S\text{-pn})]^{3+}$ (pn = propane-1,2-diamine = $H_2N\text{–}CH_2\text{–}CH(CH_3)\text{–}NH_2)]^{3+}$, where there is one chiral carbon in the chelated diamine (carrying the methyl substituent). As a consequence of the two contributions, the size of $\Delta\varepsilon$ in the CD bands is enhanced. For example, the CD spectrum of the above *S*-pn complex (Figure 7.9, left) has two features under the lower energy absorption envelope at 508 nm ($\Delta\varepsilon$ +0.11 M^{-1} cm^{-1}) and at 460 nm ($\Delta\varepsilon$ −0.32 M^{-1} cm^{-1}). However, in the CD spectrum of an analogue with only a monodentate chiral amine ligand bound, $[Co(NH_3)_5(H_2N\text{–}CH(C_2H_5)\text{–}CH_3)]^{3+}$, the single CD band is markedly smaller ($\Delta\varepsilon$ −0.03 M^{-1} cm^{-1}) at 485 nm. For complexes such as $[Co(en)_3]^{3+}$ (en = $H_2N\text{–}CH_2\text{–}CH_2\text{–}NH_2$), where chirality arises from the arrangement of three chelate rings about the metal ion (and hence where the complex may be resolved into its two optical isomers, Δ and Λ), the CD spectrum is significantly more intense (Figure 7.9, right) as the chirality source is closer to the metal ion chromophore and associated with the *configurational effect* from the spiral arrangement of the three chelated ligands about the metal ion.

CD spectra can be observed with any metal complex that is sufficiently inert to optical isomerisation to a racemic mixture, or else resistant to rapid dissociation if a chiral ligand is coordinated. For example, $\Delta-$ and $\Lambda-[Ni(en)_3]^{2+}$ cannot be resolved in solution, so no CD spectrum is possible due to rapid racemisation of the two enantiomers in accord with the labile nature of Ni(II) complexes. The above discussion has also been restricted to complexes with the common octahedral shape, but it is important to recognise that other coordination geometries may yield optical isomers, provided they lack any plane of symmetry or centre of inversion. Four-coordinate square-planar complexes have a plane of symmetry in the sense that the metal centre and four donor atoms lie in a plane and hence do not have optical isomers, but tetrahedral complexes lack a plane and a centre of symmetry, meaning they may potentially be resolvable, depending on the ligand set and complex stability.

Metalloproteins and metallopeptides exhibit chirality, and the CD spectrum of these compounds is sensitive to their three-dimensional arrangement. For example, rubredoxins, the simplest iron–sulfur proteins (also discussed in Chapter 8), have a tetrahedral FeS_4 environment and exhibit strong CD spectra due to the four chiral cysteine side chains coordinated to the metal. Although CD spectroscopy has found extensive use in the study of biomolecules, it remains valuable in coordination chemistry where chirality is involved – and a vast number of metal complexes are dissymmetric and/or include chiral ligands.

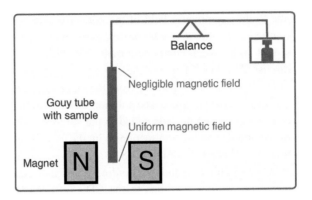

Figure 7.10

A schematic drawing of the simple equipment used in the Gouy method. A permanent magnet (raised or lowered) or an electromagnet (switched on or off) is employed to provide a magnetic field difference, with weight change from the absence to the presence of the magnetic field being recorded.

7.3.4   Magnetic Moment Measurement: The Gouy Method

Measurement of weight change of a compound in the presence and absence of a magnetic field, the **Gouy method** (proposed theoretically by physicist Louis Georges Gouy in 1889), is one of the simplest physical methods available. It allows the determination of the *magnetic susceptibility* of a compound (related to the magnetic dipole moment, basically the torque experienced in a magnetic field) by measuring the force on a sample in a magnetic field via its weight change in and out of the magnetic field (Figure 7.10).

If the strength of the magnetic field (supplied by either a permanent magnet or an electromagnet) is known, the magnetic susceptibility can be calculated; further, the *effective magnetic moment* can also be calculated via the temperature dependence of the magnetic susceptibility. The magnetic susceptibility can be viewed as the ratio of magnetisation (magnetic moment per unit volume) to the applied magnetic field intensity. A spin-only magnetic moment occurs due to an electron rotating (or precessing) about its own axis, generating a very small magnetic field; in addition, a contribution from an orbital magnetic moment arises due to electron motion around a nucleus, but this is a minor contributor for first-row transition metal ions, and hence, frequently, reference is simply to the *spin-only magnetic moment*.

As discussed earlier in Chapter 2.3, if we examine the set of d-electrons for d^4–d^7, there is choice available as regards the arrangement of electrons in the octahedral d subshells. These configurations display options for electron arrangement, namely *high spin* and *low spin*. The differentiation depends on the size of the energy gap (Δ_o) compared with the amount of spin pairing energy (P), with a large Δ_o favouring low-spin arrangements. As covered in Section 7.3.2, the size of Δ_o is ligand dependent, especially depending on the type of donor atom, and is influenced appreciably by the charge on the metal ion; P varies dominantly only with the metal centre and its oxidation state. In general, where $P > \Delta_o$, the complex will be high spin, whereas where $P < \Delta_o$, the complex will be low spin (Table 7.6). Note how, as the ligand changes from F^- to NH_3 for d^6 Co(III), and from OH_2 to CN^- for d^6 Fe(II) the spin state switches; if this model is correct, this should be experimentally observable. It is the magnetic properties that most clearly illustrate this behaviour. For d^6 octahedral complexes, low-spin compounds have all six d-electrons spin-paired in the three

Table 7.6 Comparison of pairing energy (P, in kJ mol^{-1}) versus ligand field splitting energy (Δ_o) for some metal ions with d^4–d^6 electronic configurations and consequences in terms of observed spin state.

d^n	Ion	Ligands	P	Δ_o	Spin state
d^4	Cr^{4+}	$(OH_2)_6$	282	166	HIGH $(P > \Delta_o)$
	Mn^{3+}	$(OH_2)_6$	335	252	HIGH $(P > \Delta_o)$
d^5	Mn^{2+}	$(OH_2)_6$	306	94	HIGH $(P > \Delta_o)$
	Fe^{3+}	$(OH_2)_6$	358	164	HIGH $(P > \Delta_o)$
d^6	Fe^{2+}	$(OH_2)_6$	211	125	HIGH $(P > \Delta_o)$
		$(CN^-)_6$	211	395	LOW $(P < \Delta_o)$
	Co^{3+}	$(F^-)_6$	252	156	HIGH $(P > \Delta_o)$
		$(NH_3)_6$	252	272	LOW $(P < \Delta_o)$
		$(CN^-)_6$	252	404	LOW $(P < \Delta_o)$

t_{2g} orbitals, whereas high-spin compounds have electrons in both t_{2g} and $e_g{}^*$ orbitals with four unpaired electrons (see Chapter 2); the difference is readily distinguished.

In terms of magnetism, there are two classes of compounds, distinguished by their behaviour in a magnetic field: **diamagnetic** – *repelled from* the strong part of a magnetic field; and **paramagnetic** – *attracted into* the strong part of a magnetic field. The other common form, *ferromagnetism*, can be considered a special case of paramagnetism, with domains of parallel spin in the material that lead to strong permanent magnetism. *All* chemical substances exhibit some form of diamagnetism, since this effect is caused by the interaction of an external field with the magnetic field produced by electron movement in filled orbitals (with paired electrons). However, diamagnetism is a much smaller effect than paramagnetism, in terms of its contribution to measurable magnetic properties. Paramagnetism arises whenever an atom, ion or molecule possesses *one or more unpaired electrons*. It is not restricted to transition metal ions, and even non-metallic compounds can be paramagnetic; dioxygen is a simple example. However, this definition is significant in terms of the variable number of unpaired electrons that can be present in a transition metal complex. Since we are dealing with five *d*-orbitals, the maximum number of unpaired electrons that can be associated with one metal ion is 5 – simply, there can be only from 0 to 5 unpaired electrons, or 6 possibilities.

The apparent *weight change* measured between the absence and presence of a strong inhomogeneous magnetic field (Figure 7.10) leads to the number of unpaired electrons based on theories that we shall not develop fully here. However, we will explore the factors that contribute to the magnetic properties, usually expressed in terms of an experimentally measurable parameter called the *magnetic moment* (μ).

We are dealing in our model with electrons in orbitals, which are defined to have both orbital motion and spin motion; both contribute to the (para)magnetic moment. Quantum theory associates quantum numbers with both these motions. The spin and orbital motion of an electron in an orbital involve quantum numbers for both *spin angular momentum* (S), which is actually related to the number of unpaired electrons (n), as $S = n/2$, as well as the *orbital angular momentum* (L). The magnetic moment (μ) (which is expressed in units of Bohr magnetons, μ_B) is a measure of the magnetism and is defined by Eq. (7.3), involving both quantum numbers.

$$\mu = [4S(S + 1) + L(L + 1)]^{1/2} \tag{7.3}$$

Table 7.7 Comparison of predicted and actual experimental magnetic moments (in Bohr magnetons) for octahedral transition metal complexes with from 0 to the maximum possible 5 unpaired electrons.

N^o. unpaired e^-	Calculated μ	Experimental μ	Typical metal ion
0	0	~0	low-spin Co^{3+}
1	1.73	~1.7	Ti^{3+}, Cu^{2+}, low-spin Fe^{3+}
2	2.83	2.75–2.85	V^{3+}, Ni^{2+}
3	3.88	3.80–3.90	V^{2+}, Cr^{3+}, Mn^{4+}
4	4.90	4.75–5.00	high-spin Cr^{2+} and Co^{3+}
5	5.92	5.65–6.10	high-spin Mn^{2+} and Fe^{3+}

The introduction of an equation involving quantum numbers may be daunting, but we are fortunately able to simplify this readily. Firstly, for *first-row* transition metal ions, the effect of L on μ is small, so a fairly valid approximation can be reached by neglecting the L component, and then our expression reduces to the so-called 'spin-only' approximation, Eq. (7.4).

$$\mu = [4S(S + 1)]^{\frac{1}{2}} \tag{7.4}$$

where the only quantum number remaining is S. Now, we may readily replace this quantum number by using the $S = n/2$ relationship, the substitution then leading to the 'spin-only' formula (Eq. (7.5)) for the magnetic moment.

$$\mu = [n(n + 2)]^{\frac{1}{2}} \tag{7.5}$$

Thus, we have reduced our expression to one simply involving the number of unpaired electrons (n). Using this 'spin-only' Eq. (7.5), the value of μ can be readily calculated and predictions compared with actual experimental values (Table 7.7).

It is clear that the experimental values can be used effectively to define the number of unpaired electrons, and thus the spin state of a complex. For example, Co(II) is a d^7 system, which will have three unpaired electrons if high spin and one unpaired electron if low spin, with calculated magnetic moments of 3.88 and 1.73 μ_B, respectively. If a particular complex has an experimental magnetic moment of 3.83 μ_B, then it must, by comparison with the two options, be high spin. The technique can very effectively distinguish between high- and low-spin states. For Co(III) (d^6), for example, $[Co(NH_3)_6]^{3+}$ (low spin, $n = 0$) has an experimental magnetic moment close to zero, whereas $[CoF_6]^{3-}$ (high spin, $n = 4$) has an experimental magnetic moment of nearly 5 μ_B, readily differentiated. The simple ligand field model that we have developed has triumphed again, and it is its applicability for dealing simply with an array of experimental data such as magnetic behaviour that has led to its longevity.

Yet is there any other basic evidence to support the theory that we have dwelt on so much? One of the key pillars of the theory is that there should be some inherent stabilisation of compounds as a result of introducing a field of ligands around the central metal ion, expressed in terms of the so-called *ligand field stabilisation energy*. Recall that in an octahedral field, electrons in diagonal orbitals are stabilised (by $-0.4\Delta_o$) and those in axial orbitals are destabilised (by $+0.6\Delta_o$) relative to the spherical field situation. By summing up the energy values for all d-electrons in the set, we arrive at

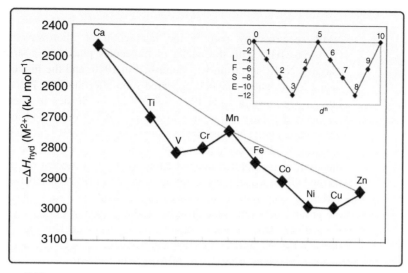

Figure 7.11

Predicted LFSE for *high-spin* d^0 to d^{10} (inset), and variation in hydration energy for first-row M^{2+} transition metal ions, which mirrors this trend, superimposed on a general increase anticipated to occur with increasing atomic number.

the overall stabilisation or *ligand field stabilisation energy* (LFSE), which will thus vary with n in d^n, as shown in Table 7.8.

For example, for high-spin d^5, we have three electrons in the t_{2g} level ($3 \times -0.4\Delta_o$) and two in the e_g level ($2 \times +0.6\Delta_o$), for an overall LFSE of zero; however, for low-spin d^5, all five electrons lie in the t_{2g} level ($5 \times -0.4\Delta_o$) for an overall LFSE of $-2.0\Delta_o$. The concept is simple, and for high-spin systems produces a pattern as shown in the inset of Figure 7.11.

There is experimental evidence that supports this trend. For example, hydration energies for M(II) ions of the first-row transition metals exhibit a W-shaped pattern (Figure 7.11), as a result of crystal field stabilisation superimposed on the expected periodic increase, consistent with this model. Likewise, the measured lattice energies of transition metal fluorides (MF_2) display a very similar pattern. The predominance of low-spin d^6 complexes is also consistent with the model, as there is a significant difference in LFSE between spin states in favour of low spin in the model. However, there are concerns with the model that have attracted criticism; for example, one might anticipate low-spin d^5 more often than is observed. Nevertheless, it is a simple and sufficient model to use at a basic level.

Table 7.8   Ligand field stabilisation energies (LFSE, given in Δ_o) for various d^n configurations.

Configuration	d^0	d^1	d^2	d^3	d^4	d^5	d^6	d^7	d^8	d^9	d^{10}
low-spin					−1.6	−2.0	−2.4	−1.8			
high-spin	0	−0.4	−0.8	−1.2	−0.6	0	−0.4	−0.8	−1.2	−0.6	0

7.3.5   Nuclear Magnetic Resonance (NMR) Spectroscopy

NMR spectroscopy records the interaction of microwave/radiowave radiation (100–900 MHz) with the nuclei of molecules placed in a strong magnetic field. The first NMR spectrum was published in 1946, and the first commercial machine (using a permanent magnet) appeared soon after in 1950. Felix Bloch and Edward Purcell were awarded the 1952 Nobel Prize in Physics for their work in the field. Elements that have either an odd atomic number or an odd mass number possess non-zero nuclear spin and hence can produce an NMR spectrum. When an external magnetic field is applied to a sample, its nuclear spin aligns either with or against the direction of the applied magnetic field (there is a modest relationship to the Gouy method discussed earlier), resulting in an energy difference between the excited state and the ground state. Following excitation with radiowaves, the return of the spin to the ground state emits the absorbed radiofrequency, effectively the NMR signal of the nucleus, detected with a radiofrequency receiver. This radiofrequency is proportional to the magnitude of the applied magnetic field; early machines with permanent magnets operated typically at 60 MHz, but modern instruments (Figure 7.12) with superconducting magnets (which require cooling to liquid helium temperatures) operate at between 300 and 1,000 MHz, providing much improved signal resolution.

The local magnetic field surrounding an atom in a molecule influences the resonance frequency, and since every molecular compound is chemically and structurally different, this produces slight variation in resonance frequencies, resulting in an NMR spectrum unique to the compound. Put more simply, NMR relies on the change in nuclear spin through the absorption of energy in the MHz range, where each atom of a particular element in a compound in a unique molecular environment will yield a signal in a unique position, reported as a chemical shift (δ) relative to a given standard compound signal. Moreover, atoms may interact with close neighbours in many cases, to produce a splitting (*coupling*) pattern superimposed on the gross chemical shift that is characteristic of this neighbouring environment. This makes the technique powerful as a weapon in determining structure, particularly in solution. Many elements (and isotopes) can produce an NMR spectrum, but the traditional element targeted is hydrogen, because ^1H is of high isotopic abundance, is in essentially all

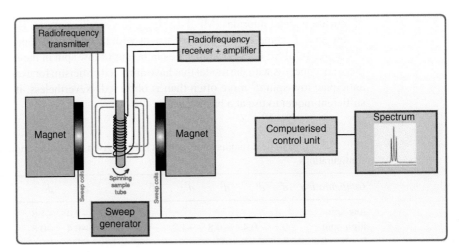

Figure 7.12

A simple schematic of a modern nuclear magnetic resonance spectrometer.

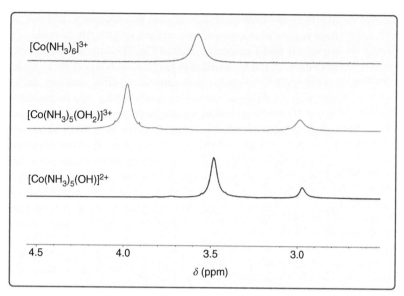

Figure 7.13

^{1}H NMR spectra (400 MHz) of $[Co(NH_3)_6]^{3+}$ (top), $[Co(NH_3)_5(OH_2)]^{3+}$ (centre, pH 5.5) and $[Co(NH_3)_5(OH)]^{2+}$ (bottom, pH 8.5) measured in H_2O with solvent suppression.

organic compounds, has simple spin characteristics, and highest sensitivity. Another popular element used, ^{13}C, has a relative sensitivity of only ~1.6% compared with ^{1}H, and a significantly lower isotopic abundance (only ~1%), so that its detection is consequently more demanding; the dominant isotope ^{12}C with zero nuclear spin is NMR silent. Given this utility, NMR is the 'go-to' physical method for all organic chemistry where carbons and hydrogen are always both present. Metals that are NMR active within complexes can be examined directly but may suffer from low sensitivities and/or inappropriate isotopic abundance, as well as other limitations that can lead to very broad peaks and extreme chemical shifts. Nevertheless, modern NMR instruments offer from adequate to excellent determination of spectra of a vast number of elements, although ^{1}H, ^{13}C, ^{19}F and ^{31}P remain most commonly available in commercial instruments, and dominantly the former two are utilised simply because they usually occur in compounds chosen for study. This means that, for coordination complexes, it is usually the organic ligands that are being probed by this technique rather than the metal or even the ligand heteroatoms. The magnetic resonance imaging (MRI) technique used for whole-body scanning in medicine is a type of NMR instrument but focussed on the variation of the properties of water molecules in different environments as the basis of its operation. This is covered in more detail in Chapter 9.

To illustrate the NMR technique very simply, we shall draw on some simple complexes as examples. The NMR method is most applicable to diamagnetic metal complexes, and we shall restrict examples to these systems. To avoid ^{1}H NMR signals from the solvent's H atoms, spectra are often measured in a deuterated solvent in which the less common isotope ^{2}H (or D) atoms replace all ^{1}H atoms, e.g. D_2O, $CDCl_3$ or CD_3CN. In aqueous solutions, the use of D_2O comes at a cost, as any H atoms associated with 'acidic' N- or O-atoms on the complex will be exchanged for D-atoms and their ^{1}H NMR signals are lost. Nowadays, NMR experiments can be designed to suppress signals from otherwise dominant solvents like H_2O, so deuterated solvents are not necessary, and this is important if signals from readily exchangeable protons are sought. An example is shown in Figure 7.13 for the ^1H NMR spectra of a set of Co(III) ammine complexes measured in H_2O; in D_2O these would be featureless (as $NH_3 \rightarrow ND_3$). If we consider highly symmetrical $[Co(NH_3)_6]^{3+}$, which is low-spin d^6 Co(III) with no unpaired d-electrons and therefore diamagnetic, all six ammonia ligands are in equivalent environments, and so the ^1H NMR spectrum should, and does, yield a single peak with one specific chemical shift at 3.55 ppm versus the reference, TMS (tetramethylsilane). If we turn to $[Co(NH_3)_5(OH_2)]^{3+}$ (Figure 7.13, centre) the four ammonia ligands around the equatorial plane of the metal are symmetrically related, but the axial ammine ligand *trans* to the aqua ligand is unique, so two peaks of different chemical shifts result, in an intensity ratio of 4:1. The chemical shifts are also sensitive to other ligands that are present. The simple deprotonation of the aqua ligand in $[Co(NH_3)_5(OH_2)]^{3+}$ (pK_a 6.4) to afford $[Co(NH_3)_5(OH)]^{2+}$ also gives a pair of ammine ligand signals in a ratio of 4:1 but the equatorial protons appear at lower chemical shift ('upfield') due to the different electronic influence of the hydroxido ligand compared with aqua. The ^1H NMR signals from the H_2O and ^-OH ligands are broadened and not visible due to rapid exchange with the solvent.

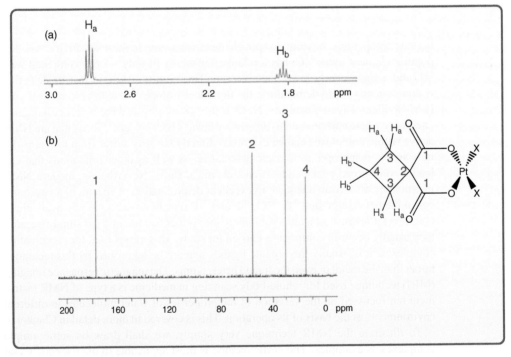

Figure 7.14

The (A) ^1H and (B) ^{13}C NMR spectra of carboplatin, with peak assignments. Coupling (between H_a and H_b types) is present in the ^1H spectrum in addition to different chemical shifts for the two types, whereas the ^{13}C spectrum is a simpler proton-decoupled version where each single peak can be assigned to carbon centres in different environments (identified by red numbers).

For coordination complexes including organic molecules as ligands, the ^1H and ^{13}C NMR spectra in combination present a powerful technique for probing complex stereochemistry and ligand character. In a simple sense, every chemically distinct carbon will have a unique ^{13}C NMR chemical shift, so that a simple count of peaks provides a good start to structural assignment, drawing on symmetry considerations to assist. The NMR technique is highly advanced, with modern spectrometers offering a range of methodologies designed to assist in structural elucidation; many of these are more often met in organic chemistry.

Beyond simply focussing on ^1H NMR chemical shifts, it is important to recognise that coupling between adjacent H atoms (spin–spin coupling) invariably exists. This is most obvious in the ^1H NMR spectra of complexes that contain coordinated organic molecules, where the coupling seen in the 'free' organic molecules is carried over to the coordinated form, albeit with some variations due to extra rigidity introduced should chelation be involved (which we will not explore at this level). This is illustrated in the ^1H NMR spectrum of the anti-cancer drug carboplatin (Figure 7.14), where the multiplicity of peaks equals the number of neighbouring H-atoms plus one e.g. three for the proton set on C3 and five for C4 in Figure 7.14. Also shown in this figure is the ^{13}C NMR spectrum, where the four different types of carbons in the complex yield four distinct ^{13}C NMR resonances; this is a so-called proton-decoupled spectrum and is the conventionally recorded ^{13}C NMR experiment.

The NMR signals of metal ions within complexes can exhibit large chemical shifts and broad spectra, but still display sensitivity to the coordination environment and oxidation state. For example, the broad ^{59}Co NMR signals for the biomolecules vitamin B_{12} (at 4,650 ppm), B_{12} coenzyme (4,480 ppm), methylcobalamin (4,215 ppm) and an ester of dicyanocobyrinic acid (4,095 ppm) display distinctive chemical shifts. As with ^1H and ^{13}C NMR for which chemical shifts are cited versus TMS, these ^{59}Co chemical shifts are relative to a 'standard', which is usually $[Co(CN)_6]^{3-}$ (defined as 0 ppm). Interpreting the NMR spectra of metals is not routine, and we shall not elaborate further. The important message here is that most elements (or specific isotopes) can theoretically display NMR signals, provided a tuneable multinuclear probe is fitted to the instrument. Effectively, all isotopes with high abundance and an odd atomic number (such as ^1H, ^{19}F and ^{31}P) or isotopes with an even atomic number but an odd atomic mass (such as ^{13}C, ^{17}O, ^{29}Si) have the properties required for NMR, although natural abundance (only 4.7% for ^{29}Si and an even lower 0.04% for ^{17}O, for example) does influence the signal intensity and calls on techniques such as isotopic enrichment, which is non-trivial chemistry.

7.3.6 Electron Paramagnetic Resonance (EPR) Spectroscopy

EPR (also occasionally called electron spin resonance, ESR) spectroscopy is used to examine compounds with unpaired electrons. EPR and NMR have many similarities but in EPR it is electron spin transitions that are probed providing information on the region around the unpaired electron(s), including aspects of metal–ligand arrangements, through detecting interactions between electron and nuclear spins in the vicinity. Unlike most spectroscopic techniques, EPR spectroscopy holds constant the frequency of the radiation employed (which is in the microwave gigahertz range)

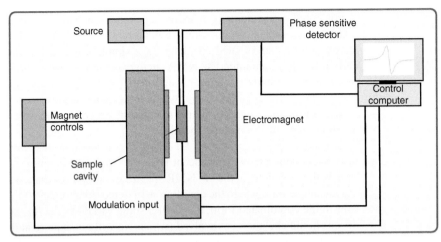

Figure 7.15
Simplified schematic block diagram for a typical EPR spectrometer.

while the magnetic field is varied linearly to produce an absorption spectrum. In the most common commercial continuous wave X-band instruments (shown schematically in Figure 7.15), the microwave radiation source typically operates at about 9.5 GHz and the magnetic field is around 250–350 mT (millitesla). The sample (solid or liquid) sits in a resonant cavity (resonator) located in the middle of an electromagnet. Absorption followed by detection is direct evidence of the presence of unpaired electrons. This occurs when the separation of the EPR energy levels matches the incident microwave photon frequency. The integrated signal intensity is directly proportional to the sample concentration.

The basis of the technique is tied to the spin of an electron and its associated magnetic moment. The mathematics behind the technique is of a high level (employing things called spin Hamiltonians – look them up if you wish!), and so we will provide only basic concepts. For a single electron spin (with spin $S = \frac{1}{2}$) in an applied magnetic field (B_o), the two possible electron spin states ($m_S = \pm\frac{1}{2}$) have different energies (the electron *Zeeman effect*). This occurs because the alignment of the magnetic moment of the electron (μ) *with* the magnetic field produces a lower energy state than where alignment is *against* the magnetic field. The two states, defined by the projection of the electron spin (m_S) on the direction of the magnetic field, are termed the parallel state ($m_S = -\frac{1}{2}$) and the antiparallel state ($m_S = +\frac{1}{2}$). For a molecule with a single unpaired electron in a magnetic field, and ignoring nuclear spin contributions, the two energy states of the electron (Figure 7.16) split and are separated by ΔE, and an EPR transition is observed when the microwave energy ($h\nu_{mw}$) matches the splitting energy according to Eq. (7.6):

$$\Delta E = h\nu_{mw} = g\mu_B B_o m_S = \pm\frac{1}{2}g\mu_B B_o \qquad (7.6)$$

where g is the proportionality factor (g-factor, indicating the field position for resonance), μ_B is the Bohr magneton (a constant), B_o is the external magnetic field (an experimental variable) and m_S is the electron spin quantum number. Simplifying Eq. (7.6), the g-value can be computed using the formula $g = 71.44775\nu_{mw}/B_o$, where the microwave frequency ν_{mw} is in GHz and the applied magnetic field B_o is in mT. Although g is a constant for a particular paramagnetic compound, the magnetic

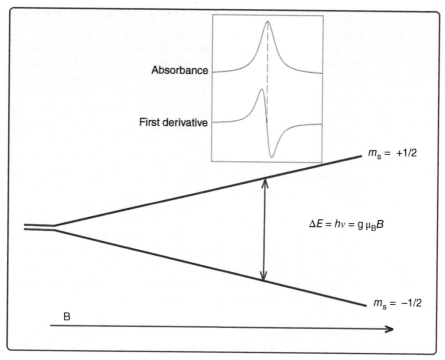

Figure 7.16
Influence of the magnetic field on the energy of the two electron spin states. Energy is absorbed when the frequency of the radiation matches the resonance condition. The inset shows an absorption spectrum and the more commonly measured first harmonic (or first derivative) that is recorded using a CW-EPR spectrometer. Here the single line arises from the *electron* Zeeman interaction. A very similar diagram applies for NMR spectroscopy involving the *nuclear* Zeeman effect, and then the chemical shift δ instead of g would be used to characterise the resonance position.

field position for the EPR signal is directly proportional to the microwave frequency, as the energy difference between the two spin states increases linearly with increasing magnetic field strength (Figure 7.16).

When the resonance condition is reached, microwave radiation is absorbed, and a signal is recorded. However, as continuous wave (CW) EPR spectrometers employ field modulation and a phase-sensitive detector that improves the signal-to-noise ratio, the CW-EPR spectrum is recorded as the first harmonic of the absorption spectrum (essentially the first derivative of the absorption spectrum), and in this case the absorption maximum corresponds to the point where the spectrum passes through zero (Figure 7.16). The first derivative accentuates features of interest in EPR spectra.

Electronic and structural information is obtained from the EPR spectrum by determining interactions with the electron spin. The proportionality factor (or g-value) describing the electron Zeeman interaction (the intrinsic spin and orbital angular momentum of the spin) is 2.00232 for a free electron, ranges only from 1.99 to 2.01 for organic radicals, but for transition metal compounds varies from 1.4 to 3.0 as a result of spin-orbit coupling and zero-field splitting effects (the latter existing for an $S > \frac{1}{2}$ spin systems such as a high-spin Fe(III) compounds with a $S = \frac{5}{2}$ spin). We shall not address the theory behind these interactions at this basic level. There are analogies between the *g*-value in EPR and the chemical shift δ in NMR as they are both sensitive to their local environment, neighbouring atoms.

Apart from the electron Zeeman effect which scales with applied magnetic field, unpaired electrons are sensitive to magnetic nuclei in their local environments. This is encompassed in the *hyperfine interaction*, which arises between the electron spin and the nuclei of neighbouring atoms that have a magnetic moment (a similar effect, often called spin–spin coupling, applies in NMR spectroscopy, and has been illustrated in Figure 7.14). Observed hyperfine interactions in a paramagnetic compound provide important information about the number and identity of coupled nuclei in addition to the distance of nuclei from the unpaired electron. The spin system energy level equation is extended to include a term associated with the hyperfine interaction, yielding (to first order)

$$E = g\mu_B B_o m_S + a m_S m_I \tag{7.7}$$

where the new terms of relevance are *a* (the hyperfine coupling constant) and m_I (the nuclear spin quantum number of the neighbouring and interacting nucleus). Nuclei such as ^1H, ^{14}N, ^{15}N or ^{13}C possess spin, and hence spin angular momentum; the spin is again defined by the spin quantum number, *I*. Not all nuclei possess spin; for example, the common isotopes ^{12}C and ^{16}O do not ($I = 0$) and are hence NMR silent as well. A single nucleus of spin $\frac{1}{2}$ further splits each EPR energy level into two (in a strong magnetic field), and as a consequence two EPR transitions are observed, with the energy difference between these two equal to the hyperfine coupling constant (Figure 7.17). The hyperfine coupling is usually much smaller than the electron Zeeman interaction and in the mega-hertz (MHz) range. Coupling patterns obey the same rules in EPR and NMR spectra.

The two possible spin states for an electron spin ($m_s = +\frac{1}{2}$ or $m_s = -\frac{1}{2}$) differ in energy in a magnetic field; this is the *Zeeman* term introduced earlier that, alone, yields a single absorbance. In addition, as introduced earlier, smaller splittings from interaction between the electron spin and a nearby nuclei (the *hyperfine* term) may operate. For a single coupled nucleus of spin $I = \frac{1}{2}$, the nuclear spin quantum numbers are $m_I = +\frac{1}{2}$ or $m_I = -\frac{1}{2}$, and in the magnetic field this electron-nuclear coupling leads to further splitting of each of the two electron spin states into two levels, so overall there are now four energy levels (Figure 7.17). EPR transitions ($\Delta m_s = \pm1$) are allowed (by more *selection rules*) only between like nuclear spin levels, here from

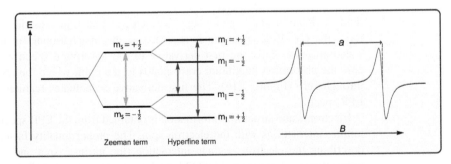

Figure 7.17

Hyperfine splitting patterns involving a single nucleus with the spin quantum number $I = \frac{1}{2}$ (with the resulting spectrum and the hyperfine coupling constant, *a*, also identified in this case). The electron spin-magnetic field interaction that produces the *Zeeman term* appears at left, with the additional *hyperfine term* arising from the interaction of the unpaired electron spin with nuclear spin shown at right.

$m_I = +\frac{1}{2}$ to $m_I = +\frac{1}{2}$ (or $-\frac{1}{2}$ to $-\frac{1}{2}$), i.e. $\Delta m_I = 0$ (as well as $\Delta m_s = \pm 1$) applies. As a consequence, only two absorbances are observed, as shown in Figure 7.17.

As it is common to see coupling to nuclei with spin $> \frac{1}{2}$, it is necessary to define the number of lines (absorbances) which result in such circumstances. This formula is simply:

$$\text{number of lines} = 2I + 1 \tag{7.8}$$

where I is the nuclear spin. From the above formula, coupling of the single electron in $^{51}V^{4+}$ to its own nucleus ($I = \frac{7}{2}$, 99.8% abundance) will result, from the above formula, in an EPR spectrum of eight lines, theoretically of equal intensity (Figure 7.18a), whereas for a single manganese nucleus (^{55}Mn, $I = \frac{5}{2}$, ~100% abundance) six lines occur (Figure 7.18b).

All discussion up to here relates to samples that are *isotropic*, i.e. have no orientation dependence, conditions usually met in solution due to rapid tumbling, or else the result of molecular shapes that are inherently isotropic (such as may occur in complexes with octahedral and tetrahedral symmetry). However, you should be aware that, more typically, EPR interactions of transition metal complexes are *anisotropic* when measured as frozen solutions or solids where the molecules cannot move; i.e. the spectrum is dependent on the orientation of the molecule relative to the spectrometer magnetic field direction. Each interaction with the electron spin is now described by three principal values, e.g. the electron Zeeman interaction has three principal g values along the three (x, y and z) axis directions of this interaction (which are obviously orthogonal to each other). For a paramagnetic transition metal complex, for example, the g values and axes are determined mostly by the d-orbital contributions (i.e. x, y and z directions) and may have different values along the different directions depending on the local ligand field symmetry of the metal ion. Complexes with axial symmetry such as square planar, for example, have $g_x = g_y$ (denoted $g_\perp$) $\neq g_z$ (denoted $g_\parallel$) where the parallel and perpendicular symbols relate to the (in this case z) axis of symmetry. A rhombic spectrum arises when the coordination environment differs along each axis such that $g_x \neq g_y \neq g_z$. This is illustrated in Figure 7.19 (left) which shows the EPR spectrum of a low-spin $Fe^{III}N_4S_2$ (d^5, $S = \frac{1}{2}$) complex showing distinct g_x, g_y and g_z values arising from when the magnetic field is aligned with the x, y and z g-value axis directions, which are all different due to the low

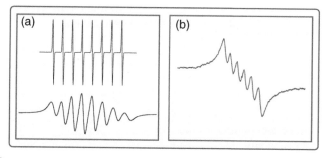

Figure 7.18

Hyperfine splitting patterns involving (a) calculated and experimental eight-line spectra for a single nucleus of ^{51}V in $[VO(OH_2)_5]^{2+}$ of spin quantum number $\frac{7}{2}$ and (b) experimental six-line spectrum for a high-spin $S = \frac{5}{2}$ Mn(II) complex of a Schiff base ligand (here only the central EPR transition is shown, for $m_s = -\frac{1}{2}$ to $m_s = +\frac{1}{2}$).

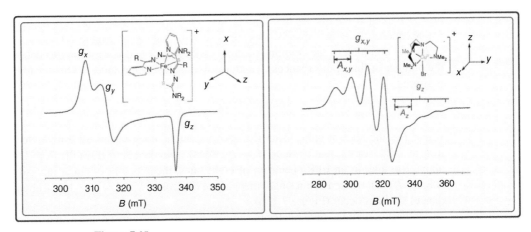

Figure 7.19

Examples of anisotropic EPR spectra. *At left*, the anisotropic spectrum of a low-spin $Fe^{III}N_4S_2$ complex, which displays three different g values when the magnetic field B_0 is oriented along the x, y and z axes of the complex. *At right*, the frozen solution anisotropic EPR spectrum (135 K) of a 5-coordinate Cu(II) complex with tetraamine and bromido ligands; copper hyperfine coupling is evident $(I = {}^3/_2)$ and parameters are indicated.

symmetry of the complex. The three different g values show that the normally degenerate non-bonding t_{2g} orbitals (d_{xy}, d_{xz}, d_{yz}), in which the unpaired electron resides in the low-spin d^5 configuration have slightly different energies. Few other spectroscopic methods can detect minor changes such as this to electronic structure.

The presence of a metal with a non-zero nuclear spin results in anisotropic spectra rich in information. A good example is Cu(II), where the nuclear spin is $I = {}^3/_2$. In accord with axial symmetry $(x = y \neq z)$, two g values are expected and hyperfine coupling splits each of these lines into four $(2I + 1)$. Note that the same consideration applies to the hyperfine coupling constants, where $a_x = a_y$ (denoted $a_\perp$) $\neq a_z$ (denoted $a_\parallel$). An example of this is shown in Figure 7.19 (right) for a trigonal bipyramidal Cu(II) complex. A wealth of information lies in the spectrum, although more in-depth interpretation lies outside the scope or needs of this book.

Overall, EPR is a powerful but still under-represented spectroscopic method in coordination chemistry, and, unlike its ubiquitous relative NMR, it is not a technique found in every chemistry research institution. However, it is of sufficient importance that it warrants this albeit limited introduction.

7.3.7   Voltammetry

In electrochemistry, reduction–oxidation (redox) reactions between two chemical species in solution involve the loss of electrons from one reactant (the reductant) and the addition of electrons to the other (the oxidant). With metals in complexes commonly able to exist in more than one oxidation state, the central metal is a prime source of, or sink for, electrons, and the study of metal-centred electrochemical processes features in coordination chemistry. Voltammetry (or polarography, nomenclature specific to the use of a mercury working electrode), where change in current as a function of change in applied potential is measured, is the dominant technique employed. There is a 'window' of applied potential in which voltammetry experiments may be conducted, influenced by the choice of electrodes, solvent and other cations or anions present, since these may undergo redox processes leading to signals that 'swamp' those from dilute solutions of a target complex. Metal-centred oxidation or reduction of a coordination complex is also influenced by the set of ligands present, leading to the potential at which this process occurs being unique for a particular complex, which assists in its identification and characterisation. Further, there are two basic outcomes resulting from an oxidation or reduction step with a metal complex: the oxidation state change may occur with full retention of the ligand environment and thus is *reversible*; or else rapid partial or complete ligand dissociation occurs and hence the process is *irreversible*. Should the latter process occur and lead to a new complex that is stable at least in the experimental timeframe, any following reduction or oxidation of the metal centre is then taking place on a complex with a different ligand environment and hence will occur almost certainly at a different potential to that of the original complex. It is possible that one or more of the ligands may also be electroactive and undergo oxidation or reduction steps not associated with the metal centre; dominantly, these ligand-centred processes are irreversible.

The redox potentials of complexes are usually probed by the experimental methods of *voltammetry*, which is a sensing rather than a complete conversion technique; the full reduction or oxidation of a solution of a complex is achieved by electrolysis *(coulometry)*, using large surface area electrodes and stirring to enhance mass transport and thus speed up the process. Of available electrochemical techniques, *cyclic voltammetry* is a simple technique commonly employed to examine the redox behaviour of metal complexes in solution. For a solid electrode and a solution of a complex met in this technique, it is only near the surface of the electrode that a redox process can occur, so both the rate of transport to an electrode surface, as well as the rate at which an electron is transferred between solid surface and dissolved complex ion, are of relevance.

For cyclic voltammetry, the voltage is changed in one direction then immediately reversed to return to the starting potential, with current monitored throughout. This may be set to repeat the process multiple times. The rate of the scanning process may be varied, but usually occurs in the seconds to minutes timeframe. The resulting current–voltage curve (a voltammogram) may show current peaks associated with reduction or oxidation processes in each direction, but reduction and oxidations waves will appear on opposite sides of the zero-current line. The technique provides rapid results, yielding voltammograms that report the presence of electroactive species. A simple experiment can provide information on the number of oxidation or reduction steps accessible in the solution environment chosen for the study, and the potential at which each occurs. It also defines, in the experimental timescale, the reversibility of the various processes, which provides information about the kinetic stability of the product of the redox reaction.

Commercial equipment for voltammetry consists of a potentiostat for voltage and scan rate control, a solution cell that holds the electrodes as well as a facility for inert gas (N_2) flushing of the solution under study to remove electroactive oxygen, and a computer for recording the output. Modern instruments employ a three-electrode system (Figure 7.20). The *working electrode* (**W**) where current is measured and potential controlled, is typically formed of carbon, gold or platinum discs or, rarely now, a hanging mercury drop. A *reference electrode* (**R**, often Ag/AgCl) provides a known and constant electrochemical potential (effectively an internal standard). This is important as it is the potential *difference* (Volts) between the working electrode and

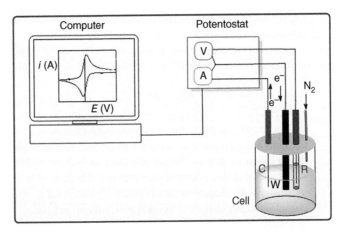

Figure 7.20

A standard three-electrode cell for cyclic voltammetry, with potentiostat to manage the current–voltage scans and a computer for overall control and data analysis. Potential is varied via the potentiostat, with switching at a selected voltage to return to the initial potential; a current–voltage curve with both forward and reverse waves (such as the example shown) results.

reference electrodes, controlled by the potentiostat, that defines the horizontal axis of the voltammogram. A *counter (auxiliary) electrode* (**C**, typically a platinum wire) completes the circuit and the current measured by the potentiostat (Amps) is equal and opposite to that at the working electrode.

Experiments may be conducted in aqueous or non-aqueous solution, with the outcome on the complex possibly influenced by solvent choice, particularly where a solvent (often potentially a ligand itself) promotes ligand dissociation reactions concomitant with reduction or oxidation steps. The experiment provides readily interpreted information on the presence of reversible or irreversible processes, as in the former situation a pair of symmetrical waves of opposite 'sign' and equivalent peak height is observed, whereas in the latter case only a single wave is seen (Figure 7.21). Another wave may sometimes be observed on the reverse scan for an irreversible process, but it is usually displaced significantly in potential, evidence for dissociative processes leading to a different complex.

Apart from an expected linear dependence of peak size with concentration of electroactive species, and variation with electrode type and surface area, voltammograms show a dependence on scan rate. For a reversible reaction, ideally the ratio of cathodic (i_{pc}) and anodic (i_{pa}) peak currents is 1 (Figure 7.21 (left)), and the peak current displays a linear dependence on the square root of the scan rate provided fast electrode kinetics operate and diffusion to the electrode surface is sufficient throughout. Separation of forward and back wave maxima, 57 mV for a perfect one-electron reversible process, may in many cases increase with increasing scan rate, due to so-called *quasi-reversible* behaviour related to a slow electron-transfer rate at the electrode surface; the behaviour can be used to determine heterogeneous rate constants for such processes. Moreover, reactions that are irreversible at slow scan rates may display reversibility at very high scan rates, as the 'dwell time' for a reduced (or oxidised) complex before it is re-oxidised (or re-reduced) becomes sufficiently short to prevent dissociative reactions being dominant.

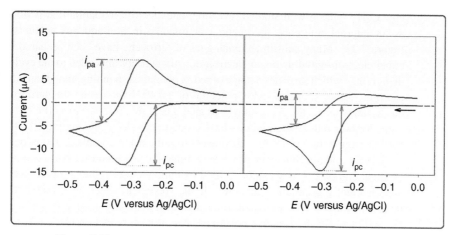

Figure 7.21

Cyclic voltammograms for (at left) a reversible single electron reduction ($i_{pa} = -i_{pc}$), and (at right) an irreversible single electron reduction ($i_{pa} \ll -i_{pc}$). The scan rate is $0.1 \, V \, s^{-1}$ in both cases. The horizontal arrows show the initial direction of the scan.

The rate at which the potential is swept is the most important variable in cyclic voltammetry as it determines the timescale of the experiment. The asymmetric irreversible curve shown in Figure 7.21 (right) can be made symmetric by increasing the scan rate as shown in Figure 7.22. At a scan rate of $0.5 \, V \, s^{-1}$, the experiment runs for 2 s in total. After reduction, the complex is unstable and dissociates to a product that cannot be reoxidised (rate constant $10 \, s^{-1}$, or half-life 69 ms). However, as the scan rate is increased, the ratio i_{pa}/i_{pc} approaches unity. For the experiment at $10 \, V \, s^{-1}$, the total time is now 100 ms and, more importantly, the time that this unstable reduced form is dominant (in the range -0.3 to $-0.5 \, V$) is only about 40 ms so most of the reduced complex is reoxidised before it can dissociate.

Since electron-donating or electron-withdrawing groups on a molecule influence the potential at which redox processes occur, cyclic voltammetry provides a facile means for illustrating the effect. The organometallic complex ferrocene ($[Fe(cp)_2]$,

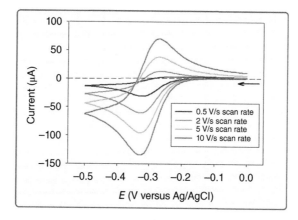

Figure 7.22

Cyclic voltammograms at different scan rates for a single electron reduction where the reduced product dissociates in a first-order process ($k_{diss} \, 10 \, s^{-1}$). The arrow shows the initial direction of the scan.

cp $= C_5H_5^-$) has a well-studied reversible one-electron oxidation that is often used as an internal reference for cyclic voltammetry in non-aqueous solvents (as in Figure 7.23). Many substituted analogues of ferrocene have been prepared bearing either electron-withdrawing or electron-donating groups appended to the cyclopentadienyl rings which provides a range of redox potentials spanning more than 0.5 V. For example, ferrocene has a redox potential of +0.40 V versus the standard hydrogen electrode, whereas the decamethylated analogue, often referred to as [Fe(cp*)$_2$], has a much lower redox potential of −0.01 V due to the electron-donating effects of the methyl groups. It is usual for substituents to influence a complex's redox behaviour in this way, and thus it may be possible to 'tune' the potential for a particular purpose.

Metal complexes may exhibit several oxidation and/or reduction processes in the experimentally accessible 'window'. If the solvent is water, then redox reactions involving water set the potential limits and these typically involve its ionised forms H^+ and OH^-. In the negative potential range, proton reduction ($2H^+ + 2e^- \rightarrow H_2$) eventually becomes the dominant reaction given the high concentration of solvent (~55 M for water). In the positive direction, water (hydroxide) oxidation becomes the dominant process at the electrode ($2OH^- \rightarrow H_2O_2 + 2e^-$). The potentials at which proton reduction and water oxidation occur are very dependent on the electrode (C, Pt, Au, Hg), but we will not explore this here.

The restrictions placed on the potential domain in cyclic voltammetry in water can be removed by carrying out the experiment in non-aqueous (organic) solvents such as MeCN, CH_2Cl_2 and dimethylformamide (DMF) which are much more resistant to redox reactions. An example of this enhanced potential range is shown in Figure 7.23 in the cyclic voltammogram of a copper complex where it can be observed in three different oxidation states (Cu(I), Cu(II) and Cu(III)).

Whereas the redox processes followed in voltammetry are limited to the region very close to the electrode surface (<1 mm), exhaustive (bulk) electrolysis of the whole solution can proceed in the experiment called *controlled potential coulometry* where the total charge passed at a fixed working electrode potential is measured over time. This requires a much larger working electrode (with much larger currents),

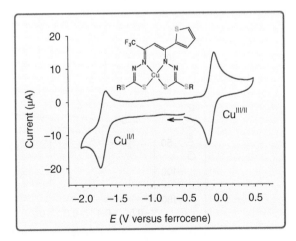

Figure 7.23

Cyclic voltammogram of a Cu complex with sequential Cu$^{III/II}$ and Cu$^{II/I}$ one-electron reversible couples. In this case, the solvent was MeCN, and the potentials are cited versus the ferrocene/ferrocenium couple.

separation of the working electrode from the counter electrode and stirring to accelerate mass transport. The amount of a species transformed during electrolysis is directly measured from the amount of charge (in Coulombs) passed over a given time, which is often hours as opposed to seconds for a typical cyclic voltammetry experiment. This method can be scaled up from the milligram to gram and even kilogram scale, enabling preparative scale redox transformations of compounds (*electrosynthesis*) without the need for chemical oxidants and reductants.

There is an array of other electrochemical methods available, but we have explored perhaps the simplest and most widely used techniques. As it happens, *in situ* electrochemical oxidation or reduction in the sample compartments of a variety of spectroscopic instruments provides access to information on electrochemically generated complexes that may otherwise not be accessible due to their instability or reactivity. This tandem technique is known as *spectroelectrochemistry* and can incorporate various spectroscopic methods already mentioned including UV–vis absorption, IR and EPR. Special cell designs are required to enable the spectroscopic measurement to be made under potential control while also ensuring that the cell solution being electrolysed does not diffuse away from the region being examined. A popular design for UV–vis spectroelectrochemistry is the so-called optically transparent thin layer electrochemical (OTTLE) cell.

7.3.8 X-Ray Crystallography

In solid-state structure determination, the ultimate characterisation method is X-ray crystallography, as it provides the three-dimensional structure of any compound or element at atomic precision. The technique has one requirement; the sample must be a crystalline solid, which essentially means it must be ordered at the atomic scale into what is called a lattice comprising repeating 'boxes', called *unit cells*. In principle, *any* element or compound can be crystallised; even things we normally associate with the gaseous or liquid state will crystallise if cooled to sufficiently low temperatures. There are six different unit cell shapes (called *crystal systems*), all having eight vertices (corners) and six faces that pack together in three dimensions to form a lattice without leaving any gaps. That there are but six shapes is a geometric (mathematical) condition, and they are shown in Figure 7.24.

To the untrained eye, they may look similar, indeed they are all six-sided objects with eight vertices. They are distinguished by six lattice parameters, which are the lengths of the three unique cell edges (a, b, c) and the angles between them (α, β, γ). The only difference is their symmetry (number of equidistant sides and internal angles), which becomes increasingly higher going from triclinic to cubic.

In 1913, only 18 years after Wilhelm Röntgen had discovered X-rays, William Henry Bragg and William Lawrence Bragg, father and son, used X-ray diffraction to determine three-dimensional crystal structures of KCl, NaCl (both cubic) and ZnS (hexagonal), including the first measurement of interatomic distances (i.e. bond lengths). The Braggs duly won the 1915 Nobel Prize in Physics for their efforts. The rest is history, and now literally millions of crystal structures have been determined of organic and metal-organic (coordination) compounds as well as all naturally occurring elements and several unnatural (radioactive) ones. Regardless of their origin or chemical form, they all must conform to one of the six shapes in Figure 7.24 when they assemble into a lattice in the solid state.

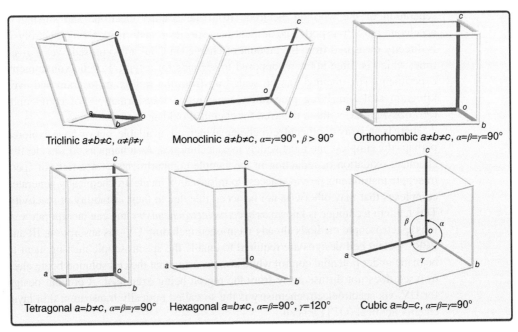

Figure 7.24
The six different shapes of the three-dimensional unit cells (crystal systems).

Beyond defining the unit cell, another primary characteristic of a structure determination is the *space group* which defines the internal symmetry of the unit cell. In other words, atoms and molecules may be related by symmetry operations within each cell which can involve the point symmetry elements in Appendix 2, and also translations in a certain direction, or combinations of both. It was independently proven by mathematicians E.S. Federov and A. Schoenflies in 1891 (before X-ray diffraction or even X-rays had been discovered) that there are (only) 230 arrangements of these symmetry operations that are unique. This sounds daunting, but the reality is that most of these theoretical space group possibilities are rarely ever seen. In fact, 80% of all crystal structures in the Cambridge Structural Database (comprising organic and metal-organic coordination compounds) fall into just five space groups! One example is shown in Figure 7.25, being in the most popular space group of all and named $P2_1/c$ (34% of structures in the CSD), comprising an octahedral Cr(III) complex with three acetylacetonato (acac) ligands. The internal symmetry elements highlighted by the orange dots are centres of inversion. The point in the centre of the cell relates the complex at top left with that at bottom right and also pairs the molecules at bottom left and top right. Although all Cr complexes in this structure have the same Cr–O bond lengths and O–Cr–O bond angles, the centre of inversion symmetry element means that half of the complexes are the Λ-enantiomer while the other half are the Δ-enantiomer. This space group has other symmetry elements which are not shown here. The presence of a centre of inversion is very common, though, and is found in 78% of all known crystal structures.

The choice of X-rays for single crystal diffraction, as opposed to other types of electromagnetic radiation (UV, vis, IR, microwave, which we met earlier in this chapter), is deliberate. The wavelengths of X-rays (around 0.1 nm, or one Angstrom, Å) are close to the separation of atoms in a crystalline lattice, leading to diffraction

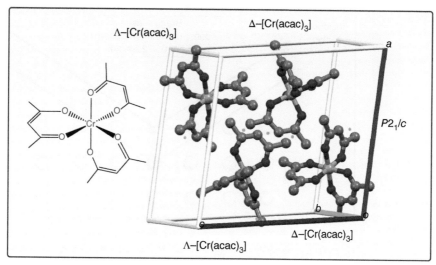

The arrangement of symmetry-related [Cr(acac)$_3$] molecules within the common monoclinic space group $P2_1/c$ (H atoms omitted for clarity). The centres of inversion are shown as orange dots.

(reflection) of part of a beam as they interact with the electron clouds of atoms in the crystal; much longer or shorter wavelengths will not do this. The assembled unit cells that make up the lattice form planes from which the X-rays are reflected. An incident source of X-rays can be considered as a stream of parallel waves of the same wavelength (monochromatic) and all in phase, which will produce an intense signal if directed onto an X-ray detector. We are more interested in the diffracted (reflected) X-rays as they leave the crystal (Figure 7.26). The first point is that the angle of incidence (θ) equals the angle of reflection (like in a game of snooker), but remember we are dealing with many parallel waves that penetrate the crystal and are reflected, at the same angle θ, by planes deeper in the crystal so they must travel further before reemerging and recombining with the other waves. Some may still be in phase (with wave height enhancement, or constructive interference) and will give a bright 'spot' on the detector or they may be out of phase and destructively interfere and no signal will be seen at this angle of reflection. Bragg's law (named after the pioneers of X-ray crystallography) is based on elementary trigonometry, which shows for a reflection to be observed from parallel planes within a crystal lattice, the angle of reflection θ is related to the interplanar spacing d by the relationship

$$2d \sin \theta = n\lambda \tag{7.9}$$

where λ is the X-ray wavelength (a constant) and n is an integer. When this diffraction condition is met, the extra path length that the lower X-ray wave in Figure 7.26 travels is $n\lambda$ ($= 2d \sin\theta$) and the recombined X-rays are still in phase. In three dimensions, the problem is more challenging, and we will not explore the mathematical details of this; there are many textbooks devoted entirely to X-ray crystallography for the interested reader.

The early years of crystallography relied on laborious photographic film images as the method for measuring X-ray diffraction. Each image would be taken at a different precisely determined orientation of the crystal and the film, and each image would require a new film, which was then developed as in standard photographic images.

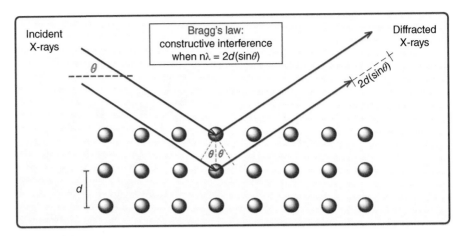

Figure 7.26

Graphical representation of Bragg's law showing the path difference between the two adjacent diffracted (reflected) X-ray waves is $2d \sin\theta$. When this equals an integral number of wavelengths ($n\lambda$), a bright spot will result from constructive interference of the recombined waves.

Just as camera technology moved beyond photographic film and paper, so did the measurement of X-ray diffraction intensities, which are now digitised. They both have the same origin; the charge coupled device (CCD) detector, invented by Willard S. Boyle and George E. Smith, who shared the 2009 Nobel Prize in Physics, has revolutionised the fields of X-ray crystallography, photography, medicine, astronomy and many other areas. The uptake of CCD technology in the 1980s greatly accelerated X-ray crystallographic data collection from days to hours. Nowadays with even faster readout times, more intense X-ray sources, and highly sensitive detectors, data collection and structure determination can now be completed in minutes compared with years of work not so long ago.

All X-ray diffractometers comprise three essential components (Figure 7.27 (left)), an X-ray source, a crystal mounted on a movable support (a *goniometer*) and an X-ray

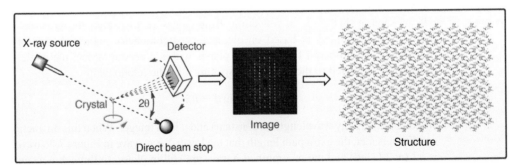

Figure 7.27

The sequence of steps involved in X-ray crystal structure determination with: (left) schematic drawing of an X-ray diffractometer; (centre) a single diffraction image showing many reflections of different intensity, and; (right) the final result, the entire three-dimensional structure of the crystal showing its regular repeating pattern.

detector that also can be moved to different positions. Large area detectors quantify the intensities of dozens of different diffracted X-rays in a single diffraction image. An immediate outcome of an X-ray diffraction experiment is that the *symmetry* of the diffraction image (Figure 7.27 (centre)) reports on the packing symmetry (in the case shown, orthorhombic) of the lattice that created it.

The details of how the diffraction images, comprising thousands of 'spots' of differing intensity measured at different orientations of the crystal and detector, becomes the overall three-dimensional crystal structure need not concern us here. This is a well understood and computer-facilitated process that effectively 'refocuses' X-rays to create a three-dimensional electron density map, which we interpret as the molecular structure of our compound. We will concentrate on the results and what they mean. When the structure determination is complete, the results of a successful analysis are the atomic coordinates (x, y, z) of every atom in the unit cell and indeed the whole crystal if all unit cells are considered. Detailed geometrical parameters, both intramolecular and intermolecular, are calculated from these atomic coordinates. Common statistical procedures are then introduced to determine standard uncertainty, so that the precision or reliability of every derived numerical parameter has an associated estimated standard deviation.

Using the same lattice shown in Figure 7.27 (right), we can zoom in to see more details of the structure in Figure 7.28. The repeating pattern of the lattice is still apparent with the unit cell highlighted. This unit cell contains four copies of the compound of interest, a Co(III) complex with two chloride and one perchlorate counter-ions which are related by rotational and translational symmetry. Most crystal structure

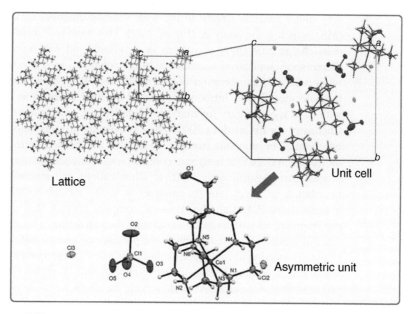

Figure 7.28

Different perspectives on the results of a single crystal X-ray diffraction experiment. The asymmetric unit (bottom) is the smallest unique component of the structure, the unit cell shows how four of these components are related by symmetry and the lattice shows how these unit cells are assembled.

solutions reported in the scientific literature include images of just the asymmetric unit, but it should never be forgotten that the entire structure, including the unit cell and the lattice, is what was actually determined.

Several factors govern the precision and quality of the crystal structure. The major contributory factor is simply the quality of the single crystal sample, as crystallinity directly affects the quality and amount of diffraction data collected. Other important factors of concern include the precision and accuracy of the measured diffraction pattern intensities (random errors dominate for the former and systematic errors for the latter), the number of measured reflections (as fewer data leads to reduced precision), and the appropriateness of the structural model used in the refinement including the space group. For larger molecules such as proteins, X-ray crystallography may also be applied, although with less precision than in 'small molecule' crystallography. It is not practicable to display all connectivity, and rather structural drawings focus on a site of interest (such as a metal centre) with the gross structure of the macromolecule represented by ribbons which show the secondary structure (shape) mapped out by chains of amino acids that may include so-called helices and sheets. Examples of metalloprotein crystal structures are postponed until Chapter 8, where they will be shown in the context of biological chemistry.

7.3.9   A Tandem Technique: Electrospray Ionisation Mass Spectrometry

Electrospray ionisation mass spectrometry (ESI–MS) is a tandem technique; it employs an electrospray ionisation device (ESI) to produce 'bare' ions in the gas phase from a supplied (usually aqueous) solution and supplies these ions as a very dilute gaseous stream to the high vacuum chamber of a mass spectrometer (MS), which is the analyser (Figure 7.29). This analyser 'sorts' and detects ions by mass/charge ratio (m/z), as cations or as anions, and will report either on request. Electrospray ionisation is termed a 'soft' technique, as it produces gas phase ions often without fragmentation of potentially thermally labile compounds, including biomolecules. It was introduced in 1989 by John B. Fenn as a tandem technique with mass spectrometry for studying biomolecules, and he shared the Nobel Prize in Chemistry for his work in 2002. The electrospray process employs dilute solutions which may be aqueous, making it suitable for both proteins and other macromolecules and also for inorganic complex ions. For coordination complexes, the ESI process can be sufficiently 'soft' to allow cations and anions of complex species in solution to travel into the analyser in their original complex form, providing a method for detecting the mass of complex species that exist in the solution environment, which can assist in defining their structure. This capacity to define the mass of complex ions, and identify the presence of several different complexes in an originally usually dilute aqueous solution, is a key application of ESI–MS. Moreover, where ESI conditions lead to some decomposition of complex species, the pattern of products observed can assist in identifying the character of the parent species and the way it undergoes change. Where a metal exists as a mixture of several isotopes, a set of peaks for a particular complex species related in pattern to the different isotopic metal centres will result, and this characteristic pattern helps confirm the presence of the metal in a complex species, and even of several metals in an oligomer.

The electrospray process expels pre-existing ions from the solution to the gas phase as droplets. These droplets undergo drying and 'shrinking', become unstable

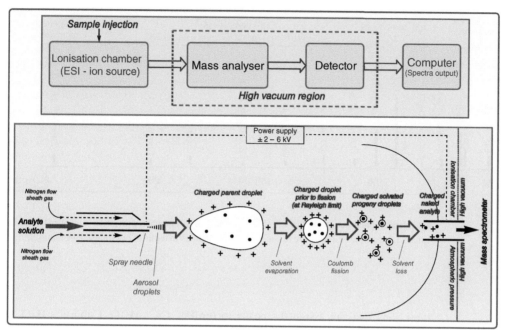

Figure 7.29
Schematic views of an ESI–MS (top) and the process of droplet formation in the ESI unit.

and consequently fragmenting into smaller droplets. This process continues until, effectively, individual ions (both unsolvated and solvated) result in the gas phase. Ions exit the spray chamber through a capillary, attracted by it being held at a high potential, leaving a one-atmosphere environment and entering a sequence of significantly pressure-reduced regions, with choice of conditions governing whether the ions undergo collision-activated dissociation or not. A series of chambers of increasingly low pressure are created by pumps and separating skimmers, with the final quadrupole chamber of the mass spectrometer (where detection occurs) held at 10^{-9} bar. Complexes can undergo oxidation or reduction reactions in the process. The ESI–MS process seems to favour low-charged ions, at least for low molecular weight species. Of course, by its nature, ESI–MS has one obvious shortcoming – it detects ions; neutral molecules are not observed. Fortunately, protonation–deprotonation processes and even weak complexation to simple ions allow some neutral species to be seen.

To illustrate the ESI–MS technique, we will look at a simple example, involving complexes formed in solution between Cu_{aq}^{2+} and ethane-1,2-diamine (en). We know from potentiometric and spectrophotometric titrations that two complex species form dominantly in solution during reaction, the 1:1 $[Cu(en)]^{2+}_{aq}$ and 1:2 $[Cu(en)_2]^{2+}_{aq}$ complex ions, and even know the stability constants for these species. In fact, careful and slow addition of en to Cu_{aq}^{2+} in a beaker allows one to see the colour changes as we step from dominantly Cu_{aq}^{2+} to successively the 1:1 and then 1:2 Cu:L species. Thus, this is a well understood system and suited to illustration and evaluation of the ESI–MS method. When a dilute solution of a 1:2 mixture of copper ion and en ligand is passed into the ESI–MS tandem instrument, a set of cations of different masses are detected (Figure 7.30). From the masses of

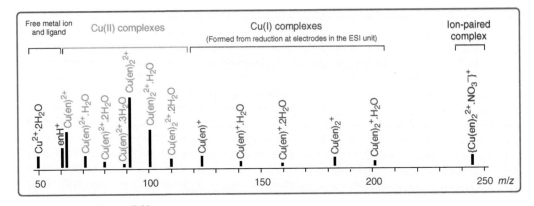

Figure 7.30

Schematic representation of the ESI–MS spectrum of compounds present in a copper(II) ion and ethane-1,2-diamine (1,2) mixture in water. Peaks are defined in terms of *m/z* (mass/charge ratio).

these ions, we can assign the peaks to (en)H$^+$, Cu^{2+}·xH$_2$O, Cu(en)$^{2+}$·xH$_2$O and Cu(en)$_2$$^{2+}$·$xH_2$O (with peaks for various x values). In addition, as a nitrate salt of copper(II) was employed, an ion-paired species {[Cu(en)$_2$]$^{2+}$·(NO$_3$$^-$)}$^+$ is observed. Further, reduced complexes [Cu(en)]$^+$·xH$_2$O and [Cu(en)$_2$]$^+$·xH$_2$O, formed by electrochemical reduction that can occur at electrodes present in the ESI unit, occur as complicating peaks. What this spectrum tells us is that the solution contains various amounts of free ligand (detected as protonated (en)H$^+$), free copper ion (detected as Cu^{2+}·xH$_2$O), 1:1 complex (detected as [Cu(en)]$^{2+}$·xH$_2$O) and 1:2 complex (detected as [Cu(en)$_2$]$^{2+}$·xH$_2$O), and an ion-paired species ({[Cu(en)$_2$]$^{2+}$·(NO$_3$$^-$)}$^+$). This is consistent with our interpretation of the solution chemistry from simple observations and from titration experiments. Hence the ESI–MS has provided additional firm evidence for these species occurring in solution. Moreover, it does not detect any [Cu(en)$_3$]$^{2+}$, consistent with the extremely low stability of this due to Jahn–Teller distortion, nor does it find any dinuclear bridged species, even for different *M:L* ratios that could support such speciation.

The shortcomings of the ESI–MS technique are first, that it can only detect ions and thus any neutral species remain invisible, and secondly, the observed peak height is not a direct measure of the relative amount of a species, so that it is not strictly valid for defining relative concentrations of species present. However, it does provide a very simple method for probing solution species, based on molecular mass/charge. There is one additional caveat; because the method 'strips' solvent molecules from droplets to form ions, it can produce and hence detect 'bare' species such as [Cu(en)]$^{2+}$ which obviously would prefer to exist as [Cu(en)(OH$_2$)$_2$]$^{2+}$ in solution. Solvent 'stripping' may be incomplete, so a sequence of solvated species, with peaks separated by 18/z (the mass of water divided by the overall charge on the ion), may be seen in some cases. Lastly, there is one additional bonus; where the metal has more than one significant isotope (as occurs with ^{63}Cu/^{65}Cu), the presence of the metal will, in a high-resolution instrument, lead to a characteristic pattern associated with that isotopic ratio whenever the metal is present, allowing easy assignment of metal-containing and metal-free species. Overall, it is a useful addition to the armoury of coordination chemists.

7.3.10 Computational Methods

One important change in recent decades has been the growth in the use of computer simulation to assist in solving chemical problems, now widely used methodology in most subdisciplines of chemistry. Computational methods, employing sophisticated computer programs, find application for modelling and predicting bonding, structure and properties. There are several methods employed: *ab initio* (applying approximations to solve Schrödinger's equation directly without the need for empirical [experimental] data, mainly applicable to smaller molecules due to high computer capacity demands), *semi-empirical* (a variation of the *ab initio* approach for larger molecules that includes various approximations in the methodology while drawing on empirical data; this combination of theory and empirical data leads to a substantial reduction in computational time); and *density functional theory* (a widely used quantum mechanical simulation and modelling approach for computing a range of properties, with several well-tested software packages available).

Computer modelling of transition metal centres presents several challenges; for example, the electronic structure arising from the 'open-shell' d^n configurations with unpaired electrons is complicated. Although researchers commonly employ the sophisticated density functional theory (DFT), the methodology requires large numbers of calculations that demand simplifications and assumptions to reduce the number of 'essential' atoms in a model and make the calculation feasible. This demand on computer time was particularly concerning in earlier times and allowed a window of opportunity for perhaps the simplest modelling approach used in coordination chemistry, *molecular mechanics* (or conformational analysis), an extension of the methodology originally developed for the conformational analysis of organic compounds. Although DFT has effectively taken over now, it presents a level of complexity inappropriate for this book. However, classical molecular mechanics (MM) is more approachable and will suffice here to give a flavour of the expanding computational field. MM still has some merit, and modest computational demands.

Molecular mechanics treats atoms and their electron sets as hard spheres that repel each other unless bonded, and treats bonds as springs. This approach, based on Hooke's law (dating back to the 17th century), has proven both resilient and surprisingly useful. Describing compounds classically in terms of strain as a result of effects such as bond stretching or compression and bond angle deformation, strain minimisation is pursued with an appropriate software package to predict the 'best' structure. Because molecules have preferred bond distances and angles, all deformations away from this optimum costs energy, and it is the sum of all these distortions and other unfavourable non-bonded interactions that add up to an overall energy associated with stability. Comparison of calculated energies for different isomers of the same molecule, for example, allows the prediction of the most stable isomer. This is a valid application, as is the calculation of energy for different conformations of a molecule. However, comparisons of calculated energy between different molecules are inappropriate, since they will contain different numbers of atoms and bonds and thus will differ inherently, so that their relative energies offer little real meaning.

The total strain energy of a system can be considered to arise from the sum of bond length distortion (stretching or compressing a bond away from its ideal; E_{str}), bond

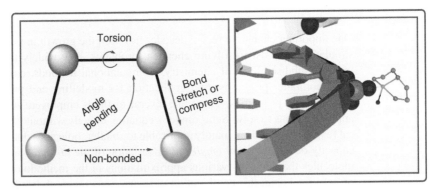

Figure 7.31
Representation of effects operating in molecular mechanics (at left). An example of binding of a copper complex to a phosphate group on the backbone of a DNA strand, optimised using molecular mechanics methodology, appears at right.

angle distortion (bending bonds to open out or close up an ideal angle; E_{bend}), torsional effects (twisting of groups around a particular bond relative to each other; E_{tor}) and non-bonded contributions (van der Waals attraction, steric repulsion, and electrostatic attraction or repulsion; E_{nb}). Each of the contributions may be represented by simple classical equations; the origins of these effects are represented simplistically in Figure 7.31. Software packages for molecular mechanics work by varying atom locations by small amounts, performing a calculation, and comparing it with the previous calculation. A stepwise process leads to a minimum. An array of sophisticated approaches is employed in software packages, but, in the end, this is still really a classical ball-and-stick modelling method, with inherent limitations. Despite this, the approach is efficient and the outcomes, with 'training' of the system through the use of appropriate force field parameters, are surprisingly good if using packages developed with suitable parameters for metal ions as well as non-metallic elements.

Overall, computational chemistry has wide practical potential. It can already be employed to calculate the vibrational spectra and normal vibrational modes for simple molecules, and assists in understanding adsorption processes, reaction mechanisms, and metal-catalysed asymmetric synthesis, among several applications. Yet there are clear challenges. The elephant in the room is calculation complexity and hence high computational cost (in both central processor time and memory), currently addressed via the use of composite semi-empirical methods. Predicting thermochemical properties to an acceptable level of certainty is also a challenge. Greater computer power is usually proposed as the saviour of the field, yet success in dealing with molecules as diverse as those met in coordination chemistry is not yet universal.

Concept Keys

Coordination chemistry relies heavily on the capacity to separate and/or isolate individual complexes from reaction mixtures, and subsequently to employ an array of physical methods to determine the nature and structure of such complexes.

Instrument-based methods commonly employed in coordination chemistry are ultraviolet–visible–near-infrared (UV–vis–near-IR) spectroscopy, infrared (IR) spectroscopy, nuclear magnetic resonance (NMR) spectroscopy (for diamagnetic complexes) and electron paramagnetic resonance (EPR) spectroscopy (for paramagnetic complexes), although a wide range of other methods exist and find ready application. A set of methods is often used in concert to define structures in both solution and the solid state.

The colour and spectra of *d*- and *f*-block complexes are intimately tied to the number and arrangement of electrons in their valence shells and to the complex stereochemistry. Further, the splitting of the *d* subshell in transition metal complexes is dependent on the type of ligands coordinated, with variation sufficient to lead, for some configurations, to high-spin and low-spin options that produce different spectroscopic and magnetic properties.

Single crystal X-ray crystallography is a key physical method, as it can provide a three-dimensional picture of crystalline complexes in the solid state at atomic precision.

Mass spectrometry (MS), commonly used for studying organic molecules, has been successfully adapted for studying aqueous solutions of metal complexes through the attachment of an electrospray ionisation (ESI) unit to produce the tandem ESI–MS technique that allows assignment of ionic species via molecular mass. However, neutral species are MS-silent and thus undetected.

Since many metal ions can exist in complexes in several oxidation states, the use of electrochemical techniques, particularly cyclic voltammetry, to probe this behaviour is often employed.

Solid-state and solution structures need not be identical, and an array of less definitive physical methods must be employed to probe complex structure and chemistry in solution.

Apart from physical methods, computational methods continue to grow in sophistication and application in coordination chemistry; molecular mechanics is the simplest technique and can model the geometry and relative stability of complexes fairly well, but is giving way to more sophisticated density function theory (DFT).

Further Reading

Cole, R.B. (Ed.) (1997). *Electrospray Ionization Mass Spectrometry: Fundamentals, Instrumentation and Applications*. Wiley-Interscience.

A lengthy but comprehensive book with sections on fundamentals and instrumentation; useful to students, despite its age.

Ebsworth, E.A.V., Rankin, D.W.H., and Cradock, S. (1987). *Structural Methods in Inorganic Chemistry*. Oxford: Blackwell.

An older, but still valid, account that may help reveal the mysteries of crystallography and three-dimensional structure determination of complexes.

Henderson, W. and McIndoe, J.S. (2005). *Mass Spectrometry of Inorganic and Organometallic Compounds: Tools, Techniques, Tips*, Wiley.

Apart from providing students with a description of how a mass spectrometer works, this textbook provides a fine coverage of various techniques, expected outcomes and their application; coverage of both coordination and organometallic compounds appears.

Hill, H.A.O. and Day, P. (Eds.) (1968). *Physical Methods in Advanced Inorganic Chemistry*. London: Interscience.

An old but readable coverage of an array of physical methods; the fundamentals have not changed, so it can be helpful for students.

Kettle, S.F.A. (1998) *Physical Inorganic Chemistry: A Coordination Chemistry Approach*. Oxford: Oxford University Press.

A venerable but readable more advanced student textbook that covers some physical methods and their underlying theory in good depth and clarity.

Moore, E. (1994) *Spectroscopic Methods in Inorganic Chemistry: Concepts and Case Studies*, Open University Worldwide.

A very brief introduction that students should find useful, set at an appropriate level.

Nakamoto, K. (2009). *Infrared and Raman Spectra of Inorganic and Coordination Compounds. Parts A and B*, 6th ed. New York: Wiley.

Fully revised editions of the classic work on infrared spectroscopy of coordination and organometallic compounds, now with bioinorganic systems. Advanced in nature, but a student may find aspects valuable and informative.

Parish, R.V. (1990). *NMR, NQR, EPR and Mossbauer Spectroscopy in Inorganic Chemistry*. Hemel Hempstead: Ellis Horwood.

An ageing but still valid advanced coverage of some important resonance methods used in coordination chemistry; the title says it all.

Scott, R.A. and Lukehart, C.M. (Eds.) (2013). *Applications of Physical Methods to Inorganic and Bioinorganic Chemistry*, Wiley.

An introductory coverage of a wide range of techniques used in coordination chemistry; provides guidance on the selection of appropriate physical methods for tasks.

Sienko, M.J. and Plane, R.A. (1963). *Physical Inorganic Chemistry*, 2nd ed., W.A. Benjamin.

Another old but readable text covering core physical methods; again, as the fundamentals remain with us, students may find some value herein.

8 A Complex Life

8.1 Metal Ions in the Natural World

It is fair to say that there is a historical bias towards the importance of organic chemistry in living systems. From the perspective of elemental abundance, the elements carbon, hydrogen, nitrogen, oxygen, phosphorus and sulfur, the key ingredients of organic chemistry, are undoubtedly dominant (making up >95% of our body mass), and all of these are essential for any life form. They are often referred to as the 'big six' (with the collective acronym CHNOPS) as they make up all the naturally occurring amino acids (the constituents of all proteins) and the genetic code (DNA and RNA) that defines all forms of life.

However, the vital roles played by metals in biology have been long recognised. Research in the 19th century conclusively showed that plants and yeasts depend on the metals iron, manganese, copper and zinc for survival. From this time, extensive studies have led to an understanding of which elements are essential for life, even though they are usually present in only trace amounts. The current state of knowledge is illustrated in Figure 8.1. This covers all life forms. The first thing that is apparent is that most chemical elements are not essential at all! Some of these exogenous elements are found in drugs to treat diseases (see Chapter 9), but the focus here is on the elements that have been shown to be genuinely essential for survival. The green and yellow squares collectively highlight the elements that are essential specifically for humans (and plants), whereas the pink squares show elements that more primitive life forms also need (but we don't).

The importance of metals in biology clearly cannot be ignored, yet textbooks often teach biological chemistry from an organic chemistry perspective, neglecting the essential role of metals and coordination chemistry. The very name of the broader field in which metals reside – 'inorganic' chemistry – is pejorative and implies irrelevance to the living world. In life, of course, a separation between organic and inorganic chemistry does not exist, and we shall attempt to counter this historical bias and focus (given the nature of this textbook) on metals that are essential for life and why this is true.

Taking humans as an example, the most abundant metals in our body are from the s-block, namely the *bulk metal ions* Ca^{2+}, K^+, Na^+ and Mg^{2+}. As we have seen in previous chapters, these classically 'hard' metal ions have their own unique coordination chemistry characterised by complexes of low stability relative to their aquated ions that undergo very rapid ligand exchange. Against this background, it is not surprising that these s-block metal ions in solution participate in reactions where they rapidly exchange ligands as part of their function. The passage of Na^+ and K^+ across the membranes of nerve cells (neurons) via the sodium-potassium pump underpins nerve impulses and the ions must rapidly exchange aqua ligands with their local environment. There are so-called protein channels specific for certain ions such as Na^+, K^+ and Ca^{2+} that are also essential for the controlled passage of these ions which create a concentration gradient either side of the channel.

Introduction to Coordination Chemistry, Second Edition. Paul V. Bernhardt and Geoffrey A. Lawrance.
© 2025 John Wiley & Sons Ltd. Published 2025 by John Wiley & Sons Ltd.
Companion website: www.wiley.com/go/coordinationchemistry2e

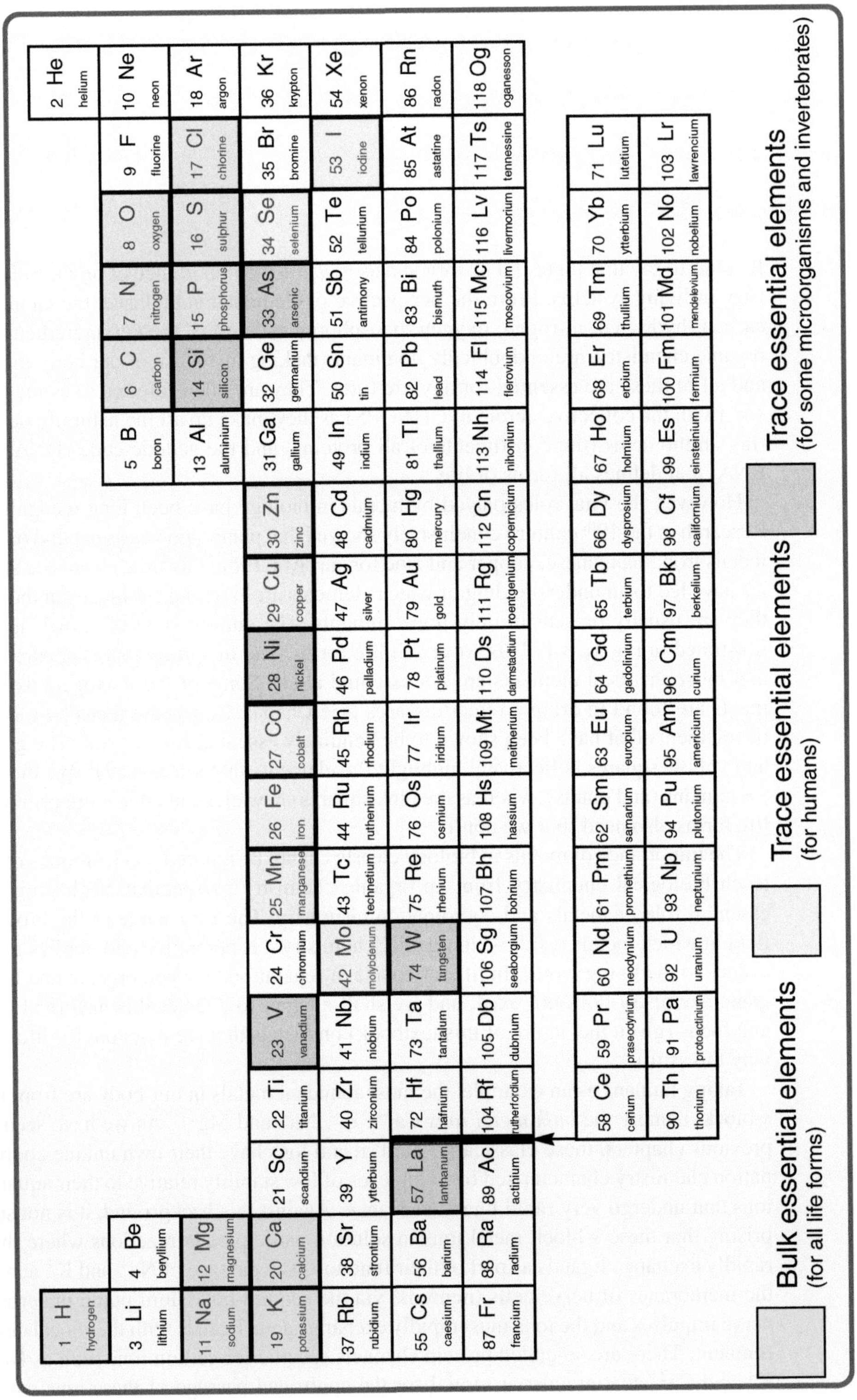

Figure 8.1

The periodic table, highlighting elements that are essential for different forms of life.

After the abundant *s*-block metals, come the *trace metals*, mostly from the *d*-block, which are present at concentrations that are thousands of times lower. However, a little bit goes a long way. Most biologically essential transition metals are found tightly bound to a protein where they carry out a specific function as part of a *metalloprotein*. It is not unusual that a solitary transition metal ion is present amongst literally thousands of C, H, N, O and S atoms within a metalloprotein, yet if that metal is removed all biological function is lost. We will explore the different ways metals associate with a protein in the following sections.

Of course, when thinking about natural systems, this also includes our environment. Apart from oxygen, the most abundant elements in the Earth's crust by mass are Si and the metals Al, Mg, Ca and Fe. These elements are the foundation of the ground we walk on and are present in various chemical forms within rocks, sands and soils. Not all of these are essential for life though; Al has no known biological role in any life form.

The three most commonly found trace metals in biological systems (including humans) are iron, zinc and copper. Other essential *d*-block elements for humans are manganese, cobalt and even the second-row transition metal molybdenum. Other *d*-block elements such as nickel, vanadium and tungsten are utilised by other forms of life. In all cases, the metal has a defined coordination sphere of ligands, and we should not forget that the same structural, bonding and electronic principles described in Chapters 2 and 3 apply equally to the coordination chemistry of metals in biological and synthetic systems. The ways in which they are complexed in biological systems are now considered.

8.1.1 Biological Ligands

As we know from earlier chapters, metal ions are never found 'naked'. In the absence of competing ligands, the coordination sphere of a transition metal ion in solution will comprise the solvent (perhaps with its counter anions as co-ligands), and in biology this will mostly be water. The coordination number and its geometry will be determined by the metal oxidation state, and this in turn determines the donor atoms that will be preferred. Transition metal ions in biology are always complexed within a protein so it is appropriate before we move on to introduce some fundamentals on how this might happen.

8.1.1.1 *Proteins as Ligands*

Proteins comprise (typically) hundreds of amino acids linked head-to-tail in a specific sequence to generate a polypeptide (polyamide) backbone. In Figure 8.2, a small fragment is shown comprising the parent amino acids alanine, glycine and serine, which undergo condensation reactions between carboxylic acid and amine groups to become linked covalently as amides. In terms of metal-binding capacity, amides are poor ligands. In Figure 8.2 (right), the amide bond is represented by two resonance structures, which is consistent with experiments that show the amide functional group is flat. In other words, the NH and CO groups all lie in a plane together, and the C–N bond exhibits significant double bond character. The N-atom is conjugated with the adjacent carbonyl group, nullifying its lone pair of electrons that would be needed for metal coordination. The carbonyl O-atom is more capable of metal coordination, but again, it is unusual for it to bind on its own in preference to water.

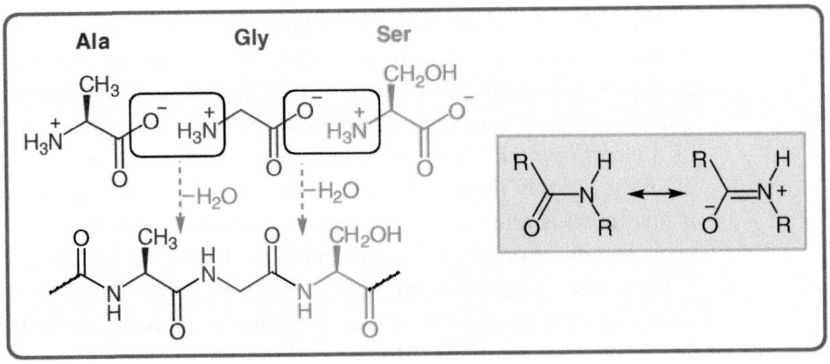

Figure 8.2

(Left) A section of a polypeptide constructed by condensation of the amino acids alanine, glycine and serine, and (Right) the two resonance forms of the amide bond illustrating the poor donor capacity of the amide NH group.

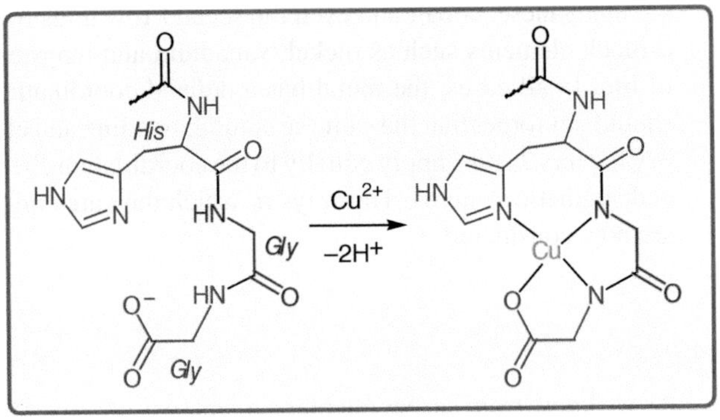

Figure 8.3

Copper(II) coordination at the C-terminus of the Prion protein, which includes binding to two deprotonated amide nitrogens. The three amino acid residues of the terminus employed are identified on the left-hand drawing.

The only way that the amide N-atom can coordinate is if the NH group of the secondary amide is deprotonated, and this is uncommon as amides are very weak acids with pK_a values greater than 13. However, metal ions can induce change, and a well-studied example is Cu^{II} coordination to the C-terminus of the Prion protein, which is associated with severe neurodegenerative disorders in humans (Creutzfeldt–Jacob disease) and other mammals (bovine spongiform encephalopathy or 'mad cow' disease). In this case (Figure 8.3), two of the amide N-atoms (from consecutive glycine residues) deprotonate and coordinate to the Cu^{II}, in addition to coordination by the terminal carboxylate group and an adjacent imidazole present as part of a histidine side chain. Here the protein is an effective tetradentate ligand as the proximity of the donor atoms enables chelation.

More commonly, the protein ligands are provided by the amino acid side chains that branch from the peptide backbone. As just demonstrated, histidine possesses such a coordinating side chain. It is one of a select grouping of natural amino acids that can coordinate to a metal ion via their side chain (Figure 8.4). These offer groups that are well known to bind to metal ions in simple coordination complexes. The functional groups in Figure 8.4 should already be familiar from Chapter 3, with

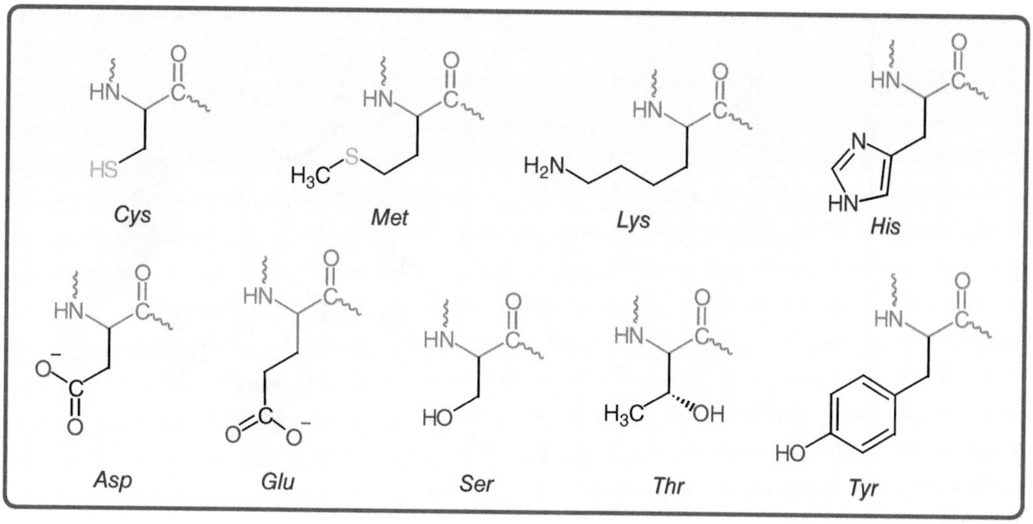

Figure 8.4

Amino acid residues cysteine, methionine, lysine, histidine, aspartate, glutamate, serine, threonine and tyrosine (within a polypeptide chain) which bear metal-coordinating side chains. The peptide backbone is shown in grey, and the coordinating group is in colour.

N-, O- and S-donor groups featuring. The side chains with −OH or −SH groups can deprotonate, which strengthens their donor capacity and also enables them to bridge more than one metal ion.

One important aspect of metalloprotein coordination chemistry that is not immediately obvious from simple chemical structure diagrams (Figure 8.2) is that the polypeptide is far from linear but instead has a preferred three-dimensional shape (or fold) which is maintained by many hydrogen bonds between its NH and C=O groups along the polypeptide backbone. This leads to a variety of familiar forms (secondary structural elements) that include α-helical patterns where the polypeptide coils and almost flat segments called β-sheets (see Figure 3.26). While we will not explore these subtleties of structural biology in any detail, the key point is that the protein is conformationally rigid, and the amino acid side chains do not have the freedom to move far from where they are within the folded protein structure. For a metal to be coordinated by two or more side chains at once, the amino acid sequence must be such that these side chains all are in proximity before the metal can coordinate.

A good example of this is the bacterial protein rubredoxin, a small metalloprotein bearing only 52 amino acids, where 4 of its 5 cysteine side chains combine to bind a single iron in a tetrahedral FeS_4 geometry (Figure 8.5, left). Unlike the example shown in Figure 8.3, the four cysteine side chains are not on consecutive amino acid residues. The pairing of Cys6 and Cys9 near the N-terminus constrains their side chains to be in proximity as they are only separated by two amino acids. Cys39 and Cys42 are also adjacent, but they are at the opposite end of the polypeptide chain, near the C-terminus. The 52 amino acids in this protein form one continuous polypeptide chain, which is seen more clearly in the ribbon representation shown in Figure 8.5 (right). If this protein was unfolded and elongated into an extended linear conformation, the pairings of Cys6/Cys9 and Cys39/Cys42 would be nowhere near each other, and simultaneous coordination to a single Fe ion would not be possible.

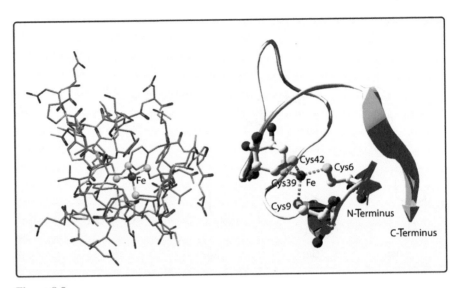

Figure 8.5

The X-ray crystal structure of the metalloprotein rubredoxin: (left) stick representation with the four cysteine S-donors and Fe ion highlighted and (right) ribbon representation with the N-terminal pairing (Cys6/Cys9) and C-terminal pairing (Cys39/Cys42) indicated.

8.1.1.2   *Nucleotides as Ligands*

Nucleotide 'building blocks' of DNA/RNA polymers contain a phosphate diester, a cyclic sugar (ribose) and an N-heterocyclic base (guanine, cytosine, adenine and thymine/uracil) and thus contain a number of potential O- and N-donor groups (Figure 8.6). The RNA single chains and DNA duplex chains are joined by linked phosphate diester groups R–O–PO$_2$–O–R, and the two terminal O atoms may in principle bind suitable metal ions (Figure 8.6, left). An alternative is for the metal to bind to the purine N-donor ligand (Figure 8.6, right). This is not a natural function of the base, as the coordinating N-atom is involved in H-bonding within the DNA double helix. However, this reaction is of great importance in drug design for

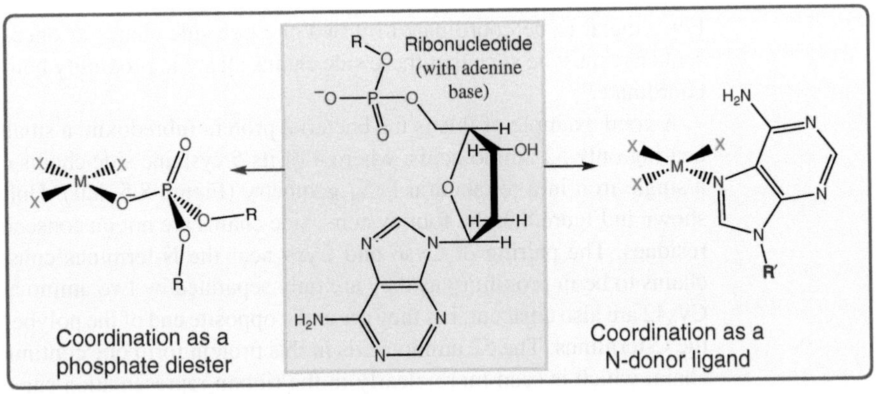

Figure 8.6

Examples of potential nucleotide coordination to a metal ion. Both oxygen and nitrogen donors have the ability to bind to metal ions.

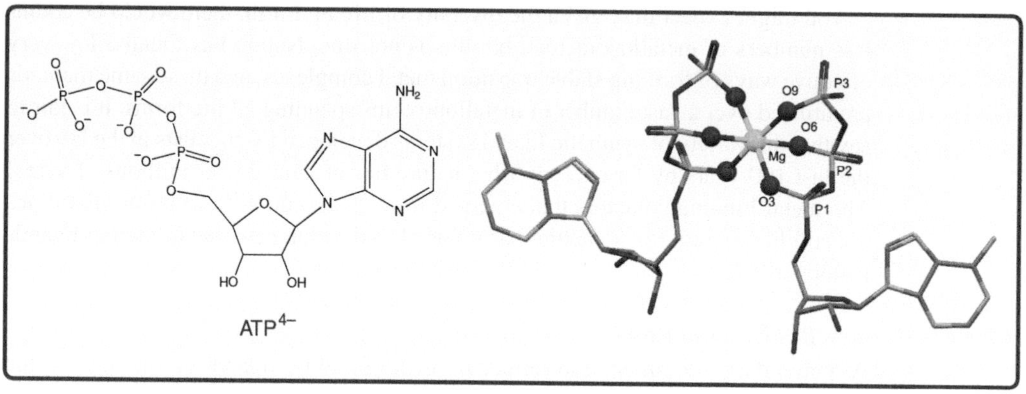

Figure 8.7

Structures of (left) the adenosine triphosphate (ATP) tetraanion and (right) its complex with Mg^{2+}. The first coordination sphere, essentially six-coordinate octahedral, is highlighted.

treating cancers, where the cancer cell DNA becomes a target. This discussion will be deferred until Chapter 9.

Returning to the terminal O atoms of the phosphate ester, coordination of Mg^{2+} to these donor atoms is necessary for cleavage of the polynucleotide backbone by enzymes known as restriction endonucleases. Similarly, in polymerases (which catalyse extension of the polynucleotide backbone), Mg^{2+} binding is also crucial. Given its relatively weak complexation and high lability, it is not straightforward to isolate relevant Mg^{2+} complexes as proof of binding, but biochemical work has shown that enzyme activity is dependent on its presence.

Adenosine triphosphate (ATP) is another nucleotide of great importance. All life forms utilise this molecule principally as a source of energy that is released upon hydrolysis to form hydrogen phosphate and adenosine diphosphate (ADP). ATP has a high affinity for Mg^{2+}, and indeed complexation with this metal is often a prerequisite for interaction with other biomolecules. An example is shown in Figure 8.7 where two ATP tetraanions complex a single Mg^{2+} ion. Complexation in a 1:1 stoichiometry is more common *in vivo*, with other ligands completing the coordination sphere. Given that Mg^{2+} is a typical 'hard' metal ion (high charge density), the negatively charged O-donors of the triphosphate unit are preferred to the N-donors of the adenine group. ATP is regenerated by various metabolic pathways, such as respiration.

8.1.1.3   Metal-Binding Cofactors

Although there are many naturally occurring protein-derived donor atoms and functional groups available for metal binding (see Figures 8.3 and 8.5), in many cases extra help from other co-ligands is needed to form a stable complex. This combination of ligand with metal produces the so-called cofactor (or metallocofactor in this case). The ligand component of the cofactor does not contain any amino acids and is formed by a separate biosynthetic pathway from the metal-free polypeptide (or apoprotein). The cofactor is introduced during the final assembly and folding of the protein. As we have seen in earlier chapters, ligand substitution can bring about significant changes to the chemistry of a metal ion, and natural systems have employed this feature in evolving over time to optimise their performance towards a particular function.

You might expect that, given the diversity of life on Earth, there would be countless numbers of metallocofactors, but this is not true. Nature has found a few very effective ways of forming stable transition metal complexes, and these same methods are utilised over a vast number of metalloproteins spanning all life forms. It is fair to say that the variety of synthetic ligand systems developed by chemists in the lab over the last 100 years by far exceeds what nature has produced over millions of years! The metal-binding cofactors that emerged through evolution from primordial times are still fit for purpose, so nature has no need to develop new metal-binding ligands to support life.

8.1.1.3.1 *Macrocyclic Metal Cofactors*

A particularly common and effective metal-binding ligand in biology is the 16-membered macrocyclic porphyrin ring whose parent structure is shown in Figure 8.8. The most common metal complexed by the porphyrin ring in biology is Fe, and when combined this unit is known as a heme.

In natural systems, substituents are always present around the periphery of the ring (see heme *b*, Figure 8.9), but the core structure and its tetradentate N_4 mode of coordination remain the same. A notable feature is that two of the pyrrole rings deprotonate upon Fe(II) coordination, which maintains charge neutrality. As we will see, many iron proteins contain a heme cofactor and the macrocyclic ligand imparts exceptional stability to the complex. This is reminiscent of complexes of synthetic macrocyclic ligand systems discussed in Chapter 4 which typically show exceptional resistance to dissociation. As Fe is often found in five- and six-coordinate geometries, extra ligands are provided by the protein or other small molecules, but these are not shown here. This aspect will be explored later.

Although iron is the dominant metal coordinated to the porphyrin ring in biological systems, there are other metals that are known to bind to closely related 16-membered macrocyclic ring systems. Chlorophylls are Mg^{2+} complexes that give plants their green colour, and they are principal participants in photosynthesis. The ring system is structurally related to the protoporphyrin IX ligand that makes up the heme *b* cofactor (Figure 8.9). Indeed, this was noted a century ago, and Hans Fischer was awarded the 1930 Nobel Prize in Chemistry for his work identifying a common biosynthetic intermediate of the heme and chlorophyll cofactors (blood and leaf pigments). This is a good example of nature efficiently using the same coordination chemistry strategy

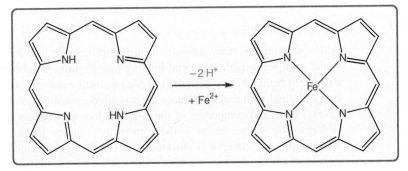

Figure 8.8

The parent porphyrin ligand, and its coordination to Fe(II) to form a heme cofactor (axial ligands or ring substituents are not shown).

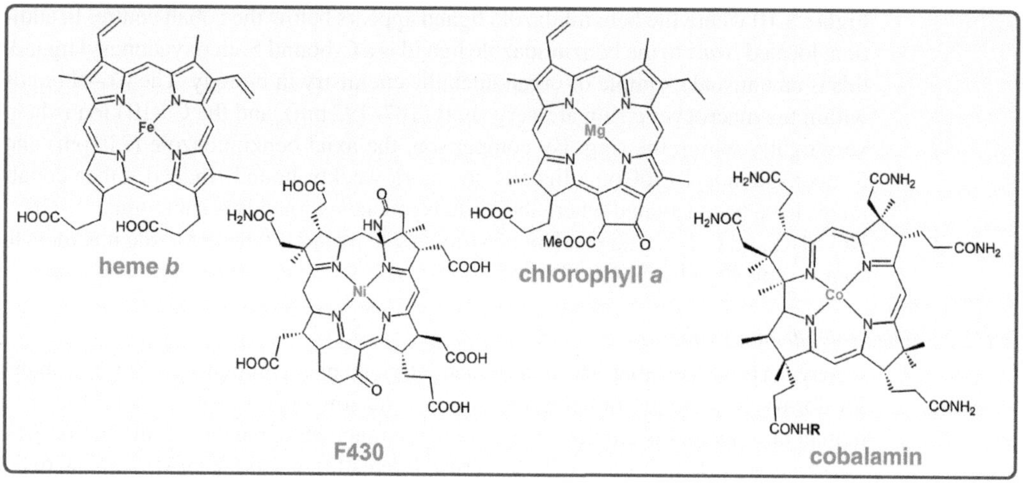

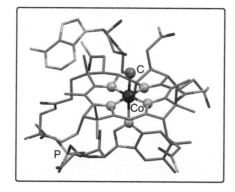

Figure 8.9

Common macrocyclic complex cofactors found in metalloproteins.

for metal complexation across various forms of life for different purposes. No point in reinventing the wheel!

Nickel is found in methane-producing archaea within the enzyme methyl-coenzyme M reductase. Like the previous examples, it is stabilised within a 16-membered macrocycle to give the so-called F430 cofactor (Figure 8.9), but the ring system is no longer aromatic due to extensive hydrogenation. The organisms containing this enzyme live in strictly oxygen-free environments usually on the ocean floor and produce more than 90% of the Earth's atmospheric methane.

Deletion of one C-atom from the ring generates the 15-membered corrin ring system, and the best-known examples are the cobalamins (vitamin B_{12}) bearing an octahedral Co(III) ion in the centre of the ring. The side chain with the $-CONH\mathbf{R}$ group bears a coordinating benzimidazole group that occupies one of the axial coordination sites perpendicular to the corrin ring (although not shown in Figure 8.9). However, the crystal structure of one of the key biologically active cobalamins is shown in

Figure 8.10

The crystal structure of 5′-deoxyadenosylcobalamin highlighting the coordination sphere of the Co(III) ion. Note that, unlike the benzimidazole ligand bound below the cobalt ion, the C-bound 5′-deoxyadenosyl ligand bound above the cobalt ion is not attached to the macrocycle.

Figure 8.10 where the benzimidazole ligand appears below the cobalt centre. In addition, located *trans* to the benzimidazole ligand is a C-bound 5′-deoxyadenosyl ligand; this is an unusual example of organometallic chemistry in biology. The Co–N bonds within the macrocyclic ring are very short (187–192 pm), and the Co(III) ion is held very tightly within the ring. By comparison, the axial benzimidazole (219 pm) and 5′-deoxyadenosyl (200 pm) ligands are more weakly bound. Indeed, other cobalamins have been isolated where the 5′-deoxyadenosyl ligands is substituted by CN^-, NO, H_2O and CH_3^-. The pendent benzimidazole ligand is tethered to the macrocycle so it is typically always coordinated.

8.1.1.3.2   *The Molybdenum Cofactor*

It may surprise you that Mo is an essential element for humans and in fact many life forms. It is unique in being the only second-row (4*d*) transition metal found in human biology and is employed despite being much less abundant than the first-row (3*d*) transition elements in the Earth's crust. Interestingly, it is the *most* abundant of all transition elements in the ocean, which is attributed to its high solubility at neutral pH as the tetraoxidomolybdate(VI) anion $[MoO_4]^{2-}$, while most other transition metals precipitate as their hydroxides at pH 7 without stabilising co-ligands. Given the vast amount of ocean-based evolution, this may be the clue to its use in biology.

Nature has furnished Mo with its own custom-made ligand (Figure 8.11, left) and the two are virtually inseparable in biology. The free ligand, comprising a pterin heterocycle with a fused pyran ring bearing a dithiolene chelating group, has been given various trivial names (including the unhelpful name 'molybdopterin', given that it contains no Mo!) but when combined with Mo it becomes the molybdenum cofactor, or Moco. In this form, it is delivered to one of four essential enzymes that are present in humans (five in plants and fungi) where it undergoes various ligand exchange reactions involving the two equatorial hydroxido ligands, but the metal never leaves its dithiolene ligand. In prokaryotes (bacteria and archaea), there are dozens of other Mo enzymes that carry out a variety of roles. Tungsten, the third transition series congener of Mo, also uses the same pyranopterin dithiolene ligand framework, and there are many parallels between the reactions of Mo and W enzymes.

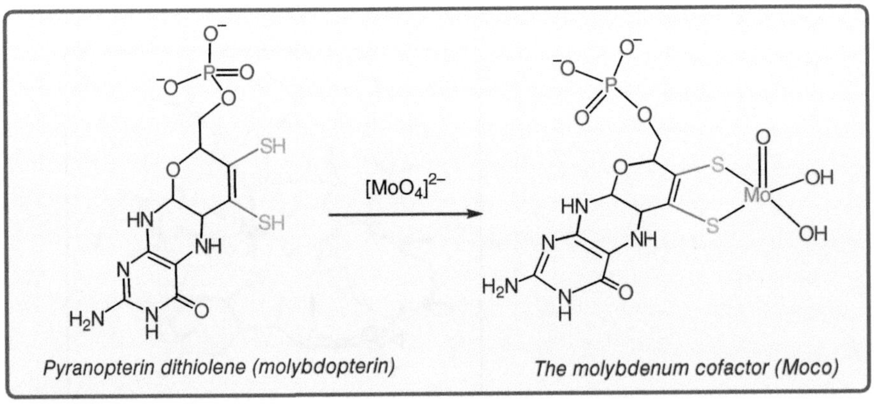

Pyranopterin dithiolene (molybdopterin)   The molybdenum cofactor (Moco)

Figure 8.11

The pyranopterin dithiolene ligand and its Mo(VI) complex (the molybdenum cofactor).

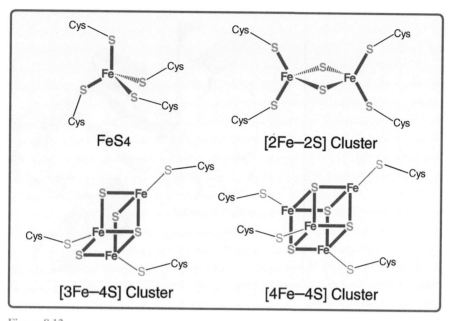

FeS4

[2Fe–2S] Cluster

[3Fe–4S] Cluster

[4Fe–4S] Cluster

Figure 8.12

The tetrahedral FeS$_4$ unit in mononuclear Fe–S proteins, the [2Fe–2S] cluster, the [3Fe–4S] cluster and the [4Fe–4S] cluster. No charges are shown here for simplicity.

8.1.1.3.3   *Iron–Sulfur Clusters*

As already seen in Figure 8.5, iron can be stabilised in a tetrahedral FeS$_4$ coordination geometry by four cysteine S-donor ligands. The same core structure can be extended to incorporate more than one Fe ion by the introduction of the simple sulfide anion as a bridging ligand, which then leads to a so-called iron–sulfur cluster (Figure 8.12). To retain the four-coordinate tetrahedral geometry, this requires that two of the 'terminal' cysteine S-donors are replaced by S^{2-} ions, which are more effective bridging ligands, but this does not bring a dramatic change to the coordination environment as the donor atom and geometry remain the same. With each additional Fe ion, the cluster grows. The formal oxidation states of the individual Fe ions are combinations of Fe(III) and Fe(II), so the overall charge of the cluster will change as electrons enter or leave. An example of this is provided by the 2Fe–2S protein spinach ferredoxin. Its structure is shown in Figure 8.13 with the 2Fe–2S cluster highlighted. Discussion of the oxidation states of the metals and the associated charges of the Fe/S clusters is deferred to Section 8.1.2.1.

8.1.1.4   *Siderophores*

Having established that transition metals are essential for life and that these elements are present in trace amounts, we can see that the acquisition of these metals becomes an important task for all life forms. Humans acquire essential metals from their diet, and there are sophisticated regulatory mechanisms in place to ensure a balance between deficiency and overload. Bacteria are equally reliant on essential trace metals such as Fe to thrive, especially pathogens that cause disease. However,

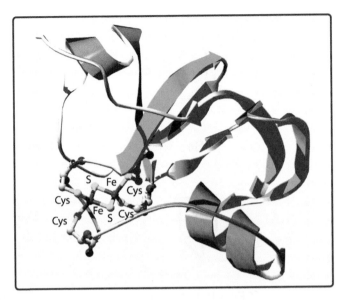

Figure 8.13
Structure of the Fe/S protein spinach ferredoxin with its 2Fe–2S cluster highlighted and the peptide back-bone represented in ribbon form. Note that four cysteine amino acid side chains function as terminal ligands, while the S^{2-} ligands bridge the two Fe centres.

under normal physiological conditions (aerobic aqueous solutions at pH 7.4), Fe(III) is completely insoluble (as Fe_2O_3), which limits its bioavailability severely. To this end, bacteria synthesise and release potent small molecular weight organic ligands known generally as *siderophores* (Greek for iron-carrier) which complex the target metal ion before being reabsorbed and stored by the organism. These bacteria are able to synthesise these siderophores in response to low-iron conditions and can switch this off when iron supply is plentiful.

Streptomyces species utilise ligand systems based on hydroxamic acids to chelate Fe(III). One of the best-known examples is the linear tris-hydroxamic acid desferriox-amine B (Figure 8.14, left) which wraps around the metal to form a very stable and soluble Fe(III) complex. The same chelator has been used successfully for decades as a drug (Desferal®) to treat people with iron overload disorders. Deposition of excess (uncomplexed) Fe in vital organs such as the heart is potentially fatal for humans if untreated. The reaction between aquated ferrous ions and hydrogen peroxide gener-ates reactive oxygen species such as hydroxyl radicals via so-called Fenton chemistry (Eq. (8.1)).

$$Fe_{aq}^{2+} + H_2O_2 \rightarrow Fe_{aq}^{3+} + {}^{\cdot}OH + OH^- \tag{8.1}$$

Desferrioxamine B captures this free iron, and it is subsequently excreted from the body as a stable, soluble and redox inactive Fe(III) complex, thus nullifying its toxicity.

Enterobactin (Figure 8.14, right) is an N-functionalised cyclic triester of the amino acid serine produced by several species of bacteria from the family *Enterobacte-riaceae* (*E. coli* is one of these). The pendent catechol (*ortho*-dihydroxybenzene) groups deprotonate readily in the presence of Fe(III), and the three catecholates func-tion as didentate chelates to form FeO_6 complexes of exceptionally high stability. The binding mode is shown in Figure 8.14 (right), which in this case illustrates V(IV)

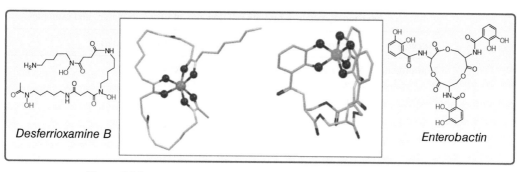

Figure 8.14

The bacterial siderophores (left) desferrioxamine and (right) enterobactin, with crystal structures of octahedral complex they form. In both cases, Fe(III) complexes of high stability are formed, with the ferrioxamine complex shown on the left. Other stable metal complexes are known; the V(IV) complex of enterobactin is shown as the example structure on the right.

complexation as a model system, as surprisingly the crystal structure of its Fe(III) complex has not yet been determined!

8.1.2 Metals in Biology – Why Are They Needed?

The preceding sections have shown that only a small number of transition and main group metals are of biological relevance. Transition metals, when carrying out a biological function, are always associated with a protein or enzyme (i.e. a protein that specifically catalyses a chemical reaction). Metals occur in several forms in biology, largely in environments that we would recognise as those of coordination compounds. Functionally the most advanced class met are the metal-dependent enzymes which catalyse a specific chemical reaction. There are literally thousands of enzymes known, which tend to be classified by their function, with seven 'families' typically recognised: oxidoreductases, transferases, hydrolases, lyases, isomerases, ligases and translocases. Of these, a considerable number (about 40%) are *metalloenzymes*, with one or several metal ions present coordinated to donor groups that are part of a biopolymer. The metal ions in metalloenzymes and other metallo-biomolecules are commonly the lighter, more abundant and more reactive metals of the *s*- or *d*-block of the periodic table, but not exclusively. Rarer and heavier elements (such as molybdenum) can be used, as shown earlier. The metal ion is also not present just by accident – it has a specific role. In many cases, it is found coordinated at the active site in enzymes (i.e. the site of catalysis), where the chemical reaction catalysed by the enzyme occurs.

In Table 8.1, a high-level summary of the functions of different types of metal-dependent proteins and enzymes is given. This list is not exhaustive but illustrates the amazingly diverse roles metals fulfil.

When we recall the features of the *d*-block elements compared with elements in the rest of the periodic table, a couple of properties stand out. The first is their ability to exist in multiple stable oxidation states, which is a feature in biological coordination chemistry as well. The ability to reversibly exchange electrons *one at a time* is a special property of the *d*-block elements and usually outside the domain of organic chemistry where radicals (with unpaired electrons) are seldom encountered as stable entities. This function is routine in metalloproteins bearing Fe(III/II) and Cu(II/I)

Table 8.1 Division of metals in biology (all domains of life), sub-classes and examples.

Category	Function	Examples (metals involved)
Non-protein	Metal transport	siderophores (Fe)
	Structural	skeletal minerals (Ca)
Protein	Electron transfer	cytochromes (Fe); ferredoxins (Fe); blue copper (Cu)
	Metal storage/transport	ferritin (Fe); transferrin (Fe); ceruloplasmin (Cu); metallothionein (Zn, Cu)
	Structural	zinc fingers (Zn)
	Oxygen storage/transport	myoglobin (Fe); hemoglobin (Fe); hemerythrin (Fe); hemocyanin (Cu)
	Neurotransmission	ion channels (Ca, K, Na); ion pumps (Ca, K, Na)
Enzyme	Hydrolysis	carboxypeptidases (Zn); aminopeptidases (Mg, Mn); phosphatases (Mg, Zn, Cu, Fe, Mn)
	Redox	oxidoreductases (Fe, Cu, Mo, W); hydroxylases (Fe, Cu, Mo); oxygenases (Fe, Cu); superoxide dismutases (Mn, Fe, Ni, Cu, Zn); nitrogenases (Fe, Mo, V); hydrogenases (Fe, Ni)
	Photo-redox	Photosystems I and II: light harvesting complex (Mg), photosynthetic reaction centre (Mg); oxygen evolving complex (Mn, Ca)
	Isomerisation	aconitase (Fe) vitamin B_{12} coenzymes (Co)

centres, and there are many Fe- and Cu-dependent proteins that do exactly this as their sole function. Oxidation and reduction reactions in biology are a team effort requiring the participation of many electron transfer relays (proteins) that can ferry electrons between the terminal oxidant (e.g. O_2 in respiration) and the reductant, so the individual 'stations' along the way must be stable in both their oxidised and reduced forms for this to be sustainable.

A second and distinct feature is the electron-withdrawing influence the metal has on the reactivity of its ligands, particularly aqua (water) ligands. While the proportion of hydroxide ions at pH 7 is insignificant (<0.0000001%) in comparison to its conjugate acid H_2O, when water coordinates to a transition metal, its acidity is enhanced by many orders of magnitude such that the coordinated hydroxido ligand can be the dominant species at pH 7. The coordinated hydroxido ligand is a highly effective nucleophile that can attack a variety of organic functional groups in synthetic systems (Figure 6.14), and we will show biological examples of this here.

Below, we highlight some of the functional roles played by metalloproteins by choosing examples that illustrate the reactivity of the central metal (or metals) as well as its first coordination sphere of donor atoms, while not forgetting the organic polypeptide which provides the three-dimensional shape of the protein as well as the local environment of the metal or metals. All these cases are founded on the concepts covered in earlier chapters.

8.1.2.1 Electron Transfer Proteins

As mentioned earlier, the shuttling of electrons, one at a time, along a series of relays requires strategic placement of electron transfer proteins whose function is only to accept or donate electrons. This may occur over several different 'stations' along the way, and thermodynamics dictates that electrons will flow in one direction and not the other, gravitating towards centres with the highest redox potential.

The cytochromes (named from the intense colours they bring to cells) are a large family of heme-dependent proteins that conduct electron transfer reactions. There are various categories of cytochromes (*a*, *b* and *c*-type, based on their spectral properties), but we will not explore these subtleties here. Cytochrome *c* is one of the most heavily investigated members of this family and is found in many life forms (humans included). It serves several functions depending on the cells in which it is found, but in humans, it is a source of electrons during respiration where it is oxidised from its Fe(II) to Fe(III) state by the enzyme cytochrome *c* oxidase. As shown in Figure 8.15 (left), the heme cofactor is covalently linked to the polypeptide through C–S bonds via two cysteine residues (Cys14 and Cys17). In addition, two amino acid side chains (histidine and methionine) occupy the axial coordination sites, with the N_4 donor set of the porphyrin ligand completing the coordination sphere. The redox potential of this protein (+0.26 V versus the standard hydrogen electrode) means that the Fe(III) form of the protein is an effective oxidant, but importantly this electron transfer reaction occurs very quickly and without any change to the coordination sphere. There are many other cytochromes that bear different axial ligands (and redox potentials).

The Fe/S proteins, such as ferredoxin (Figure 8.13), are also electron transfer proteins where again they can mediate single electron reactions. Neglecting the charges of the coordinated cysteine residues, the formal charge of the ferredoxin cluster is 2×3 (Fe^{3+}) $- 2 \times 2$ (S^{2-}) $= +2$, represented as $[2Fe–2S]^{2+}$. There are numerous 2Fe–2S proteins spanning all life forms, and their typical role is to transfer electrons by undergoing a reversible one-electron reduction where formally one Fe is then Fe(III) and the other is Fe(II), i.e. $[2Fe–2S]^{2+} + e^- \rightarrow [2Fe–2S]^+$. Interestingly, the two metals in this 'mixed valence' reduced form are indistinguishable, as this electron is delocalised over both metal centres and there are barely any structural changes after electron transfer. Moreover, there are equally many examples of 4Fe–4S proteins, and the most common forms that participate in redox reactions are the $[4Fe–4S]^{2+}$ ($2 \times$ Fe(III), $2 \times$ Fe(II)) and $[4Fe–4S]^+$ ($1 \times$ Fe(III), $3 \times$ Fe(II)) forms. The 3Fe–4S proteins are less common, but their typical redox states are $[3Fe–4S]^+$ ($3 \times$ Fe(III)) and $[3Fe–4S]^0$ ($2 \times$ Fe(III), $1 \times$ Fe(II)).

The so-called Blue Copper Proteins constitute the other major type of single electron transfer proteins. They are found in plants, algae and bacteria where they mediate electron transfer in processes such as photosynthesis and respiration. Like the cytochromes, they are effective electron acceptors in their Cu(II) form as their Cu(II/I) redox potentials are unusually high; in fact, much higher than the standard $Cu_{aq}^{2+/+}$ redox potential (of +0.17 V versus SHE). The example shown in Figure 8.15 (right) is the bacterial protein rusticyanin which bears a CuN_2S_2 (two histidines, a cysteine and

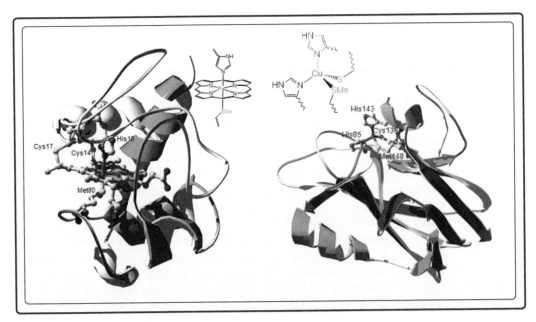

Figure 8.15

(Left) The structure of cytochrome *c* (from horse heart), with the heme *c* cofactor and the axial ligands highlighted. (Right) The bacterial blue copper protein rusticyanin with its coordinating amino acids highlighted.

a methionine) coordination sphere which is typical of this family. Rusticyanin possesses a Cu(II/I) redox potential of +0.69 V versus SHE at pH 6, making its Cu(II) form one of the strongest single electron oxidants in biology. This is an excellent example of how ligand substitution (from all O-donors to a N_2S_2-donor set) can dramatically change the preferred oxidation state, with in this case Cu(I) being now preferred to Cu(II). The blue colour that gives these proteins their name is due to an intense S(Cys)-to-Cu^{II} charge transfer transition in their oxidised state. In their reduced Cu(I) forms, these proteins are colourless.

8.1.2.2   *Metal Transport and Storage*

Given the tendency of aquated transition metal ions to precipitate at physiological pH, nature has had to devise proteins that can stabilise these essential metal ions as complexes that can then be transported to various parts of the body before release. Given their importance and abundance, the three main transition metals, Fe, Cu and Zn, all have different proteins responsible for their transport and storage. It is important (even essential) that complex organisms (like us) have ways of storing reserves of essential metals for future use.

Most of the excess Fe reserves in our body reside in the storage protein ferritin, which is found in many different organs. Ferritin comprises 24 identical subunits (with a combined molecular weight close to 470,000 Da (g mol^{-1})). Each subunit is a four-helix bundle (Figure 8.16) that self-assembles into a spherical protein with a large internal cavity. Within this cavity, ferric ions are stored in the form of hydrated iron oxide ($Fe_2O_3 \cdot nH_2O$), which in the macroscopic world we know as rust. Phosphate ions are also present, thought to act as terminal ligands on the outside of a cluster

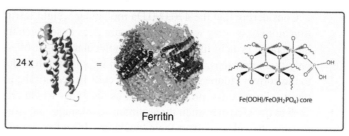

Figure 8.16

The crystal structure of apoferritin (no Fe) showing the shape of the monomeric four-helix bundle (left) and the structure of its core of hydrated iron oxide (centre). Note that for the hydrated iron oxide in the internal cavity, each Fe is in an octahedral FeO_6 environment, as shown at right.

that comprises something in the range of 4,000–5,000 Fe atoms! Serum ferritin levels are often used as a biomarker for healthy Fe levels in medicine.

In mammals, the protein transferrin (whose name is self-explanatory) is tailor-made for transporting ferric ions in serum and depositing it in cells that require Fe for metalloprotein synthesis. Transferrin binds two ferric ions in identical coordination sites (Figure 8.17, left). The ligands are an interesting mixture of amino acid side chains (two tyrosines with deprotonated phenolate groups, one histidine, and one aspartate) and a didentate carbonate to give a FeO_5N coordination sphere (Figure 8.17, right).

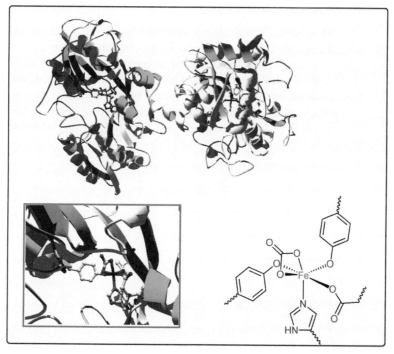

Figure 8.17

The crystal structure of porcine transferrin showing the two separate Fe-binding sites and a zoomed in view of the Fe coordination sphere with the didentate carbonate co-ligand a feature of the FeO_5N complex.

Considering that the 470,000 Da molecular weight ferritin can store thousands of iron atoms, you may wonder why nature goes to so much trouble constructing a similarly large protein like transferrin (molecular weight of ~77,000 Da) that can bind only two ferric ions. Transferrin has an important job to do, as Fe must be delivered to the correct 'address', which will be cells in need of the element. On the surface of these cells are other proteins known as receptors which recognise transport proteins such as transferrin by their three-dimensional shape and structure and only allow them to enter the cell to deliver their 'package'. Intruders are not allowed to enter! There are several other proteins employing elements like Cu and Zn that conduct similar functions.

Metallothioneins contrast with transferrin in being small molecular weight proteins replete with cysteine residues where they can comprise nearly a third of all amino acids in the protein. These proteins (of which there are many types) have a particular affinity for d^{10} metal ions such as Cu(I) and Zn(II), but they also are known to complex toxic metals such as Cd(II) and Hg(II). They bind several metals at once with groupings of cysteines, typically comprising pairs separated by one amino acid (i.e. a so-called CXC motif). The metal-binding stoichiometry is metal-dependent and variable. This makes sense, as metals with lower coordination numbers require fewer (cysteine) S-donor atoms. For instance, in some metallothioneins, seven Zn(II) ions are accommodated, whereas in the structure shown in Figure 8.18, eight Cu(I) ions are included in an intricate pattern comprising bridging cysteine thiolate S-donors. Note that the Cu(I) ions are only two- or three-coordinate, although it is more common for Zn(II) in analogues to be four-coordinate (tetrahedral).

8.1.2.3   *Oxygen Transport and Storage*

Most forms of life, except anaerobic bacteria and archaea, depend on dioxygen as the terminal electron acceptor in the electron transport chain where it is reduced to water as part of respiration. This is central to metabolism and energy production. By analogy with metal transport and storage just mentioned, we need to have readily available stores of dioxygen that can be quickly utilised when needed. In mammals, this is achieved with two heme-dependent proteins myoglobin (for oxygen storage) and hemoglobin (for oxygen transport). In fact, most of the 4 g or so of iron in our

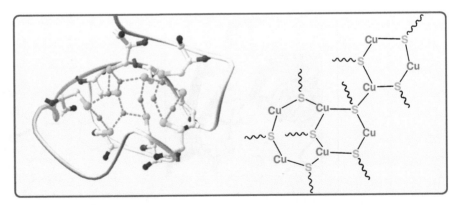

Figure 8.18

The crystal structure of Cu-loaded metallothionein from yeast, highlighting the Cu cluster comprising eight 2- and 3-coordinate Cu(I) (grey) bridged by cysteine S-donor (yellow) ligands.

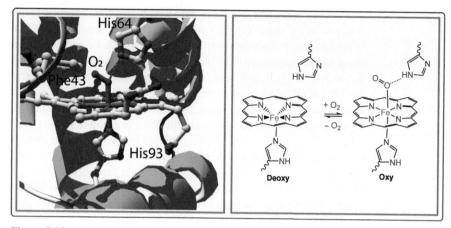

Figure 8.19

The active site of oxymyoglobin (from sperm whale) and the coordination geometry changes at the Fe centre associated with reversible dioxygen coordination including hydrogen bonding to the distal histidine His64.

bodies is found in these two proteins. Myoglobin and hemoglobin have been intensively studied and were one of the early successes of protein X-ray crystallography. Max Perutz and John Kendrew were awarded the 1962 Nobel Prize in Chemistry for determining the crystal structures of both myoglobin and hemoglobin.

Myoglobin is a 'small' (M.W. ~17,000 Da) protein with a single heme cofactor non-covalently embedded within the polypeptide chain that features several α-helices (Figure 8.19). It stores oxygen in muscle tissue until it is required for oxidative phosphorylation, thus providing a ready reserve of oxygen to cope with intense respiration demands. Hemoglobin has the primary role of transporting oxygen (although its full function is somewhat more complicated and also includes transporting carbon dioxide) and is found in red blood cells. It is essentially a tetrameric version of myoglobin, although it has other distinctive features due to its four subunits working together (cooperatively) to bind each successive O_2 molecule more strongly.

In both myoglobin and hemoglobin, dioxygen coordinates to Fe(II), but not Fe(III), within the heme cofactor. The only connection between the heme and the polypeptide is a histidine side chain (His93 in Figure 8.19, left) which occupies an axial coordination site; this is commonly known as the *proximal* position. The heme is embedded in a crevice where it is surrounded by hydrophobic groups such as an aromatic ring from the phenylalanine (Phe43) side chain. Another imidazole group (from His64) is on the same (*distal*) side of the heme as Phe43. It is at this protected site that dioxygen coordinates as a monodentate ligand, taking the iron(II) from a five-coordinate to a preferred six-coordinate octahedral complex.

In the deoxy form, a five-coordinate square pyramidal geometry is found, and the Fe ion lies out of the macrocycle plane closer to the axially coordinated imidazole of His93 (Figure 8.19, right). When oxygen binds at the 'vacant' sixth site, *trans* to the imidazole, the Fe changes to a six-coordinate octahedral geometry and the metal becomes coplanar with the porphyrin ring. In achieving this, the protein is required to adjust its conformation, to retain the required Fe–N(histidine) bond distance. The release of oxygen allows relaxation back to the five-coordinate square-based pyramidal shape around the iron. No permanent change in the protein is involved, and hence this process happens reversibly and rapidly.

The role of the distal 'pocket' into which dioxygen enters is twofold. It provides a water-resistant (hydrophobic) environment which repels polar (charged) species, so it protects the coordinated O_2 ligand from attack by such other species. Secondly, the distal histidine (His63) imidazole ring forms a strong hydrogen bond (NH ... O) to the bound O_2 ligand, which stabilises coordination. It is interesting to note that the O_2 ligand coordinates in a bent configuration (Fe—O—O angle of 125°), whereas other diatomic ligands such as CO, CN^- and NO bind in a linear disposition to the heme. The formal oxidation state of the heme in oxymyoglobin is not straightforward. The oxyheme cofactor may be represented by two resonance structures, ferrous-oxy and ferric-superoxy forms (Eq. (8.2)), showing that delocalisation of electron density from Fe to O_2 must be considered. The current view is that the ferric-superoxy form (Eq. (8.2), right) is the dominant contributor and consistent with the bent coordination mode. When O_2 dissociates, it leaves the electron behind to regenerate the ferrous deoxy form.

$$(8.2)$$

Not all life forms bind and carry dioxygen in this way. Hemerythrin is a non-heme iron protein used by *sipunculid* and *brachiopod* marine invertebrates for oxygen transfer and/or storage. The complete protein has a complicated structure including a large number of subunits, each of which is an active site for oxygen addition. Each subunit has a molecular weight of around 14,000 Da, contains two iron atoms and binds one molecule of oxygen. This diiron oxido protein contains octahedral iron(III) centres linked by one oxido (O^{2-}) and two carboxylato (–COO⁻) bridges. The oxygen-free form, deoxyhemerythrin (deoxyHr), is colourless with two high-spin d^6 Fe(II) centres bridged by a hydroxide ion; one Fe is six-coordinate (bound to three histidine residue N-donors, two peptide carboxylate groups and the bridging hydroxide ion), the other one is five-coordinate (bound to two histidine residue N-donors, two peptide carboxylate groups and the bridging hydroxide ion) and has the one 'vacant' site where dioxygen can bind. The dioxygen-bound form, oxyhemerythrin (oxyHr) is red-violet in colour, and now contains two six-coordinate low-spin d^5 Fe(III) centres (Figure 8.20). The O_2 binds to the five-coordinate iron(II) centre and abstracts an H-atom from the bridging hydroxido group, forming a FeOOH group that remains strongly hydrogen-bonded to what is now a bridging oxido group. Oxygen uptake is also accompanied by one-electron oxidation of *both* Fe(II) centres to Fe(III), meaning that the dioxygen undergoes two-electron reduction to bound peroxide ion.

8.1.2.4   *Oxidoreductases*

This family includes all enzymes that catalyse either the oxidation or reduction of a substrate. In most cases, this is formally a two-electron redox reaction with accompanying atom transfer. In many cases, protons accompany electrons in the overall reaction. Oxidoreductases are the largest group of enzymes, and transition

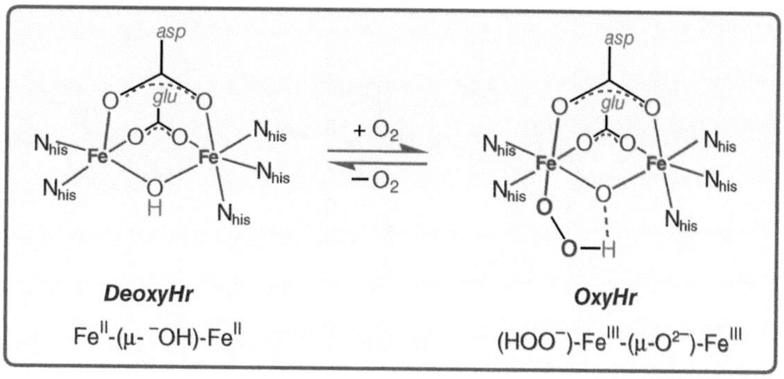

Figure 8.20

The active site of the non-heme Fe-protein hemerythrin. The geometry at the iron centres of the bridged dinuclear unit in its dioxygen-free *deoxy* form (left) and oxygenated *oxy* form (right) form is illustrated.

metal-dependent enzymes are dominant. Now a key point to emphasise is that enzymes are catalysts; that is, they only accelerate reactions but are not consumed in the process. This means that any oxidoreductase-catalysed reactions must have three components: (a) the enzyme; (b) the substrate which will be oxidised (or reduced) and (c) an oxidant (or reductant) to accept or donate the electrons provided or taken by the substrate. Given their abundance and diversity, it is worthwhile to show a few examples of how transition metal-dependent oxidoreductase enzymes work.

8.1.2.4.1 *Molybdenum-Dependent Oxidoreductases*

As mentioned earlier, all mammals (and many bacteria) require Mo enzymes to survive. In humans, one of the most important Mo enzymes is sulfite oxidase which catalyses the reaction shown in Eq. (8.3). Sulfite is reactive and extremely neurotoxic, so we rely on human sulfite oxidase (abbreviated as SO) to convert it to inert and benign sulfate. Mutations leading to sulfite oxidase deficiency in mammals and also plants are catastrophic.

$$2\text{cyt } c(\text{Fe(III)}) + \text{SO}_3^{2-} + \text{H}_2\text{O} \xrightarrow{\text{SO}} \text{SO}_4^{2-} + 2\text{cyt } c(\text{Fe(II)}) + 2\text{H}^+ \quad (8.3)$$

The role of the Mo ion is to transfer an O-atom from its oxido-Mo(VI) form to the sulfite substrate whilst simultaneously accepting two electrons. The immediate product is an O-bound sulfato-Mo(IV) complex which then releases SO_4^{2-} before transferring the two electrons taken from sulfite to cytochrome *c*, which we met in Figure 8.15. The main steps in this enzyme-catalysed oxidation are shown in Figure 8.21. There are other electron transfer steps that involve a heme cofactor in SO which relays electrons to cytochrome *c*, but this is not shown here. As its name suggests, SO can also donate electrons to O_2 but this is not the principal oxidant in this reaction.

8.1.2.4.2 *Peroxidases and Catalases*

Hydrogen peroxide is an important and high energy biomolecule that is involved in many redox reactions. The reduction of H_2O_2 has already been introduced in Eq. (8.1), where the O–O bond may be cleaved *homolytically* by a single electron reduction to release hydroxyl radicals (a Fenton reaction). Nature tries to avoid the production of reactive oxygen species such as ·OH due to their indiscriminate and

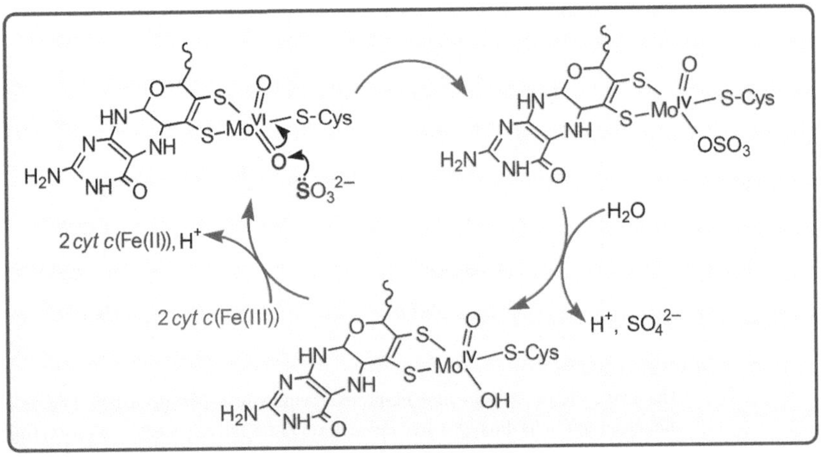

Figure 8.21

Reactions at the active site of human sulfite oxidase where sulfite is converted to sulfate.

high reactivity. The heme-dependent peroxidases and catalases both use H_2O_2 as a substrate to conduct related but distinct reactions (Eqs. (8.4) and (8.5)) which involve heterolytic cleavage of the O–O bond leading to H_2O as a product.

$$2RH + H_2O_2 \xrightarrow{\text{peroxidase}} 2R^{\cdot} + 2H_2O \tag{8.4}$$

$$2H_2O_2 \xrightarrow{\text{catalase}} 2H_2O + O_2 \tag{8.5}$$

The similarities between the two systems can be appreciated in Figure 8.22 where their heme cofactors are shown. The axial ligands are different in heme peroxidases (imidazole of His) and catalases (phenolate of Tyr), but they share common reactive intermediates formed after the reaction between the resting enzyme in its ferric-hydroxido form and H_2O_2. The intermediate H_2O_2 complex forms a two-electron oxidised species given the nebulous name 'Compound I'. This form of the cofactor is interesting in that formally one electron has been lost from the Fe(III) ion (to give an oxido-Fe(IV) species), whereas the other electron originates from the porphyrin ring. In both the peroxidase and catalase mechanisms, proton transfer involving nearby amino acid side chains as weak bases or acids is important. A histidine residue on the proximal side plays an important role in both enzymes as a proton acceptor and is highlighted in Figure 8.22 along with a generic weak acid 'HA' which can be various side chains.

From the Compound I point, the mechanisms diverge. The peroxidases carry out a single electron oxidation of their substrate (RH, which can be any of many different organic compounds such as phenolates) to give the species designated 'Compound II' and a radical R· where formally the electron has been accepted by the one-electron oxidised porphyrin ring. This is repeated in a second single electron reduction by the substrate to regenerate the resting form. The substrate that donates a single electron is not always an organic compound and may be chloride (chloroperoxidase, which has a cysteine S-donor axial ligand) or the protein cytochrome *c* (cytochrome *c* peroxidase, with a histidine axial ligand).

The reaction facilitated by catalase (Eq. (8.5)) is the disproportionation of H_2O_2 into water and dioxygen. This means that H_2O_2 is both the oxidant and the reductant,

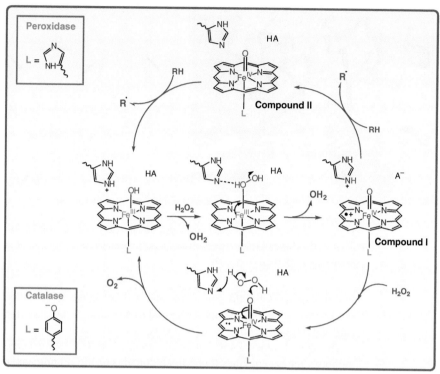

Figure 8.22

The parallels between heme-dependent peroxidases (upper cycle) and catalases (lower cycle). The different axial ligands L for each enzyme are identified in the boxes at upper and lower left.

which is unusual although not unprecedented. After the generation of Compound I, the second H_2O_2 molecule is oxidised to O_2 formally through a hydride transfer to the axial oxido ligand. Note that in catalase, Compound II is never formed, as all electron transfer steps involve two electrons. This is important, as it avoids the formation of the superoxide radical ($O_2^{-\bullet}$) as an intermediate. A large family of Mn-, Fe-, Ni- and Cu/Zn-dependent enzymes known as superoxide dismutases conduct a similar transformation on the superoxide radical ($2O_2^{-\bullet} + 2H^+ \rightarrow O_2 + H_2O_2$) to remove this potentially damaging reactive oxygen species.

8.1.2.4.3 *Oxygenases*

These enzymes bind and activate O_2 as an oxidant for various reactions with organic substrates. They are subdivided into monooxygenases (introducing a single O-atom) and dioxygenases (where both O atoms are involved). The copper enzyme tyrosinase (widely distributed in bacteria, plants and animals) catalyses the *ortho*-hydroxylation of phenolic substrates. The active site in its reduced form exhibits two Cu(I) ions in proximity each coordinated by three histidine side chains (Figure 8.23). Dioxygen binds as a bridging ligand between the two metals in a side-on ($\mu-\eta^2:\eta^2$) manner, and it has been shown that two-electron transfer to give a di-Cu(II)-peroxido complex occurs. The phenolic O-atom of the substrate (shown in Figure 8.23) coordinates to the active site, which positions the substrate specifically for *ortho*-hydroxylation. This $\mu-\eta^2:\eta^2$ Cu_2O_2 binding mode has been seen in other systems, including the oxygen transport protein hemocyanin found in some invertebrates where it serves the same

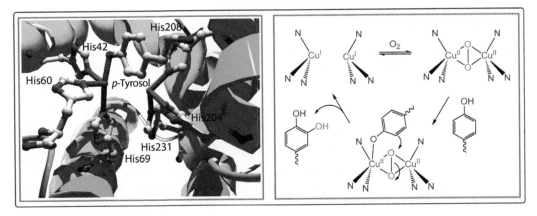

Figure 8.23

The di-copper active site structure of *Bacillus megaterium* tyrosinase with model phenolate substrate *p*-tyrosol (2-(4′-hydroxyphenyl)ethanol) coordinated and proposed mechanism, where the copper ions are in green (as Cu(I)) or pink (as Cu(II)).

role that hemoglobin does in mammals. A fine balance needs to be struck between reversible O_2 binding (hemocyanin) and irreversible O_2 binding (tyrosinase and also catechol oxidase) leading to monooxygenase activity.

The cytochromes P450 (often abbreviated as CYP) are heme-(Fe)-dependent monooxygenase enzymes found in all domains of life. In humans, they are involved in the oxidation of xenobiotic compounds (such as drugs) in the liver so that they can be more easily cleared from the body. The majority of CYP enzymes conduct hydroxylations of organic substrates or variations of this reaction (Eq. (8.6)) where they insert an O-atom into an inactivated C–H bond; this is a challenging reaction to carry out in a controlled way in the laboratory.

$$RH + 2H^+ + O_2 + 2e^- \xrightarrow{\text{CYP}} ROH + H_2O \tag{8.6}$$

CYP turnover requires two electrons, relayed by a partner protein (Fe/S or flavin-dependent protein), and originating from nicotinamide adenine dinucleotide-based reductants (NADH or NADPH). These two electrons must be delivered one after the other, first to the Fe(III) resting form of the enzyme then to the coordinated O_2 ligand. Protonation results in heterolytic cleavage of the O–O bond (releasing water) to generate yet again Compound I (see Figure 8.22) which then hydroxylates the organic substrate. There are many parallels with the chemistry of peroxidases in that Compound I is an intermediate, but this is formed via O_2 coordination (not H_2O_2). Also, dioxygen coordination in myoglobin (Figure 8.19) and hemoglobin is relevant, as a ferrous-dioxy form is also an intermediate.

8.1.2.4.4   Nitrogenases

This family of metalloenzymes continues to attract much attention. The chemical reaction they catalyse is arguably one of the most difficult to achieve in the laboratory – the six-electron (six-proton) reduction of dinitrogen to ammonia. The dinitrogen triple bond is the strongest chemical bond known, and despite N_2 being abundant on Earth (as 80% of our atmosphere), it rarely features in any chemical reactions. Ammonia is a vital agricultural feedstock for fertilisers and is also used in the manufacture of explosives. Without synthetic fertilisers to support large-scale food production and feed our planet's rising human population, we would face starvation. Industrial-scale synthetic hydrogenation of N_2 to NH_3 is performed at high gas pressures ($\sim$300 bar) and temperatures ($\sim$500°C) over an Fe catalyst according to the Haber–Bosch process developed in the early part of the 20th century. This reaction alone accounts for $\sim$2% of human energy consumption!

Bacteria that live on the nodules of plants can 'fix' atmospheric dinitrogen and reduce it into ammonia, which is then taken up by the plant in a bioavailable form for protein synthesis in a symbiotic and vital partnership, as nitrogenase is not produced by plants. This reaction occurs at atmospheric pressure and ambient temperature. The overall reaction catalysed by nitrogenase (a reductase enzyme) involves two processes running in parallel; N_2 reduction to NH_3 (NH_4^+ at physiological pH) and H^+ (H_2O) reduction to H_2, overall requiring eight electrons and a considerable amount of energy provided by ATP hydrolysis (Eq. (8.7)).

$$N_2 + 16\text{MgATP} + 16H_2O + 8e^- \xrightarrow{\text{nitrogenase}} 2NH_4^+ + H_2 + 16\text{MgADP}$$
$$+ 16HPO_4^{2-} + 6H^+ \quad (8.7)$$

Given the number of reactants and products in Eq. (8.7) and the consumption of eight electrons overall, you will not be surprised that the mechanism of how nitrogenase works is complicated and still a work in progress at this time. There are two proteins that are primarily involved. The first, referred to (perhaps unhelpfully given the omnipresence of Fe in biology) as the 'Fe-protein', has the sole task of delivering electrons for substrate reduction. It bears a single [4Fe–4S] cluster that sits at the centre of a dimeric assembly (Figure 8.24) and electrons must be delivered one at a time via its $[4Fe–4S]^{2+/+}$ redox couple. It also has a binding site for two Mg-ATP moieties (see Figure 8.7) which are necessary to make the reaction (Eq. (8.7)) thermodynamically feasible.

The other component of the nitrogenase system is more complicated. In fact, this component can be one of three related but distinct proteins that are referred to as the

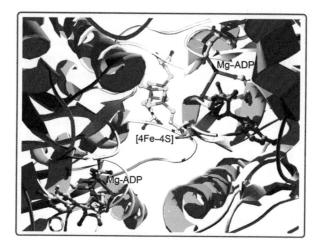

Figure 8.24

Structure of the *Azotobacter vinelandii* nitrogenase Fe-protein highlighting its [4Fe–4S] cluster which sits in the middle of the dimeric protein and two Mg-ADP units (the product of Mg-ATP hydrolysis). Only three O-donor atoms are visible for each Mg^{2+} ion (see top right Mg for a clear view). The other three sites are occupied by H_2O, not shown.

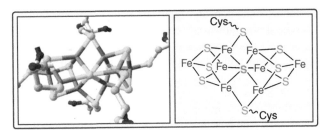

Figure 8.25

Structure (left) and line drawing (right) of the [8Fe–7S] P-cluster within *Azotobacter vinelandii* nitrogenase MoFe protein in its reduced state. The central six-coordinate S^{2-} ligand is highlighted, with the two
bridging cysteine residues above and below it in the structure. Only the two bridging cysteine residues are
shown in the line drawing for clarity.

MoFe protein, the VFe protein and the FeFe protein. As you may have already gath-
ered, this nomenclature refers to key transition metals that are present. They are quite
similar, so we will focus on the most studied member, the MoFe protein. Two cofactors are present in the MoFe protein that are both Fe/S clusters, but they are each
unique. The first, known as the P-cluster, is a [8Fe–7S] cluster that is coordinated
by six cysteine ligands and has no precedent in biology (Figure 8.25). The cluster
may be viewed as a fusion of two [4Fe–4S] distorted cubic clusters that share a central S-atom. This structure is reinforced by two bridging cysteine S atoms. Despite
its extraordinary structure, like the [4Fe–4S] cluster in the Fe-protein, this [8Fe–7S]
cofactor is only required to relay electrons (one at a time) between the Fe-protein and
the active site where N_2 is reduced. Crystal structures of this protein in its oxidised
and reduced forms have shown that there is some reorganisation of the structure with
oxidation or reduction and other amino acids (e.g. serine) may replace cysteine. The
reduced form is shown in Figure 8.25 with six cysteine ligands coordinated. Whereas
two of them bridge the two cubes with each binding two Fe ions, the other four only
bond to a single Fe ion.

The other cofactor within the MoFe protein bears the active site and is known as the
M-cluster or FeMo cofactor (FeMoco). The exact composition of the FeMoco cluster
took at least two decades of structural and spectroscopic work to resolve. In the first
crystal structure of the MoFe protein reported in 1992, a rather unusual cluster was
defined with six apparently three-coordinate Fe ions (Figure 8.26). Higher quality
crystallographic data reported in 2002 then identified a hitherto unseen 'light' atom
at the centre of the cluster that was bonded to six Fe ions in a distorted octahedral
configuration. The electron density showed that it was a (non-metallic) element from
the second period, which left only the $2p$ elements B, C, N, O and F as possibilities.
Sulfide, the most obvious candidate, was unconditionally ruled out.

The final piece of the puzzle was provided in 2011 by X-ray absorption spectroscopic measurements which showed that the atom at the centre of FeMoco was
carbon! The FeMoco cluster was then established to be [Fe_7MoS_9C] (Figure 8.26).
The only coordinating amino acids are histidine, which is bound to the six-coordinate
Mo ion, and cysteine to the Fe ion at the opposite end. Homocitrate is also a co-ligand
for the Mo ion. The cluster can be removed from the MoFe protein intact if kept
away from oxygen. Parallel biochemical studies showed that the carbide ion was
inserted into the cluster initially as a methyl group from a so-called radical SAM
(*S*-adenosyl-ʟ-methionine) enzyme NifB. Despite several atomic resolution crystal
structures of the active site, much remains to be understood about the mechanism of

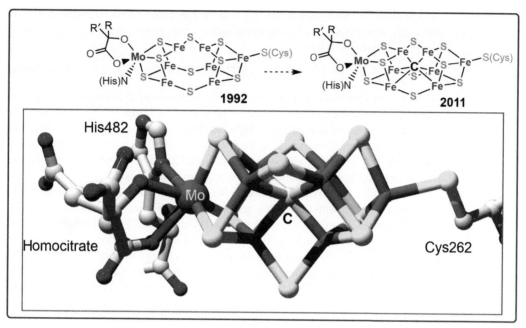

Figure 8.26

Structure of FeMo cofactor of *Clostridium pasteurianum* within the nitrogenase MoFe protein. The six-coordinate interstitial carbide is a feature (defined in 2011 but undetected in earlier studies).

dinitrogen (and proton) reduction at the FeMoco cluster. Fundamental questions such as where the N_2 molecule binds remain unresolved at this time. It is thought that, with H^+ reduction, the protons associate with the sulfido ligands as electrons are added to FeMoco and that bridging hydrido ligands are involved, spanning pairs of Fe atoms. The related VFe and FeFe proteins have either V or Fe in place of Mo but still bear the six-coordinate carbide ion. Nitrogenase catalyses many other (non-physiological) reductions. It will reduce other small molecules of similar size and or shape to N_2, including CO_2, CO, C_2H_2, SCN^- and CS_2, to mention a few. Finally, the appearance of six-coordinate carbon in the FeMoco cluster perhaps should not have been a complete surprise, as carbon atoms occupy similar interstitial sites within the structure of steel, where six-coordinate C atoms are surrounded by Fe atoms! Who would have thought that an enzyme and steel were related?

8.1.2.4.5 Oxygen Reduction

Respiration is central to all life forms and involves a series of redox reactions that generate energy in the form of adenosine triphosphate (ATP), the 'energy currency' for cellular reactions. In all species that breathe air, electrons generated by the oxidation of organic compounds (through various pathways) ultimately reach dioxygen where it is reduced to water in a concerted four-electron reaction. The enzyme that catalyses this reaction is cytochrome *c* oxidase (CcO), and it is located in the mitochondria (membrane-bound cell organelles). As its name suggests, the substrate is indeed the same heme protein (ferrous cytochrome *c*) that we have met before (Figure 8.15), and four of these deliver one electron sequentially to CcO. Cytochrome *c* is the penultimate step in a long electron transfer chain. The overall reaction is shown in Eq. (8.8). As electrons and protons often move together, an additional important outcome of

CcO turnover is proton translocation across the inner mitochondrial membrane. This process is an integral part of energy production where redox free energy is converted to a proton motive force.

$$O_2 + 4 \text{cyt } c_{Fe(II)} + 4H^+ \rightarrow 2H_2O + 4 \text{cyt } c_{Fe(III)} \tag{8.8}$$

Since dioxygen reduction provides much more energy ($\sim 400 \, \text{kJ} \, \text{mol}^{-1}$) than is needed to form a single ATP from ADP + phosphate ($\sim 30 \, \text{kJ} \, \text{mol}^{-1}$), it is to the cell's advantage to conduct the reduction stepwise. The actual process yields six ATP (i.e. $\sim 45\%$ efficient) per dioxygen. This involves intermediate oxygen reduction products, superoxide (O_2^-) and peroxide (O_2^{2-}), which are hazardous to biochemical systems if they 'escape'. Special metalloenzymes 'mop up' these ions.

The structure of CcO has been thoroughly studied in various oxidation states. It is a complex enzyme with four redox active cofactors (Figure 8.27) and some highly unusual features. A mixed-valent Cu(II)/Cu(I) cluster (Cu$_A$) with bridging cysteine ligands and a heme *a* cofactor (with two coordinated histidines) are electron transfer relays to the active site comprising heme a_3 (five-coordinate, one histidine coordinated) and the three-coordinate Cu$_B$ site (a Cu(I) centre with three histidine ligands). One of these histidines (His240) is cross-linked to a tyrosine (Tyr244) (highlighted by a red arrow) creating a uniquely modified amino acid pairing. Cu$_A$, heme *a*, heme a_3 and Cu$_B$ are all capable of accepting one electron; as well, the unusual tyrosine phenol group also participates in redox reactions.

Unsurprisingly, the multi-electron, multi-proton transfer mechanism of O_2 reduction by CcO is very complicated, and there are still aspects being investigated. It is without doubt that in its fully reduced (Fe(II)/Cu(I)) form the heme a_3/Cu$_B$ active pairing binds dioxygen in the vacant site between the two metals where, in a series of carefully orchestrated electron and proton transfer reactions, two water molecules are eventually released (Figure 8.28). The adjacent Mg^{2+} ion (Figure 8.27, centre) coordinated to three amino acid side chains binds three aqua ligands and it is from these coordinated H_2O molecules that the protons essential for turnover are delivered to the active site. The appearance of an oxido-Fe(IV) ('Compound II' type) intermediate is reminiscent of the peroxidase and cytochrome P450 enzymes and again shows nature's habit of repurposing the same machinery for different chemistry. The unique tyrosine residue also plays various parts in the mechanism as H-atom donor, electron acceptor and a proton acceptor at different points of the catalytic cycle. This is an example of some of the very sophisticated coordination chemistry going on in your body every time you draw breath.

8.1.2.5 *Isomerases*

Some enzymes catalyse the isomerisation of its (organic) substrate and are known as *isomerases*. Not all Fe/S proteins participate in electron transfer. A fascinating example is the enzyme aconitase, which is widely distributed in nature (including humans). There are different roles for aconitase enzymes, including as iron regulatory proteins. In mitochondria, aconitase functions as an isomerase to catalyse the isomerisation of citrate to isocitrate (Figure 8.29) as part of the tricarboxylic acid (TCA) cycle. The mechanistic fine details need not concern us, but this involves elimination of water then rehydration, and stereochemical control is important. The active form of aconitase bears a [4Fe–4S] cluster but only three of these Fe centres are coordinated by a cysteine side chain so the fourth is labile; when it dissociates to give a [3Fe–4S] cluster, all activity is lost. Coordination of citrate to the labile Fe centre is essential, and the structure of the product isocitrate coordinated to the complete [4Fe–4S] cluster illustrates the importance of a complete cluster for activity.

The cobalamins (vitamin B_{12}, introduced in Section 8.1.1.3) are cobalt-containing metallocofactors (like hemes) that bind to various enzymes where they carry out a variety of functions. Cobalamins are unusual, not only as one of nature's rare organometallic compounds (forming Co–C bonds), but also as the only vitamin known to contain a metal ion. Vitamin B_{12} is essential for all higher animals but is not found in plants.

One function of cobalamins is as cofactors of isomerase enzymes such as methylmalonyl-CoA mutase (Figure 8.30). The mechanism is very different from that of aconitase as it involves radical intermediates facilitated by the redox activity of the cobalt centre. Vitamin B_{12} is formally a cobalt(III) compound, but homolytic dissociation of the Co(III)–C bond generates Co(II) and an adenosyl radical; this then undergoes reversible H-atom transfer with the substrate, which rearranges to give the product. A general mechanism is shown in Figure 8.31.

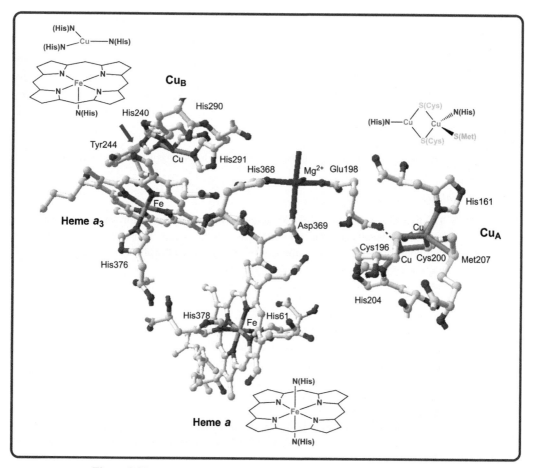

Figure 8.27

Structure of the key cofactors within bovine heart cytochrome *c* oxidase, along with line drawings.

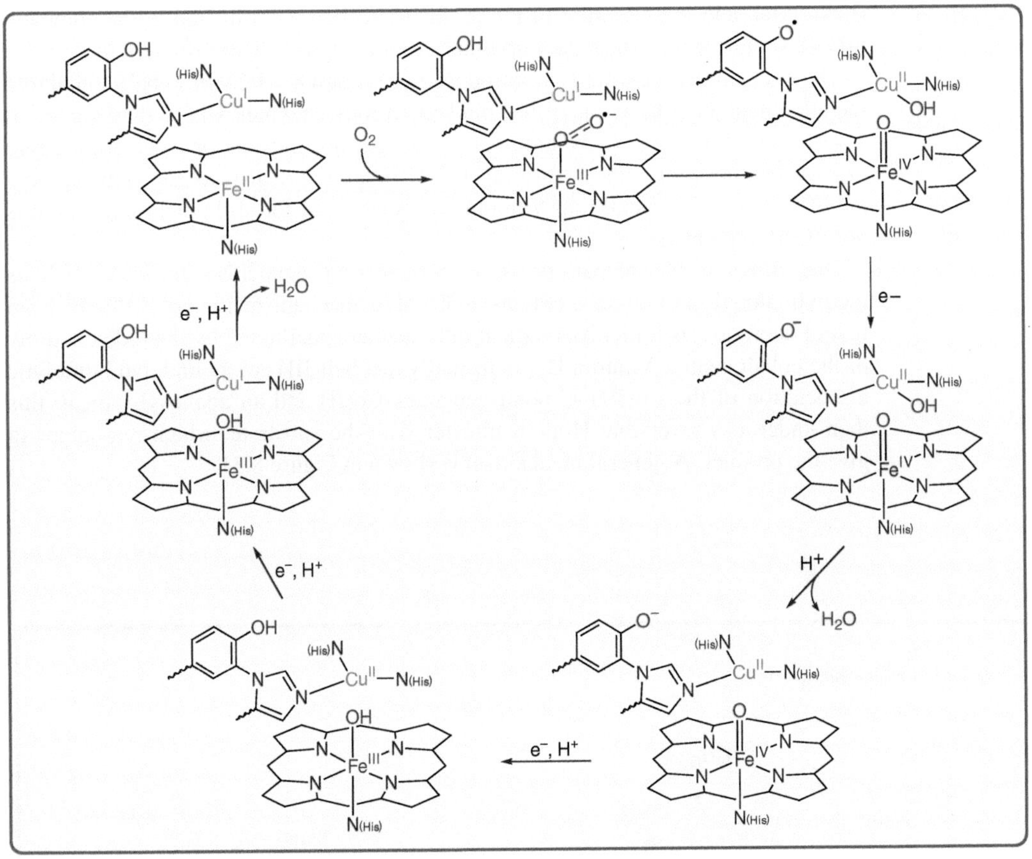

Figure 8.28

Mechanism of O_2 reduction by cytochrome c oxidase. Each electron is delivered by ferrous cytochrome c.

8.1.2.6 Hydrolases

The hydrolysis of a substrate (organic or inorganic) only requires water as a reagent, and this is usually in plentiful supply in living systems. However, the hydrolysis of compounds such as amides and esters (organic and phosphate) is generally slow, so there are enzymes (all within the hydrolase family) that can accelerate these reactions by many orders of magnitude relative to their uncatalysed rates. Some enzymes containing zinc belong to this class. Zinc proteins are common, but the absence of any accessible oxidation states apart from d^{10} Zn(II) means that they have no redox activity. There are many Zn(II) proteins that have been structurally characterised now, and their structural chemistry shows the expected distribution of four-, five- and six-coordinate geometries as found in related synthetic complexes. Although it does not have the redox activity of Fe or Cu, the most significant role of Zn(II) ions within enzymes emerges as catalytic centres for hydrolysis reactions of organic and inorganic substrates.

It is well known that base-catalysed hydrolysis reactions are accelerated by hydroxide being a better nucleophile than water. However, at physiological pH (around 7.4), the concentrations of hydroxide are negligible. As we have seen in earlier chapters, the coordination of water to a transition metal ion dramatically enhances its acidity. Lying at the end of the first transition metal series, the Zn(II) ion

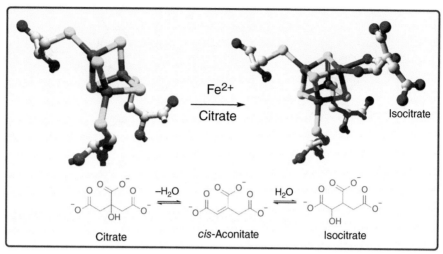

Figure 8.29
Interconversion between the inactive (3Fe–4S) and active (4Fe–4S) cluster forms of aconitase with the product coordinated on the right-hand side.

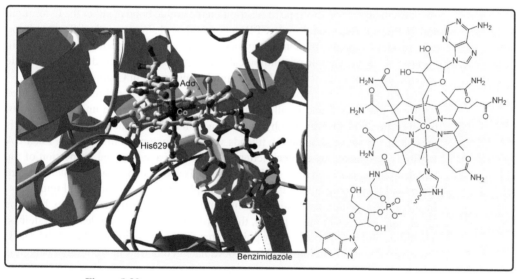

Figure 8.30
Active site of *Mycobacterium tuberculosis* methylmalonyl-CoA mutase with the cobalamin cofactor shown. Note dissociation of the benzimidazole pendent arm by a histidine reside from the protein. The other axial ligand is C-bound 5′-deoxyadenosyl.

has the highest effective nuclear charge and is best placed of all divalent metal ions to polarise an O–H bond of water. When Zn(II) ions are embedded within the active sites of proteins, the pK_a of their aqua ligands fall in the region of 7–8 so the concentrations of the aqua and hydroxido complexes at physiological pH are similar (Figure 8.32).

An additional feature of Zn(II) ions related to the above point is that they are labile centres and effective Lewis acids; consequently, they can rapidly bind other ligands with high affinity. As illustrated in the generic scheme shown in Figure 8.33,

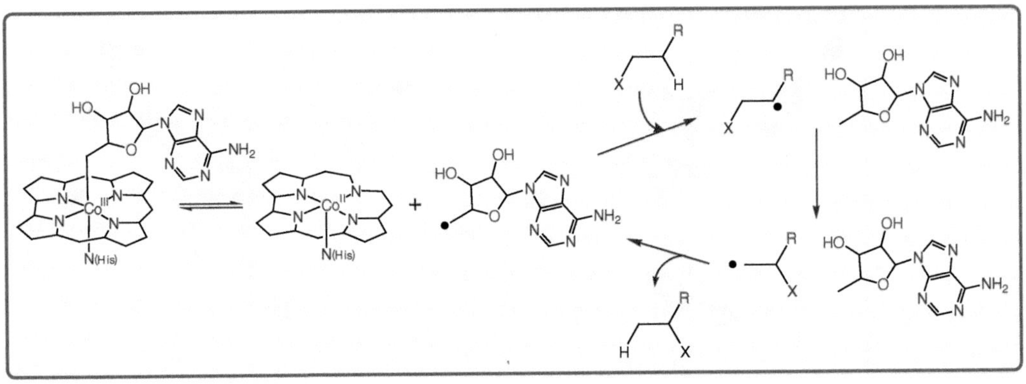

Figure 8.31

The isomerisation reaction facilitated by the octahedral adenosylcobalamin cofactor, showing radical intermediates.

coordination of an amide via its carbonyl O-atom further polarises the C=O bond, making the C-atom more electrophilic, and the adjacent hydroxido ligand is perfectly positioned to attack, leading to irreversible amide hydrolysis and release of the free primary amine formed. The Zn-dependent carboxypeptidases (CP) are digestive enzymes that perform this type of reaction on the amide bond at either the C-terminal end (CPA) or N-terminal end (CPB) of peptide substrates, with different amino acid residues targeted by some enzymes over others. The mechanism depicted in Figure 8.33 is an example of a biological reaction involving a coordinated hydroxide nucleophile, reminiscent of synthetic examples discussed in Chapter 6.

The Zn-dependent carbonic anhydrases (CA) are a large and intensively studied family of enzymes found in all forms of life. They catalyse the interconversion of bicarbonate and CO_2; this (de)hydration reaction regulates the local pH. Zinc(II) is present at the active site, but the ligands vary from one class to the other. Each CA family is designated by a Greek letter and currently a substantial proportion of the Greek alphabet has been used to classify them. Two of the most studied members are the α- and β-carbonic anhydrases (CA). Examples of each are shown in Figure 8.34. In the α-CAs (which are also found in humans), three histidine residues bind the Zn(II) ion, whereas in many β-CAs, the coordination sphere includes two cysteines, a histidine and an aspartate. A key point is that the coordination number of the Zn(II) ion is variable and both four- and five-coordinate geometries interconvert easily. It is relevant that Zn(II) is a d^{10} ion, and consequently has no real preference (ligand field

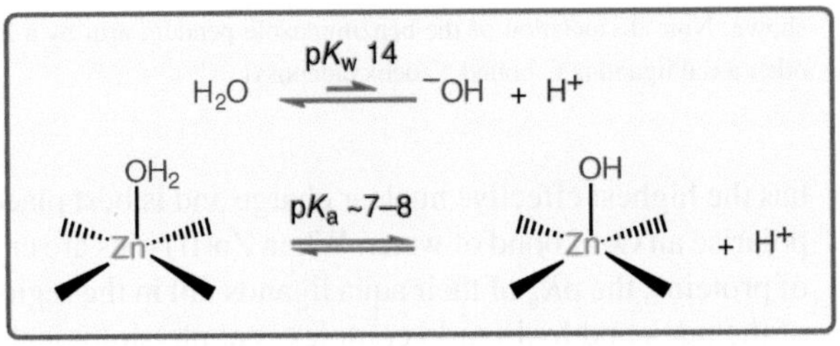

Figure 8.32

Ionisation equilibria of free and Zn(II)-coordinated water showing the electron-withdrawing effect of the metal cation.

stabilisation energy) for any geometry. The CA mechanism shown in Figure 8.35 has similarities with that of carboxypeptidases in Figure 8.33.

As seen in earlier sections, sometimes two metals are better than one. The same applies to dinuclear hydrolase enzymes. There are many examples where the metals may be the same or different, but one example is the family of purple acid phosphatases (PAPs) which are found in plants and mammals and catalyse the hydrolysis of phosphate monoesters. Their active sites contain two metal ions in proximity, one being Fe(III), while the other is a divalent metal that is always Fe(II) in mammals but can be Mn(II), Fe(II) or Zn(II) in plants. The coordinating amino acids are highly conserved across all known PAPs including a bridging aspartate and a bridging hydroxido ligand in the absence of substrate (Figure 8.36). The PAP enzymes catalyse phosphate ester hydrolysis under mildly acidic conditions and have been found to be involved in a diverse range of biological processes. In humans, they participate in bone metabolism. Their purple colour originates from the tyrosine phenolate ligand coordinated to Fe(III) which leads to an intense metal-to-ligand ($O^- \rightarrow$ Fe(III)) charge transfer transition.

The mechanism of PAP has been well studied and is illustrated in Figure 8.37. Several features that are particular to this dinuclear active site warrant mention. The phosphate ester coordinates in a bridging mode which straddles the two metal centres and increases their coordination number from 5 to 6. The adjacent bridging hydroxido ligand rapidly attacks the electrophilic P-atom, and the alcohol is released, leaving phosphate coordinated in a tridentate mode. Water enters the active site, and the phosphate reverts to a monodentate coordination before being displaced altogether by protonation and re-establishment of the bridging hydroxido ligand. The two metals have very different Lewis acidities. The Fe(III) centre is a classic 'hard' metal ion, and five of its six ligands are O-donor carboxylates, phenolates or hydroxide. The 'soft' divalent site includes two N-donor histidines and an O-bound amide (from asparagine). The phosphate product naturally will have a higher affinity for the Fe(III) site, so it gravitates to this metal before finally departing as HPO_4^{2-}.

So where would we be without metals in biology? It is hard to imagine, given the vital roles they play with every breath (dioxygen absorption, transport, storage and reduction), in addition to nerve cell function and energy production. The material presented here is just the tip of the iceberg, but what should be clear is that every

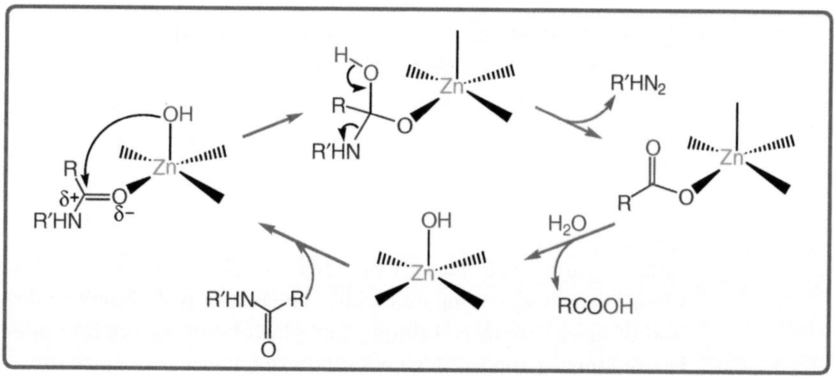

Figure 8.33

General catalytic cycle for Zn-dependent peptidase enzymes, which perform amide hydrolysis to an amine and a carboxylic acid.

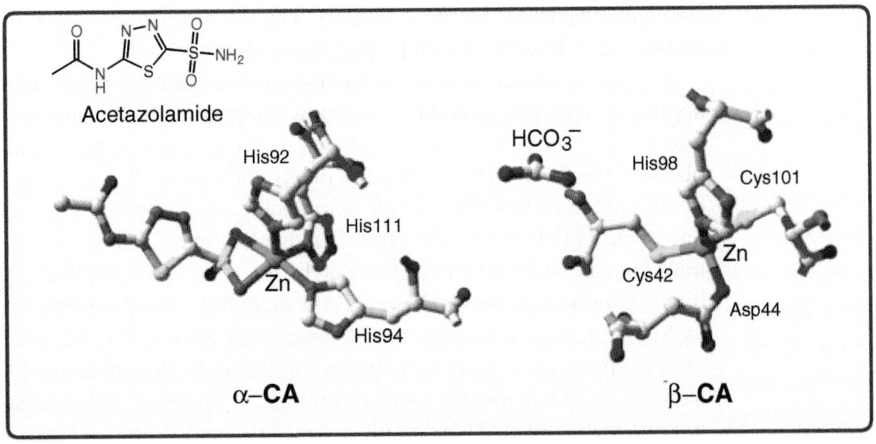

Figure 8.34
Active sites of (at left) α-carbonic anhydrase from the bacterial pathogen *Neisseria gonorrhoeae* complexed with the inhibitor acetazolamide, and (at right) β-carbonic anhydrase from *Hemeophilus influenzae* with bicarbonate at the active site.

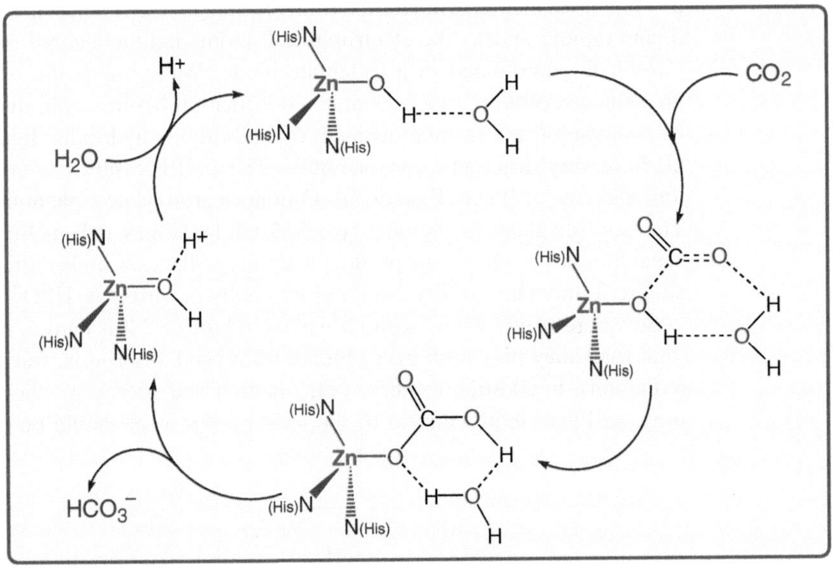

Figure 8.35
A mechanism featuring the ligand environments for zinc(II) in α-carbonic anhydrase II during catalysis.

system discussed in this chapter is founded on the fundamental principles laid out in earlier sections. The rules are the same. Nature just has a way of doing things in very sophisticated ways compared with synthetic chemists, but to be fair to us it had a head start of more than three billion years of evolution on Earth to optimise its chemistry before humans appeared, so we still have a fair way to catch up!

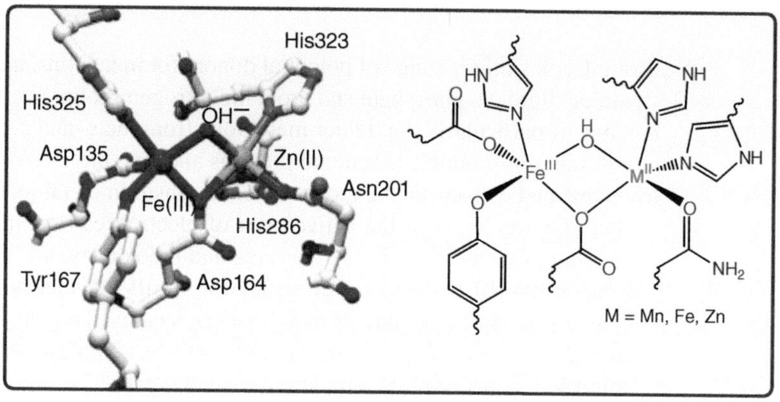

Figure 8.36

The active site of the dinuclear Fe/Zn enzyme purple acid phosphatase (from red kidney bean). The ferric ion present in all PAPs is highlighted in magenta. The chemical structure drawing on the right shows known variations at the divalent metal centre site.

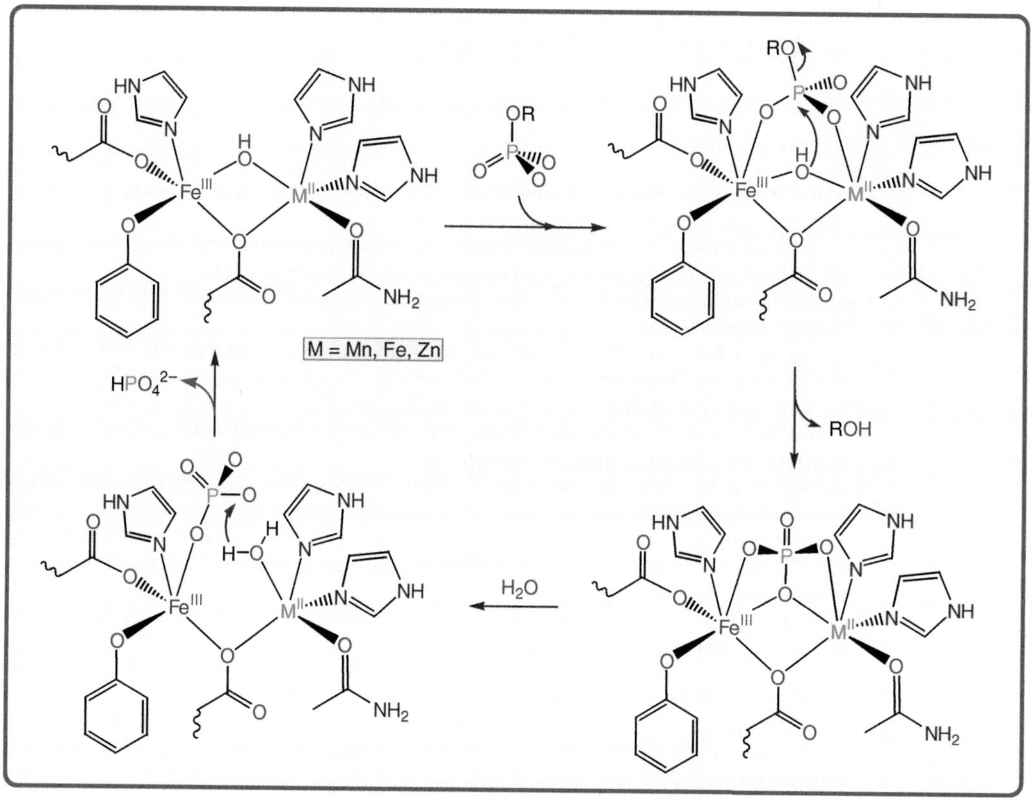

Figure 8.37

The proposed mechanism of phosphate ester hydrolysis for purple acid phosphatase.

Concept Keys

Biomolecules offer a range of potential donors for metal ions, including carboxylate, amine, thiolate, phosphate and aromatic nitrogen groups.

A range of particularly the lighter metal ions from the *s*- and *d*-block find biological roles in non-proteins, functional proteins and enzymes.

Iron, zinc and copper are the most common transition metal ions in the human body and play key roles at the active sites of electron carrier, structural and oxygen binding proteins as well as in cleavage and redox enzymes.

Overall, a limited number of key structural motifs are met in the molecular components acting as ligands in metallobiomolecules. The oxygen binding proteins *myoglobin* and *hemoglobin* are examples of iron complexes of one class, involving unsaturated macrocyclic tetraamines called hemes.

Metal complexes lie at the heart of proteins involved in vital life processes of oxygen storage, transport and reduction in all life forms.

Polynuclear metal units are common in biomolecules as exemplified by the ferredoxins, which contain up to four iron centres in an iron–sulphur cluster.

Mixed-metal polynuclear units are also met in enzymes, with each metal ion usually playing a different role in the chemistry performed by the enzyme.

The chemistry in metalloenzymes usually occurs at the metal centre(s) and effectively involves reactions of coordinated ligands.

Further Reading

Bertini, I., Gray, H.B., Stiefel, E.I., and Valentine, J.S. (2006). *Biological Inorganic Chemistry: Structure and Reactivity*. University Science Books.

    A broad account of bioinorganic systems, from a perspective that provides good exposure of students to metal environments and active sites in metalloproteins.

Lippard, S.J. and Berg, J.M. (1994). *Principles of Bioinorganic Chemistry*. Mill Valley: University Science Books.

    An aged but accessible introduction for students; since fundamentals don't change markedly, it is still relevant.

McDowell, L.R. (2002). *Minerals in Animal and Human Nutrition*, 2nd ed. Elsevier.

    For those wanting to stray more deeply into the role of metals in nutrition, this is a useful resource, with good coverage of core aspects.

Roat-Malone, R.M. (2020). *Bioinorganic Chemistry: A Short Course*, 3rd ed. Hoboken: Wiley.

    A student-focused coverage of bioinorganic chemistry with good readability. Apart from introductory coordination chemistry, it covers metals in proteins, metalloenzymes, and includes coverages of both applicable computational methods, metals in medicine, and nano bioinorganic chemistry.

Trautwein, A. (1997). *Bioinorganic Chemistry: Transition Metals in Biology and Their Coordination Chemistry*. Weinheim: Wiley-VCH.

    An overview of biological systems, with a coordination chemistry perspective consistent with the approach in this chapter.

Williams, F. (Ed.) (2022). *Principles of Bioinorganic Chemistry*. ML Books International.

    A recent, more advanced book effectively covering all key topics that may serve as a reference source for those seeking to pursue more detail. A good example from a growing range of new books in the field.

9 Complexes and Commerce

It would be inappropriate, before concluding, to leave the subject without some mention of the roles that coordination complexes play in the working world. Otherwise, one could be left with the impression that the field is little more than an academic plaything. Of course, we have seen in Chapter 8 how Nature has made fine use of coordination complexes, but this goes on essentially without human intervention. What is important to appreciate is that coordination chemistry and coordination complexes lie at the core of an array of important applications in industry, medicine and other fields. Applications turn up in surprising ways. For example, the formation of a coordination complex of iron(II) sulfate with the natural compound gallic acid (3,4,5-trihydroxybenzoic acid) is the source of the blue colour of iron gall ink, used widely from at least the 4th to the early 20th century. Coordination complexes even play roles in the food and agriculture industries; for example, inert ferrocyanide ion, $[Fe(CN)_6]^{4-}$, is widely used as an anti-caking agent for common salt, and has found use in citric acid and wine production.

Given the inventiveness of humans, it should come as no real surprise to find that the development of new compounds often leads to an analysis of their potential applications, and these can be surprisingly diverse. For example, who could have thought that the simple palladium complex $[PdCl_2(PPh_3)_2]$ would play a key role in the synthesis of the well-known anti-inflammatory drug ibuprofen? Here, we shall touch on a few well explored areas and examples, to give but a flavour for the field; a full serving can be pursued in specialist texts, if desired.

9.1 Metals in Medicine

Medicine is an area of high activity and interest, partly because health issues are of strong interest to us all as we seek to achieve and retain the best possible quality of life. Recovery from and or control of disease often involves the use of drugs, which may be natural products, synthetic organic compounds and, though less often met, synthetic coordination complexes. To distinguish the latter from the more common organic drugs, those containing metal complexes are sometimes defined as *metallodrugs*. The use of metals in medicine has a history thousands of years old, at least back to the ancient Egyptians (who used copper compounds in potions) and Chinese (who used gold compounds), and perhaps beyond. In the distant past, some treatments may have caused more damage than provided a cure, but to be useful in modern medicine, a compound must provide a measured beneficial effect while displaying low toxicity. The *therapeutic index* (TI) defines relative toxicity versus

Introduction to Coordination Chemistry, Second Edition. Paul V. Bernhardt and Geoffrey A. Lawrance.
© 2025 John Wiley & Sons Ltd. Published 2025 by John Wiley & Sons Ltd.
Companion website: www.wiley.com/go/coordinationchemistry2e

benefit, expressed as LD_{50}/ED_{50}, or the ratio of the median dose required to kill 50% of a host (clearly undesirable!) versus the median dose needed to produce an effective therapeutic response in 50% of the host; ideally TI ≫ 1. Before any new drug can be introduced, it must go through a series of clinical trials covering, for human use, mostly three phases; needless to say, given the expense and time involved in this process, a new drug has to be markedly better than any others already on the market to succeed commercially. Phase I screens for human toxicity, Phase II for activity against the medical condition of concern, and Phase III benchmarks activity against other existing medications, and these data largely determine if further development will be economically viable.

Over time, a wide range of essential and non-essential elements have found application in medicine, as exemplified in Figure 9.1. Almost all of the *s-*, *p-* and *d-*block elements have been probed, and even some *f-*block elements. Naturally, many metal ions feature in medically-related applications, partly as a result of their dominance of the periodic table. Whereas at least ten metals (as well as a similar number of non-metals) are considered essential for human life (Chapter 8), at present about 50 other non-essential metals and radionuclides have found use in therapy, diagnosis, nuclear medicine and physical implants. Trace levels of many metals appear in the human body, although a good number of these are best viewed as environmental contaminants and non-essential. The essential/non-essential divide, as well as medical applications, is constantly disturbed by a vast amount of on-going research, so Figure 9.1 is certainly a 'work in progress', but nevertheless a useful guide.

Pharmaceutical companies have appeared somewhat resistant to the development of metallodrugs in the past, perhaps concerned about toxicity, specificity and metal accumulation in patients over extended periods of use but also reflecting relative unfamiliarity with the background coordination chemistry. They have also benefited from serendipity as regards discovery of some current metallodrugs, allowing them to put aside rational design in approaches to specific targets. Metallodrugs are often considered pro-drugs since they commonly undergo activation by ligand substitution or redox reactions *in situ*, which usually complicates establishing structure–activity relationships, and this influences their pursuit. Changes are afoot, however, with growing recognition in recent decades that metallodrugs can be sophisticated options to conventional pure organic drugs, tied not only to reactivity and stereochemical differences but also to access to nuclear, magnetic and optical properties simply not available with organic drugs.

9.1.1   Introducing Metallodrugs

Metal-containing coordination compounds that show a capacity to cure or control a disease have grown remarkably in number and range of applications in recent decades. They may alleviate symptoms and regulate the immune system in humans in various ways; these include interrupting infections, affecting the process of virus replication by inhibiting critical enzymes involved, and serving to create a stronger immune response and hence enhance the efficacy of antiviral drugs and vaccines. Their form covers a wide range of metal ions, ligands and stereochemistries, their accessible geometries going well beyond the limited geometries enforced at carbon centres. Some examples appear in Table 9.1; at the beginning of the 2020s, specific complexes of about ten *d-*block metals (and about the same number of other metals) have been approved as drugs in the EU, United States, and many other countries;

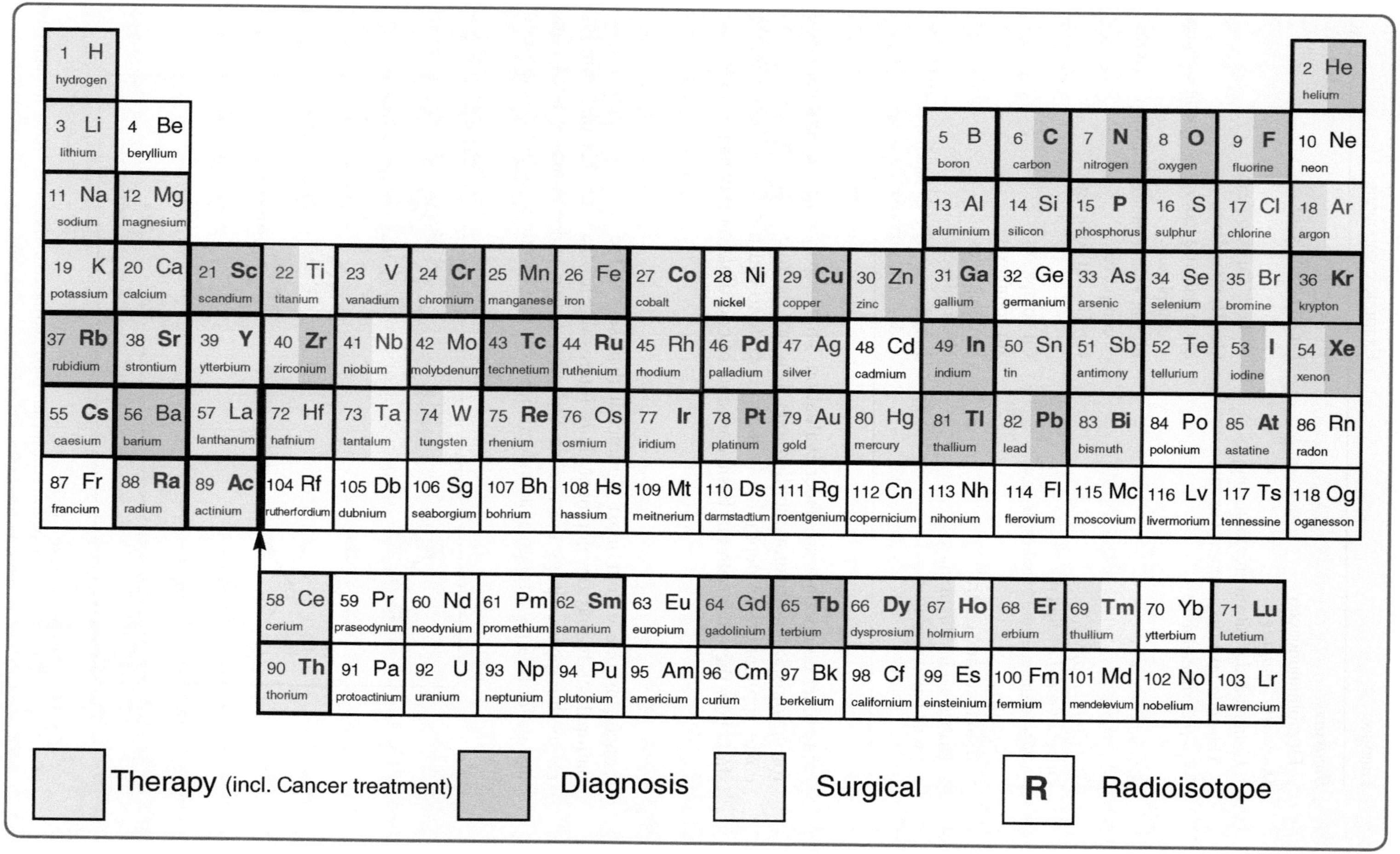

Figure 9.1

A medicinal periodic table. Elements in coloured boxes are present in compounds and/or materials in clinical trials or approved for either therapeutic use (light blue), including cancer treatment, diagnostic use (pink), surgical applications (yellow) or several of these. Elements in red possess potentially medically useful radioisotopes.

Table 9.1   Some medically-related applications of metal coordination complexes.

Function	Example compounds
Bioassay	
Fluoroimmunoassay	Eu(III) complexes
Diagnostic imaging	
Gamma and positron emitting isotopes	^{99m}Tc, ^{64}Cu complexes
Magnetic resonance contrast agents	Gd(III) complexes
Radiopharmaceuticals	Short half-life radioactive metal salts and complexes
Anticancer drugs	Pt, Ru, Ti and photosensitive porphyrin complexes
Antiviral therapy	Co, Ga, Au complexes
Other selected treatments	
Wilson disease and Menkes disease	Cu complexes
Arthritis	Au complexes
Blood pressure control	Fe complexes
Diabetes—insulin mimics	V complexes
Metal poisoning/overload	Polydentate ligands (such as EDTA, desferrioxamine)

many other compounds are under evaluation. This is apart from the compounds used for diagnostic and imaging applications. It is not the role of an introductory text to cover applications in depth, but it is appropriate to illustrate their applications and show how they link with our basic concepts of coordination chemistry.

9.1.2   Anti-cancer Drugs

Cancer, as one of our most significant diseases, is the focus of intensive international study and development. It is interesting to discover that platinum-based chemotherapeutic drugs are among the most active and commonly used clinical agents for treating a range of advanced cancers. *cis*-Diamminedichloridoplatinum(II)

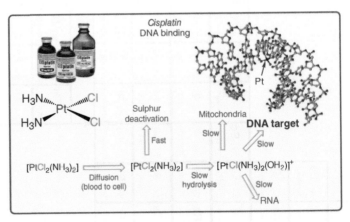

Figure 9.2

The drug *cisplatin*, its reaction pathways upon diffusion into the cell, where inactivation through binding to sulphur donors and to mitochondria or RNA compete with efficacious binding to the DNA target. Also shown is an example of coordination to DNA, leading to distortion of the double helix.

(*cis*-[PtCl$_2$(NH$_3$)$_2$], called *cisplatin* in medical use; Figure 9.2) was the first to be used clinically, and remains one of the largest-selling drugs on the market. This is an old coordination compound, first reported by Michele Peyrone in 1845, with its square planar structure defined by Alfred Werner. The potential anti-cancer properties of cisplatin were discovered serendipitously by Barnett Rosenberg in 1965, and the drug was introduced clinically in the late 1970s. It is routinely widely used, including for treating ovarian, testicular, bladder, head, neck, cervical and small-cell lung carcinomas. In particular, as a drug cisplatin has made testicular cancer eminently curable, with more than 90% of sufferers now cured. A high-profile beneficiary was American cyclist Lance Armstrong, a seven-time *Tour de France* winner, who ironically was duly stripped of all titles for taking performance enhancing drugs. Efforts to enhance specificity of attack at cancerous cells and minimise side effects in patients has led subsequently to the development of a range of related complexes, discussed below.

Substitution reactions of coordinated chlorido ligands are the key chemistry in reactions of cisplatin in a human cell. Most cisplatin circulates in the blood unchanged over a short timeframe, as the high chloride ion concentration in blood suppresses chlorido hydrolysis and hence further reaction. Once across the cytoplasmic membrane of a cell, however, it meets a much reduced chloride concentration (as low as 4 nM) and undergoes a series of hydrolysis reactions, forming species including [PtCl(NH$_3$)$_2$(OH$_2$)]$^+$, where one chlorido ligand has been substituted by water. Although the biological target of the complex is DNA in the cancer cell, it is not particularly discriminating in its chemistry, and can participate in coordination chemistry with sulfur-containing groups and other functionalities that lead to its 'capture' and inactivation; it binds so strongly to these that it is no longer available for further chemistry. The interaction sought (Figure 9.2) is with the heterocyclic nitrogen bases that form part of DNA — cytosine, guanine, adenine and thymidine. These types of N-donors are excellent ligands for platinum(II); guanine is believed to be a particularly favourable donor. Binding is covalent, but as mentioned earlier, chlorido ligand hydrolysis is necessary first to provide a more readily substituted site (an aqua ligand) that facilitates the introduction of the N-donor DNA base into the inner coordination sphere. The cationic aqua complexes formed following initial chloride hydrolysis (particularly [Pt(NH$_3$)$_2$(OH$_2$)$_2$]$^{2+}$) enhance the activity of the drug as a cytostatic agent due to an ionic attraction to the negatively-charged DNA helix, which has an anionic phosphate backbone. However, it is the structural changes arising from coordination of N-donor DNA bases in place of aqua ligands that are the key to their activity.

In addition, cisplatin may bind to other proteins and biomolecules, sometimes together with binding to DNA. In achieving binding to two adjacent DNA bases, it is using both *cis* coordination sites originally occupied by chloride ions so that hydrolysis of both chlorido ligands is necessary. Possible binding modes are shown in Figure 9.3. Which of these is most important is under continued debate, but intrastrand processes are favoured. Cisplatin can form intrastrand crosslinks between adjacent bases, preferring two guanine bases (a cisplatin–guanine–guanine intrastrand DNA adduct is depicted in Figure 9.3). This coordination mode cannot be adopted by biologically inactive *trans*-[PtCl$_2$(NH$_3$)$_2$], suggesting it is indeed of importance. This cross-linking has been shown to cause unwinding and duplex bending, perhaps sufficient to attract some of the high mobility group family of proteins able to recognise defective DNA, which bind and further inhibit replication.

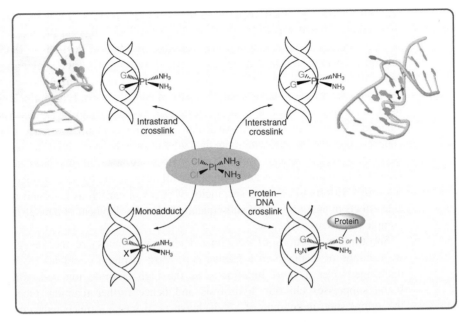

Figure 9.3
Various modes of coordination of the platinum(II) drug cisplatin to bases in DNA strands, following initial chloride hydrolysis; a guanine N-donor (G) is favoured. Models based on structural studies of intra-strand (at left) and inter-strand (at right) coordination to DNA oligomers are also shown.

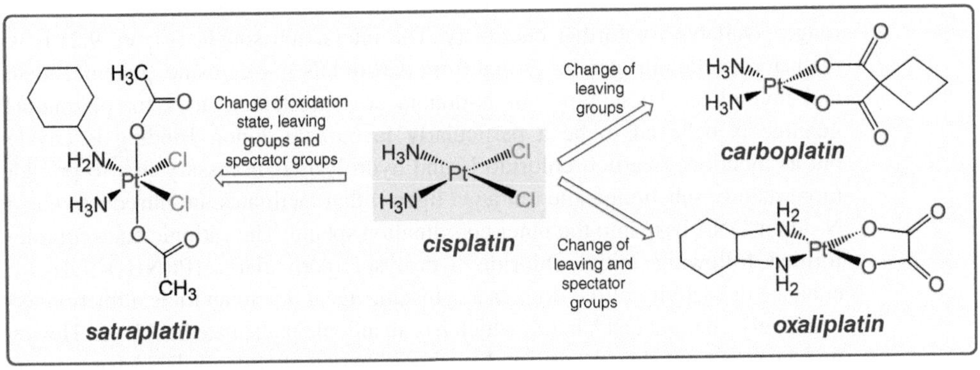

Figure 9.4
Evolution of the first-generation platinum drug cisplatin into later generation drugs.

The core problem with cisplatin is that dose-related toxicity occurs, leading to an array of problems for patients, such as joint pain, ringing in the ears and hearing difficulties, and general weakness. The maximum tolerated dose of cisplatin is around 100 mg/day for up to only five consecutive days. While the drug displays high activity, these unfortunate side-effects led to a search for a better analogue (Figure 9.4). Subsequently, carboplatin [*cis*-diammine(1,1-cyclobutanedicarboxylato)platinum(II)] was introduced as a second-generation drug in 1990. It shows about a fourfold reduction in side effects to the kidneys and nervous system. Because the drugs must circulate in the bloodstream before finding their target, slower degradation *in vivo* is believed to be one factor of importance.

Both cisplatin and carboplatin suffer from the emergence of acquired resistance with time. Subsequently, oxaliplatin [(1,2-cyclohexanediamine)(oxalato)platinum(II)] was introduced as a third-generation drug. It is active in all phases of the cell cycle and binds covalently to DNA guanine and adenine bases, dominantly via intrastrand cross-linking. It is thought that DNA mismatch repair enzymes are unable to recognise oxaliplatin–DNA adducts in contrast with some other platinum–DNA adducts, as a result of the bulkier nature of the complex. However, the complex is inactivated by thiol protein binding, alteration in cellular transport and increased DNA repair enzyme activity. Development of a new-generation drug to address inactivation has led to satraplatin, the first orally-administrable platinum drug—in this case a platinum(IV) complex. A mode of action that involves reduction to Pt(II) *in vivo* has been proposed. This drug shows no neurotoxicity or nephrotoxicity, with myelosuppression (decreased bone marrow activity) the dose-limiting adverse effect; it is under consideration for approval as a drug in prostate cancer treatment. There are metal-free competitors for the latter treatment; enzalutamide is an organic benzamide nonsteroidal chemical targeting the androgen receptor which is currently used in the treatment of prostate cancer.

Platinum complexes are amidst the best-selling anticancer drugs in the world, with billion-dollar annual sales, although they meet strong competition from an array of patented organic compounds, several of which lead the best-selling list. Three complexes, cisplatin, carboplatin and oxaliplatin, are in international use, with almost half of cancer patients worldwide receiving platinum drugs; an additional three (nedaplatin, lobaplatin and heptaplatin) have been approved for use in some countries, but these are traditional Pt(II) complexes with *cis*-N_2O_2 donor sets that differ only in the ligands employed. Nevertheless, the search for better platinum drugs goes on, including development and trial of some polynuclear complexes, and a focus on complexes that operate via mechanisms of action that differ from those employed by the approved drugs (such as *trans* rather than the conventional *cis* isomers) or else involve targeting of different species such as sugars, steroids and peptides. Apart from the array of platinum(II) compounds examined, several platinum(IV) complexes are under investigation as prodrugs following the promise shown by satraplatin. Chemotherapy with platinum drugs can be effective but may have severe side effects because the drugs are toxic and cannot discriminate sufficiently between normal and cancer cells. Development of new drugs centred on the introduction of components that deliver the drugs specifically to cancer cells will assist in specificity and reduction of side effects.

Although platinum drugs have a well-established place in the array of treatments used for cancer sufferers, they are not the only coordination complexes that exhibit activity. Side effects and resistance development with platinum drugs have been two aspects associated with the drive to examine complexes of other metals. Overall, at this time, the alternative metal complexes for anti-cancer therapy include (in alphabetical order) antimony, arsenic, bismuth, copper, gallium, gold, iron, rhodium, ruthenium, titanium and vanadium, although most of these are not approved for use as drugs yet. In effect, complexes of most of the accessible metals in the periodic table have been the subject of some level of study as potential drugs, driven by the profits to be made from a successful candidate as much as by a simple desire to develop compounds with enhanced performance.

After platinum, the most interest in metallodrug research is arguably in its coinage metal neighbour ruthenium. Two Ru(III) compounds stand out in terms of clinical studies to date, which currently have the rather nebulous names NAMI-A and

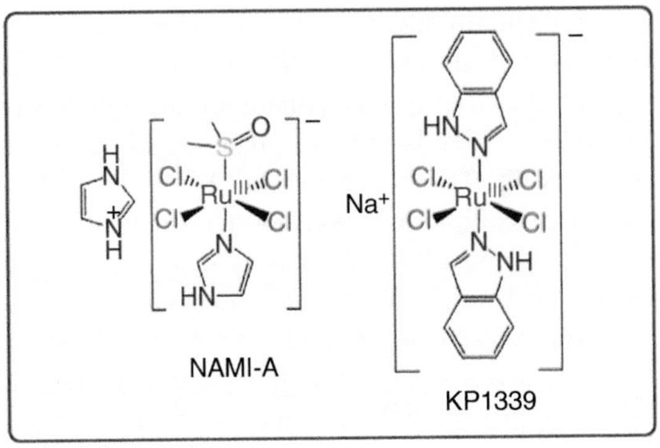

Figure 9.5
The Ru(III) anticancer drugs NAMI-A and KP1339.

KP1339 (Figure 9.5). The ligands are not sophisticated, and their structures are obviously similar. NAMI-A includes an S-bound dimethyl sulfoxide (DMSO) ligand. The preference here for S-coordination of DMSO is unusual, as O-coordination is more common especially with 'hard' transition metals. The remaining co-ligands are chloride and N-heterocycles. While a deep understanding of the mode of action and metabolism of any metallodrug is work in progress, it is believed that both NAMI-A and KP1339 are activated by reduction to the Ru(II) oxidation state which then leads to ligand substitution. In this context they, like cisplatin, are *prodrugs* which are converted *in vivo* to a more active form. An important distinction, though, is their mode of action; KP1339 is cytotoxic (killing cancer cells), while NAMI-A is antimetastatic (inhibiting cancer cell growth spreading to other tissues).

Another family of complexes under examination as anti-cancer drugs are those of titanium, some complexes of which show a range of anti-tumour activity. One titanium(IV) anticancer drug, budotitane [Ti(bzac)$_2$(OEt)$_2$] (Figure 9.6), was the first non-platinum complex assessed in clinical trials, with activity assigned to a DNA intercalating mechanism associated with the presence of the aromatic groups in the ligand since the analogue without these groups is inactive. Briefly, intercalation in this context involves insertion of the flat aromatic ring of the metallo-intercalator between adjacent base pairs in a 'sandwich' configuration, again disrupting the structure of the double helix. Decomposition in biological environments is a challenge with this class of compound, although at least the final decomposition product, titanium dioxide, is safe and readily excreted. Recently, the titanium(IV) complex of a N$_2$O$_4$-hexadentate ligand formed from two phenolato and two alkoxido groups attached to an ethane-1,2-diamine core has shown high efficacy, low toxicity and enhanced stability likely to be related to the high denticity of the ligand chosen. Inclusion in this list possibly reflects the level of research activity as much as their capacity to intercept and eliminate cancer cells. A range of organometallic compounds such as titanocene dichloride (Figure 9.6) have also shown promise.

9.1.3 Other Metallodrugs

Apart from an ability to also act on cancer cells, complexes of gold(I) and gold(III) serve several other purposes; in fact, gold compounds are amidst the oldest remedies used by humans and have been employed since about 2,500 BCE. Given this

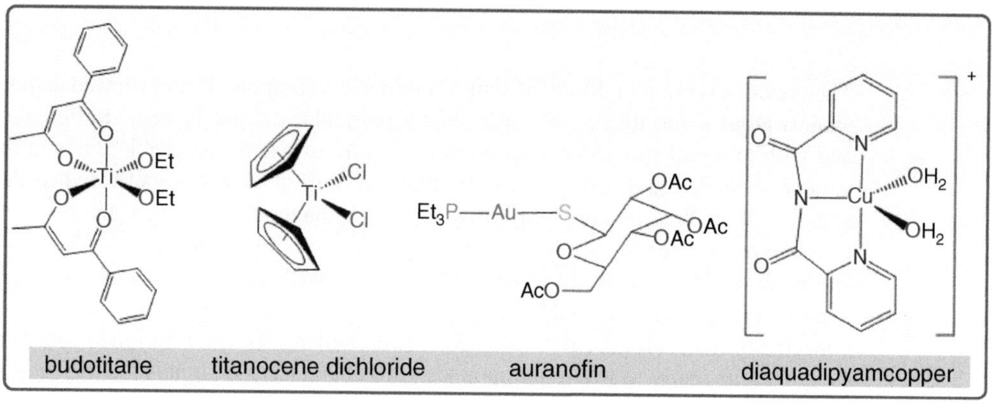

Figure 9.6

The structures (from left to right) of two titanium(IV) potential anti-cancer drugs, the gold(I) arthritis drug auranofin and a copper(II) protease inhibitor of the lentivirus HIV-1.

venerable history, it is hardly surprising that modern-day science has probed their applicability. The orally-administered gold(I) complex auranofin was developed for the treatment of rheumatoid arthritis. This is a typical linear two-coordinate complex of gold(I), with 'soft' S and P donors to match the 'soft' metal ion (Figure 9.6). One of its ligands is an acetylated glucosethiolate, whereas the other is a phosphine. The bulky ligands here promote compatibility in the bio-environment, and activity is influenced by the substituents on the RS^- and R_3P ligands. Auranofin is versatile and has also been investigated both as an anticancer drug and an antimicrobial agent targeting antibiotic resistant pathogens. Toxicity issues have limited its clinical use.

Employing an expensive element such as gold (and platinum or ruthenium) is not a requirement for bioactivity, with even the relatively inexpensive copper finding success. For example, a copper(II) complex (Figure 9.6) is effective as a competitive inhibitor of the lentivirus HIV-1 which may lead to AIDS over time. Vanadium is also relatively common, and its compounds, mainly square pyramidal complexes of the type $[O{=}V(O{-}R{-}O)_2]$, function as insulin mimetics with potential in diabetes control if toxicity issues can be overcome; one is in use as an orally-administered agent in animals. Manganese is the second most abundant transition metal and an element essential for humans, so it is hardly surprising that it has also attracted interest. For example, a pentagonal bipyramidal complex of manganese(II) with a pentaaza macrocycle acts as a superoxide dismutase mimetic.

Despite the promise shown by the utilisation of metals in medicine and their likely novel modes of action, they are still very much the exception than the rule in the marketplace compared with purely organic synthetic and natural products. To date, very few have been approved for use and there remain many hurdles to overcome in terms of metabolism and toxic side effects. The mechanisms of organic drug metabolism in the human liver principally using cytochrome P450 enzymes (see Chapter 8) is a mature field. Much less is known about the fate of metal ions in a metallodrug once it breaks down. After all, there are many competing ligands in the biological milieu, and this leads to potential risks of unintended toxic metal accumulation during a course of treatment. Nevertheless, the variety of complexes under study as potential drugs is astounding and this suggests that more advances will be reported in the future.

9.2   Medical Imaging and Diagnostics

The previous section highlighted some coordination compounds that showed activity against a particular disease. In order that a particular disease is treated effectively and with minimal side effects, an accurate picture of where the problem lies in the patient is needed and a way to ensure that the drug of choice is delivered to that site. The long-time use of X-ray imaging in the diagnosis of bone fractures is an example of how highly accurate information can be obtained to assist in diagnosis and treatment with negligible harm to the patient. With diseases such as cancers or conditions affecting heart and brain function, higher resolution imaging of the areas of concern are critical for accurate diagnosis. This challenge is now referred to generally as *molecular imaging* where compounds with characteristic signals can be tracked *in vivo*.

9.2.1   Nuclear Imaging

Due to its inherent sensitivity and zero-background response, the use of radioactive isotopes as signalling units is now a well-established field. The fact that the *d*- and *f*-block metals outnumber the *s*- and *p*-block elements means that there are many more options for suitable radioisotopes in these blocks with suitable coordination chemistry and radiochemical properties (half-life, decay pathways) than there are in the main group, which is basically limited at the moment to only a couple of elements and isotopes (^{11}C and ^{18}F). Furthermore, the incorporation of ^{18}F ($t_{1/2}$ 2 h) into a drug must be done through organic chemistry, which can take too much time. Coordination chemistry offers a solution here, and coordination compounds that include a metal radionuclide have made, and continue to make, great inroads in the radiochemistry field due to the potentially facile complexation reactions that can occur rapidly at room temperature under mild conditions.

While we will not go into great detail about the multifarious pathways by which a radioactive isotope may decay to a more stable daughter element, there are a couple of important points to highlight. Although radioactive elements can decay by emitting alpha particles (helium nuclei), beta particles (β^-, high energy electrons), positrons (β^+) or gamma radiation, not all pathways are suitable for medical imaging due to the degree of tissue penetration and or detector sensitivity. When injected into a patient, the radionuclide can be detected by its characteristic decay and, by collecting data from different directions, a 3D image of the affected tissues can be reconstructed, a process called *tomography* which effectively reconstructs 2D data into 3D shapes.

Detection sensitivity is paramount, and the imaging techniques of choice are positron emission tomography (PET—for positron emitters) and single photon emission computed tomography (SPECT—for gamma emitters). The positron, once emitted, travels only a short distance before it annihilates by combining with an electron to produce two gamma photons that travel in exactly opposite directions. Cameras at opposite sides of the patent detect these signals simultaneously, which provide accurate information about the origin of the radiation. This method, of course, relies on positron emitters and subsequently is less common than SPECT detection which only requires gamma emitters.

9.2.1.1   SPECT Imaging

Technetium is unique in being the only transition element in the first three rows without a stable isotope. Its longest-lived isotope (^{97}Tc) has a half-life of (a mere) four million years so if any Tc-containing ores were ever present in the Earth's crust 4.5 billion years ago, then they decayed a long time ago. Technetium was first isolated by Emilio Segrè and Carlo Perrier from a sample of molybdenum that had been bombarded with radiation. It was the first 'synthetic' element, and its name (from the Greek for 'artificial') reflects this. Of course there are now many synthetic elements. All 26 *trans-uranium* elements plus the lanthanoid promethium (like Tc an oddity in being radioactive, yet surrounded by stable elements) have only been synthesised with nuclear chemistry.

Nowadays ^{99m}Tc, the clinically used isotope, is produced by nuclear fission of ^{235}U (via ^{99}Mo) and used routinely around the world. The final step in its synthesis is beta decay of radioactive $[^{99}MoO_4]^{2-}$ ($t_{1/2}$ 3 d) to $[^{99m}TcO_4]^-$ (the clinically used metastable isotope, $t_{1/2}$ 6 h) that undergoes gamma emission to long lived $[^{99}TcO_4]^-$ ($t_{1/2} > 10^6$ y). Therefore, transportation of the $[^{99}MoO_4]^{2-}$ source must be rapid from reactor to clinic, which typically involves international air travel. After ^{235}U fission, the chemically stable molybdate anion $[^{99}MoO_4]^{2-}$ is separated from other fission products and then loaded onto an alumina column (also referred to as a generator) from which $[^{99m}TcO_4]^-$ can be eluted. In its 'as prepared' form, as pertechnetate $[^{99m}TcO_4]^-$, it has no tissue specificity, so it must be complexed to steer it towards its intended target. Two of the most used Tc complexes in the clinic are shown in Figure 9.7. The technetium(I) complex Sestamibi (or MIBI) is used extensively for imaging blood flow in the heart. Its specificity stems from it being both lipophilic and cationic and it is taken up by transport proteins and has a high specificity for heart tissue. The technetium(V) complex Mertiatide is employed for renal imaging, and this highly soluble compound remains in the bloodstream and passes rapidly through the kidneys of healthy subjects, whereas slow passage is an indication of renal impairment.

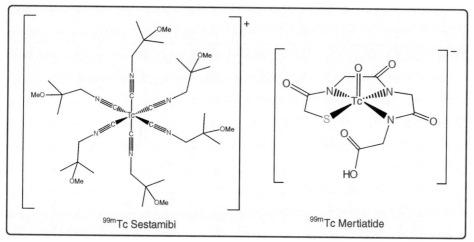

^{99m}Tc Sestamibi ^{99m}Tc Mertiatide

Figure 9.7

Two widely employed ^{99m}Tc compounds for SPECT imaging.

9.2.1.2 *PET Imaging*

The feature that makes PET radionuclides so attractive is the sensitivity (limit) at which they can be detected; this can be as low as picomolar concentrations, which means that the radiochemical dose can be low, thus minimising harmful side effects. PET detectors are more expensive and less common than SPECT, but the advantages in accuracy and sensitivity are driving the development of new positron emitting compounds. The main group positron-emitting isotope ^{18}F is typically incorporated in a glucose molecule to give fluorodeoxyglucose (FDG), which then is taken up by tissues with an abnormally high glucose demand that may indicate tumour growth, inflammation or heart disease. The isotope ^{11}C ($t_{1/2}$ 20 min) has also been intensely investigated due to the ubiquitous nature of carbon in anything living. It can be introduced via a ^{11}C-methylation or delivered as ^{11}C-acetate where it is incorporated into well-understood biochemical cycles. Regardless of the objective, the incorporation of ^{18}F or ^{11}C within a compound requires organic chemistry which is inherently slower and requires more forcing conditions than we find necessary in coordination chemistry.

There is a wealth of emerging positron-emitting radiometals that can be incorporated within a complex rapidly and under mild conditions; the isotope ^{64}Cu ($t_{1/2}$ 13 h) is one such example. As discussed in earlier chapters, Cu^{2+} is a labile metal ion and undergoes complexation reaction in most cases during mixing time. The downside, however, is that an effective PET imaging complex must be thermodynamically stable and kinetically inert, otherwise dissociation and ^{64}Cu^{2+} release during its passage to the site of interest will be compromised, with 'off target' imaging the result. The most effective ^{64}Cu chelators to date are macrocyclic polyamines such as the pendant arm amino-carboxylate tetraazacyclododecane tetraacetate (or DOTA for short) and the macrobicyclic hexaaamine cage MeCOSar, which both rapidly form highly stable ^{64}Cu(II) complexes for PET imaging (Figure 9.8).

Another promising PET isotope is ^{89}Zr ($t_{1/2}$ 3 d). This element is exclusively found in its tetravalent oxidation state, which presents challenges in terms of its hydrolysis and formation of insoluble ZrO$_2$. In fact, the recent emergence of ^{89}Zr for PET imaging has brought about renewed interest in its relatively unexplored

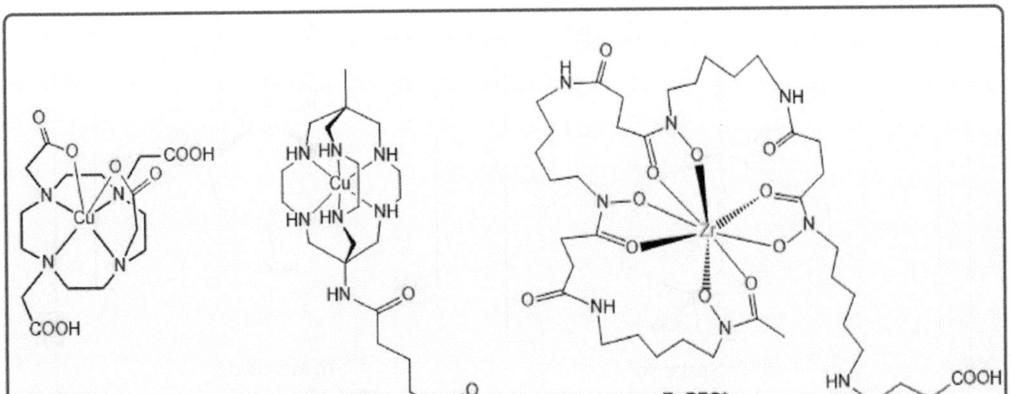

Figure 9.8
^{64}Cu and ^{89}Zr complexes that have shown promise as PET imaging agents.

aqueous coordination chemistry! The octadentate tetrahydroxamate DFO* has been found to be an effective 'hard' ligand to offset the charge of Zr^{4+} and forms a very stable complex (Figure 9.8) that is a promising lead for ^{89}Zr PET imaging. The inspiration for this ligand stemmed from the natural Fe(III)-binding siderophore desferrioxamine B (DFO, which we met in Chapter 8, Figure 8.14). The present example DFO* bears an extra hydroxamate chelating group which thus satisfies the preferred eight-coordinate geometry of Zr^{4+}. It is important to note that all the ligand systems in Figure 9.8 have a free carboxylic acid functional group that enable their attachment to targeting groups such as peptides or even antibodies.

9.2.2 Magnetic Resonance Imaging (MRI)

In Chapter 7, we saw how NMR spectroscopy can be applied to coordination compounds much in the same ways as it is employed in organic chemistry. Although less commonly seen, ^1H NMR spectra can also be recorded with *paramagnetic* compounds, with associated large paramagnetic shifts of the expected resonances from the chemical shifts that would normally apply in a diamagnetic spin-paired analogue. The technique of magnetic resonance imaging (MRI) utilises paramagnetic complexes, but it is not the NMR signals of the probe that are of interest but rather the effect the paramagnetic centre has on its environment, namely, water. In MRI, the more unpaired electrons, the better. In terms of paramagnetic candidates, the $4f^7$ metal ion gadolinium(III) selects itself as the only stable metal ion with seven unpaired electrons, and hence Gd(III) complexes dominate MRI to this day. The mechanism by which they act relies on interaction between the slowly relaxing electrons spins of Gd(III) and the H-atom (proton) nuclear spins of water molecules within the first coordination sphere (an aqua ligand) and neighbouring water molecules in the second coordination sphere that are associated only through H-bonding. Both effects are important, but second sphere interactions are difficult to control. However, the inner sphere component can be influenced by multidentate ligands bound to Gd(III). Overall, the target is a complex that enhances *proton relaxivity*. MRI is especially effective in imaging soft tissue, and various pathologies such as tumours and spinal or ligament damage can be highlighted through abnormalities in the local water environment. It should be noted that so-called uncomplexed Gd_{aq}^{3+} (actually a complex bearing only aqua ligands) is quite toxic and accumulates in various organs due to its chemical similarity to Ca^{2+}, although this is still poorly understood. Chelating ligands are thus required, and high thermodynamic stability of the Gd(III) complex is paramount to ensure little free Gd^{3+} is ever released.

Like all lanthanoid ions, high coordination numbers (8 or 9) are the norm for Gd(III). Multidentate ligands bearing O- and N-donors dominate the field. Indeed, many of these ligands have already been met earlier in this book (and this chapter). Two of several examples currently in clinical use are shown in Figure 9.9. The octadentate H_4DOTA ligand has also been used to complex Cu in PET imaging (Figure 9.8) and has a long history in coordination chemistry. In this case, the ligand occupies eight of the nine coordination sites of Gd(III), leaving one available for an aqua ligand. Another ligand family is based on the acyclic diethylenetriamine pentaacetate (DTPA); a benzyl ether analogue H_5BOPTA is illustrated in Figure 9.9. The presence of at least one aqua ligand is essential for an effective Gd(III) MRI contrast agent. Although more aqua ligands would theoretically enhance the proton relaxivity, this comes at a cost of lower thermodynamic stability. Furthermore, the

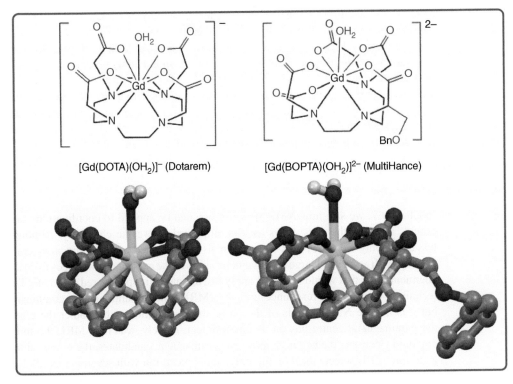

Figure 9.9

Two Gd(III) MRI contrast agents currently in clinical use comprising octadentate amino-acid ligands. Both include one aqua ligand in the coordination sphere. Their crystal structures are shown with H-atoms omitted except for the aqua ligand. Dotarem is used in brain and spinal imaging, whereas MultiHance is used for liver imaging.

possibility that competing ligands *in vivo* (phosphate, citrate, carbonate) might replace these aqua ligands exists, which would cause a decrease in relaxivity. The rate of aqua ligand exchange has a major effect on proton relaxivity; the faster the better. Through intensive research on a large family of complexes, it has emerged that Gd(III) complex anions and a sterically crowded environment surrounding the aqua ligand are both desirable features for enhancing water exchange within the coordination sphere. Also, the rate of rotation (tumbling) in solution is a factor, as slower rotation rates increase relaxivity. In this introductory text, the fine details that must be optimised for an ideal MRI agent need not concern us, but it is clear that an ideal match between metal and ligand in terms of coordination number and structure is central.

9.3 Analysing with Complexes

Complexation, as we have seen already, causes change in the properties of a metal ion. Where such a change is reflected in a property that varies linearly (or at least in a known manner) with respect to complex concentration, there is the makings of an analytical method. This may be applied to the detection of a metal ion or alternatively to the detection of a ligand species that binds the metal ion effectively. It may even be possible to have complexation as a sensor of another process or species that is 'switched on' by a changed non-covalent interaction tied to the complexation event (such as an ion-pairing interaction). The two examples given below relate to changing luminescence intensity associated with complexation, but in principle, any measurable process can be employed. Luminescence (fluorescence or phosphorescence) is a useful sensor, where it occurs in compounds, as it provides usually high sensitivity due to low background 'noise' levels. This therefore permits detection of the luminescent species down to very low concentrations, with instrumentation that is able to detect luminescence now widely available at reasonable cost.

9.3.1 Fluoroimmunoassay

The discovery by Samuel I. Weissman in 1942 that europium(III) complexes of β-diketone O-donor chelates absorb light in the UV region but emit light in the visible region (called a Stokes' Shift) set in train research on the analytical applications of fluorescent *f*-block complexes that continues to this day. These complexes work by light being absorbed by the ligand (at ~360 nm) when it carries a strongly absorbing chromophore which acts as an 'antenna', before passing this energy to a lower energy Eu^{3+} metal 4*f*-orbital. This 4*f*-excited state then relaxes to its ground state by emitting light at a longer wavelength (~615 nm). The lanthanoid ions that promote readily measurable fluorescence are Sm^{3+}, Eu^{3+}, Tb^{3+} and Dy^{3+}, because the excitation energy level of these four metals lies slightly lower than the excited levels of the 'antenna' ligands and thus are able to readily accept the energy transfer. With aqua ligands in particular, a mechanism for rapid decay of the excited state other than by fluorescence operates, and so it is only in the presence of appropriate ligands that the lifetime of the excited state is sufficient to lead to strong fluorescence, with applicability assisted by a sharp emission peak. The emission process for the complexes is relatively slow, and the complex is said to have a long fluorescent lifetime (albeit still in the sub-second range), attributed in part to the 4*f* orbital being shielded by outer orbitals like the 5*s*. This long lifetime is a key to analytical applications, since a process that waits a short time following an excitation pulse to let other unwanted but short-lived fluorescence die away is employed, allowing the lanthanide-based fluorescence to be measured essentially free of interference; this is called a *time-resolved fluorescence analysis* method. Typically, a 'wait time' of ~200 μs followed by a measurement time of ~400 μs is employed, permitted because the fluorescent life span of the lanthanide complex is hundreds of microseconds, making it possible to detect the target fluorescence with good sensitivity after fading away of the short-lifespan background fluorescence of impurities and organic species.

Fluoroimmunoassay makes use of the above behaviour. One of the common commercial methods is dissociation-enhanced fluoroimmunoassay (DELFIA). In this approach, a non-fluorescent Eu(III) EDTA-like complex is attached by a simple chemical reaction to an antibody or antigen, in a process called labelling. An immunoreaction is next initiated to bind the biological target, and then a mixture of a β-diketone and trioctylphosphine oxide (TOPO) is added to the formed immunocomplex at pH ~3 to promote the release of the Eu(III) from the antibody and its subsequent complexation as the strongly fluorescent complex $[Eu(\beta\text{-diketonate})_3(TOPO)_2]$, which is then measured by time-resolved fluorescence methods. The signal size relates to the amount of europium complexed, which in

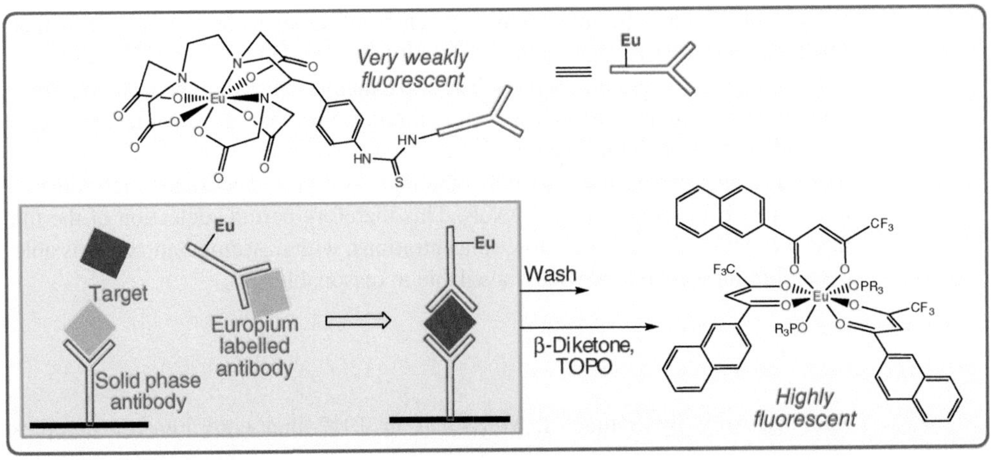

Figure 9.10

Chemistry involved in the fluoroimmunoassay process. Europium ion is bound to an EDTA-like pendant on an antibody, which traps the target pathogen at a second surface-anchored antibody with high specificity. Following washing to remove the test solution, the europium in the trapped adduct is released on chelate addition to form a highly fluorescent complex in solution that is sensed by fluorescence spectroscopy. Signal intensity is directly related to target pathogen concentration.

turns relates directly to the amount of the specifically formed target immunocomplex. This process is represented schematically in Figure 9.10.

The ease of handling tied to the high detection sensitivity of fluorescent lanthanide probes has led to their use in base sequence analysis of genes, DNA hybridisation assay, and fluorescent bioimaging, apart from in immunoassay. They are rapidly replacing organic fluorescent dyes, which have the disadvantage of short fluorescent lifetimes so consequently they are compromised by interference signals. This lanthanide-based technology is a fine example of simple coordination compounds at work.

Understanding the photochemistry and photophysics of coordination complexes resulting from the interaction of complexes with light provides a fundamental base for development of other applications. Apart from biomedical imaging and therapy applications of the type exemplified above, applications under active development or the subject of preliminary research include the use of coordination complexes in photocatalysis, optical information transfer and storage, optical computing, analytical sensing, and harvesting of solar energy. The penultimate of these fields is exemplified below.

9.3.2 Fluoroionophores

Fluoroionophores are large ligands that are described as supramolecular systems; they consist of two components joined together covalently—an *ionophore* ('ion-lover') molecule linked to a light-sensitive *fluorophore*. The role of the first unit is to capture metal ions, whereas the role of the second unit is to capture light and emit it at a different wavelength as fluorescence, which can then be detected and measured. It is the impact of a captured metal on the fluorescence process that is at the core of the analytical application. These fluoroionophores are designed to sense metal

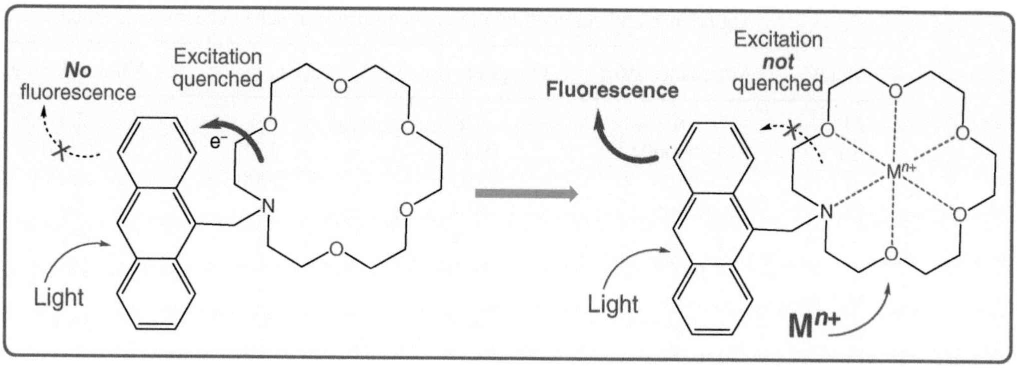

Figure 9.11

A scheme of a fluorescent 'switch' turned on or off (quenched) depending on the presence or absence of a metal ion, respectively. The ionophore (the cyclic polyether) is the metal-binding component, the fluorophore (the fused-ring aromatic unit) is the component activated by light. Complexation stops electron transfer that otherwise quenches fluorescence.

ions selectively; ion recognition involves the ionophore binding the metal ion through coordinate covalent bonds, whereas the signalling process of the fluorophore depends on photophysics. Typically, the process involves metal ion control of photo-induced electron transfer (Figure 9.11). Selective metal-ion detection using a fluoroionophore arises because of the variation of colour (or fluorescence wavelength) with metal ion. Even the various alkali and alkaline earth ions can be distinguished.

Phosphorescence or fluorescence of complexes in the visible region has also permitted other developments of note. For example, Pt(II), Rh(III) and Ir(III) complexes of polydentate ligands such as aromatic N-donors, and mixed N- and P- or C-donor compounds produce emissions in the visible region and are being developed as organic light emitting diodes (OLEDs) or phosphorescent organic light emitting diodes (PHOLEDs). Clearly, this is a very liberal definition of the word 'organic', given that the process relies on a fluorescent transition metal centre to work! Phosphorescent metal complexes were the second generation of emitters used in OLED panel displays, following earlier purely organic fluorescent compounds. Ruthenium complexes used to lead research in photochemistry of metal compounds, but iridium complexes have recently overtaken them as the key target compounds due to their applications in OLEDs. This is a lively and ever-changing field; for example, over 90% of luminescent iridium(III) complexes have been reported only since 2005. With their luminescence 'tuneable' through ligand choice, iridium complexes are firm candidates for optical display applications. Recently, QLED panel displays have become available, which use CdSe- and InP-based colloidal quantum dots for which vibrant colour is dependent on dot size.

9.4 Profiting from Complexation

9.4.1 Metal Extraction

An ore body is a commercially viable concentration of one or usually several metals in a solid matrix. This concentration reflects the amount of an element in the Earth's crust; for a common element like iron, an ore body could be ~50% iron, whereas

Table 9.2   Abundance and commercial ore concentrations of selected economic metals.

Metal	Abundance (%)[a]	Ore grade (av. %)[b]	Conc. factor [c]	Value ($US/kg)[d]
Al	8	28	3.5	2.3
Au	0.0000004	0.0001	250	70,000
Cr	0.01	30	3,000	12.5
Cu	0.005	0.4	80	8.6
Fe	5	25	5	0.1
Li	0.002	0.01	5	25.7
Mn	0.09	35	390	2.4
Ni	0.007	0.5	70	23.1
Pb	0.001	4	4,000	2.1
Pt	0.0000005	0.0001	200	35,000
Sn	0.0002	0.5	2,500	26.0
Zn	0.007	4	570	2.6

[a] Average abundance in the Earth's crust.
[b] Approximate minimum exploitable grade of deposits, as a percentage of the ore.
[c] Concentration of an ore grade deposit relative to the earth's crustal average.
[d] Value in 2023—varies regularly; the value also affects the concentration factor acceptable for a commercial mining operation.

for a rare element like gold, it could be much less than 1% (Table 9.2). However, digging an ore out of the ground is but the first step in the release of the metal and its conversion into commercially useful forms. Whereas some common metals are recovered by very simple redox reactions that do not involve coordination chemistry to any marked extent, some metals do rely heavily on coordination chemistry for their extraction and recovery.

Ores can be classified into several 'families', related to the reactivity of the metal and ore type. The least reactive metals, such as gold, platinum and silver, can exist in nature as free elements, albeit in most cases dispersed at very low concentration in the body of an ore; consequently, they require mechanical and or chemical treatment to permit their recovery. Some metals of relatively low reactivity, such as mercury, can exist as sulfide ores and require further chemical treatment. As element reactivity rises to what are sometimes termed medium reactive metals (those less reactive than carbon, which includes zinc, tin and iron), the range of commercial ores increases to include oxides and carbonates in addition to sulfides. Highly reactive metals (such as sodium) are often met in ores as halides or nitrates; electrolysis of chloride salts is the usual route to the element.

Prior to reactions directed towards recovery of the elemental metal, crushing and ore concentration to separate the desired material from the unrequired and essentially valueless remainder of the ore body (called gangue) is undertaken. Concentration processes employed included *gravity separation* (such as by using a stream of water to simply wash off lighter gangue material), *froth flotation* (where air is sparged through a mixture of crushed ore, water and a frothing agent to 'float' the desired ore for collection), *magnetic separation* (relying on magnetic properties of ore or gangue, with separation achieved on magnetic rollers) or *leaching* (of batches of crushed ore in tanks or else in large heaps using an aqueous solution containing a chemical leaching agent).

The ore concentrates recovered are subject to further processing steps. Sulfide ores undergo *roasting* in the presence of air to form oxides or, in a few cases, the metal. Such high temperature solid-state reactions are examples of *pyrometallurgy*. For example, mercury sulfide is converted on roasting to the oxide (Eq. (9.1)), and in this particular case, further heating yields the element (Eq. (9.2)).

$$2\,HgS + 3\,O_2 \xrightarrow{\Delta} 2\,HgO + 2\,SO_2 \qquad (9.1)$$

$$2\,HgO \xrightarrow{\Delta} 2\,Hg + O_2 \qquad (9.2)$$

Carbonate ores undergo *calcination*, effectively loss of CO_2 to form an oxide, by heating in the absence or near-absence of air, such as occurs for $ZnCO_3$ (Eq. (9.3)).

$$ZnCO_3 \xrightarrow{\Delta} ZnO + CO_2 \qquad (9.3)$$

Reduction of oxide ores to the metal follows, producing a crude metal that generally requires further refining to produce the pure metal. Reduction of an oxide ore can be achieved in a redox reaction with a chemical reducing agent, such as carbon (employed to produce zinc; Eq. (9.4)), or hydrogen (used to produce copper; Eq. (9.5)).

$$ZnO + C \xrightarrow{\Delta} Zn + CO \qquad (9.4)$$

$$CuO + H_2 \xrightarrow{\Delta} Cu + H_2O \qquad (9.5)$$

Electrochemical methods have great historical significance in the isolation of metallic elements, especially highly reactive ones. Humphrey Davy pioneered the approach, and in the early 1800s, he was the first to isolate Na, K, Mg, Ca and Ba in their pure metallic forms. Electrochemical reduction of active metals such as sodium is achieved by passing a current through molten salt. For sodium chloride, sodium is formed in a reduction reaction at the cathode, and chlorine gas is formed in an oxidation reaction at the anode. The above examples of what are now traditional methods do not rely much on coordination chemistry, but there are several processes where coordination chemistry plays a key role, and we shall concentrate on these henceforth.

Pure titanium metal is made by chlorination of the oxide TiO_2 (from ores such as ilmenite) at about 1,000°C in the presence of carbon (which forms CO and some CO_2) to form the tetrahedral complex $TiCl_4$ (Eq. (9.6)). This is purified by distillation and then reduced in a redox reaction with magnesium (or else sodium) metal in a sealed vessel at from 800 to 1,000°C to yield free titanium metal as a powder (Eq. (9.7)).

$$TiO_2 + 2\,C + 2\,Cl_2 \xrightarrow{\Delta} TiCl_4 + 2\,CO \qquad (9.6)$$

$$TiCl_4 + 2\,Mg \xrightarrow{\Delta} Ti + 2\,MgCl_2 \qquad (9.7)$$

To provide a continuous loop of reagents, the $MgCl_2$ (or NaCl) also formed in this reduction step is recovered and electrolysed to produce chlorine and magnesium (or sodium) metal, with the chlorine and magnesium (or sodium) re-used fully. Although impure titanium metal was first isolated in 1825 by Jöns Jacob Berzelius, it was not until 1910 that metal of 99% purity was isolated, with subsequent process improvement leading to the modern method described above. This is a good example of an economical industrial process with few waste products. Industrial waste is costly to dispose of due to environmental protection laws, so using a process that avoids it in the first place is the best approach. Of course, energy is consumed, so titanium recovery still carries an environmental impact. Titanium metal represents only 5% of all titanium consumption annually, although it plays a very important role in medical implants and also as an alloying agent in aircraft and spacecraft. The vast majority of titanium is used as its oxide in the pigment industry (such as in paints). It also is an active ingredient in sunscreen lotions.

Gold recovery is a simple example where complexation plays a key role. Gold ore, where the gold is present as finely-divided metal that cannot be successfully recovered mechanically, is slurried in the presence of air and added cyanide ion. The elemental gold is oxidised by dioxygen and complexed by cyanide to form a linear two-coordinate gold(I) cyanide complex (Eq. (9.8)).

$$Au + 2\,CN^- + \tfrac{1}{2}\,O_2 + H_2O \rightarrow [Au^I(CN)_2]^- + 2\,{}^-OH \qquad (9.8)$$

This water-soluble complex is then captured following filtration by adsorption onto carbon, and subsequently recovered by further processing. The coordination complex of Au(I) shown in Eq. (9.8) is at the core of the process. The Au(I) cyanide complex is very stable, and this lowers the $Au^{I/0}$ redox potential, which enables oxidation by O_2. Without appropriate ligands present, oxidation of elemental gold requires much harsher conditions. Other ligands are being developed to replace the environmentally problematic and highly toxic cyanide, such as thiourea ($S=C(NH_2)_2$), which also acts as a good soft monodentate ligand and forms a stable linear two-coordinate complex. A gold extraction level of 99% is achievable under optimal conditions, close to that achieved with cyanide ion, although loss of thiourea during processing due to oxidation is a disadvantage. Interestingly, household cleaning liquids to remove silver sulfide tarnish have thiourea as their active ingredient. The coordination chemistry is the same as with its congener gold, in this case forming a soluble linear bis(thiourea)silver(I) complex.

The chemistry of gold recovery discussed above is an example of what is called *hydrometallurgy* or the treatment of ores in aqueous solution. It is in this medium that most of the coordination chemistry of metallurgical processes can be found. A decrease in the grade and increase in the complexity of available ores of some metals has increased the need to develop new hydrometallurgical processes. The two key steps for ore treatment in hydrometallurgy are ore *leaching* to dissolve the target metal and *recovery* (often involving a precipitation reaction) of the purified dissolved target metal. Although simple acid or base leaching can be successful, some ores require the addition of complexing agents (ligands) to assist, such as the dissolution of copper oxide in ammonia/ammonium chloride (Eq. (9.9)).

$$CuO_{(s)} + 4\,NH_4^+ + 2\,{}^-OH \rightarrow [Cu(NH_3)_4]^{2+} + 3\,H_2O \qquad (9.9)$$

This is only a complexation reaction, as the copper is already in its desired oxidation state. By contrast, the gold recovery process described above involves both oxidation and complexation. Another example of such a reaction is nickel sulfide pressure leaching, where it is the sulfide anion that is oxidised, the nickel remaining in its Ni(II) form throughout (Eq. (9.10)).

$$NiS_{(s)} + 2\,O_2 + 6\,NH_3 \rightarrow [Ni(NH_3)_6]^{2+} + SO_4^{2-} \qquad (9.10)$$

This process also extracts cobalt(II), often present to a lesser extent in the ore. Reduction to the metal follows; one technique employs hydrogen at elevated temperature and pressure, although reduction electrochemically is growing in use. Obtaining nickel metal of a very high purity can be achieved through reaction with carbon monoxide to form volatile $[Ni(CO)_4]$, which is then distilled and thermally decomposed to recover the carbon monoxide and pure nickel powder.

Complexation also plays a role in the commercial process for the separation of the platinum group metals, involving formation of chloride and ammonia complexes; this separation of so many related elements that occur naturally in the ore bodies is a demanding procedure. Platinum metal concentrates are usually dissolved in the strong mixed acid *aqua regia* (1:3 concentrated HNO_3:HCl) as the first step, to separate the soluble fraction (Au, Pd and Pt species) from the insoluble fraction (Rh, Ir, Ru, Os and Ag species). Solvent extraction of the soluble fraction with bis(2-butoxyethyl)ether then separates gold from palladium/platinum species. The latter water-soluble species are the square planar complex $H_2[PdCl_4]$ and the octahedral complex $H_2[PtCl_6]$. Addition of ammonium chloride precipitates sparingly soluble $(NH_4)_2[PtCl_6]$, leaving the palladium complex alone in solution for subsequent isolation. From the separated complexes, the free metals are recovered through different redox reactions.

Solvent extraction of a metal ion in aqueous solution with another immiscible solvent containing ligands, introduced above, allows the target metal ion to be separated from other unwanted metal ions in the original leach solution; it is used in several hydrometallurgical processes. One example is the recovery of uranyl ion (UO_2^{2+}), which uses a phosphate diester dissolved in kerosene to complex and extract the uranyl ion into the organic phase, from where it can be recovered. The extraction chemistry involves a reaction (Eq. (9.11)) where two didentate $R_2PO_2^-$ ligands bind to the uranyl cation to form a neutral complex that is highly soluble in the organic phase, into which it shifts.

$$UO_2^{2+}{}_{(aq)} + 2\,R_2PO_2H_{(org)} \rightarrow [UO_2(O_2PR_2)_2]_{(org)} + 2\,H^+{}_{(aq)} \qquad (9.11)$$

Similar chemistry, but with a different chelate ligand, is used to purify copper(II). The copper complex is loaded into the organic phase at a pH of between 2 and 4, then back-extracted into water as the Cu_{aq}^{2+} ion using dilute sulfuric acid at pH 0. It is recovered as either copper sulfate or, with following electrochemical reduction, copper metal. Solvent extraction involving complex formation is also employed in the separation of cobalt–nickel mixtures.

9.4.2 Coordination Complexes as Industrial Catalysts

Coordination chemistry offers many examples of applications in industry beyond those addressed above. Historically, one may point to the development and use of transition metal phthalocyanine complexes as dyes and in paints and inks since the 1930s, but the industrial use of coordination compounds now extends well beyond that early example. Recent developments in chemistry promise greater applications in the future. Here, a number of examples are presented to give a sense of current use and opportunities for coordination complexes in commercial roles, with a focus here on their role as *catalysts*, given that at least 90% of manufactured products are produced with the aid of at least one catalyst. Some catalysts employ metal ions in simple compounds such as oxides rather than formal coordination complexes; vanadium oxide is used in the manufacture of over 200 million tonnes of sulfuric acid annually, for example. While not discounting the obvious importance of catalysts such as metal oxides, here our interest will lie dominantly with coordination complexes, which are the central focus of this book.

The chemical industry as we currently know it would be markedly different without transition metal catalysts, as these play key roles in a wide range of processes in the organic chemical industry. Metal catalysts may be *homogeneous* (dissolved in a solvent) or else *heterogeneous* (present as a solid, possibly dispersed on the surface of an inert material), and both types find application in industry. The key task of a catalyst is to accelerate a reaction by effectively lowering the activation barrier for the reaction; a catalyst must also be regenerated so that it can promote reaction at the molecular level over and over, as well as robust, allowing many 'turnovers' to occur without undergoing decomposition. Apart from acceleration, a catalyst may also be able to selectively direct a reaction towards a particular product amidst a number of options or induce optical activity in an organic product if the catalyst includes a chiral ligand. The success of an asymmetric catalyst is defined by the *enantiomeric excess*, which is the difference in percentage yields of the major and minor enantiomers of the product. If 90% of one optical isomer forms and 10% of the other, the enantiomer excess is 80%; obviously, the closer that this value is to 100% (at which stereo electivity is achieved) the better. Asymmetric synthesis in industry depends fully upon transition metals as the active site of the catalysis.

One early development in metal catalysts was the discovery in 1954 that polymerisation of alkene monomers could be promoted under relatively mild conditions using a mixture of $TiCl_3$ and $Al(C_2H_5)_3$. While not formally well defined complexes due to changes in the coordination sphere, it was not long before complexes such as $[PdCl_4]^{2-}$ and $[Co(CO)_4H]$ found use for promoting the oxidation of ethylene to acetaldehyde and hydroformylation of olefins to form aldehydes, respectively. One important advance in homogeneous catalysis was the establishment of the Wacker Process in 1959, the result of a cooperative development between two German companies, Wacker and Hoechst. This process produced acetaldehyde from oxidation of ethene. Nominally, the catalyst was a simple mixture $PdCl_2 \cdot CuCl_2$, but this belies the chemical complexity of the reaction. Overall, the reaction is

$$C_2H_4 + \tfrac{1}{2}O_2 \rightarrow H_3C\text{---}CHO \tag{9.12}$$

thought to proceed via the redox process

$$C_2H_4 + H_2O + Pd^{II}Cl_2 \rightarrow H_3C\text{---}CHO + Pd^0 + 2\,HCl \tag{9.13}$$

$$Pd^0 + 2\,Cu^{II}Cl_2 \rightarrow Pd^{II}Cl_2 + Cu^I{}_2Cl_2 \tag{9.14}$$

$$Cu^I{}_2Cl_2 + 2HCl + \tfrac{1}{2}O_2 \rightarrow 2\,Cu^{II}Cl_2 + H_2O \tag{9.15}$$

Although this appears to involve only simple salts, there is strong evidence that the active catalyst is actually the complex anion $[PdCl_4]^{2-}$, which goes on to form the intermediate $[Pd(C_2H_4)(OH_2)Cl_2]$ in which the coordinated water is involved in an intramolecular nucleophilic attack on the coordinated ethene with β-elimination of a H atom to form a vinyl alcohol. This is converted in a series of steps to the aldehyde and palladium(II) hydride, with the latter being unstable in water and decomposing to HCl and Pd(0), the latter being re-oxidised by $CuCl_2$. Overall, a complicated but satisfying piece of coordination chemistry. Synthetic reactions with most other complex catalysts involve a series of steps that draw on traditional concepts of coordination chemistry. Unfortunately for the original Wacker process, potentially hazardous chlorinated by-products were a concern, but improvements using Pd/C catalyst and

$KBrO_3$ oxidant have been developed. This is typical of an ongoing focus on seeking improvements in efficiency, selectivity and safety of catalysed industrial processes.

Epoxides, or olefin oxides, are compounds with a three-membered C–O–C ring; they tend to be reactive and hence find plenty of use in the organic chemicals industry, often featuring metal catalysts. A Mo(VI) complex catalyses the epoxidation of propene with an organic hydroperoxide, apparently involving a coordinated peroxide intermediate. The epoxide produced is employed in the synthesis of certain alcohols, ethers and glycols. One of the best-known coordination complexes that acts as a catalyst of chiral epoxidations is Jacobsen's catalyst, developed in the 1990s and named after its developer, Harvard professor Eric Jacobsen. This is a manganese(III) complex of a sterically demanding chiral ligand (Figure 9.12), which is used in conjunction with the oxidant hypochlorite (ClO^-) in asymmetric epoxidation. The oxidant initially oxidises the complex to form an oxidomanganese(V) compound, which is then able to deliver the oxygen to an alkene to form an epoxide, returning to the Mn(III) state. The epoxidation occurs with high stereospecificity, with enantiomeric excesses of usually greater than 95% achieved routinely. It is believed that the substrate is bound to the metal ion in the activated state, with oxygen transfer occurring in the chiral environment of the tetradentate N_2O_2 ligand leading to chirality in the epoxide product. Since the Mn(III) complex is regenerated, it is able to function as a catalyst, undergoing many 'turnovers' before degradation through ligand dissociation reactions. This is another example of a *homogeneous catalyst*, which is one that is dissolved in solution for the reaction.

Another catalyst type mentioned earlier is the *heterogeneous catalyst*, which remains as a solid and promotes chemistry at the surface. To function well, these catalysts require high surface areas per unit mass. Metal oxides and hydroxides are simple common examples. A vanadium(V) oxide is employed in the formation of ammonia from nitrogen and hydrogen under elevated temperature and pressure, for example. Synthetic accessibility and structural diversity of polyoxometallate metal clusters (described in Chapter 6.7.2), which are oxido-ligand coordination complexes employing dominantly O^{2-} and HO^- as ligands, makes them attractive for some catalytic roles. One particularly topical area of research is their application as catalysts or co-catalysts in CO_2 conversion to useful products. Embedding these

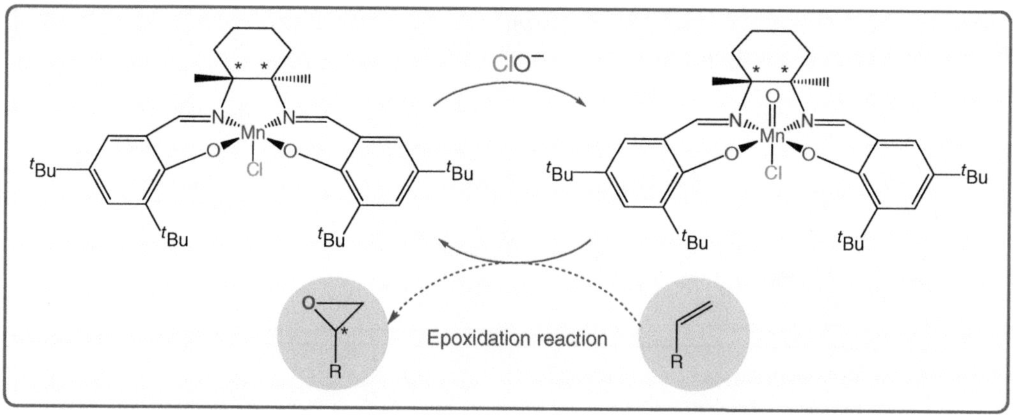

Figure 9.12

Chiral epoxidation with Jacobsen's catalyst. Chiral centres in the ligand of the catalyst and the product are marked (*).

clusters in metal-organic frameworks provides a higher level of sophistication that may enhance catalysis and is under investigation.

The two traditional types of catalysts (homogeneous and heterogeneous) merge with the development of what are called 'tethered' catalysts, where a homogeneous catalyst is amended so that it can be attached covalently to an inert surface, such as silica. This is also called a 'supported' catalyst. Having the active component available as part of a solid can assist processes where carrying the catalyst forward in solution to another stage of the process may lead to contamination or catalyst destruction. Furthermore, surface attachment also can alter catalytic activity favourably in certain cases.

A wealth of industrial syntheses of organic compounds employs metal compounds as catalysts (see Figure 9.13 for examples); several processes appear briefly below:

Hydrocyanation, the conversion of alkenes to nitriles, is a reaction used to prepare the monomer adiponitrile (hexanedinitrile) used for the production of *Nylon 66*, of which well over a million tonnes is prepared annually. This monomer is hydrogenated to hexane-1,6-diamine for nylon production. Synthesis employs a $[Ni^0L_4]$ catalyst with organophosphate ligands (L). Oxidative binding of HCN to the catalyst forms an intermediate $[NiL_3(CN)H]$, which apparently coordinates to the C=C bond of the butadiene precursor prior to hydrocyanation.

Olefin metathesis, a catalytic reaction that involves an exchange of double bond locations between carbon atoms, is broad in scope. It requires metal catalysts, with *heterogeneous catalysts* (dispersed solids) generally used in commercial processes, although homogeneous catalysts were developed more recently. The active heterogeneous catalyst is usually supported on an inert solid such as alumina;

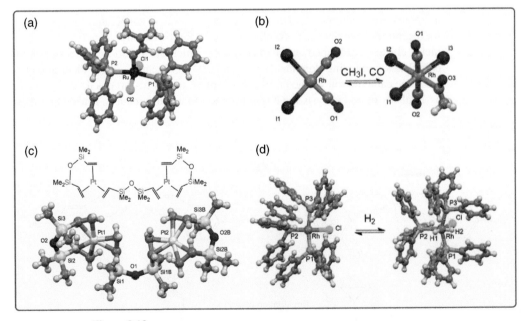

Figure 9.13

Examples of metal complexes employed as catalysts in various organic syntheses. (a) A carbenoid complex $RuRCl_2(PPh_3)_3$ (Grubbs' catalyst for olefin metathesis); (b) *cis*-$[Rh(CO)_2I_2]^-$ and $[Rh(CO)_3(COMe)I_3]$ (carbonylation reaction); (c) Karstedt's catalyst (hydrosilylation); and (d) $[Rh(PPh_3)_3Cl]$ and $[Rh(H)_2(PPh_3)_3Cl]$ (Wilkinson's catalyst for hydrogenation of terminal alkenes).

octahedral complexes $[Mo(CO)_6]$ and $[WCl_6]$ are two catalysts employed. Alternative homogeneous catalysts more elaborate in structure are Grubbs Catalysts (named after Robert H. Grubbs who shared the 2005 Chemistry Nobel Prize), which are five-coordinate ruthenium(II) carbenoid complexes featuring a Ru=C double bond (Figure 9.13a). Applications include the preparation of fuels with high octane rating, the conversion of ethylene and 2-butene to propylene, and synthesis of certain pharmaceutical drugs.

Carbonylation is the reaction where CO is attached to an organic compound, as exemplified by the carbonylation of methyl alcohol to acetic acid, catalysed by the *cis*-$[Rh(CO)_2I_2]^-$ ion (Figure 9.13b), formed *in situ*. Oxidative addition of this square planar Rh(I) complex is facile and a mechanistically related example is also shown in Figure 9.13b, where methyl iodide adds to give an octahedral Rh(III) complex with the iodido ligand coordinating while the methyl group attacks one of the carbonyl ligands forming an acetyl ligand, the latter reaction being formally a carbonyl insertion. Iridium and palladium complexes have also been patented for use in various other processes. For example, $[Pd(PPh_3)_2Cl_2]$ is used in the production of the anti-inflammatory drug ibuprofen.

Olefin polymerisation is a large-scale industrial process, with millions of tonnes of products manufactured annually, usually employing $TiCl_4$ deposited on small $MgCl_2$ grains as the heterogeneous catalyst. One example of an organometallic homogeneous catalyst is the Zr(IV) complex $[Zr(CH_3)(\eta^5\text{-Cp})_2X]$, which operates by binding alkene monomers to the metal prior to addition to a growing carbon chain. Atom transfer radical polymerisation, conducted in water, also makes use of copper and iron or titanium complexes as catalysts to mediate controlled reactions, whereas radical polymerisation of acrylates uses cobalt, iron and chromium complexes as catalysts. The ability of these metals all to undergo facile *single electron* redox reactions is key to their success in catalysing radical formation.

Hydrosilylation reactions are, obviously, directed towards the wealth of silicon-containing compounds, such as silicones, which find wide application in the modern world as adhesives and polymers. A wide array of metal complexes (of Pt(0), Pt(II), Pd(II), Co(II), Ni(II), Rh(II), Ru(I) and Ir(I)) are in use as catalysts. The example shown in Figure 9.13c is the dinuclear formally Pt(0) compound known as Karstedt's catalyst, which is the most widely employed industrial catalyst for hydrosilylation reactions. Three alkenes bind in a side-on manner to each 16-electron Pt centre. Interestingly, two of the ligands function as chelates while the third bridges the two metals.

Hydroformylation involves reaction of olefins with hydrogen and carbon monoxide to form straight-chain and branched-chain olefins, for example:

$$R - CH = CH_2 + CO + H_2 \rightarrow R - CH_2 - CH_2 - CHO + R - CH(CH_3) - CHO$$

$$(9.16)$$

Although $[Co(CO)_4H]$ was the first catalyst used, this was supplanted by the breakthrough development of rhodium-based Wilkinson's catalyst (named for the British joint 1973 Chemistry Nobel laureate Geoffrey Wilkinson, who illustrated and popularised its use). It has about a 1,000-fold higher catalytic activity than the first cobalt catalyst. This, one of the best understood and long-established coordination complexes acting as a catalyst, is the simple square-planar Rh(I) complex $[Rh(PPh_3)_3Cl]$ (see Figure 9.13d) and is used for the hydrogenation of terminal alkenes. An oxidative addition reaction of H_2 generates the octahedral *cis*-dihydrido complex $[Rh(H)_2(PPh_3)_3Cl]$ (Figure 9.13d). Alkene insertion into the coordination sphere generates an intermediate where both hydrogen and the alkene are coordinated, allowing an intramolecular reaction between these to form an alkane product—effectively a reaction of coordinated ligands, but one where the product can depart so that the original complex is reformed, allowing the process to occur again, a key requirement for catalysis.

Hydrogenation is the general name for reaction between molecular hydrogen and an organic compound, and a metal catalyst is employed to drive what is a thermodynamically unfavourable process. The reaction has been in extensive use since the end of the 19th century. Although heterogeneous catalysts of, for example, finely divided nickel, platinum or palladium are often employed, homogeneous metal complexes as catalysts can enhance regioselectivity and stereoselectivity. Wilkinson's catalyst can also be used as a general hydrogenation catalyst, but an array of other complexes (of Ru, Rh and Ir, for example) are in use. Oxidative addition of dihydrogen to the transition metal catalyst is again thought to be the first step (exemplified in Figure 9.13d). Hydrogenation of alkenes and alkynes to alkanes is one example of a technology widely employed in the food, petrochemical, agricultural and pharmaceutical industries.

9.4.3 Complexes as Nanomaterials

Nanotechnology is one of the more recent frontiers of chemistry, focussed on the development of new materials with well-defined structure within the size range of from 1 to ~100 nm. In particular, the properties of the nanomaterial should differ from those of both conventional molecular compounds and bulk solids, and it this difference that is the key to their attraction and potential applications. This definition can include metal complexes, particularly larger polymetallic systems (see Chapter 6.7). Of course, many biomolecules fall within the definition of nanomaterials in terms of size, but the interest in the field at present lies in the development of synthetic more than natural materials.

Fabrication of nanomaterials divides into two approaches—'top down', which relies on starting with bulk materials and processing them to yield nanoscale materials, and 'bottom up', which employs atomic and molecular species aggregated to form larger nano-scale materials. It is the latter of these two approaches that may involve coordination chemistry. Addressed already in Chapter 6, a simple example here may suffice to display this in application. It is possible to produce (cation)$^+$[AuCl$_4$]$^-$ as a finely-divided particulate by choosing an appropriate cation that limits complex solubility in the chosen solvent. In the presence of a thiol (RSH), substitution of chlorido ligands on the gold(III) by thiolate (RS$^-$) anions can occur. When the resultant thiolato-gold(III) complex is treated with a suitable reducing agent, Au(III) is reduced to form nano-scale metallic gold particles that remain coated with thiol, of type {Au$_x$(RS)$_y$} (see Figure 6.15). This represents a new form of gold particles with special properties, whose size and surface character can be controlled through manipulation of reaction conditions and reagents. Metal nanoparticles may also form as a result of a coordination complex catalyst undergoing reduction during the reaction process or may be pre-formed; controlled reduction of [PdL$_2$Cl$_2$] complexes leads to a Pd metal nanocatalyst, for example.

Building polymetallic clusters into new nanomaterials through combination of particularly shaped ligands and monomer complex components is an area of growing development. This is an extension of concepts we have already developed in Chapter

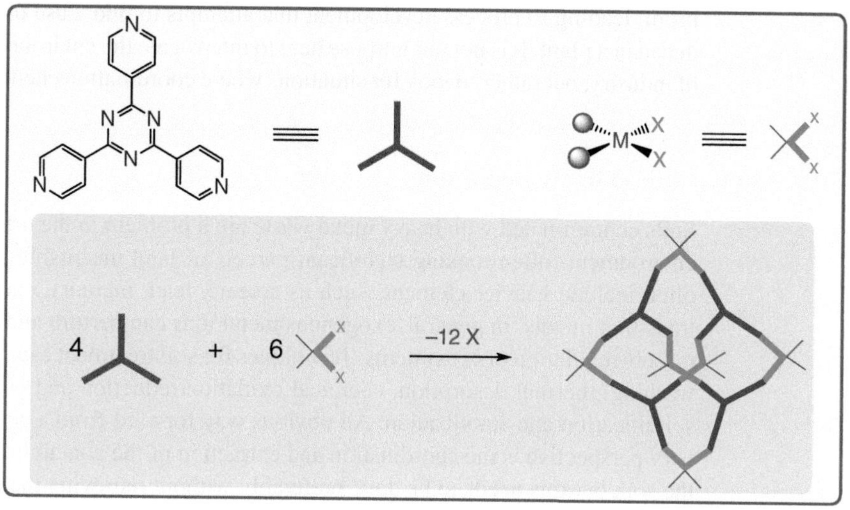

Figure 9.14
Assembly of rigid precursor units into a larger assembly (even to nano-scale) is directed by the shape of components.

6 (see Figure 6.9). This is really molecular *Lego*®—there is only a limited number of ways ligand and complex precursors of particular and usually rigid shapes can combine as larger units, just like *Lego* blocks can only fit together in certain ways. A simple example of the concept is illustrated in Figure 9.14.

Substitution of the two *cis* X-groups from the complex allows coordination of terminal pyridine groups from the Y-shaped ligands in their place. However, as a result of the rigid ligand shape, any ligand can only attach as a monodentate to a single metal, and so a polymeric network of attachments is built up to satisfy the demands of the Y-shaped ligand for tridentate coordination and the metal for filling two adjacent coordination sites. A large number of complex, shape-directed nano-sized clusters have been developed using this type of approach.

Coordination compounds attached to nanomolecules, either covalently or through strong ion pairing, are being examined with regard to medical applications. The hope is that biological activity of complexes with potential or known efficacy as drugs can be improved, and as a consequence the number of doses required can be reduced in addition to reduction in the severity of side effects. The aim is for nanotechnology to deliver an active compound to the desired location, where conventional treatments do not succeed, with specificity and in sufficient concentrations for efficacy. Examples of systems probed include cyclodextrins, gold clusters and oligomers.

9.5 Being Green

If you are familiar with the *Muppet* philosopher Kermit the Frog, you will probably be aware of his view that it is not easy being green; however, it seems that chemists are developing some solutions to this problem. Key new driving forces in applied chemistry are environmental sensitivity, social responsibility and energy saving processes. Some may say that greenness has been thrust upon industry by government, but for some time, there has been recognition in industry that a waste product is also wasted profit, leading to process development that attempts to find a use for everything produced in a plant. It is not our purpose here to interrogate the environmental credentials of industry, but rather to look for situations where coordination chemistry plays a role.

9.5.1 Complexation in Remediation

Soils contaminated with heavy metal waste are a problem in the urban industrialised environment, often making significant parcels of land unsuitable for re-use. These often include heavier elements such as arsenic, lead, mercury, cadmium and some transition metals. In general, exogenous metal ions can perturb and disrupt the natural self-regulation of ecosystems. Techniques for soil treatment examined include soil washing, thermal desorption, chemical oxidation/reduction and even cement-based solidification and stabilisation. An obvious way forward from a coordination chemistry perspective is the coordination and extraction of the contaminating metals from the soil, but this needs to be done preferably without removing the benign and dominant natural metal ions in the soil. Selective complexation offers a way forward; in effect, one can adopt the approaches used and already developed for metal extraction from commercial ore bodies—the contaminated soil can be 'mined' for its unwanted contaminating metals. This may be best achieved by the employment of molecules as ligands that are strongly selective for the target metal or metals, allowing others to remain undisturbed. Polydentate chelating amino acids such as EDTA have been examined in recent years; while very good complexing agents, their cost for large-scale use may be prohibitive, and selectivity may not be high enough. Consequently, alternatives need to be found.

An alternative to removal is immobilisation *in situ*, but this may not be a long-term solution if subsequent slow on-going reactions lead to heavy metal availability being re-established. Environmentally-friendly materials such as biochar and natural minerals such as zeolites have been employed in immobilisation studies. Remediation of contaminated sites is a pressing problem internationally; in Europe alone, there are estimated to be well over a million sites requiring treatment.

9.5.2 Better Ways to Fine Organic Chemicals

Many reactions directed towards syntheses of organic compounds are conducted in organic solvents. There are several approaches to removal of the inherently dangerous and harmful organic solvents under development: water-based reactions, use of non-volatile and more benign ionic liquids as solvents, and solid-state reactions using microwave energy are three popular areas of development. Although these need not draw on coordination chemistry, they represent the likely future of synthesis of the so-called fine organic chemicals (i.e. relatively limited amounts of high purity compounds). Where coordination chemistry can play a role lies in the development of water-based and ionic liquid-based chemistry, since ionic coordination complexes are inherently soluble in water and many ionic liquids. Thus, employing metal-directed reactions (see Chapter 6.5), where metal ions template the syntheses of larger organic molecules from small components, offers a way forward. In processes where metal complexes are currently employed as catalysts of organic reactions, many operate in organic solvents because they are simply most stable in such solvents, rather than any inherent insolubility of organic substrates in water; the development of more catalysts that are stable and active in an aqueous environment is a clear target. Ionic liquids are low melting point and very low vapour pressure salts comprising typically an organic cation and an inorganic anion; reactions using these as solvents is one area attracting strong interest. One example developed for use in a pyridinium ionic liquid is the methoxycarbonylation of iodobenzene to a benzoic acid methyl ester, where the complex [PdCl$_2$(cod)] (cod = 1,5-cyclooctadiene) is employed as catalyst; a related process described is reaction with ethylene to form methyl propionate. Since the first small-scale commercial process set up by the Eastman Chemical Company in 1996, more industrial processes employing ionic liquids have been developed, with the multinational company BASF showing in 2002 that ionic liquids could be employed on a large scale and efficiently recycled. At this stage, examples in industry of coordination complexes as new catalysts in non-traditional processes are limited and commercialisation not yet widely achieved, so opportunities exist for the entrepreneurial coordination chemist.

9.6   Complex Futures

If there is one certainty—apart from death and taxes—it is that the human race cannot predict the future well. Society generally has a very poor record of prediction, perhaps partly because not enough effort is put into developing a dependable model on which to base prediction. Sometimes advances simply come out of left field; a now classical example was the belief, extrapolating from the growth in horse-drawn traffic in New York, London and other large cities in the late 19th century, that the streets could eventually become at least waist-deep in manure. (*The Times* predicted in 1894 that "in 50 years, every street in London will be buried under nine feet of manure".) However, this proposal failed to predict the rapid development of the automobile (so that now the streets are merely shoulder-deep in cars). Scientists have had greater success at prediction because they are prepared to develop a model, evaluate and refine the model, and then apply it. Where we get it wrong also is in trying to predict beyond the limitations of our models and available information; for example, 60 years ago, one could feel comfortable based on then-known chemistry in saying "all cobalt(III) complexes are six-coordinate" because no other coordination number had been reported, whereas today we know that this is not true. So it is with some temerity that one can even tackle something called 'complex futures'; but of course, consistent with human frailty, that's not going to stop us trying—and setting some challenges for future coordination chemists at the same time.

Metals, at least the accessible ones, are naturally limited in number, but the chemistry of some is as yet not well explored. For example, it is not too many years ago that examples of organometallic complexes of *f*-block elements were very few in number; while some people could apparently explain this from a theoretical standpoint, laboratory chemists simply beavered away and found their way well into the field. Ligands, the very body of coordination chemistry, have evolved through advances in synthesis to the level where it is practicable to routinely produce the 'designer' ligand—made for a particular purpose. While molecular synthesis lies at the heart of the field, there is growing interest in the '*in silico*' complex—modelled on a computer chip, but not necessarily made in the laboratory. Classical molecular modelling, which is essentially based on Hooke's Law and treats atoms as solid spheres joined by springs, gives a surprisingly valid view of shape and isomer preference, even as more sophisticated approaches, such as density functional theory (DFT), are overtaking it.

Complexes are finding their way into a wide field of endeavour, including medicine, solid state chemistry and nanotechnology. For a field that some thought was 'all done' by the 1970s, coordination chemistry continues to expand and surprise. One good reason for this continued growth is that, despite all our sophistication, we really know very little still; there is plenty remaining out there to be discovered. The wealth of research papers that continue to use words such as 'remarkable', 'unusual' and even 'unprecedented' suggest that our voyage of discovery has hardly begun. So where could this voyage take us? Perhaps some of the speculation one meets seems a little unlikely, but one word that should be used in science with great caution is 'impossible'. Coordination chemistry has thrown up some wondrous outcomes in the past century of effort, so who can know where it is going to take us. Just book your ticket and get on board.

Concept Keys

Coordination complexes find wide commercial applications, from being used in medicine to serving as the mode by which metals are extracted from ores.

Medically-related applications of coordination complexes include use in bioassay, diagnostic imaging and as drugs.

Metal complexes find important roles in the fight against cancer. In particular, cisplatin and its homologues are used a great deal and to beneficial effect, and initiate their activity through binding to the aromatic amine bases of DNA in cancer cells.

Fluoroimmunoassay makes use of the fluorescent properties of europium complexes and is one example of how a strong spectroscopic response has been adopted to provide an analytical method, in this case, in bioanalysis.

Hydrometallurgy makes use of redox and complexation reactions for the recovery of valuable metals from ores. This is exemplified by the dissolution of gold metal in ores through oxidation by air and complexation by cyanide.

The synthetic organic chemical industry is heavily reliant on the use of metal complexes as catalysts, including for the preparation of optically pure chiral compounds.

Metal complexes can function as a starting point for what is termed 'bottom up' synthesis of nanomaterials, through commencing with simple complexes that aggregate into large nano-scale products.

Environmentally sensitive 'green' chemistry is a developing commercially relevant field, with roles for coordination chemistry included in current research.

Further Reading

Arpe, H.-J. and Hawkins, S. (Translator) (2010). *Industrial Organic Chemistry*, 5th ed., Wiley. A lengthy but popular, comprehensive and authoritative coverage of industrial organic chemistry in one volume. Apart from production aspects, important precursors and intermediates, as well as the many processes employing metal catalysts, are addressed.

Bellussi, G., Bohnet, M., Bus, J., Drauz, K., Greim, H., Jäckel, K.-P., et al. (2011) *Ullmann's Encyclopedia of Industrial Chemistry*, 40 vols, 7th ed. Wiley-VCH. This is the premier source book for all facets of industrial chemistry, with detailed but accessible chapters giving coverage of many processes and key chemical reagents.

Benvenuto, M.A. (2015). *Industrial Inorganic Chemistry*, De Gruyter.

A moderately short book focused on major inorganic industrial chemistry processes. While limited in coordination chemistry aspects, it does include discussion of efforts to improve processes and 'green' processes.

Dabrowiak, J.C. (2017). *Metals in Medicine*, 2nd ed., Wiley.

Following an introduction to basic coordination chemistry, an engaging coverage of metallodrugs and prodrugs, platinum and other anti-cancer complexes, complexes for treating arthritis and diabetes, parasites, bacteria and viruses follow. Further, metal ion imbalance in the body, as well as roles for metal complexes in bioanalysis are reported.

Evans, A.M. (1993) *Ore Geology and Industrial Minerals: An Introduction*, 3rd ed. Oxford: Blackwell Science Publications.

For those who wish to read beyond this text and deeper into geochemistry, this is an ageing but accessible and student-friendly book.

Gielen, M. and Tiekink, E.R.T. (Eds.) (2009). *Metallotherapeutic Drugs and Metal-based Diagnostic Agents: The Use of Metals in Medicine*, Wiley.

Provides a comprehensive and advanced account of medicinal uses of metals, for those who yearn for a more detailed coverage.

Hadjiliadis, N. and Sletten, E. (Eds.) (2009). *Metal Complex – DNA Interactions*. Oxford: Wiley-Blackwell.

Overviews metal–DNA interactions and mechanisms, metallodrugs and toxicity; includes a deep coverage of *cisplatin*.

Jones, C.J., Thornback, J.R., Sadler, P.J., Dilworth, J.R., and Williams, D.R. (2007). *Medicinal Applications of Coordination Chemistry*. London: RSC Publishing.

A readable, though now somewhat dated, account of the use of coordination compounds for pharmaceutical and medical use, such as in cancer therapy and diagnostic imaging.

Shiflett, M.B. (Ed.) (2020). *Commercial Applications of Ionic Liquids*, Springer.

Although fairly advanced, this book exemplifies one particular area of emerging industrial technology where metal catalysts may play a role. It outlines applications using ionic liquids that have been commercialized or are the subject of pilot studies.

Steed, J.W., Turner, D.R., and Wallace, K. (2007). *Core Concepts in Supramolecular Chemistry and Nanochemistry*, Wiley.

Fundamentals of the frontier research fields of supramolecular chemistry and nanochemistry are covered in a fairly concise manner, explaining the early evolution of the fields from more basic foundations. Suitable for students who want a clear and accessible introduction to these areas, of value despite its age.

Storr, T. (Ed.) (2014). *Ligand Design in Medicinal Inorganic Chemistry*, Wiley.

This advanced book reports key aspects of ligand design for a diversity of complexes directed towards targeting a range of disease sites, and should guide those keen to explore such aspects.

Swaddle, T.W. (1997). *Inorganic Chemistry: An Industrial and Environmental Perspective*, Academic Press.

A refreshing book aimed at the undergraduate chemist that deals with a key question – what is inorganic chemistry good for? Although topics stray away from coordination chemistry, there is sufficient of relevance to make this a worthwhile read, despite its age.

Appendix One

Nomenclature

From very early times, alchemists gave names to substances, although these names gave little if any indication of the actual composition and or structure, which is the aim of a true nomenclature. This was eventually addressed in the early days of 'modern' chemistry in the late 18th century, and modern nomenclature evolved from that early work. Since nomenclature evolved along with chemistry, it was far from systematic even up to the beginning of the 20th century. In large part, our current approach in coordination chemistry derived from nomenclature concepts introduced by Werner to represent the range of new complexes that he and contemporaries were developing, providing both composition and structural information. His system of leading with the names of ligands followed by the metal name, as well as also employing structural 'locators', is still with us today. Although an international 'language' for organic molecules commenced from a meeting in 1892, it was sometime later that a systematic international inorganic nomenclature developed, and it was as late as 1940 before a full systematic nomenclature was developed. Constant development in the field has demanded evolution of nomenclature, and the international rules were revised or supplemented in 1959, 1970, 1977, 1990 and again in 2005; like all languages, chemical language continues to evolve.

In describing chemical substances, we are dealing with a need for effective communication using an appropriate language. In a sense, chemical nomenclature is as much a language as is Greek or Mandarin, albeit a restricted one with a very specific purpose; it has an organized structure, 'rules of grammar', conventions, and undergoes continuous evolution. One advantage is that it is a universal language, governed by rules set in place by the International Union of Pure and Applied Chemistry (IUPAC). This body produces 'dictionaries' for chemical nomenclature that serve in much the same way as a conventional dictionary, thesaurus or grammar rule book, and which are updated regularly. For coordination compounds it is known as the IUPAC Red Book (Nomenclature of Inorganic Chemistry).

A.1 Nomenclature Basics

The object of the nomenclature adopted is to provide information on the full stoichiometric formula and shape of a compound and its isomeric form in a systematic manner, without the need for a structure diagram. For coordination chemistry, we need to deal with a number of aspects – the ligands (of which there may be more

Introduction to Coordination Chemistry, Second Edition. Paul V. Bernhardt and Geoffrey A. Lawrance.
© 2025 John Wiley & Sons Ltd. Published 2025 by John Wiley & Sons Ltd.
Companion website: www.wiley.com/go/coordinationchemistry2e

than one type), the central metal (or metals, in some cases), metal oxidation state(s), ligand distributions around the metal(s) and counter-ions (if the compound is a salt). Collectively, these place a great deal of demand on the nomenclature system, to the point where it has become both sophisticated and difficult to use. We shall try and provide just a basic and introductory nomenclature here, even neglecting some more advanced aspects of naming for the sake of brevity and clarity.

Molecules can be described in terms of a structural drawing, a written name or a formula. These are in a fashion all representations of the same thing – a desire to express the character of a chemical compound in a manner that will be understandable to others. Let's examine these options for two very simple examples (Figure A1.1).

The three representations in Figure A1.1 of the same thing serve in different circumstances. The structural drawing of the molecule perhaps gives the clearest view of the complex under discussion since it can be fairly easily 'read' by people with modest chemical training. The molecular formula is the most compact representation, and the molecular name provides a word-like form of representation for use in running text; both carry some higher demands in terms of rules for full interpretation, however.

Let's look at the first example in the Figure in terms of information that can be 'read' from the representations.

The compound drawing:

The set of square brackets around the cobalt-centred species separates it from the remainder of the drawing, defining it as the coordination complex unit – square brackets are the standard delineator of a complex unit, though these are not always included, particularly for isolated charge neutral complex entities. From the drawing, note:

- the metal ion is obviously six-coordinate, and from the shape of the drawing apparently octahedral – it is also a 3+ cation from the charge placed outside the square brackets at top right;
- it has six NH_3 (ammonia) molecules bound, apparently all uniformly via the N atom;
- there are additional molecules, not bound to the metal, present – their presence in this representation suggests the complex carries these as counter-ions, as it is a salt.

Compound drawing	Compound formula	Compound name
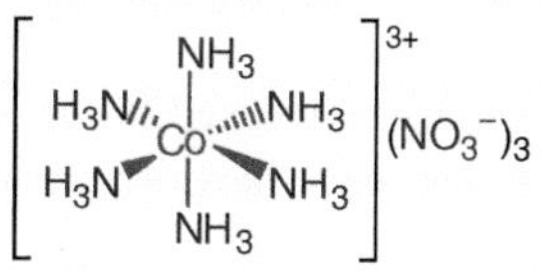 $[Co(NH_3)_6]^{3+}(NO_3^-)_3$ structural drawing	$[Co(NH_3)_6](NO_3)_3$	hexaamminecobalt(III) nitrate
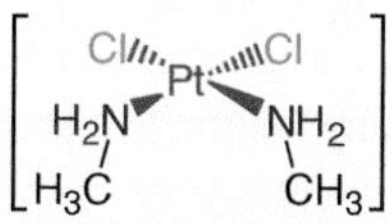 cis-$[PtCl_2(NH_2CH_3)_2]$ structural drawing	cis-$[PtCl_2(NH_2CH_3)_2]$	cis-dichloridobis(methanamine)platinum(II)

Figure A1.1

The three methods for describing a coordination compound: a structural drawing (left), chemical formula (centre) and systematic compound name (right).

The charges on the cation and anion have been inserted above, but do not invariably appear. In such circumstances, some additional interpretation is required, namely:

- in the absence of ion charges, it is necessary to deduce that '$(NO_3)_3$' means three NO_3^- anions, which must be inferred from chemical knowledge sufficient to assign 'NO_3' as the nitrate monoanion;
- if no charges were shown (which is more conventional) then we would infer that, if three NO_3^- anions are present, the complex unit within the square brackets must have an overall charge of +3 to balance these anionic charges;
- further, if one draws on chemical knowledge that ammonia is neutral, this means the 3+ charge lies on the cobalt centre, and, by further extension, that a +3 charge on a metal equates with an oxidation state of three, or we are dealing with a cobalt(III) ion (with the oxidation state conventionally written in Roman numerals).

The compound formula:

- importantly, the set of square brackets here still defines the presence of a complex unit, and everything within the square brackets 'belongs' to the one complex unit, linked by coordinate (covalent) bonds – the central atom (metal) which is the focus of bonding invariably comes first in the formula of the ion or neutral species in which it is contained;
- like the compound drawing, we can subsequently establish (using arguments as for the compound drawing) that we are dealing with a $[Co(NH_3)_6]^{3+}$ cation and three NO_3^- anions, with the metal being cobalt(III);
- the number of each ligand present appears as a subscript after the ligand's molecular formula (if more than one ligand type are present, the ligands are arranged in alphabetical order).

The compound name:

- in this case, we notice there are two 'words', one for the cation and the other for the anions – a single word means the complex is charge neutral;
- the metal and its oxidation state are defined clearly, and come last in the name of the metal-containing unit, with the metal name written without alteration;
- the ligand name is defined clearly, along with a (Greek) numerical prefix that says how many ligands are present (the same convention is followed in organic chemistry);
- the anion name is defined clearly, although the number must be worked out from the charge on the cation, demanding a knowledge of the charge (or lack of charge) of the ammonia ligands.

What is hopefully clear from all of the above is that all three approaches require some basic tacit knowledge of aspects of coordination chemistry to extract the full information from the names; this is, of course, an inevitability of any language – some concepts of context and 'unspoken' rules are necessary. You can show your knowledge by interpreting the second of the two examples in Figure A1.1.

However, let's move on to see how we can interpret both molecular formula and names generally, and then step forward to some simple rules of nomenclature and examples. How we can 'read' inorganic nomenclature is exemplified in Figure A1.2.

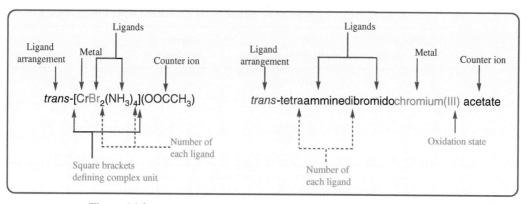

Figure A1.2

The conventional ordering of each component in the molecular formula (left) and systematic name (right).

Note how the metal is placed first in the formula representation of the complex unit and last in the written name; some other aspects are obviously common to both representations.

Some basic rules for naming compounds are given below. These may not make you an expert, but will allow some understanding of how names are generated; only extensive use (as with any language) brings expertise. Fortunately, structural formulae are now frequently met and allow an escape from the more demanding naming methodology.

A.2 Ligand Polydenticity

Not all ligands are simple and coordinated at only one site. When dealing with polydentate ligands, as is common in coordination chemistry, the concept of *denticity* is important in assisting development of the broader nomenclature. Denticity simply refers to the number of donor groups of any one ligand molecule that are bound to the metal ion; they can be (but are not always) separately identified in nomenclature, but knowledge of denticity assists in name development. Where coordination is by just one donor group, we are dealing with monodentate ligands. The terminology is outlined in Table A1.1, and in effect represents the form of nomenclature for discussing the way ligands are bound to metal ions; it is simply easier to say 'a tridentate ligand' rather than 'a ligand bound through three donor groups'. The same Greek numerical prefix applies as for multiple ligands bound to the same complex unit.

Table A1.1 Class names of polydentate ligands in terms of bound donor group numbers.

Number of donor groups coordinated	Name	Number of donor groups coordinated	Name
0	*Free ligand*	6	hexadentate
1	monodentate	7	heptadentate
2	didentate	8	octadentate
3	tridentate	9	enneadentate
4	tetradentate	10	decadentate
5	pentadentate	*Many*	polydentate

Table A1.2   Prefixes used in defining the number of a particular type of bound ligand.

Number of monodentates	Prefix to ligand name	Number of polydentates[a]	Prefix to ligand name
1	*No prefix*	1	*No prefix*
2	di	2	bis
3	tri	3	tris
4	tetra	4	tetrakis
5	penta	5	pentakis
6	hexa	6	hexakis

[a]Also used for large and or complicated monodentate ligands, or where ambiguity may arise.

To assist in recognition of polydenticity in a written name, the number of any poly-dentate ligands of denticity ≥ 2 bound to a metal ion is represented itself by a sequence of different prefixes (inserted immediately in front of the relevant ligand name) to those used for simple monodentate ligands, as given in Table A1.2. The need to use these only arises for written names and to avoid ambiguity (for example, in a case where there are two (*bis*) triamine (*tri*) ligands bound).

A.3   Naming Coordination Compounds

An abbreviated, simple set of rules for naming coordination complexes follows. The complete set of rules for nomenclature is very extensive and to be found in the >300 page tome that is the IUPAC Red Book and the primary source for the determined reader.

A.3.1   Simple Ligands

The names of *neutral* ligands are usually unchanged in naming a coordination com-pound containing them. For example, the molecule urea is still called urea as a ligand. There are but four important exceptions for the following very common ligands:

- water (H_2O) becomes **aqua**;
- ammonia (NH_3) becomes **ammine**;
- carbon monoxide (CO) becomes **carbonyl**;
- nitric oxide (NO) becomes **nitrosyl**.

Coordinated *anions* are simply named by replacing the final letter 'e' of the free anion by an 'o'. Examples of common anionic ligands are given in Table A1.3.

Table A1.3   Names of common coordinated anionic ligands.

Free anion	Coordinated anion	Free anion	Coordinated anion
Cl^-, chloride	chlorido	HO^-, hydroxide	hydroxido
O_2^-, superoxide	superoxido	CN^-, cyanide	cyanido
SCN^-, thiocyanate	thiocyanato	CO_3^{2-}, carbonate	carbonato
NO_3^-, nitrate	nitrato	NO_2^-, nitrite	nitrito
N_3^-, azide	azido	NH_2^-, amide	amido
ClO_4^-, perchlorate	perchlorato	SO_4^{2-}, sulfate	sulfato

Nomenclature always has tradition and habit as a barrier to change. In organometallic chemistry, formally anionic ligands like CH_3^- are often referred to as a substituent in the same way they are in organic chemistry; i.e. a *methyl* ligand rather than *methanido*, which does not roll off the tongue as easily.

There are very few examples of cationic ligands given that their partner the metal is also often cationic. The most important example is NO^+, and confusion between this and its neutral NO (nitrosyl) ligand can arise due to an interpretation of metal oxidation state. IUPAC recommends oxidonitrogen(1+) for this formally cationic ligand, and we'll leave it there!

A.3.2   Complexes

A.3.2.1   *Complex Name*

Neutral complexes have only a single 'word' name.

Salts comprising complex cations and or anions are written as two 'words', with one word name for the cation and one word name for the anion. For *all* salts, invariably, the *cation* is named *first*, and the anion last. (This follows the convention for simple salts like sodium chloride.) These basic rules mean all components of a single neutral, cationic or anionic complex entity are run together without any breaks, inspired by German (non-chemical) compound nouns such as the now infamous *Donaudampfschifffahrtsgesellschaftskapitän*.

These rules for naming the complex entity will now be defined.

- Ligands are arranged *first*, in *alphabetical order of their names* (numerical prefixes do not affect the ordering) with the metal atom at the end. The *oxidation number* of the central metal atom may be included at the end of the name, following the metal name, as a *Roman numeral* in parentheses.
- Alternatively, it is permitted to enter the overall *formal charge* on the complex ion in parentheses at the end which includes the charges of the ligands as well as that from the metal oxidation state (e.g. for the complex cation $[CrCl(NH_3)_5]^{2+}$ pentaamminechloridochromium(2+) or pentaamminechloridochromium(III) are both acceptable). There is no suffix appended to the metal name in *neutral or cationic* complexes.
- In *anionic* complexes an *-ate* suffix is included which typically involves contraction of the elemental name for (the vast majority of) metals whose names end in '*um*' (e.g. molybdenum to molybdate) while metals such as zinc just become *zincate* as its complex anion. As just mentioned above, it is now acceptable to write the name of a complex anion by only referring to its formal charge, e.g. tetraoxidochromate(2−) for $[CrO_4]^{2-}$ while the formal oxidation state (chromium(VI)) is not stated. In an acknowledgement of their venerable history, complex anions of metals first isolated in ancient times have symbols that reflect their original Latin names (e.g. Fe *ferrum*, Cu *cuprum*, Sn *stannum*, Au *aurum*) and their anionic complexes often use their Latin prefix, e.g. tetrachloridoferrate(1−) for $[FeCl_4]^-$.

A.3.2.2   *Complex Formula*

Note that in writing the formula representation of a complex, for complex cations or neutral complexes the metal is written first at left after the opening square bracket, and then the *identical* approach of *alphabetical order* to that reported above for the 'word' name is used. Ligands are listed in alphabetical order of the first letter of the formula or standard abbreviation that is used for each ligand (thus, for example, Cl before NH_3 before OH_2, and (en) before NH_3). Any bridging ligands in polynuclear complexes are listed after terminal ligands, and, if more than one, in increasing order of bridging multiplicity (i.e. number of bonds to metals). The formula is completed with a closing square bracket. If there are any counter-ions, their standard formulae are written before (if cations) or after (if anions) the complex formula within the square brackets, e.g. $K[MnO_4]$, $[Ni(CO)_4]$ and $[CoCl(NH_3)_5](SO_4)$.

A.3.2.3 Multipliers

Commonly, complexes contain more than one of the single or several ligand types attached to a central metal ion/atom. The *number* of a given ligand needs to be defined in the name, and is indicated by the appropriate Greek prefix, as tabulated in Table A1.2. The rule, in summary, is di (2), tri (3), tetra (4), penta (5), hexa (6) *if* the ligand is monoatomic or one of the simple ligands above. *Polyatomic* (and multidentate) ligands are placed in parentheses, and their number is indicated by a (different) prefix outside the parentheses: bis (2), tris (3), tetrakis (4), pentakis (5), hexakis (6). Apart from also using these latter number prefixes for bulky substituted organic ligands acting as monodentates, even simpler ligands may use this prefix set if it helps avoid ambiguities.

In formula representations, the number of each ligand is included as a subscript after the symbol (except for the case where there is only one of a ligand).

Simple cations and anions serving as counter-ions to balance charge on the complex are simply named and their number is not defined; you will see examples of this convention below.

Examples:

$[Co(NH_3)_6]Br_3$	hexaamminecobalt(III) bromide (*or* hexaamminecobalt(3+) bromide *using charge instead*)
$Li_2[PtCl_6]$	lithium hexachloridoplatinate(IV) (or lithium hexachloridoplatinate(2−) *using charge instead*)
$[RhCl_2(NH_3)_4]NO_3$	tetraamminedichloridorhodium(III) nitrate
$K_3[Cr(C_2O_4)_3]$	potassium tris(oxalato)chromate(III)
$[IrBr(H_2NCH_2COO)_2(OH_2)]$	aquabromobis(glycinato)iridium(III)
$Na[PtBrCl(NH_3)(NO_2)]$	sodium amminebromidochloridonitritoplatinate(II)
$[Pt(PPh_3)_4]$	tetrakis(triphenylphosphane)platinum(0)

A.3.2.4 Locators

Structural features of molecules associated with stereochemistry and isomerism are indicated by prefixes attached to the name, and *italicized*, and separated from the rest of the name by a hyphen.

Typical of this type are the geometry indicators *cis-* (same side), *trans-* (opposite), *fac-* (facial, three mutually *cis* donors), *mer-* (meridional, spanning a pair of *trans* coordination sites). Note that a more elaborate but highly functional form of stereo-chemical descriptors exist in the nomenclature system; we shall avoid its use here.

Examples:

cis-[PtCl$_2$(py)$_2$]	*cis*-dichloridobis(pyridine)platinum(II)
mer-[IrH$_3$\{P(C$_6$H$_5$)$_3$\}$_3$]	*mer*-trihydridotris(triphenylphosphine)iridium(III)

A.3.2.5   *Organic Molecules as Ligands*

These take *exactly* the names they carry as an organic molecule in general; there are two simple examples immediately above. The exception is that anionic molecules are usually altered slightly to carry the usual **-o** ending as in Table A1.3. Thus, an understanding of organic nomenclature is vital as well. Some organic molecules have well accepted and approved 'trivial' (non-IUPAC) names, and these are also able to be carried forward into the complex name in this form.

Examples:

H$_2$N–CH$_2$–CH(CH$_3$)–NH$_2$	propane-1,2-diamine
[Co(H$_2$N–CH$_2$–CH(CH$_3$)–NH$_2$)$_3$](ClO$_4$)$_3$	tris(propane-1,2-diamine)cobalt(III) perchlorate
H$_2$N–CH$_2$–CH(CH$_3$)–COOH	alanine (*or* alaninate *as the anion*)
[Cu(H$_2$N–CH$_2$–CH(CH$_3$)–COO$^-$)$_2$]	bis(alaninato)copper(II)

The sole variation of nomenclature you may meet where a ligand is coordinated is the addition of κ nomenclature in the molecular formula, which is done to provide more information by defining more exactly the way a ligand is bound, that is the number of donors attached and their type. As an example, instead of just writing [Co(H$_2$N–CH$_2$–CH(CH$_3$)–NH$_2$)$_3$](ClO$_4$)$_3$ as above, you may occasionally find it written as [Co(H$_2$N–CH$_2$–CH(CH$_3$)–NH$_2$-κ^2-*N,N*)$_3$](ClO$_4$)$_3$ where the κ^2 identifies that two (the superscript number) unconnected groups are acting as donors, and the *N,N* identifies them both as nitrogen donors. In a simple case like this example, most people do not really need the additional information, and so it tends to be left out. However, for mixed-donor polydentate ligands where there are options for coordination, such as an excess of donor groups over sites available to choose from, the additional specification is necessary. Given that you will meet only simple examples at this level, this is an extension that is best simply set aside at present.

A.3.2.6   *Bridging Ligands*

Some ligands, with more than one lone pair of electrons, can bind two metals simultaneously, and are then termed *bridging* ligands, leading to polynuclear complexes.

These ligands are identified by the Greek letter mu (μ) to indicate the ligand is bridging, together with a superscript *n* to indicate the number of metal atoms to which the ligand is attached (but only if $n > 2$).

Examples:

$[Fe_2(CN)_{10}(\mu\text{-CN})]^{5-}$	μ-cyanido-bis(pentacyanidoferrate(III))
$[Fe_2(CO)_6(\mu\text{-CO})_3]$	tri-μ-carbonyl-bis(tricarbonyliron(0))
$[\{Co(NH_3)_4\}_2(\mu\text{-Cl})(\mu\text{-OH})]^{4+}$	μ-chlorido-μ-hydroxido-bis(tetraamminecobalt) (4+)

This simplified group of rules clearly does not by any means cover the full set of naming rules that apply to deal with modern coordination chemistry. However, they go some of the way to allowing you to 'navigate' around nomenclature for the relatively simple complexes you will likely meet. It is a difficult task to name complicated compounds – which is why, even in the chemical research literature, people sometimes choose to avoid it as much as possible. For some, nothing compares with a drawing of the complex molecule and a trivial name(s) for the ligand(s) involved!

A.4   Organometallic Compounds

The same basic rules apply to the naming of organometallic compounds as to traditional Werner-type coordination compounds.

However, met more commonly in this field is a nomenclature for dealing with connected donor atoms bound to a metal, the *hapticity* (η) nomenclature, which defines the number of atoms involved in bonding of a ligand. For example, the ligand $H_2C=CH_2$ bound side-on so that both connected carbons atoms are effectively equally linked to a metal would be designated as (η^2-C_2H_4). This device allows for the commonly met feature in organometallic compounds of a ligand like cyclopentadienyl ($C_5H_5^-$), which can involve one, three or all five connected carbon atoms in close (bonding) contact to the metal. The distinction is clearly defined in terms of η, which reports, for example, one (η^1), three (η^3) or all five (η^5) of the carbon atoms of the ligand coordination as a prefix to the ligand representation (e.g. η^5-$C_5H_5^-$). This concept is analogous to the use of κ, outlined earlier, to designate the number of donors of a polydentate ligand actually bound in Werner-type complexes.

Further Reading

The International Union of Pure and Applied Chemistry (IUPAC) publishes a number of books and reports on nomenclature. Some are available with free access on-line through their website. IUPAC published the last update of *Nomenclature of Inorganic Chemistry*, at the time of writing this book, in 2005, and an online version is freely available (https://iupac.org/what-we-do/books/redbook/, ISBN 0-85404–438-8) for greater detail.

Appendix Two

Molecular Symmetry: The Point Group

Molecular shape plays an important role in the spectroscopic properties of complexes. Even if we compare two simple two-coordinate linear complexes, one with two identical ligands and one with two different ligands, it should be immediately apparent that they do not look the same. For one, the left- and right-hand sides of the molecule as drawn are equivalent, for the other they are different (Figure A2.1). Turn one around by a rotation of 180° about an axis perpendicular to the bond direction, and it looks identical to what you had in the first place; do this for the other and it is clearly in a different position. Further, consider the situation where the linear molecule with two identical donors is compared with an analogue where the X–M–X unit is bent. Now, consider a 180° rotation about an axis containing the X–M–X unit. For the linear molecule, the outcome is indistinguishable from the starting situation, whereas for the bent molecule this is not the case. In each of these examples, the linear molecule has undergone what is called a *symmetry operation*. The non-symmetrical and bent molecules have clearly yielded a different view, due to their different shape and connectivity. What we are seeing is that the process we have undertaken has distinguished between what are clearly different molecules. In effect, we can use any molecule's behaviour upon applying a series of operations like those presented below to define the molecular shape (or symmetry). Molecules can be described in terms of a symmetry *point group*, effectively a combination of a limited number of what are termed *symmetry elements* based on considering a set of interrelated operations like those introduced above. There are a limited number of both individual symmetry elements and combinations (called the *point group*).

The only symmetry element that is common to every molecule is the *identity* (E), which simply means 'no change' (and hence appears trivial but does have a role that mathematicians interested in *group theory* will explain to you, but we won't).

The only symmetry elements that can operate are defined below in Table A2.1. A legitimate symmetry operation employing these elements occurs when the molecular views before and after the operation are indistinguishable. Operations occur relative to an x, y, z coordinate system arranged around a particular point in the molecule, selected with regard to defining symmetry elements and maximizing operations in such a way that the z-axis forms the principal axis, passing through the molecular centre and being the axis around which the highest order (n) operation occurs. To summarise, a symmetry *operation* is the act of moving an object in some way such that it is indistinguishable from its original position, while the symmetry *element* is an object (a plane, an axis or a point) in three-dimensional space about which the symmetry operation is performed. It is a good idea to not mix them up.

Introduction to Coordination Chemistry, Second Edition. Paul V. Bernhardt and Geoffrey A. Lawrance.
© 2025 John Wiley & Sons Ltd. Published 2025 by John Wiley & Sons Ltd.
Companion website: www.wiley.com/go/coordinationchemistry2e

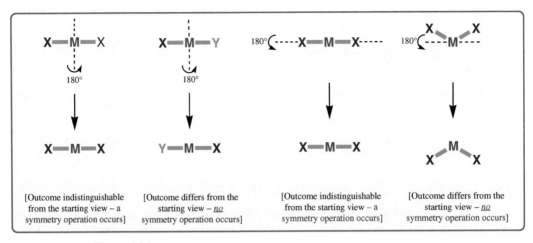

Figure A2.1
Two operations with linear MX_2 and MXY molecules and linear and non-linear MX_2 that distinguish their differing symmetry.

Table A2.1   Symmetry elements and operations.

Element	Description
E	**Identity**
σ	**Plane of symmetry**. The two types involve reflection in a plane containing the principal axis (σ_v (vertical) or σ_d (dihedral)), and reflection in a plane perpendicular to the principal axis (σ_h, horizontal)
C_n	**Rotation axis**. The subscript n denotes the *order* of the axis, which is the angle of rotation ($360°/n$) to achieve an indistinguishable image
S_n	**Improper rotation axis**. This involves sequential steps of rotation by $360°/n$, followed by a reflection in a plane perpendicular to the rotation axis
i	**Inversion**, which occurs through a centre of symmetry

Determination of all possible symmetry elements provides the ability to then assign the *point group* for the element, which (as the name suggests) is based on the symmetry around a point in the molecule that either coincides with the central atom or the geometric centre of the molecule. A flow chart (Figure A2.2) is frequently used to assist in the assignment of all symmetry operations for a molecule. Once all symmetry operations have been identified, the point group evolves. Point group types are classified into 12 categories (identified in Table A2.2), although given that there is an infinite number of positive integers (n) there is actually no limit to the theoretical number of point groups. In chemistry the geometric constraints of chemical bonding do place practical limits on these possibilities.

To exemplify the concepts, we shall examine two systems: the trigonal bipyramidal MX_5 and the square planar MX_4. The shapes, axes and operations involving these are shown in Figure A2.3 and A2.4.

For MX_5, the principal (z) axis passes through the M and the two axial X groups; the M is at the point about which the point group is defined. The three X groups coplanar with the metal lie in the *xy* plane; one axis (designated *x*) passes through one M–X bond, which then requires the other to pass through M but not include the

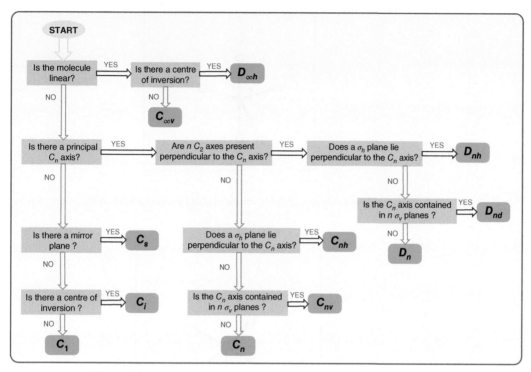

Figure A2.2

A simplified flow chart for assignment of a symmetry group. Special high symmetry groups (T_d, O_h, I_h) are not included in this chart, and need to be identified at the start of the flowchart; the chart applies to other and generally lower symmetry molecules, which are more often met in reality.

Table A2.2 Point groups.

Point group	Symmetry elements involved
C_1	Only the identity operation applies
C_s	One plane of symmetry
C_i	A centre of symmetry (inversion)
C_n	One n-fold rotation axis
D_n	One n-fold rotation axis (about the principal axis) and n horizontal twofold axes
C_{nv}	One n-fold rotation axis (about the principal axis) and n vertical mirror planes
C_{nh}	One n-fold rotation axis (about the principal axis) and one horizontal reflection plane
D_{nh}	One n-fold rotation axis (about the principal axis, as for D_n), one horizontal mirror plane, and n vertical planes containing the horizontal axes
D_{nd}	One n-fold rotation axis (about the principal axis, as for D_n), and vertical planes bisecting angles between the horizontal axes
S_n	Systems with one S_n axis (alternating $n = 4, 6, 8, \ldots$)
$C_{\infty v}$, $D_{\infty v}$	Linear systems with an infinite rotation axis
T_d, O_h, O, I_h, I	Special high symmetry (cubic) groups: tetrahedral, octahedral, cubic and icosahedral

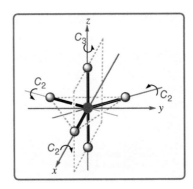

Figure A2.3

The coordinate system and symmetry operations for trigonal bipyramidal MX_5, of D_{3h} symmetry point group.

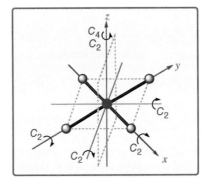

Figure A2.4

The coordinate system and symmetry operations for square planar MX_4, of D_{4h} symmetry point group. Note that there is a C_2 operation colinear with a C_4 around the z-axis.

other X groups. Around the z-axis, rotation by 120° leads to an indistinguishable arrangement; this is thus a C_3 axis. Rotation by 180° about each of the three M–X bonds in the xy plane leads to an indistinguishable arrangement; thus, there are three C_2 elements. These operations alone define the molecules as belonging to a D point group. However, there are also three vertical planes of symmetry each containing one M–X bond in the xy plane and the C_3 axis (only one of which is shown as a dotted square in the figure, for clarity), as well as one horizontal plane that is the xy plane containing the central equatorial MX_3 part of the molecule (shown as a dotted triangle in the figure). Overall, then this means the molecule belongs to the D_{3h} point group.

The square planar MX_4 has some similarities to the trigonal bipyramidal MX_5 as along the z-axis there is a C_4 operation (rotation by 90° leads to formation of an indistinguishable arrangement), and there are four C_2 elements (two about each in-plane X–M–X, and two about imaginary axes bisecting the x and y axes in two directions). There are four vertical planes containing the C_4 axis (only one of which is shown as a dotted square in the figure, for clarity) and one horizontal plane of symmetry (shown as a dotted square incorporating the four X groups in the figure). Following the flow chart, this leads to the D_{4h} point group.

It is possible with the flow chart and careful examination of drawings and or three-dimensional models of a complex to assign the point group with reasonable

Table A2.3 Point groups for several coordination geometries with identical ligands.

Coordination number	Stereochemistry	Point group
2	linear	$D_{\infty h}$
3	trigonal planar	D_{3h}
	trigonal pyramidal	C_{3v}
	T-shaped	C_{2v}
4	tetrahedral	T_d
	square planar	D_{4h}
5	trigonal bipyramidal	D_{3h}
	square-based pyramidal	C_{4v}
6	octahedral	O_h
	trigonal prismatic	D_{3h}

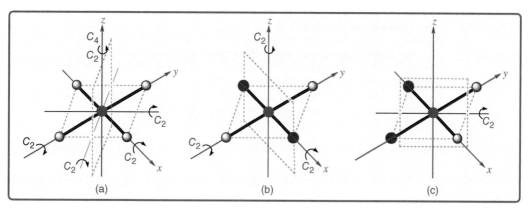

Figure A2.5

The coordinate system and symmetry operations for the family of square planar complexes (a) MX_4, (b) *trans*-MX_2Y_2 and (c) *cis*-MX_2Y_2.

rapidity and success after some practice. To assist further, a list of the point groups of basic higher symmetry structures are collected in Table A2.3. Those shown assume identical ligands in all sites. You should assume that these shapes with a mixture of different ligands will be of lower symmetry and have a lower symmetry point group. This aspect is illustrated in Figure A2.5 for moving from square planar MX_4 (D_{4h}) to square planar *trans*-MX_2Y_2 (D_{2h}) and *cis*-MX_2Y_2 (C_{2v}).

Distortions of common geometric shapes lead necessarily to changes in the point group. For example, an octahedron (O_h) can undergo three types of distortion. The first is *tetragonal distortion*, involving elongation or contraction parallel to a single C_4 axis direction, leading to D_{4h} symmetry. The second is *rhombic distortion*, where changes occur along two the C_2 axes perpendicular to this C_4 (which becomes C_2) so that no two sets of bond distances along each of the three axes are equal, leading to D_{2h} symmetry. The third is *trigonal distortion*, involving contraction or elongation along one of the C_3 axes to yield a trigonal antiprismatic shape, reducing the point group to D_{3d}. Similarly, other common stereochemistries may distort, leading to shapes with different point groups. On the basis of its point group, we can predict if a molecule will be polar or chiral, for example.

Just as molecules have certain symmetry, molecular orbitals likewise have symmetry (developed in Chapter 2). Orbital labels such as σ and π relate to the rotational symmetry of the orbital, whereas the labels *a*, *e*, e_g, *t* and so on met in complexes arise from considering orbital behaviour in the context of all symmetry operations of the point group of the molecule. A *character table*, which defines the symmetry types possible in a particular point group, provides a way of assigning these labels, but this will not be pursued here.

Further Reading

Ameta, R. and Ameta, S.C. (2016) *Chemical Applications of Symmetry and Group Theory*. Waretown: Apple Academic Press.

Opening with chapters on symmetry and point groups, it avoids complicated mathematics. It also covers applications to various spectroscopic methods.

Kettle, S.F.A. (2007). *Symmetry and Structure: Readable Group Theory for Chemists*, 3rd ed. New York: Wiley.

An accessible revised and updated introduction for students, relying on a diagrammatical and non-mathematical approach.

Ogden, J.S. (2023). *Introduction to Molecular Symmetry*. Oxford: Oxford University Press.

A readable and well-presented book suitable for students at any level.

Vincent, A. (2013). *Molecular Symmetry and Group Theory: A Programmed Introduction to Chemical Applications*, 2nd ed. New York: Wiley.

This popular textbook may serve the more advanced reader well, with a slightly deeper mathematical base but a well-staged programmed learning approach.

Index

Introduction to Coordination Chemistry, Second Edition. Paul V. Bernhardt and Geoffrey A. Lawrance.
© 2025 John Wiley & Sons Ltd. Published 2025 by John Wiley & Sons Ltd.
Companion website: www.wiley.com/go/coordinationchemistry2e

Printed and bound by CPI Group (UK) Ltd, Croydon, CR0 4YY

18/12/2024

14613872-0001